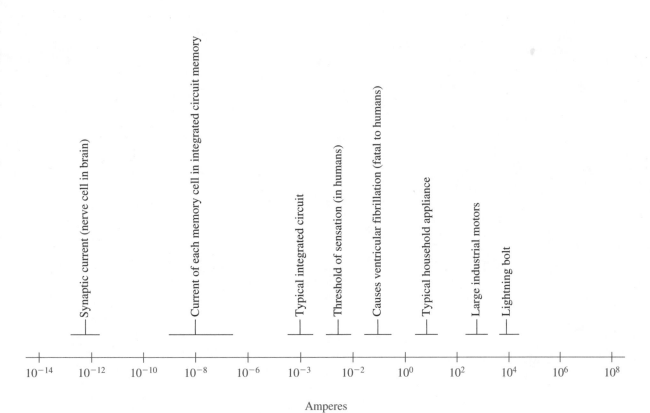

The Range of Currents

Introduction

to Electrical Engineering

Introduction
to Electrical Engineering

J. David Irwin
Auburn University

David V. Kerns, Jr.
Vanderbilt University

Prentice Hall
Englewood Cliffs, New Jersey 07632

Irwin, J. David.
    Introduction to electrical engineering / J. David Irwin, David V. Kerns, Jr.
        p. cm.
    Includes bibliographical references and index.
    ISBN 0-02-359930-8
    1. Electric engineering.   I. Kerns, David V.   II. Title.
TK145.I78   1995
621.3—dc20   94-15502
                                          CIP

Acquisitions Editor: Alan Apt
Buyer: Lori Bulwin
Production Supervisor: bookworks
Production Coordinators: Aliza Greenblatt/Bayani Mendoza de Leon
Editorial Assistant: Shirley McGuire
Cover Designer: Blake Logan
Cover Art: Photo courtesy of Intel; Pentium is a trademark of Intel Corporation.

 © 1995 by Prentice-Hall, Inc.
A Simon & Schuster Company
Englewood Cliffs, New Jersey 07632

The following figures are reproduced with the permission of Macmillan College Publishing Company from BASIC ENGINEERING CIRCUIT ANALYSIS 4/E by J. David Irwin. Copyright © 1993 by Macmillan College, Publishing Company, Inc.: P2.4–P2.7, P2.9–P2.13, P2.15–P2.16, P2.18–P2.19, P2.24, P2.26–P2.31, P2.33–P2.36, P2.39, P2.48–P2.49, 3.2, 3.5, P3.2–P3.4, P3.9, P3.15–P3.16, P3.20, P3.26–P3.27, P3.32, P3.33, 4.1–4.2, 4.6, D4.6, 4.8–4.11, D4.6, 4.19, 4.21–4.22, P4.9, P4.11–P4.18, P4.20–P4.25, P4.27, P4.32–P4.33, P4.35, P4.38, P4.49, 5.1–5.3, D5.1–D5.4, 5.5, 5.7 A and B, 5.16–5.21, 5.23–5.24, P5.4–P5.9, P5.11–P5.12, P5.14, P5.35–P5.36, 6.4 A–D, 6.7–6.11, D6.4, 18.8, D18.2–D18.3, 18.10, 18.11, P18.13, P18.19–P18.20, 21.1, 21.3–21.10, P21.2–P21.3.

Printed in the United States of America

10 9 8 7 6 5 4 3 2 1

ISBN   0-02-359930-8

PRENTICE-HALL INTERNATIONAL (UK) LIMITED, London
PRENTICE-HALL OF AUSTRALIA PTY LIMITED, Sydney
PRENTICE-HALL CANADA INC., Toronto
PRENTICE-HALL HISPANOAMERICANA, S.A., Mexico
PRENTICE-HALL OF INDIA PRIVATE LIMITED, New Delhi
PRENTICE-HALL OF JAPAN INC., Tokyo
SIMON & SCHUSTER ASIA PTE. LTD., Singapore
EDITORA PRENTICE-HALL DO BRAZIL LTDA., Rio de Janeiro

TRADEMARK INFORMATION:

IBM PS/2, Personal System 2, PC, PC/XT and PC/AT are trademarks of International Business Machines. Intel 386SX, 386DX, 486DX, 486DX2, 486DX4 and Pentium processors are trademarks of Intel Corporation. MSDOS and Microsoft Windows are trademarks of Microsoft Corporation. R2000 and R3000 are trademarks of MIPS Technologies Inc. SPARC, SPARCstation and Open Windows are trademarks of Sun Microsystems, Inc. UNIX is a trademark of UNIX Systems Laboratories.

DEDICATION

To Edie and Sherra

and our far-flung, still growing families

ACKNOWLEDGMENTS

The authors wish to express their thanks to a number of their collegues and other individuals who have contributed to the preparation of this text. We are especially indebted to Professor Charles A. Gross of Auburn University for making substantial contributions to the material on machines and preparing the sections of the text that deal with ac residential power circuits and transformer applications. Professor C. H. "John" Wu of Auburn University made major contributions to the chapter on microprocessors, complimented by suggestions from Professor Lloyd W. Massengill of Vanderbilt. Professor Lloyd A. "Pete" Morley of the University of Alabama and Professor S. Mark Halpin of the Mississippi State University made a number of contributions to the problem sets; the critique of the chapters on electronics by Professor Robert Fox of the University of Florida is greatly appreciated. We also recognize the help of Ashwin Matta, Randall Milanowski, and Manish Pagey all of Vanderbilt University for their assistance with homework problems and solutions.

Contents

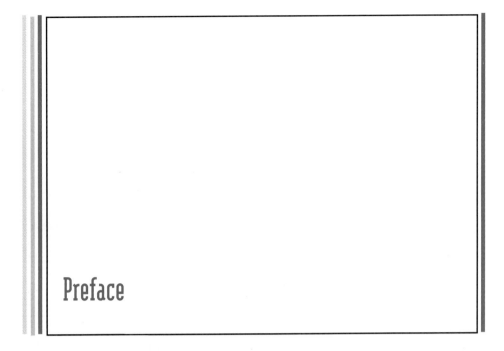

Preface

P.1

Purpose

This book is intended to serve as a text for an introductory course or course sequence in electrical engineering. It is a broad and comprehensive introduction to the field designed to provide the student with an understanding of the fundamental concepts and a practical working knowledge of the application of these basic principles to real-world, up-to-date problems.

The fundamental strength of the text is the clear, solid and concise presentation of the essential components of the analysis of electric circuits, electronic circuits and systems, and electromechanical systems and safety. Furthermore, the book contains a number of other related topics outlined below, that greatly enhance the flexibility of presentation. The material is organized and presented in a manner that permits the instructor great latitude in the selection of topics and their order of presentation.

When used with a survey course comprising one or two semesters, typically only selected chapters or portions of chapters can be covered. In general, because of the breadth of the electrical engineering field, there simply is insufficient time to cover all the important elements of the discipline. However, this text is designed to assist the skillful instructor to empower students with the confidence and understanding that they are capable of solving electrical engineering problems outside of those specifically addressed in class. By becoming familiar with the entire organization of this book and treating it not only as a text, but as a reference as well, the student can learn one of the most crucial points about any survey course—how to access the information they need when they need it in the future. In this way the experience of a survey course provides a life-long benefit.

All the major disciplines within electrical engineering are represented in this text. Therefore, if students retain this book as part of their permanent library, they will be able to address throughout their careers the many electrical and electronic issues that literally permeate the engineering problems that are encountered in our modern technological world. In addition, the book is designed to be a very useful reference for students who wish to take the Fundamentals of Engineering exam (see Appendix C) that is a prerequisite for becoming licensed as a Professional Engineer.

This text is also suitable for the reader who wishes to use a self-study approach to learn the fundamentals of electrical engineering. The many worked examples, drill problems, and careful explanations within each chapter test the student's understanding, clarify key points, and provide necessary feedback.

For students who may later study many of these subjects in more detail, this text provides a clear presentation of fundamentals including the relationships among the various disciplines within electrical engineering and a comprehensive foundation for further advanced study in any selected area.

P.2
Features

This book contains a number of unique features that enhance its use as a survey text and valuable reference.

- The first chapter sets the stage for the study of electrical engineering topics by outlining the many disciplines within the field, their interrelationships, and the work of an electrical engineer including the value of operating as a member of a team consisting of many professionals with diverse backgrounds.

- The book is organized around six major parts, each of which represents a principle area of study within the field. The outline of topics within each part provides a logical grouping of the subject material as well as the relationship of these subjects to the field as a whole.

- After a brief study of the first few chapters, the book is organized in such a way that the topics may be selected for presentation in almost any order. Chapters are designed, to the extent possible, to be independent; therefore, the instructor has great flexibility in the selection of chapters, or sections within a chapter, to meet the special objectives of any particular course.

- A large number of examples, strategically placed, reinforce learning. In addition to these examples, which follow essentially every topic after it's introduced, there are a number of drill problems designed to quickly test the readers' understanding of the material. The combination of these features enhances both learning and retention by providing immediate demonstrations of the concept under discussion.

- The frequent use of analogies is a valuable aid to the newcomer in developing intuition for the fundamental principles. The use of fluid flow and mechanical analogies have been shown to be particularly helpful in quickly providing new students or those from other engineering disciplines an intuitive grasp of the principles that govern electrical engineering subjects.

- The subject of digital electronics and systems is presented prior to analog circuits, reflecting the growing importance of digital systems in today's world. Simple "switch" models for the transistor provide an easily understood, and very practical introduction to electronic circuits.

  In addition to these general features, a number of specific features, not generally found in other texts, have been included to enhance the learning experience for the student and provide significant latitude to the instructor.

- A compact introduction to the topics of both digital and analog electronics is covered in a single chapter (Chapter 7) for use in courses where time does not permit the luxury of a more detailed analysis.

- Although the diode is covered in some detail in Chapter 8, this chapter is organized such that the device physics (sections 8.1 to 8.5) can be easily skipped and the diode circuit models used directly to analyze practical circuits containing these elements.

- The microprocessor is covered in a relevant and practical manner in Chapter 12. This chapter emphasizes the Personal Computer, its specification, selection, and operation.

- An introduction to microelectronics design and manufacturing (Chapter 13) is presented in such a way that it can be easily understood by someone unfamiliar with the area. This material is a very useful reference for any engineer associated with the design of an electronic system in which integrated circuits must be procured. The importance of this unique presentation is the fact that the economic competitiveness of today's world in electronic systems of all types requires that some familiarity with chip technology options be included in an engineering education.

- Electrical safety, a critically important topic which unfortunately is often neglected, is a consideration throughout the text and is presented in a very practical manner in Chapter 21. The fundamentals of electrical safety are discussed in the context of numerous illustrations of potentially hazardous conditions together with suggestions for avoiding injury.

- The topic of communications, an area of growing importance in our world where machines and human beings are interconnected throughout the globe, is presented in a way that fosters the student's understanding and appreciation of the subject. Many of the modern topics, which they encounter in their everyday lives, are explained in a simple and straightforward manner in Chapter 23.

P.3
Using This Text

No particular background in electric circuits, electronics, or electrical systems is assumed. This text is intended to be a first course in this field. There is, however, sufficient advanced material and depth to support the needs of students who wish to study some aspect in more detail.

This text is divided into six major parts. The first part includes a quantitative and intuitive definition of the terms and parameters used in the text. The fundamentals of circuit analysis are also presented, and while portions of the second and fourth chapters are prerequisite for much of the material that follows, the remaining topics can be skipped or included as desired without adversely affecting a carefully planned study of the remaining subjects.

Part II of the text is organized such that coverage of Chapter 7 and a selection of other material to fit particular course needs, can provide a basis for an introduction to the area of electronics. For example, this chapter together with selected topics on diodes and transistors would permit a quick introduction to the field.

The presentation of Digital Systems, Part III, provides great flexibility in its use. Topics may be selected from any of the chapters in this section consistent with the instructor's priorities without losing continuity with subsequent chapters.

One of the most important topics in Part IV is the operational amplifier presented in Chapter 14. This chapter is virtually independent and may be presented at any point in a course. The coverage of this topic, together with a selection of topics in small-signal and large-signal analysis, would provide an excellent introduction to analog transistor circuits.

Part V of the text presents the traditional topics in electromechanical systems. A brief presentation of these topics, which includes some very practical examples readily identified by the students, can provide them with a basic understanding of electric machines. The topics in the chapter on safety can be of enormous value to them throughout their lives.

The topics in control contained in Part VI are traditional and cover important fundamentals while introducing the importance of this subject. The material on communications provides historical perspective and an up-to-date discussion of many elements of today's modern communication technologies. These subjects are presented in an easily understood format and emphasize the importance and usefulness of this critically important area.

It is important to note that whatever route through this text is chosen for a particular course, the instructor may well provide the student the greatest long-term benefit by communicating the breadth of the field of electrical engineering, the interrelationship of its topics, and the confidence to address related problems as needed in their future careers.

Finally, for teachers employing this text in their courses, a manual that contains solutions to all drill problems and end-of-chapter problems, is available as a Laboratory Manual.

P.4
System of Units

The system of units used in this text is the international system of units, the Systéme International des Unités, which is normally referred to as the SI standard system. This system, which is composed of the basic units meter (m), kilogram (kg), second (s), ampere (A), degree Kelvin (K), and candela (cd), is defined in all modern physics texts and therefore will not be defined here; we will, however, discuss the units in some detail as we encounter them in our subsequent analyses.

The abbreviations and symbols used to represent the various quantities studied in this text follow standard practice. Table P.1 may be a useful reference if you encounter unfamiliar symbols.

Table P.1 Standard Abbreviations and Symbols

ac	Alternating Current
A	Ampere
C	Coulomb
db	Decibel
dc	Direct current
F	Farad
H	Henry
Hz	Hertz
J	Joule
m	Meter
N	Newton
N-m	Newton-meter
Ω	Ohm
PF	Power factor
rad	Radian
RLC	Resistance-inductance-capacitance
rms	Root-mean-square
s	Second
S	Siemens
V	Volt
VA	Voltampere
W	Watt

The standard prefixes that are employed in SI are shown in Fig. P.1. Note the decimal relationship between these prefixes. These standard prefixes are employed throughout our study of electrical engineering.

Only a few decades ago, a millisecond, 10^{-3} s, was considered to be a short time in the analysis of electric circuits and devices. Advances in technology, however, have led to a state in which we now think of doing such things as performing calculations in nanoseconds or even picoseconds. The remarkable increases in speed and functional performance have been accompanied by phenomenal decreases in the physical size of electronic systems. Miniaturized integrated circuits are commonplace in calculators, computers, and other electronic equipment. A single such chip (typically about 1 cm on each side) can contain millions of devices, while the size of each device or circuit element on the chip is measured in fractions of a micron (1 micron = 10^{-6} m). As an example, consider the advanced integrated circuit chip shown in Fig. P.2.

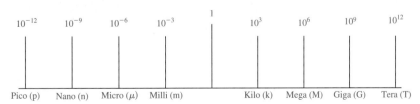

Figure P.1 Standard SI multiplier prefixes.

Figure P.2 An example of an advanced integrated circuit, the Intel Pentium™ Processor, contains 3.1 million transistors. (Courtesy of Intel; Pentium is a trademark of Intel Corporation)

P.5
Notation

Two of the most commonly used symbols in the study of electrical engineering are those for voltage (V) and current (I). The use of lowercase or uppercase letters and the particular subscript used with the letter provide information about the voltage or current as illustrated in the following example:

V, V_1, or V_A = a dc (direct current) value. The use of a capital V or I with number or capital subscript following indicates a constant, dc value.

$v(t)$, $v_1(t)$, $v_a(t)$, or v_A = the instantaneous value. The use of lowercase v or i with "(t)," or with a capital subscript, indicates the total instantaneous value "as a function of time."

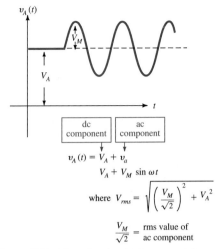

Figure P.3 Notation for Voltage and Current

V_M, V_m, I_M, or I_m = the amplitude or maximum value of a sinusoidally varying voltage or current.

V_{rms} or I_{rms} = the RMS or root-mean-square value of a sinusoidally varying voltage or current.

v_a or v_1 = the instantaneous value of the time-varying component (zero average).

The use of these quantities is illustrated graphically in Fig. P.3.

J. David Irwin
David V. Kerns, Jr.

INTRODUCTION

Introduction

1.1
Overview

The Electrical Engineer

In today's world, many technological problems are so complex that their scope spans a number of engineering and scientific disciplines. Teams of individuals are required to address the seemingly endless number of issues that impact the solution. As a result, many engineers find themselves working closely with individuals whose technical specialties are quite diverse from their own.

Each branch of science and technology has its own areas of specialization, and students in these areas are typically somewhat familiar with the breadth of their discipline. For example, in civil engineering, some typical subdisciplines are hydraulics, structures, and waste water treatment. In mechanical engineering, typical areas of specialization are fluid mechanics, solid mechanics, and thermodynamics. In fact, all engineering disciplines can be subdivided into a number of fairly distinct areas. Therefore, as we begin our study of electrical engineering, it is helpful to present an overview of the discipline and attempt to provide an appreciation of the vast array of technological areas addressed by electrical engineers.

Electrical engineering involves the conception, design, development, and production of the electrical or electronic products and systems needed by our technological society. Electrical engineers are professionals who play an essential role in creating and advancing this high-technology world of computers, lasers, robots, space exploration, communications, energy, and many other applications of electronic devices and systems. Electrical engineers (EEs) contribute with other engineers in areas as diverse as research, development, design, manufacturing, operations, sales, technical management, and education. Engineering projects today are often complex and require expertise from several engineering disciplines. EEs team with engineers from other areas (e.g., chemical, mechanical, civil, etc.) to design, develop, and help produce a great variety of services and products such as energy distribution systems, desktop computers, microprocessor-controlled systems, artificial hearts for the disabled, satellite systems, hand-held radios, radar systems, process control systems, electric cars, and almost anything else you can imagine that involves electrical or electronic components.

Good electrical engineering designs and products perform their intended function and do so safely, reliably, and cost effectively, and are easy to repair if broken; this text will assist in developing sensitivity to all these issues.

Many engineering students take the Fundamentals of Engineering (FE) exam administered by the National Council of Examiners for Engineering and Surveying. Successfully passing this exam is one of the first steps to becoming professionally licensed—an important consideration if you ever plan to offer your engineering services to the public. The same FE exam is taken by engineers from all disciplines. There is currently a portion of that exam dealing with electrical engineering, and one of the objectives of this book is to prepare you well for that exam.

Like graduates of other engineering disciplines, EE graduates may choose from a wide variety of career paths. Many choose to begin their careers in an industrial or government organization after completing the bachelor's degree. Some students continue their education to obtain a master's degree or a doctorate in electrical engineering. Others use their electrical engineering degree as a very strong base for graduate education in law, medicine, or management.

In almost every aspect of our everyday lives, we see electrical or electronic functions replacing older technologies. Automobiles have electronic dashboards and electronic ignitions; surveying is done with lasers and electronic range finders; manufacturing processes and other industrial processes in every field, from chemical refining facilities to metal foundries to waste water treatment, all utilize electronic sensors to detect information about the processes; electronic instrumentation systems to gather that information; computer control systems to process the incoming information, make decisions about the process flow, and send out electronic commands to actuators within the facility to correct and control the operation.

The material in this text will give you insight into the knowledge and tools required to understand how these systems operate and to be able to communicate effectively with experts in the field. It is intended to provide you with the vocabulary and the skills to assist you in specifying and purchasing electronic systems or equipment, setting it up in a laboratory or manufacturing area, and effectively utilizing this equipment or instrumentation to accomplish your intended goals. This knowledge will greatly enhance your ability to do your job, whether it be in engineering, management, or any related field.

The Field of Electrical Engineering

The field of electrical engineering involves the application of electricity to meet the needs of society. There are two primary uses of electricity: first, as a means to transfer electrical energy or power from one location to another; and second, to carry or transfer information.

There are a myriad of ways in which this is accomplished in individual systems and applications, and because electrical engineering is a dynamic field with innovative developments occurring constantly, new approaches appear rapidly. The wide scope and breadth of interest and study by electrical engineers is illustrated by the diversity of topics in the listing of the titles of the Transactions published by the Institute of Electrical and Electronic Engineers (the IEEE—often pronounced "I-triple-E"); such a list is presented in Table 1.1.

For simplicity, one might assume that the traditional core of electrical engineering study could be divided into seven major specialty areas. They are:

1. Power engineering
2. Electromagnetics
3. Communications
4. Computer engineering
5. Electronics
6. Systems
7. Controls

These broad areas are addressed in logical groups for ease of instruction in the six major parts of this text.

Power engineering, the only one in the list of seven that specifically deals with the generation and transmission of electrical energy from one location to another, is the oldest specialty within the field, and continues to be of great importance to society. We have become critically dependent upon electrical energy to power the appliances in our home and to power the tools and machines of industry.

Power engineering deals with the generation of electric power, its transmission over what are often very large distances, and the conversion of that power into forms that can do useful work. The generation of electric power most often is accomplished by the conversion of mechanical energy from a rotating shaft to electric energy in a generator. Many large power generating stations use some form of fuel to generate heat which produces steam. The steam is routed to a turbine which has blades on a shaft, much like a large fan, which rotates by the action of the steam. This rotating shaft couples the energy to the generator. Hydroelectric power generating stations use the energy of water falling by gravity to turn the turbine blades. Often the water is backed up in a river by a dam in order to create a convenient site for the water to fall a large distance. Nuclear power generating stations use the heat generated by nuclear fission to produce steam to turn the turbine blades.

More recently, solar energy has become a viable means of generating electric energy. Solar cells are carefully prepared slices of special materials fabricated in precise patterns and layers that directly convert the light from the sun into an electric current. Silicon solar cells as well as solar cells made from other materials are becoming more common as a source for electric power, particularly in remote locations.

Batteries are sources of electric power which derive their energy from a chemical reaction within the battery. Rechargeable batteries, such as the battery in your car, allow the chemical reaction to be reversed so that electric energy can be stored in the battery and then extracted at a later time.

Table 1.1 Transactions of the IEEE
The IEEE publishes the IEEE Transactions on:

Aerospace and Electronic Systems
Antennas and Propagation
Applied Superconductivity
Automatic Control
Biomedical Engineering
Broadcasting
Circuits and Systems
Circuits and Systems for Video Technology
Communications
Components, Hybrids, and Manufacturing Technology
Computer-Aided Design of Integrated Circuits and Systems
Computers
Consumer Electronics
Education
Electrical Insulation
Electromagnetic Compatibility
Electron Devices
Energy Conversion
Engineering Management
Geoscience and Remote Sensing
Image Processing
Industrial Electronics
Industry Applications
Information Theory
Instrumentation and Measurement
Knowledge and Data Engineering
Magnetics
Medical Imaging
Microwave Theory and Techniques
Neural Networks
Nuclear Science
Parallel and Distributed Systems
Pattern Analysis and Machine Intelligence
Plasma Science
Power Delivery
Power Electronics
Power Systems
Professional Communication
Reliability
Robotics and Automation
Semiconductor Manufacturing
Signal Processing
Software Engineering
Systems, Man, and Cybernetics
Ultrasonics, Ferroelectrics, and Frequency Control
Vehicular Technology

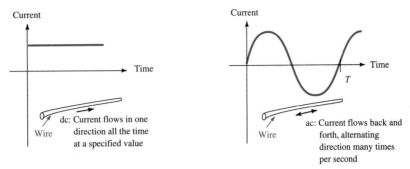

Figure 1.1   dc and ac.

Power generated in large power generation facilities is carried across many miles by high-voltage transmission lines, where it is distributed at various substations to the users of the power. This power is used for heating, generation of electric light, driving various types of electric motors, and many other applications. Electric power is classified as either dc or ac, direct current or alternating current, respectively.

The abbreviation "dc" stands for direct current, and refers to current which is flowing in a wire continuously in one direction at a specified value. On the other hand, "ac" means alternating current, in which the current goes back and forth, reversing its direction many times per second. The number of times the direction changes (and changes back) per second is called the *frequency,* and is expressed by the number of cycles per second, a unit called the Hertz (Hz). An illustration of dc and ac current is shown in Fig. 1.1.

Alternating current (ac) often follows a sinusoidal variation as a function of time. The positive and negative swing of the sine function in the previous figure represents the current first flowing one way and then the other—completing one entire "cycle" in time, *T,* the period.

So now, understanding that in ac the current goes back and forth, you ask, "What is current?" That is a good question which will be answered later in this chapter.

Power engineering has a long history. It was 1879 when Thomas A. Edison invented the first successful carbon-filament incandescent lamp. Within three years, commercial application of this wondrous new light began. On September 4, 1882, Edison started operations of the first central-station electric generating plant in New York City, Pearl Street Station. This station supplied power to 59 customers within an area of approximately one square mile; the system could supply 400 lamps, each rated at 83 watts. The same year a similar station was put into service in London (Holborn). These early systems used low voltage, less than 100 volts dc; by 1886, Edison Companies were experiencing difficulties because they could deliver power only a short distance from their generating stations. With their dc distribution, as voltage dropped at increasing distances from the generating station, there was no way to step it back up. This limitation could be overcome with the use of ac and the transformer, which was made commercially practical by Stanley in 1885. Nonetheless, Edison pressed on with his commitment to dc systems and there was heavy debate and discussion.

Within a few years the advantages of ac triumphed, primarily due to the ability to step the voltage up for transmission, and then down to the proper value for use near the customer's site. The first ac transmission in the United States was in 1889, and took place between Oregon City and Portland, 13 miles (21 km) away. A year later a major ac power station was commissioned at Deptford to supply power to central London. Today virtually all electric power distribution, worldwide, is done by ac.

Household current in the United States and Canada today operates at a frequency of 60 Hz, or the current reverses direction, flowing back and forth 60 times in each second. In Europe the standard is a frequency of 50 Hz. In the early development of electric power there was no universally accepted frequency; one of the most common in the late 1800s was 133 Hz. In 1891, a standard of 60 Hz was proposed, but for many years frequencies of 25, 50, and 60 Hz were all utilized in various parts of the United States. The city of Los Angeles operated on 50 Hz, but converted to 60 Hz when Hoover Dam power became available; the Edison Company completed its conversion to 60 Hz in 1949.

The discipline of *electromagnetics* bridges the gap between applications of electricity for energy transfer and the remaining disciplines, which are primarily associated with information transfer. Electromagnetics deals with the interaction between magnetic fields, electric fields, and the flow of current. A coil of wire carrying an electric current generates a magnetic field; a piece of iron brought in proximity to this "electromagnet" will experience a force on it which can be made quite strong. This is the origin of the force used to turn electric motors and cause motion in other electromechanical devices.

At dc, and low-frequency ac (up to several hundred Hz), the electric and magnetic interactions remain relatively localized around the current-carrying wire. As frequency is increased, however, a wonderful thing happens. Energy begins to radiate from the wire and propagate through the atmosphere as electromagnetic waves. These are the waves that make possible radio communications, television, satellite communications, radar, and all related forms of wireless communications. These are the same electromagnetic waves which are beamed into a chicken in your microwave oven to transfer energy for cooking. An alternating current applied to a conducting structure tailored to be effective in radiating the energy into the atmosphere can transmit electromagnetic waves around the world and beyond. Such a structure is called an antenna.

Electromagnetic waves propagating through space travel at the speed of light, c, which is 186,000 miles per second or 3×10^8 meters per second. Light is an electromagnetic wave within a particular range of frequencies; the exact frequency determines the color.

The frequency of an electromagnetic wave affects many of its characteristics, including the way it propagates through the atmosphere. Figure 1.2 shows the electromagnetic spectrum, with frequency plotted horizontally and the common usage of various parts of this spectrum indicated.

The field of *communications* focuses on the engineering and science required to transmit information from one place to another. Electronic communications systems often utilize unconfined electromagnetic waves for information transfer; however, it is also common for data to be transmitted by telephone wires, various cables, or optical fibers. One important question in the area of communications deals with the ways in which information is encoded on an electrical signal. The process of putting information on a carrier signal is called modulation or encoding, and the process of extracting that information at the receiving end is called demodulation or decoding. As an example, when information is carried by an electromagnetic wave, its frequency might be varied up and down slightly to encode the desired information. This form of modulation, called *frequency modulation*

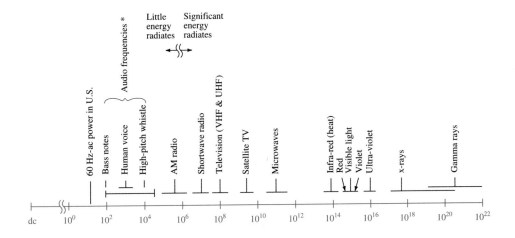

Hertz (Hz)

Figure 1.2 Frequency spectrum. *Note: We hear sounds by acoustical waves at these frequencies.

or FM, is commonly used in stereo music broadcasting today. AM is an abbreviation for *amplitude modulation,* in which the intensity or amplitude of the electromagnetic wave is varied in time to transmit the desired information.

Developments in communications engineering have produced much more sophisticated systems for the encoding of information for transmission, and for the extraction of the desired information from the received signal at the other end. During transmission the desired signal is often mixed with a great deal of interference or random noise, and communications engineers develop filtering techniques to exclude the unwanted part and isolate the desired information. Many of these techniques involve state-of-the-art computer sampling and digital processing of the information.

Computer engineering is one of the fastest-growing specialties within electrical engineering, and includes the design and development of computer hardware systems and the computer programs or "software" that control them. Computer systems range in size from the large "mainframes," which are generally used for highly complex computing tasks, to special-purpose computers for engineering, business, accounting, banking, finance and many other purposes. Personal computers (PCs) and workstations are being used at a rapidly increasing rate in industry, as well as for individual use. Engineering and science students in all disciplines will make use of the computer hardware and software developed by computer engineers in their specific field.

Electronics refers to the utilization of various materials in special configurations or structures to make devices which can valve "on" or "off" the flow of current. These devices can be interconnected to form circuits. Such electronic circuits may simultaneously control the flow of many different electric currents and be able to perform complex functions, such as the electronic computer just described. To create the complex hardware utilized in today's electronic systems, components such as transistors, diodes, and integrated circuits are utilized; these elements will be studied in Part II of this book. This electronic hardware includes everything from circuit boards for computers to instrumentation systems, from electronic pacemakers to the electronic engine monitors in new automobiles, from radio and television transmitters and receivers to complex radar systems, and many other electronic applications.

The rapid growth of the field of electronics can be largely attributed to the development of the integrated circuit or "chip." Integrated circuit technology provides a means of fabricating a very large amount of electronic circuitry in a very small volume, at low cost, with high reliability and generally low power consumption. A single silicon chip containing a four-megabyte static memory, for example, may contain over 20 million transistors, all of which must operate and be interconnected with each other properly; this is accomplished in an area of the order of one square centimeter.

The specialty of *systems engineering* utilizes mathematical principles to model and describe complex systems and predict their performance based on engineering analyses. Systems engineers might, for example, utilize mathematical principles and an engineering approach to describe the flow of airplanes in and out of an airport, or the traffic through New York City at rush hour. With such a mathematical description of the problem, the engineer can play "what if" games and attempt to optimize the system for a given set of conditions. Further, he or she might predict what would happen to the system under a rare set of conditions not generally observed, and thus determine what the limits are for proper system operation. Systems engineering principles can be applied to quantitatively address a wide variety of society's problems.

Control systems are a very important class of systems, and the field of "controls" is widely studied within electrical engineering departments. Control systems are often electronic systems designed to provide fast and accurate mechanical adjustments or placements upon some command. Such actions as the control of the horizontal stabilizer or tail of a fighter aircraft as the pilot pushes the control wheel, can be described by the application of the mathematical and engineering concepts of control engineering. Robotics and the precise control required of robotic arms in picking and placing objects has placed new emphasis on this area. Control systems are certainly not limited to mechanical adjustments; they are used to provide fast and accurate "control" of electrical systems, chemical systems, hydraulic systems, and others.

An understanding of electric *"circuits"* is essential in the study of almost all of the other specialties in electrical engineering; it is at the core of the material required for an introduction to the field. The latter part of this chapter will begin to introduce the basic elements of electric circuits; further development of these concepts and the tools one needs for analyzing circuits will be presented in the chapters of Part I. Subsequent parts of the text will make use of these circuit concepts.

The field of electrical engineering is constantly in a state of change due to technology innovations which rapidly and continuously occur. In electrical engineering we have the challenge of learning more and more about more and more new things because the devices and systems on which we work are constantly evolving; the basic rules of the game change almost daily.

There is, however, also a set of fundamental concepts and principles in electrical engineering forming a core body of knowledge and which make some understanding of all the new innovations possible. This text strives to provide a reasonable balance of fundamental core knowledge with insight into the latest and most important technologies of application.

The dynamic growth and change of this field is both a blessing and a curse. It demands continuous education and re-education, a life-long aggressive pursuit of new knowledge and understanding as new technologies and ideas emerge; but it also has its reward. That reward is the excitement of being involved in one of the greatest revolutions in the history of the world; this revolution is likely to be recorded by historians as significant as was the industrial revolution of the previous century. The products and benefits produced by engineers of various disciplines working together have the potential to change and improve the lives of people throughout society.

1.2
Basic Concepts

Charge

Electrical engineers deal primarily with charge, its motion, and the effects of that motion. *Electricity* is a word most often used in a nontechnical context to describe the presence of charge; the term *electricity* is used both to describe charge in motion (for example, through a wire as an *electric current*) and stationary charge, *static electricity.*

Charge is a fundamental property of matter and is said to be conserved—that is, it can neither be created nor destroyed. This means that if the charge moves away from one location, it must appear at another. There are two types of charge, positive charge and negative charge. Charge is the substance of which electric currents are made.

Charges near each other will attract each other or repel from each other according to the following rule: Like charges repel each other; opposite charges attract each other.

The basic structure of an atom is held together by the attractive force between unlike charges. Recall that an atom of hydrogen, with a positively charged nucleus, has an electron with a negative charge which is held in an orbit around the nucleus by the constant attraction between these unlike charges. You've probably done experiments where you rubbed a wool sweater (or a cat) on a hard rubber comb and created static electricity. The comb was originally neutral, containing an equal balance of positive and negative charges. The rubbing action strips charge off the wool and the comb in an unequal way so that there is no longer a balance; one contains more positive charge and less negative charge than the other. The unequal charges attract each other and the charged comb can be used to attract little bits of paper in order to demonstrate the attractive force between unequal charges. In this example the charge once placed on the comb doesn't move easily, and hence the designation static electricity. In case you're wondering, in the above experiment, the comb picks up extra electrons and becomes negatively charged.

Charge is designated by the symbol q, and is measured in units of coulombs (C). The negative charge carried by a single individual electron is -1.602×10^{-19} coulombs, and this is the smallest unit of charge that exists. Therefore, charge is quantized in blocks, the magnitude of the charge on a single electron.

The Force Between Two Charges

The force between two small clusters of charge (each one small enough to be considered a point) has been found to be described by the following equation:

$$F = k\frac{q_1 q_2}{d^2} \tag{1.1}$$

where q_1 is the charge at position 1 in coulombs, q_2 is the charge at position 2 in coulombs, d is the distance between the charges in meters, and k is a constant of 8.99×10^9 newton meter^2/C^2.

EXAMPLE 1.1
If $q_1 = 0.50$ coulombs and $q_2 = -0.03$ coulombs, let us calculate the separation between these two charges if they are attracted together by a force of one newton.
From Eq. (1.1) we find that

$$d = \sqrt{k\frac{q_1 q_2}{F}}$$

and hence

$$d = \sqrt{(8.99 \times 10^9)\frac{0.50 \times 0.03}{1}} = 11,612 \; meters \qquad ■$$

Note: In these calculations, the square root will always be applied to a positive number; if the charges are of opposite sign, the force, F will be negative (indicating attraction).

EXAMPLE 1.2
Two electrons drifting through space have come to a position where they are separated by a distance of one micron, that is, 1×10^{-6} meters. We wish to calculate the force between these two electrons and determine if it is attractive or repulsive.
Applying Eq. (1.1) we find that the force $F = 2.3 \times 10^{-16}$ newtons; since the two electrons each have negative charge, they repel each other. ■

DRILL EXERCISE

D1.1. A point charge of 1 C creates an attractive force of 0.2 N when placed at a distance of 0.1 m from a second charged object. Find the charge on the second object.

Ans: -2.23×10^{-13}C.

Conductors and Insulators

In order to put charge in motion so that it becomes an electric current, we must provide a path through which it can flow easily. In the vast majority of applications, charge will be carried by moving electrons along a path through which they can move easily. Materials through which charge flows readily are called *conductors*. Most metals, such as copper, are excellent conductors and therefore are used for fabrication of electrical wires and the conductive paths on electronic circuit boards.

Insulators are materials which do not allow charge to move easily. Therefore, electric current cannot be made to flow through an insulator. Charge placed on an insulating material, such as the rubber comb, just stays there as static electricity; charge has great difficulty moving through it. Insulating materials are often wrapped around the center conducting core of a wire to prevent the charge from flowing off to some undesired place if the wire inadvertently touches some other object.

Table 1.2

Conductors	Semiconductors	Insulators
Silver	Silicon	Glass
Gold	Germanium	Plastic
Copper	Gallium Arsenide	Ceramics
Aluminum		Rubber

Resistance will be defined quantitatively later; however, qualitatively a conductor has a low resistance to the flow of charge, and an insulator has a very high resistance to the flow of charge. There is a wide range of charge conducting abilities of various materials; the resistance to charge movement of copper is about 10^{25} times lower than that of a comparable single piece of quartz. *Semiconductors* fall in the middle between conductors and insulators, and have a moderate resistance to the flow of charge. Table 1.2 lists some common conductors, semiconductors, and insulators.

EXAMPLE 1.3

Based on our experience, let us classify the following materials as insulators or conductors, that is, selecting the one that fits best: iron, paper, wood, salt water.

Using either experience or intuition, we select as conductors: iron and salt water, and as insulators: paper and wood. ∎

Current and Voltage

This section defines two of the most commonly used terms in electrical engineering: *current* (measured in amperes) and *voltage* (measured in volts).

Charge in Motion. Electric *current* implies "charge in motion"; the term *current* is simply a measure of how much charge is moved per unit of time. Current is measured in *amperes,* frequently called the amp and abbreviated as A; one ampere is defined as the transfer of one coulomb in one second. Imagine for example a wire carrying four amperes of current. Now imagine a cross section cut through this wire; in one second of time, four coulombs of charge pass through the plane of that cross section. Therefore, one ampere is equal to one coulomb/second and four amperes is equal to four coulombs/second.

EXAMPLE 1.4.

Let us determine the number of electrons that pass through a cross section of a wire carrying one ampere in one second of time.

Recall, the charge on a single electron is -1.6×10^{-19} coulombs. Therefore, the number of electrons in 1 coulomb is

$$\frac{1}{1.6 \times 10^{-19}} = 6.25 \times 10^{18} \ electrons \qquad ∎$$

EXAMPLE 1.5.

A lightning bolt carries 1,000 amperes of current and lasts 38 microseconds. Let us calculate the charge that is deposited on a golf cart hit by the lightning strike.

If we let Q be the total charge deposited on the cart,

$$Q = \frac{\Delta q}{\Delta t} \times \Delta t = current \times time\ interval$$

and hence

$$Q = (1000\ amperes) \times (38 \times 10^{-6}\ seconds) = 0.38\ coulombs \qquad \blacksquare$$

As illustrated in the above example, the *average* current over a time interval is defined as the total charge transferred divided by the total time interval; we will use the symbol I to represent the average current.

$$I = \frac{\Delta q}{\Delta t} \tag{1.2}$$

where Δq is the change in charge, and Δt is the time interval over which that change occurs.

The instantaneous current, i, is defined as the time rate of change of the charge transfer at any particular instant and is defined by

$$i = \frac{dq}{dt} \tag{1.3}$$

The letter "i" is used to represent current in most texts; the letter "C" was already taken for "coulomb."

DRILL EXERCISE

D1.2. How long does it take a 20A battery charger to deliver a charge of $10^4 C$?

Ans. $500s.$

Charge can be transported by various mechanisms. As mentioned, the most common is the movement of electrons through a conductor; however, positive ions flowing the opposite direction can transfer charge, as is the case in electrochemical reactions in batteries or in electroplating. In addition, solid state electronic devices use semiconductor materials in which charge can be moved by electrons carrying negative charge, and "holes" carrying positive charge. The total current through any particular plane is the total net charge transferred divided by the time interval.

$$i = \frac{dq}{dt} = \frac{dq^+ + dq^-}{dt} \tag{1.3a}$$

where dq^+ is the incremental positive charge transferred and dq^- is the incremental negative charge transferred.

Current, a scalar quantity, requires a sign convention; if we assume that the direction we call "positive current" flow is the direction positive charge moves (which is what we will assume in this text), then Eq. (1.3a) reveals that the total current is the sum of the rate of flow of positive charge in one direction and negative charge in the opposite direction.

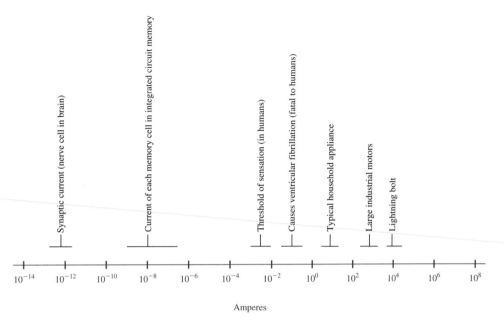

Figure 1.3   The range of currents.

The *sign convention,* which we will adopt here, is the following: Positive current will be defined as the net rate of flow of positive charge. Therefore, a wire conducting electrons to the left will be described as having a positive current to the right.

In our world we encounter currents over many orders of magnitude. Lightning strikes consist of bursts of current that can be tens of thousands of amperes, while at the other end of the scale, the current in a nerve pulse may be only picoamperes. In between is a wide range of currents. For example, typical household appliances require from 0.5 to 10 amperes; individual electronic circuits may require microamperes to milliamperes, and large industrial motors may require hundreds of amperes. A chart showing the range of commonly encountered currents is shown in Fig. 1.3.

Fluid Analogy. In beginning the study of electrical engineering, it is often helpful to have an easily understood analogy for some of the concepts. We will use such analogies throughout this book. Consider water flowing in a pipe as analogous to current flowing in a wire. We measure the quantity of water, let's say in gallons; the quantity of electricity is the charge, measured in coulombs. The flow rate of water, the amount of water flowing through a cross section of pipe per unit time, will be measured in gallons per second; the amount of electric current is the flow rate of charge measured in coulombs per second which we have given the special name ampere.

Note that current is always a measure of the flow through a conducting path, measured at a particular cross section through the path, just as flow rate of water is measured at a particular point along the pipe. In the top half of Fig. 1.4, we can see that if the conducting path, be it a wire or a pipe, has no branches or alternate paths, the flow rate at any point along the pipe must be the same.

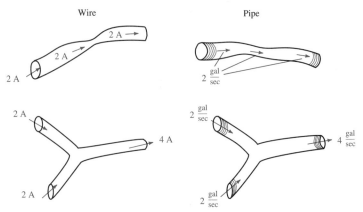

Figure 1.4   Flow rate analogy.

If conducting branches merge into a single path, as illustrated in the lower half of Fig. 1.4, the sum of the flow rates in the entering branches must equal the flow rate of the out-going branch. This is illustrated in the figure with two branches, but it could be any number. Put simply, this is an illustration of the concept "what goes in must come out" (if there's no storage device). This is an illustration of Kirchhoff's current law, which will be explained in more detail in Chapter 2.

Voltage and Energy. Continuing for a moment with the analogy of water flowing in a pipe, imagine that this pipe connects to the input of a turbine, as shown in Fig. 1.5(a). The water enters the turbine and pushes against the blades on a shaft, rotating the shaft and doing work. The water is collected at the bottom of the turbine and exits through another pipe. The water entering the turbine has a higher potential energy than that exiting; the potential energy is converted in the turbine to useful work (as energy delivered to the rotation of the shaft), and then exits at a lower potential energy. In the case of water, this potential energy is measured by the pressure. The water pressure entering the turbine is higher than that leaving and this pressure difference forces the water through the turbine. If the water pressure were the same at the entry point of the turbine as at the exit, then there would be no push or force to drive the water through the turbine. Thus, it is the pressure difference across the turbine that forces the water through it.

The fluid collected at the output of the turbine is at a lower pressure. If this output fluid is to be cycled through again to continue the process, it must be increased in pressure so that it will have the potential energy necessary to drive through the turbine again. This is accomplished by routing the collected water to a pump. The pump provides a source of energy that forces the fluid around in the closed loop.

Now consider an electric current flowing through an electric light bulb as illustrated in Fig. 1.5(b). The "force" or "pressure" that pushes the charge through the light bulb is the voltage difference supplied by the battery. The charge which flows through the light bulb does work, that is, w joules of energy for generating heat and light, and then the charge emerges at the other terminal of the bulb at a lower voltage or potential. There is a *voltage difference* across the bulb, $V_A - V_B$, that moves charge through the bulb; if the voltages were the same, no charge would move.

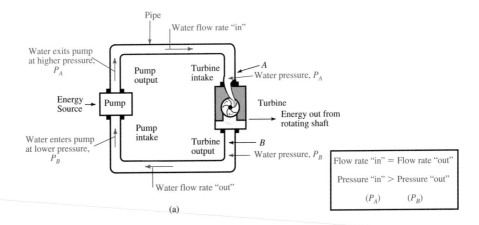

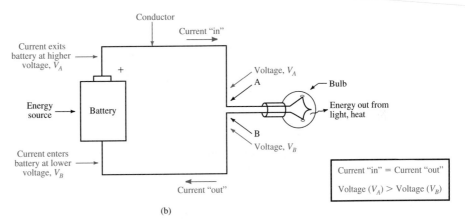

Figure 1.5   Simple fluid and electric circuit.

The voltage value represents potential energy, and the potential difference or voltage difference across the bulb reveals the energy-transfer capability of the charge flow. Voltage is measured in joules per coulomb, defined as the volt, V, and one volt is the energy in joules required to move a positive charge of one coulomb (C) through an element.

Therefore, assuming differential amounts of charge and energy:

$$v = \frac{dw}{dq} \qquad (1.4)$$

EXAMPLE 1.6.

If the battery, connected as shown in Fig. 1.5(b), supplies a voltage difference of 1.5 volts, such as a flashlight battery, let us calculate the amount of energy used to transfer 0.5 coulombs of charge through the bulb.

From Eq. (1.4)

$$dw = v\, dq$$

Integrating, we obtain

$$w = \int dw = v \int dq = vq = (1.5 \text{ V}) (0.5 \text{ C}) = 0.75 \ joules$$

∎

D1.3. A 20-volt battery delivers 0.1 J of energy. (a) How much charge was delivered by the battery? (b) If the process in (a) occurred in 1 μs, what was the current?

Ans: (a) 5×10^{-3} C, (b) 5×10^3 A.

We encounter a wide range of voltages in everyday experiences. Figure 1.6 shows some examples of voltages ranging from the extremely large voltage differences between clouds and earth that can produce lightning, to the very small voltage differences between locations on the human scalp resulting from electrical activity in the brain. Recordings of the latter are called electroencephalograms (EEGs) and are a common medical diagnostic tool for neurological brain disorders.

Circuit Analogy. Figures 1.5(a) and (b) both show single-loop "circuits," the first utilizing a fluid flow, and the second a current flow.

In Fig. 1.5(a), notice that the fluid exits the pump at a high pressure. If the pipe is sufficiently short or is large enough and of good design, we can assume that there is no pressure loss from the exit of the pump to the entry point of the turbine. We make the analogous assumption in our study of electric circuits as illustrated in Fig. 1.5(b). The conductor paths, represented by solid lines in the circuit drawing, are assumed ideal and conduct electric charge perfectly, that is, with no loss in voltage. Therefore, the voltage at the plus terminal of the battery is the same as the voltage, V_A, at the top of the bulb; this same voltage would be measured at any point along that conductor between those two points.

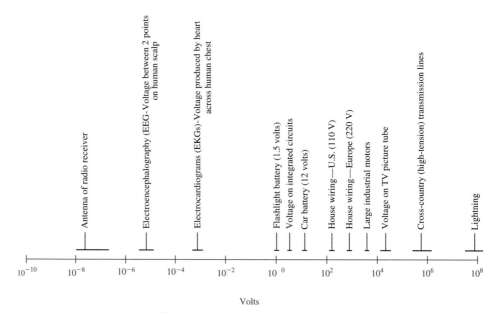

Figure 1.6 The range of voltages.

Similarly, the voltage at the bottom of the bulb is the same anywhere along the lower conducting path to the negative terminal of the battery.

The circuit drawing in Fig. 1.5(b) represents therefore an idealized "model" of the actual physical circuit. We construct models making certain simplifying assumptions about the various parts of the circuit that will enable us to easily apply mathematical techniques and predict the circuit's performance. If our predicted values agree with those actually measured in real circuits, then the validity of these simplifying assumptions is justified. In a later section in this chapter we will discuss modeling in more detail.

Unless the turbine in Fig. 1.5(a) contains some magical storage capacity, in the steady state, that is, after the system has been operating for a while, the flow rate "out" must be the same as the flow rate "in." That is, there can be no net accumulation of water in the turbine that occurs for an unlimited time. It's difficult to imagine more water flowing out than flowing in because there is no source for the extra water; and, the only way less water could flow out than flows in is if it is diverted to some other path or storage area within the turbine. This might happen initially when the system is first turned on—as the reservoir in the bottom of the turbine fills. However, in steady state, the amount of water flowing in during a given amount of time must equal the amount of water flowing out in the same amount of time, that is, their flow rates must be equal.

The analogous situation holds in the electric circuit of Fig. 1.5(b). When the circuit is first energized, there may be a brief period of time when the current in does not equal the current out due to establishing an equilibrium charge on various components within the circuit. This "transient" behavior will be studied in Chapter 3. However, after the battery has been connected in the circuit as shown in Fig. 1.5(b) for some time, the current "in" will exactly equal the current "out." The steady state behavior of dc circuits will be examined in Chapter 2 and the steady state behavior of ac circuits will be examined in Chapter 4.

In single-loop circuits such as that illustrated by Figs. 1.5(a) and 1.5(b), the flow rate, or current in the electric circuit, is the same at any point around the entire loop.

Reference Directions and Polarities

In solving problems involving mechanical systems, we generally set up a coordinate system which defines what we will assume is the positive direction; then when we calculate velocity, for example, and if the answer is a positive number, we know the motion is in the same direction as our assumed positive direction. If, on the other hand, we compute a negative velocity, then the actual motion is in the opposite direction.

We must set up an analogous reference direction or polarity in the solution of electric circuit problems. It is extremely important that the variables that are used to represent voltage between two points be defined in such a way that the solution will let us interpret which point is at the higher potential with respect to the other.

In Fig. 1.7(a), the variable that represents the voltage between points A and B has been defined as V_1. The + and − signs define a reference direction for V_1. Since V_1 has a positive value (+2 volts), the terminal with the + reference (terminal A) is at a higher potential than terminal B by 2 volts. If a unit positive charge is moved from point A through the circuit to point B, it will give up energy to the circuit and have 2J less energy when it reaches point B. If a unit positive charge is moved from point B to point A, extra en-

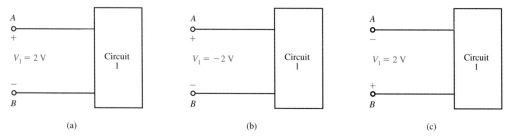

Figure 1.7 Voltage polarity relationships.

ergy must be added in the amount of 2J, and hence the charge will end up with 2J more energy at point *A* than it started with at point *B*.

If, on the other hand, V_1 had a negative sign (like $V_1 = -2$ volts), then it would imply that terminal *B* was at a higher potential than terminal *A*, and the situation above would be reversed. This is illustrated in Fig. 1.7(b); this figure tells us that terminals *A* and *B* have a potential difference (voltage difference) of 2 volts, and that terminal *B* is more positive than terminal *A*.

The same situation as that described in Fig. 1.7(b) is presented in a fully equivalent way in Fig. 1.7(c). Here the reference direction (the + and − signs) are reversed, but the polarity of V_1 is also reversed, so it still conveys the fact that terminal *B* is more positive than terminal *A*.

As you can see, it is absolutely necessary to specify both magnitude and direction for voltage and current. It is incomplete to say that the voltage between two points is 10 volts or the current in a line is 2 amperes, since only the magnitude has been given. A figure or schematic circuit drawing with a reference direction shown for each voltage and current is the most commonly used way to show the reference polarities.

Energy is another important term in circuit analysis, and the direction of energy transfer is defined by the way the signs of the voltage and current are presented. In Fig. 1.8(a), energy is being *supplied to* the element by whatever is attached to the terminals. Note that 2 A, that is, 2 C of positive charge, is moving from point *A* to point *B* through the element each second. Each coulomb loses 3J of energy as it passes through the element from *A* to *B*. Therefore, the element is absorbing 6J of energy per second. Note that *when the element is absorbing energy, a positive current enters the positive terminal*. In Fig. 1.8(b),

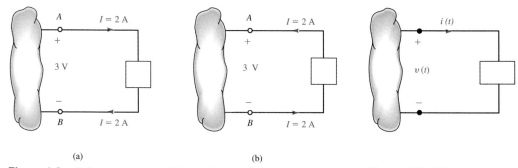

(a)   (b)

Figure 1.8 Voltage–current relationships for (a) energy absorbed and (b) energy supplied

Figure 1.9 Sign convention for power.

energy is being *supplied by* the element to whatever is connected to terminals $A - B$. In this case, note that *when the element is supplying energy, a positive current enters the negative terminal and leaves via the positive terminal.*

Power

Power is defined as the time rate at which energy, w, is produced or consumed, depending on whether the element is a source of power or a user of power, respectively. That is

$$p = \frac{dw}{dt} \tag{1.5}$$

which can be rewritten as

$$p = \frac{dw}{dt} = \left[\frac{dw}{dq}\right]\left[\frac{dq}{dt}\right] = v\, i \tag{1.6}$$

The above equation shows that power can be computed by the product of the voltage across a circuit element and the current through it. Since both voltage and current can vary with time, the power, p, is also a time varying quantity, and can be expressed as $p(t)$.

Therefore, the change in energy from time t_1 to time t_2 can be found by integrating Eq. (1.6), that is

$$w = \int_{t_1}^{t_2} p\, dt = \int_{t_1}^{t_2} vi\, dt$$

Calculation of power requires the use of a consistent sign convention. To determine the sign of any of the quantities involved, the variables for the current and voltage should be arranged as shown in Fig. 1.9. The variable for the voltage $v(t)$ is defined as the voltage across the element with the positive reference at the same terminal that the current variable $i(t)$ is entering. This convention is called the *passive sign convention* and will be so noted in the remainder of this book. The product of v and i, with their attendant signs, will determine the magnitude and sign of the power. If the sign of the power is positive, power is being absorbed by the element; if the sign is negative, power is being supplied by the element.

EXAMPLE 1.7
We wish to determine the power absorbed, or supplied, by the elements in Fig. 1.8.

In Fig. 1.8(a), $P = VI = (3\ \text{V})\ (2\ \text{A}) = 6$ watts (W) is absorbed by the element. In Fig. 1.8(b), $P = VI = (3\ \text{V})\ (-2\ \text{A}) = -6W$ is absorbed by the element, or $+6W$ is supplied by the element.   ∎

EXAMPLE 1.8
Given the two diagrams shown in Fig. 1.10, let us determine whether the element is absorbing or supplying power and how much.

In Fig. 1.10(a), the power is $P = (2\ \text{V})\ (-4\ \text{A}) = -8W$. Therefore, the element is supplying power. In Fig. 1.10(b), the power is $P = (2\ \text{V})\ (2\ \text{A}) = 4W$. Therefore, the element is absorbing power.   ∎

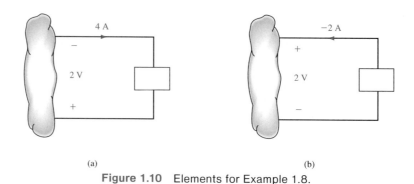

(a)   (b)

Figure 1.10   Elements for Example 1.8.

D1.4. Determine the amount of power absorbed or supplied by the elements of Fig. D1.4.

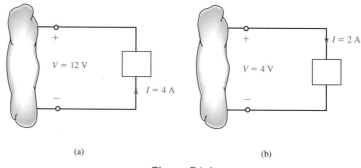

(a)   (b)

Figure D1.4

Ans: (a) $P = -48W$; supplied (b) $P = 8W$; absorbed.

Modeling

Modeling is the process of representing a real physical system with a representation of that system that under some assumptions allows us to solve for its behavior using the analytical tools we have available. Generally, we choose assumptions which make things simpler by disregarding unimportant features. We will apply various mathematical techniques to the model of our system to arrive at solutions. A "predictive model" is one which represents the actual system sufficiently well that we can impose different sets of assumptions and conditions, and accurately predict, by our calculations using the model, the expected behavior of the real physical system.

In our study of electric circuits we make a number of simplifying assumptions. First we assume that the important characteristics of the circuit can be grouped together in "lumps" or separate blocks, connected together by ideal conductors. This approach is called "lumped element circuit modeling" and is used in electrical and electronic circuit analysis up to microwave frequencies. We will use this approach in this text.

The single-loop electric circuit of Fig. 1.5(b) could be described or represented by a single loop of ideal conductors connecting a rectangular block on the left representing the

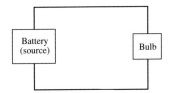

Figure 1.11 A simple circuit.

battery with a rectangular block on the right representing the light bulb as shown in Fig. 1.11.

Two terminal circuit elements are classified as either *active* or *passive* based simply on whether they supply energy to the circuit or absorb energy.

Batteries and generators are typically modeled by active elements since they supply energy to a circuit. There are three fundamental passive circuit elements: resistors, capacitors, and inductors.

Ideal Circuit Elements

To make the analysis of electrical circuits simpler, we have defined some idealized circuit elements. These elements can be completely described by knowing the mathematical relation between the voltage across and the current through the element. Idealized active elements consist of independent sources and dependent sources. Each of these comes in two types, the voltage source and the current source.

There are three circuit elements that normally are assumed as single-valued in circuit analysis, and as such, they are called ideal passive circuit elements: the resistor, the inductor, and the capacitor. All of these will be discussed in detail in later chapters; however, a brief introduction is provided here. These three passive circuit elements differ electrically in the way in which the voltage across is related to the current through each of the elements. As you will see, if v is the voltage across each element and i is the current through each element, the relationships and the commonly used symbols are given in Fig. 1.12. The constants, R, L, and C, are known as the resistance (in ohms, Ω), inductance (in henries, H) and capacitance (in farads, F), respectively.

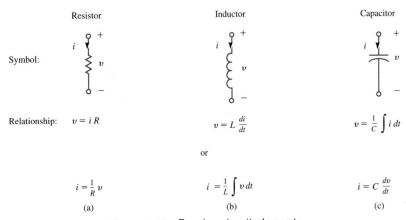

Figure 1.12 Passive circuit elements.

The resistor represents the part of a circuit component in which energy entering the element by the flow of current through it is transformed to heat. Light can be emitted also if the resistive element becomes hot enough to glow.

The inductor represents a two-terminal electric element in which energy is stored in a magnetic field. Coils of wire such as those used to make an electromagnet, or the windings of wire in an electric motor, must be modeled using inductances.

The third passive circuit element, the capacitor, stores energy in an electric field. The capacitor is often fabricated by two parallel conducting plates separated by an insulating layer.

These three circuit elements, combined with the active elements (sources), can be used to represent, model, and study a wide range of electrical and electronic systems. The next chapter describes the first of these passive elements, the resistor, in more detail.

Sources

Again consider the analogy of fluid flow in the "circuit" of Fig. 1.5(a). The pump is the source of energy that powers the flow of fluid—it forces or "pushes" the fluid by providing a fluid pressure at the exit of the pump which is higher than that at the intake of the pump.

We might imagine the pump acting in one of two ways. First, we might construct an image of a pump which increases the pressure of the fluid a predefined amount. That is, the pressure at the output of the pump is a fixed number of pressure units larger than that at the input; the pump would provide this pressure difference irrespective of the amount of fluid flowing through the pump. Now, imagine what happens if suddenly the diameter of the pipe and the turbine could somehow magically be increased, so that much more fluid could flow. In order to maintain the same pressure differential as before, the pump would have to quickly work much harder and move much more water per unit time. Conversely, if the flow around the loop were suddenly restricted so that it was more difficult to force as much volume of fluid per unit time around the circuit, the flow rate through the pump would have to quickly decrease; otherwise, there would be a sudden increase in pressure.

The second type of pump would be one which maintains a predefined flow rate through it independent of the pressure differential across it. In this situation if the loop of pipe somehow developed a restriction to flow, the pump would continue to force the same number of gallons per minute around the loop; however, in order to do this it would have to increase the pressure differential across it.

We will use these analogies in the next section.

Independent Sources. There are electrical sources with direct analogies to the situations described above. An ideal independent voltage source maintains a specified voltage across its terminals independent of the current through it.

This source corresponds to the first type of pump described in the previous section. The voltage between the terminals of the ideal voltage source is determined by the value of the voltage source, regardless of the current passing through it, and regardless of any other circuit parameters.

The specified value of voltage could be a constant value, V_1, in which case we represent this source by the symbol shown in Fig. 1.13(a), which would closely model a typ-

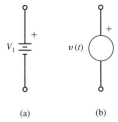

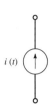

(a)   (b)

Figure 1.13   Independent voltage sources.

Figure 1.14   Independent current source.

ical battery. If the specified voltage is a predescribed function of time, $v_1(t)$, we represent this source with the symbol shown in Fig. 1.13(b).

In both of these sources the figure shows an additional and very important piece of information, the "reference polarity" or "sign" of the source. In the case of a constant voltage source, that is, a dc source, the "long-bar" end of the symbol is defined to be the positive reference side, and the "+" designates the positive reference side in the other symbol. This designation is equivalent to establishing the positive direction in a coordinate system. It does not mean that the voltage is always more positive at the "+" terminal. It means that if the value of $v(t)$ is positive, then the "+" terminal will be more positive than the other terminal.

The corresponding electrical analogy to the second type of pump described in the previous section is the ideal current source. An ideal independent current source maintains a specified current through it, independent of the voltage developed across its terminals.

The symbol for the ideal independent current source is shown in Fig. 1.14. The direction of the arrow indicates the chosen reference direction for "positive" current flow. If the value of $i(t)$ becomes negative, then at that time, positive current would actually be flowing in the opposite direction of the arrow.

The same symbol can be used to represent a dc current source by letting $i(t)$ be a constant, I.

It is important that we pause here to inject a comment concerning the application of circuit models. In general, mathematical models approximate actual physical systems only under a certain range of conditions. Rarely does a model accurately represent a physical system under every set of conditions. To illustrate this point, consider the model for the voltage source in Fig. 1.13(b). We assume that the voltage source produces $v(t)$ volts, regardless of what is connected to its terminals. Theoretically, we could adjust the external circuit so that an infinite amount of current would flow, and therefore the voltage source would deliver an infinite amount of power. This is, of course, physically impossible. Hence, the reader is cautioned to keep in mind that models have limitations and are valid representations of physical systems only under certain conditions.

Dependent Sources. In contrast to the independent sources, which produce a particular voltage or current completely unaffected by what is happening in the remainder of the circuit, dependent sources generate a voltage or current that is determined by a voltage or current at a specified location in the circuit. These sources are important because they are an integral part of the circuit models used to describe the behavior of many electronic circuit elements, such as transistors.

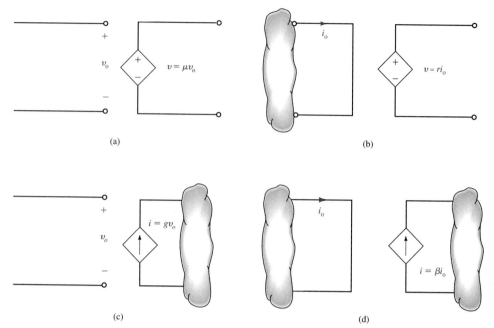

Figure 1.15 Dependent sources.

In contrast to the circle used to represent independent sources, a diamond is used to represent a dependent or controlled source. Figure 1.15 illustrates the four types of dependent sources. The "input" terminals on the left represent the voltage or current that *controls* the dependent source, and the "output" terminals on the right represent the output current or voltage of the controlled source. Note that in Figs. 1.15(a) and (d) the quantities μ and β are dimensionless constants because we are transforming voltage to voltage and current to current, respectively. This is not the case in Figs. 1.15(b) and (c); hence, when we employ these elements in a later section, we must describe the units of the factors, r and g.

EXAMPLE 1.9

Given the circuits shown in Fig. 1.16, we wish to determine the outputs.

In the first circuit, Fig. 1.16(a), the output voltage is $V = \mu V_0$ or $V = (20)\,(2\text{ V}) = 40$ V. Note that the output voltage has been "amplified" from 2 V at the input port to 40 V at the output port; that is, the circuit has an amplification factor or gain of 20.

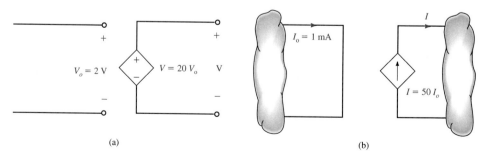

Figure 1.16 Circuits used in Example 1.9.

Figure 1.17 Ideal circuit model of simple circuit.

In the circuit of Fig. 1.16(b), the output current is $I = \beta I_0 = (50)$ (1 mA) = 50 mA; that is, the circuit has a current gain of 50. The output current is 50 times greater than the input current. ∎

Ideal Circuit Model

We can now construct an idealized circuit model of the battery and bulb shown in Fig. 1.5(b).

The battery, if ideal, would be represented as an active element, an independent constant voltage source. We could utilize either of the symbols in Fig. 1.13, the symbol for a battery, or the more general "circle" symbol for a voltage source where we let the value of the source, $v(t)$, be a constant, V_1. A real battery would likely have some internal resistance that would have to be added to the model, but we will postpone that consideration until later.

The bulb principally changes the flow of electric current into heat and light, and therefore is best modeled by a resistor, which has a symbol as shown in Fig. 1.12(a).

Therefore, the real circuit of Fig. 1.5(b) can be represented by an idealized lumped element model as shown in Fig. 1.17. This drawing is often called a circuit schematic drawing and shows the individual circuit elements connected by lines that represent ideal conductors.

Figure 1.18 summarizes the types of circuit elements used in the analysis of electric circuits. Note that the branches in this figure under "current sources" have the same format as those under "voltage sources."

1.3

Summary

- Seven major specialty areas with EE include: (1) power engineering, (2) electromagnetics, (3) communications, (4) computer engineering, (5) electronics, (6) systems, and (7) controls.
- Direct current (dc) describes charge flow in one direction all the time; alternating current (ac) repeatedly reverses direction. A cycle is defined as one direction reversal and return to the original direction.
- The number of times per second that an ac current completes a full cycle is defined as the frequency in hertz (Hz).
- Like charges repel each other; opposite charges attract.
- The force (in Newtons) between two point charges is

$$F = k \frac{q_1 q_2}{d^2}$$

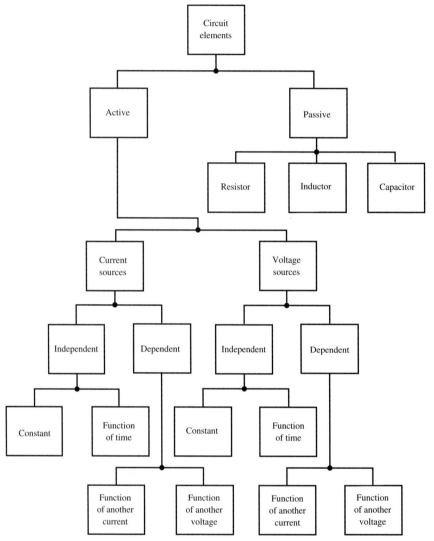

Figure 1.18   Summary of circuit element designations.

where k = constant = $8.99 \times 10^9 \ N{-}m^2/C^2$, q_1 and q_2 are the charges in coulombs, and d is the separation between the charges in meters.

• Current is the time rate of flow of charge, that is,

$$i = \frac{dq}{dt}$$

where i = the current in amperes, dq is the incremental value of charge crossing a plane in an incremental time, dt.

• Positive current is assumed to be in the direction of positive charge movement.
• Voltage difference is analogous to pressure difference and is defined by

$$v = \frac{dw}{dq}$$

where dw is the incremental energy required to move an incremental charge dq through a voltage difference of v.

- Reference directions must be assigned to voltages across and currents through elements.
- Power, p in watts, is the time rate at which energy, w, is produced or consumed; thus

$$p = \frac{dw}{dt} = vi$$

where v = the voltage across an element, and i = the current through the same element.

- Power is assumed positive if power is being absorbed by an element, and negative if power is being supplied.
- Ideal active circuit elements include both current and voltage sources; each can be one of two types: (1) independent or (2) dependent.
- Current sources maintain a prescribed current independent of terminal voltage; voltage sources maintain a prescribed voltage independent of current.
- Independent sources have a value that does not depend on any other variable in the circuit; dependent sources are a function of a circuit variable, for example, a voltage in another part of the circuit.
- Ideal passive circuit elements include:
  (1) the resistor, R in ohms, where

$$v = iR$$

  (2) the inductor, L in henrys, where

$$v = L\frac{di}{dt}$$

and (3) the capacitor, C in farads, where

$$v = \frac{1}{C} \int i \, dt$$

PROBLEMS

1.1. Categorize the following sources of electric power as either ac or dc.
1. automobile battery
2. solar cell
3. residential power outlet
4. watch battery
5. commercial power generating station

1.2. Why is ac preferable to dc for power distribution? What frequency is used today in the United States?

1.3. Edison's first commercial power generating station was capable of delivering a total of how many watts? Was it ac or dc?

1.4. What is the approximate frequency (in Hz) near the center of the portion of the electromagnetic spectrum corresponding to:
1. infrared
2. shortwave radio
3. gamma rays

1.5. Classify the following as insulators or conductors: copper, glass, rubber, platinum, salt water, silver, ceramics.

1.6. Find the force between two identical point charges each of magnitude 0.01 μC separated by a distance of 1 cm. Is this force attractive or repulsive?

1.7. Assume the radius of a hydrogen atom is 5.3×10^{-9} cm. Calculate the force between the nucleus of the hydrogen atom and an electron at this radius.

1.8. Three charges are placed in a straight line as shown in Fig. P1.8.

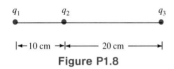

Figure P1.8

(a) If $q_1 = -2$ mC, $q_2 = 3$ mC, and $q_3 = 9$ mC, find the magnitude and direction of the resultant force on q_1, q_2, and q_3.
(b) If $q_1 = -4$ mC, find the magnitude and sign of the charge q_3 such that the resultant force on q_2 is zero.

1.9. If 5×10^{18} electrons flow uniformly through the cross section of a wire in 2 s, what is the current flowing through the wire?

1.10. How many electrons are flowing through the cross section of a wire per second if a current of 0.5 A flows through it?

1.11. What is the typical current range for the following:
(a) Household appliances
(b) Lightning bolt
(c) Memory cell of an integrated circuit memory

1.12. If a battery supplies 1 joule of energy in transferring 1/9 coulombs of charge through an element, what is the voltage across the battery?

1.13. A CRT (Cathode Ray Tube) is used for most television and computer video displays. In the back of a CRT is an electron gun emitting a beam of electrons toward a phosphor coated display screen; the beam makes a visible spot where it impacts the screen. Assuming the electron gun is made 25kV more negative than the screen, how much energy is transferred to a single electron by the time it reaches the screen?

1.14. One electron-volt (1 eV) is defined as the amount of energy gained by an electron moving through a voltage difference of 1 volt. What is this energy expressed in joules?

1.15. A 12V car battery delivers 200 A of current during startup. If it takes one second to start the car, how much energy is supplied by the battery during ignition?

1.16. Assign values to V and I in element B so that it is equivalent to element A (See Fig. Pl.16).

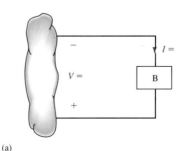

(a)

Figure P1.16 (continues)

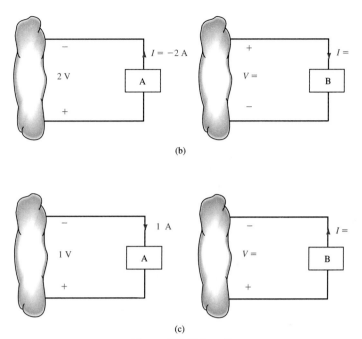

(b)

(c)

Figure P1.16 (cont)

1.17. Assign directions to currents and polarities to voltages in element B so it is equivalent to element A (See Fig. Pl.17).

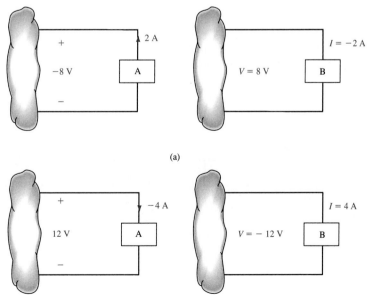

(a)

(b)

Figure P1.17 (continues)

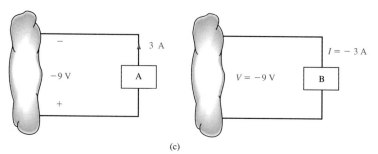

(c)

Figure P1.17 (cont)

1.18. In the elements of Fig. Pl.18, determine for each if they are supplying power or absorbing power and the magnitude of the power being transferred.

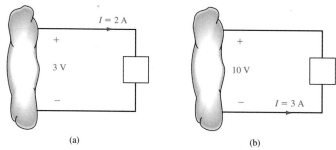

Figure P1.18

1.19. In Fig. Pl.19 determine for each element whether it is supplying or absorbing power. If all the energy absorbed is dissipated as heat, determine the heat dissipated (in joules) over a period of one hour for each of the circuits shown.

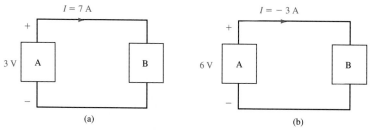

Figure P1.19

1.20. A dependent voltage source shown in Fig. 1.15(a) has $\mu = 25$. (a) What is the voltage between the terminals on the right, if $v_0 = 3.3$ V? (b) If $v_0 = 60$ V?

CIRCUITS

DC Circuits

2.1
Introduction

In this chapter we will introduce some of the basic concepts and laws that are fundamental to circuit analysis. We will also illustrate the utility of some of these concepts in practical situations.

The basic laws, such as Ohm's Law and Kirchhoff's Laws, are presented, along with a number of network theorems which sometimes help simplify and often provide additional insight into our analysis of electric circuits. This material forms the basis for most of the topics contained within the book.

2.2
Ohm's Law

Ohm's law is named for the German physicist Georg Simon Ohm, who is credited with establishing the voltage-current relationship for resistance.

Ohm's law states that *the voltage across a resistance is directly proportional to the current flowing through it.* The resistance, measured in ohms, is the constant of proportionality between the voltage and current.

A circuit element whose electrical characteristic is primarily resistive is called a resistor and is represented by the symbol shown in Fig. 2.1. A *resistor* is a physical device that can be fabricated in many ways. Certain standard values of discrete resistors can be purchased in an electronic parts store. These resistors, which find wide use in a variety of electrical applications, are normally carbon composition or wirewound. In addition, resistors can be fabricated using thick or thin films for use in hybrid circuits or they can be fabricated in a semiconductor integrated circuit.

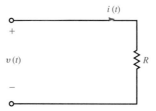

Figure 2.1 Symbol for the resistor.

The mathematical relationship of Ohm's law is illustrated by the equation

$$v(t) = Ri(t) \text{ where } R \geq 0 \tag{2.1}$$

Note carefully the relationship between the polarity of the voltage and the direction of the current. In addition, note that we have tacitly assumed that the resistor has a constant value and therefore that the voltage-current characteristic is linear.

The symbol Ω is used to represent *ohms,* and therefore

$$1 \ \Omega = 1 \text{ V/A}$$

Although we will assume here that the resistors are linear, it is important for readers to realize that some very useful and practical elements do exist that exhibit a nonlinear resistance characteristic. Diodes, which are used extensively in electric circuits, are examples of nonlinear resistors. These elements are discussed in a later chapter.

Since a resistor is a passive element, the proper current-voltage relationship is illustrated in Fig. 2.1. The power supplied to the terminals is absorbed by the resistor. Note that the charge moves from the higher (+) to the lower (−) potential as it passes through the resistor and the energy absorbed is dissipated by the resistor in the form of heat. The rate of energy dissipation is the instantaneous power, and therefore

$$p(t) = v(t)i(t) \tag{2.2}$$

which, using Eq. (2.1), can be written as

$$p(t) = Ri^2(t) = \frac{v^2(t)}{R} \tag{2.3}$$

This equation illustrates that the power dissipated in a resistor is a nonlinear function of either current or voltage and that it is always a positive quantity.

Conductance, represented by the symbol G, is another quantity with wide application in circuit analysis. By definition, conductance is the reciprocal of resistance: that is

$$G = \frac{1}{R} \tag{2.4}$$

The unit of conductance is the *siemens,* and the relationship between units is

$$1 \ S = 1 \ A/V$$

Using Eq. (2.4), we can write two additional expressions

$$i(t) = Gv(t) \tag{2.5}$$

and

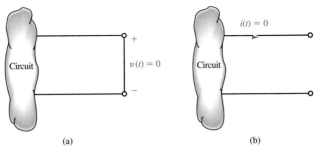

Figure 2.2   Diagrams illustrating a short circuit ($R = 0$) and an open circuit ($R = \infty$).

$$p(t) = \frac{i^2(t)}{G} = Gv^2(t) \tag{2.6}$$

Equation (2.5) is another expression of Ohm's law.

Two specific values of resistance, and therefore conductance, are very important: $R = 0$ ($G = \infty$) and $R = \infty$ ($G = 0$). If the resistance $R = 0$, we have what is called a *short circuit* illustrated in Fig. 2.2a. From Ohm's law

$$v(t) = Ri\,(t)$$
$$= 0$$

Therefore, $v(t) = 0$, although the current could theoretically be any value. If the resistance $R = \infty$, as would be the case with a broken wire, we have what is called an *open circuit* illustrated in Fig. 2.2b, and from Ohm's law

$$i(t) = \frac{v(t)}{R}$$
$$= 0$$

Therefore, the current is zero regardless of the value of the voltage across the open terminals.

EXAMPLE 2.1

In the circuit shown in Fig. 2.3, determine the current and the power absorbed by the resistor.

Using Eq. (2.1), we find the current to be

$$I = \frac{V}{R} = \frac{10}{2} = 5\,A$$

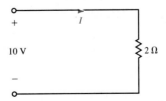

Figure 2.3   Circuit for Example 2.1.

and from Eq. (2.2) or (2.3), the power absorbed by the resistor is

$$P = VI = (10)(5) = 50 \ W$$
$$= RI^2 = (2)(5)^2 = 50 \ W$$
$$= \frac{V^2}{R} = \frac{(10)^2}{2} = 50 \ W$$

∎

D2.1. Assume a small lamp is modeled as a resistor. Measurements indicate that the voltage across the lamp is 12 V and the current through the lamp is 92.3 mA. Determine the lamp's resistance.

Ans: $R_{Lamp} = 130 \ \Omega$.

D2.2. A heater element draws 2.0 A when connected to a 120 V source. Calculate both the resistance of the element and the power absorbed in the form of heat.

Ans: $R = 60 \ \Omega$, $P = 240$ W.

D2.3. A speaker is a device which converts electrical energy into sound energy. Assume the internal resistance of a speaker is typically 8 Ω. The speaker's power rating is the maximum power that can be delivered to it without damage. Therefore, determine the maximum safe current that can be delivered to a stereo speaker with internal resistance of 8 Ω and a power rating of 200 watts.

Ans: $I_{max} = 5.0$ A.

2.3
Kirchhoff's Laws

The previous circuits that we have considered have all contained a single resistor and were analyzed using Ohm's law. At this point we begin to expand our capabilities to handle more complicated networks which result from an interconnection of two or more of these simple elements. We will assume that the interconnection is performed by electrical conductors (wires) that have zero resistance: that is, perfect conductors.

To aid us in our discussion, we will define a number of terms that will be employed throughout our analysis. As will be our approach throughout this text, we will use examples to illustrate the concepts and define the appropriate terms. For example, the circuit shown in Fig. 2.4(a) will be used to describe the terms *node, loop, mesh,* and *branch*. A *node* is simply a point of connection of two or more circuit elements. The nodes in the circuit in Fig. 2.4(a) are exaggerated in Fig. 2.4(b) for clarity. The reader is cautioned to compare the two figures carefully and note that although one node can be spread out with perfect conductors, it is still only one node.

A *loop* is simply any closed path through the circuit in which no node is crossed more than once. For example, the paths defined by the nodes *ACBA, BCDB,* and *ACDBA* are all loops. A *mesh* is any loop that does not contain within it another loop. Therefore, the first two paths are meshes; however, the third is not. Finally, a *branch* is a portion of a circuit containing only a single element and the nodes at each end of the element. The circuit in Fig. 2.4 contains five branches.

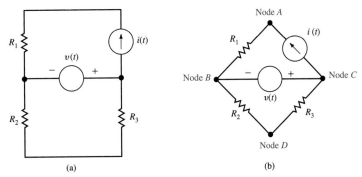

Figure 2.4   Circuits used to illustrate terms.

Given the previous definitions, we are now in a position to consider Kirchhoff's laws, named after the German scientist Gustav Robert Kirchhoff. These two laws are quite simple but extremely important. The first law is *Kirchhoff's current law,* which states that *the algebraic sum of the currents entering any node is zero.* In mathematical form the law appears as

$$\sum_{j=1}^{N} i_j(t) = 0$$

where $i_j(t)$ is the jth current entering the node through branch j and N is the number of branches connected to the node. To understand the use of this law, consider the node shown in Fig. 2.5. Applying Kirchhoff's current law to this node yields

$$i_1(t) + [-i_2(t)] + i_3(t) + i_4(t) + [-i_5(t)] = 0$$

We have assumed that the algebraic signs of the currents entering the node are positive and therefore that the signs of the currents leaving the node are negative.

If we multiply the foregoing equation by -1, we obtain the expression

$$-i_1(t) + i_2(t)] - i_3(t) - i_4(t) + i_5(t) = 0$$

which simply states that *the algebraic sum of the currents leaving a node is zero.* Alternatively, we can write the equation as

$$i_1(t) + i_3(t) + i_4(t) = i_2(t) + i_5(t)$$

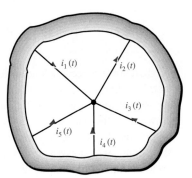

Figure 2.5   Currents at a node.

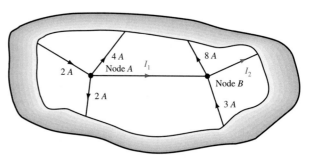

Figure 2.6   Illustration nodes for Kirchhoff's current law.

which states that *the sum of the currents entering a node is equal to the sum of the currents leaving the node.* These expressions are alternative forms of Kirchhoff's current law.

EXAMPLE 2.2

In Fig. 2.6, we wish to determine the currents I_1 and I_2.

The equations for Kirchhoff's current law at nodes A and B, respectively, are

$$2 - 4 - 2 - I_1 = 0$$
$$I_1 - 8 + 3 - I_2 = 0$$

From the first equation, $I_1 = -4$ A. Since it was assumed that I_1 was leaving node 1, the negative sign illustrates that positive current is actually entering node 1. Using this value of I_1 in the second equation, we find that $I_2 = -9$ A. ■

DRILL EXERCISE

D2.4. Find the currents I_1 and I_2 in the network in Fig. D2.4.

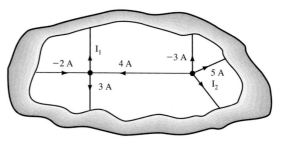

Figure D2.4

Ans: $I_1 = -1$ A, $I_2 = -6$ A.

Kirchhoff's second law, called *Kirchhoff's voltage law,* states that *the algebraic sum of the voltages around any loop is zero.*

Recall that in Kirchhoff's current law, the algebraic sign was required to keep track of whether the currents were entering or leaving a node. In Kirchhoff's voltage law, the algebraic sign is used to keep track of the voltage polarity. In other words, as we traverse the circuit, it is necessary to sum to zero the (− to +) increases and (+ to −) decreases in potential energy level as represented by the voltage. Therefore, it is important that we keep track of the voltage polarity as we go through each element.

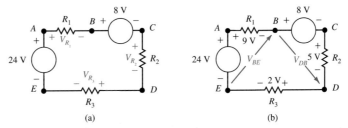

Figure 2.7 Circuits used to illustrate Kirchhoff's voltage law.

EXAMPLE 2.3

Consider the circuit shown in Fig. 2.7(a). In applying Kirchhoff's voltage law, we must traverse the circuit and sum to zero the increases and decreases in energy level. We can traverse the path in either a clockwise or counterclockwise direction, and we can consider an increase in voltage as positive and a decrease in voltage as negative, or vice versa.

Assuming that an increase in voltage is positive, starting at node A and traversing the circuit in a clockwise direction, we obtain

$$-V_{R1} - 8 - V_{R2} - V_{R3} + 24 = 0$$

Assuming that an increase in voltage is negative, using the same starting point and direction yields

$$V_{R1} + 8 + V_{R2} + V_{R3} - 24 = 0$$

Note that these two equations are identical. Furthermore, we would have obtained the same result if we had started at any node and traversed the network in a counterclockwise direction.

Now suppose that V_{R1} and V_{R2} are known to be 9 V and 5 V, respectively. Then either equation can be used to find $V_{R3} = 2$ V. The circuit with all known voltages labeled is shown in Fig. 2.7(b). This network is also used to illustrate the notation V_{AB} which denotes the voltage of point A with respect to point B. Since the potential is measured between two points, it is convenient to use an arrow between two points with the head of the arrow located at the positive reference. Note that the double subscript notation, the + and − notation, and the single-headed arrow notation are all the same if the head of the arrow is pointing toward the positive terminal and the first subscript in the double subscript notation. If we apply this notation to the network in Fig. 2.7(b), we find for example that $V_{AE} = +24$ V, $V_{EA} = -24$ V, $V_{CD} = +5$ V, and $V_{DC} = -5$ V. Furthermore, we can apply Kirchhoff's voltage law to the circuit to determine the voltage between any two points. For example, suppose we want to determine the voltage V_{BE}. Note that we have a choice of two paths; one is *BCDEB* and the other is *BEAB*. For the first path we obtain

$$-8 - 5 - 2 + V_{BE} = 0$$
$$V_{BE} = 15 \text{ V}$$

For the second path the equation is

$$-V_{BE} + 24 - 9 = 0$$
$$V_{BE} = 15 \text{ V}$$

In a similar manner we can show that

$$V_{DB} = -13 \text{ V}$$

∎

In general, the mathematical representation of Kirchhoff's voltage law is

$$\sum_{j=1}^{N} v_j(t) = 0 \qquad (2.8)$$

where $v_j(t)$ is the voltage across the jth branch (with the proper reference direction) in a loop containing N voltages. This expression is analogous to Eq. (2.7) for Kirchhoff's current law.

DRILL EXERCISE

D2.5. In the network in Fig. D2.5, V_{R1} is known to be 4 V. Find V_{R2} and V_{bd}.

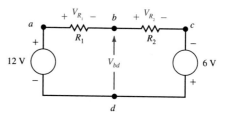

Figure D2.5

Ans: $V_{R2} = 14$ V, $V_{bd} = 8$ V.

D2.6. In the network in Fig. D2.6, find V_{R4}, V_{bf}, and V_{ec}.

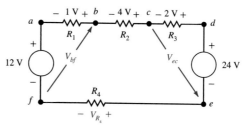

Figure D2.6

Ans: $V_{R4} = -5$ V, $V_{bf} = 13$ V, $V_{ec} = -22$ V.

2.4

Single-loop Circuits   At this point we can begin to apply the laws we presented earlier to the analysis of simple circuits. To begin, we examine what is perhaps the simplest circuit—a single closed path, or loop, of elements. The elements of a single loop carry the same current and therefore are said to be in series. We will now apply Kirchhoff's voltage law and Ohm's law to the circuit to determine various quantities in it.

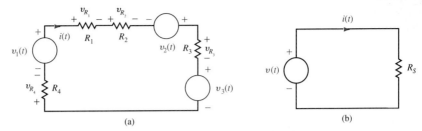

Figure 2.8 A single-loop circuit.

The circuit shown in Fig. 2.8 will serve as a basis for discussion. This circuit consists of a number of independent voltage sources in series with several resistors. We have assumed the current to be in a clockwise direction. If this assumption is correct, the solution of the equations that yields the current will produce a positive value. If the current is actually in the opposite direction, the value of the current variable will simply be negative, indicating that the current is in a direction opposite to that assumed. We have also made voltage polarity assignments for the voltage across the resistors. These assignments have been made using the convention employed in our discussion of Ohm's law and our choice for the direction of $i(t)$, that is, the convention shown in Fig. 2.1.

Applying Kirchhoff's voltage law to the network in Fig. 2.8(a) yields

$$-v_{R1} - v_{R2} + v_2(t) - v_{R3} - v_3(t) - v_{R4} + v_1(t) = 0$$

or

$$v_1(t) + v_2(t) - v_3(t) = v_{R1} + v_{R2} + v_{R3} + v_{R4}$$

However, from Ohm's law we know that

$$v_{Rj} = i(t)R_j$$

Therefore

$$v_1(t) + v_2(t) - v_3(t) = i(t)R_1 + i(t)R_2 + i(t)R_3 + i(t)R_4$$
$$= i(t)\,[R_1 + R_2 + R_3 + R_4]$$

This equation can be written as

$$v(t) = R_s i(t)$$

where

$$v(t) = v_1(t) + v_2(t) - v_3(t)$$
$$R_s = R_1 + R_2 + R_3 + R_4$$

Under the definitions above, the network in Fig. 2.8(a) is equivalent to that in Fig. 2.8(b). In other words, several voltage sources in series can be replaced by one source whose value is the algebraic sum of the individual sources. Furthermore, *the equivalent resistance of any number of resistors in series is simply the sum of the individual resistances.*

EXAMPLE 2.4
Given the network in Fig. 2.9, (a) find I, V_{BE}, V_{FD} and the power absorbed by the 3 Ω resistor, and (b) repeat part (a) if the 12 V source is changed to 72 V.

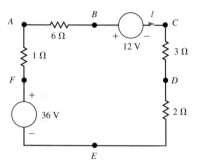

Figure 2.9   Circuit used in Example 2.4.

(a) starting at node A and traversing the network in a clockwise direction writing Kirchhoff's voltage law, we obtain

$$-6I - 12 - 3I - 2I + 36 - I = 0$$

or

$$I = 2\text{A}$$

Kirchhoff's voltage law for the path $ABEFA$ is

$$-6I - V_{BE} + 36 - 1I = 0$$

Since $I = 2$ A, $V_{BE} = 22$ V. In a similar manner using the path $FDEF$, we find that

$$-V_{FD} - 2I + 36 = 0$$

or

$$V_{FD} = 32 \text{ V}$$

The power dissipated in the 3 Ω resistor is

$$P_{3\Omega} = I^2(3)$$
$$= 12 \text{ W}$$

(b) Kirchhoff's voltage law for the network in this case is

$$-6I - 72 - 3I - 2I + 36 - 1I = 0$$

or

$$I = -3 \text{ A}$$

Once again, employing Kirchhoff's voltage law for the paths $ABEFA$ and $FDEF$, we obtain

$$-6I - V_{BE} + 36 - 1I = 0$$

or $V_{BE} = 57$ V and

$$-V_{FD} - 2I + 36 = 0$$

or $V_{FD} = 42$ V. Finally, $P_{3\Omega} = I^2(3) = 27$ W.   ∎

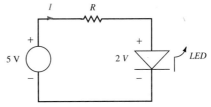

Figure 2.10   A simple LED circuit.

EXAMPLE 2.5

Light emitting diodes, or LEDs, are used as indicator lamps in a wide variety of electric equipment such as stereos, computers, and cameras. The operation of a standard LED is such that as long as there is a current through the LED, the voltage drop across the LED is approximately constant at 2 V. In the circuit shown in Fig. 2.10, determine the value of the resistor R so that the current in the LED does not exceed 20 mA.

Applying Kirchhoff's voltage law to the circuit in Fig. 2.10 yields

$$+5 - IR - 2 = 0$$

and if $I = 20$ mA

$$R = 150 \ \Omega$$ ∎

EXAMPLE 2.6

An industrial load is served by a power company generator as shown in Fig. 2.11. The transmission line is 100 miles long and has a line resistance of 0.1 Ω/mile. The load absorbs 1.2 megawatts. Determine the amount of power that must be supplied by the generator if the line voltage at the load is (a) 12 kV or (b) 120 kV.

(a) If $V_L = 12$ kV and $P_L = 1.2$ MW, then the line current I_L is

$$I_L = \frac{1.2 \times 10^6}{12 \times 10^3}$$

$$= 100 \text{ A}$$

The line losses are then

$$P_{Line} = I_L^2 R_{Line}$$
$$= (100)^2(100)\,(2)\,(0.1)$$
$$= 0.2 \text{ MW}$$

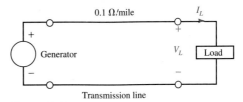

Transmission line

Figure 2.11   Circuit used in Example 2.6.

Therefore, the power that must be supplied by the generator is

$$P_{Gen} = P_L + P_{Line}$$
$$= 1.4 \text{ MW}$$

(b) If $V_L = 120$ kV, then

$$I_L = \frac{1.2 \times 10^6}{120 \times 10^3}$$
$$= 10 \text{ A}$$

The line losses are then

$$P_{Line} = (10)^2(100)\,(2)\,(0.1)$$
$$= 2 \text{ kW}$$

Therefore, the power that must be supplied by the generator is

$$P_{Gen} = P_L + P_{Line}$$
$$= 1.202 \text{ MW} \qquad\blacksquare$$

Note that the efficiency, that is, the ratio of output power to input power, is quite different. In the first case, the efficiency is 86%, while in the second case, the efficiency is 99.8%. Transmitting at high voltage and low current minimizes the I^2R losses, and therefore, the utilities operate in this mode. The issue is much more complicated than indicated in this simple approach, since as the voltage goes up, so does the cost of insulation and switchgear. Furthermore, broader right-of-ways and higher towers are needed for transmission.

Consider now the network shown in Fig. 2.12.
Applying Kirchhoff's voltage law to this network yields

$$v(t) = R_1 i(t) + R_2 i(t)$$

Solving the equation for $i(t)$ yields

$$i(t) = \frac{v(t)}{R_1 + R_2} \qquad (2.9)$$

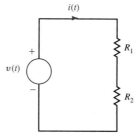

Figure 2.12   Two single-loop circuits.

Knowing the current, we can now apply Ohm's law to determine the voltage across each resistor:

$$v_{R1} = R_1 i(t)$$

$$= \frac{R_1}{R_1 + R_2} v(t) \tag{2.10}$$

Similarly

$$v_{R2} = \frac{R_2}{R_1 + R_2} v(t) \tag{2.11}$$

Note that the equations satisfy Kirchhoff's voltage law, since

$$+v(t) - \frac{R_1}{R_1 + R_2} v(t) - \frac{R_2}{R_1 + R_2} v(t) = 0$$

Although simple, Eqs. (2.10) and (2.11) are very important because they describe the operation of what is called a *voltage divider*. In other words, the voltage of the source $v(t)$ is *divided between the resistors R_1 and R_2 in direct proportion to their resistances.*

EXAMPLE 2.7

The volume control circuit for a radio or TV is shown in Fig. 2.13(a). A portion of the receiver circuit applies a small fixed voltage to the variable resistance R, which is the volume control potentiometer (pot). This is a resistor with a sliding contact that can be placed anywhere along the resistor. The volume control circuit is actually a voltage divider between V_{in} and V_{out} as shown in Fig. 2.13(b). The voltage V_{out} is amplified by a fixed amount to drive the speaker.

Now suppose that $V_{in} = 1.6$ V, $R = 24$ kΩ, $R_1 = 12$ kΩ, and $R_2 = 12$ kΩ. Under these conditions, let us find V_{out}.

The voltage divider output is

$$V_{out} = V_{in} \left(\frac{R_2}{R_1 + R_2} \right) = V_{in} \left(\frac{R_2}{R} \right)$$

$$= \frac{1.6}{24000} (12000)$$

$$= 0.8 \text{ V} \qquad\blacksquare$$

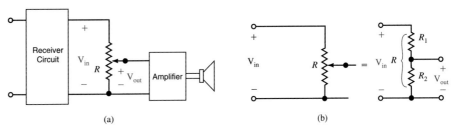

(a)   (b)

Figure 2.13   Volume control circuit.

D2.7. Find I and V_0 in the network in Fig. D2.7.

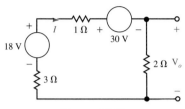

Figure D2.7

Ans: $I = -2$ A, $V_0 = -4$ V.

D2.8. Find I, V_{AC}, V_{CB}, and $P_{6\Omega}$ in the network in Fig. D2.8.

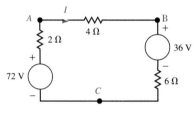

Figure D2.8

Ans: $I = 3$ A, $V_{AC} = 66$ V, $V_{CB} = -54$ V, $P_{6\Omega} = 54$ W.

2.5
Single-node-pair Circuits

An important circuit is the single-node-pair circuit, an example of which is shown in Fig. 2.14(a). In this case the elements have the same voltage across them, and therefore are in parallel. Kirchhoff's current law and Ohm's law will be used to determine the various unknown quantities in these networks.

With reference to the network shown in Fig. 2.14(a), we have assumed that the upper node is $v(t)$ volts positive with respect to the lower node. Applying Kirchhoff's current law to the upper node yields

$$i_A(t) - i_1(t) + i_B(t) - i_2(t) - i_C(t) - i_3(t) - i_4(t) = 0$$

or

$$i_At + i_B(t) - i_C(t) = i_i(t) + i_2(t) + i_3(t) + i_4(t)$$

However, from Ohm's law we know that

$$i_j(t) = \frac{v(t)}{R_j}$$

Therefore

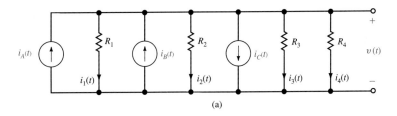

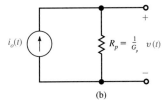

Figure 2.14   Equivalent circuits.

$$i_A(t) + i_B(t) - i_C(t) = \frac{v(t)}{R_1} + \frac{v(t)}{R_2} + \frac{v(t)}{R_3} + \frac{v(t)}{R_4}$$

$$= v(t)\left[\frac{1}{R_1} + \frac{1}{R_2} + \frac{1}{R_3} + \frac{1}{R_4}\right] = v(t)\,[G_1 + G_2 + G_3 + G_4]$$

This equation can be written as

$$i_o(t) = \frac{v(t)}{R_p} = G_p v(t)$$

where

$$i_o(t) = i_A(t) + i_B(t) - i_C(t)$$

$$\frac{1}{R_p} = \frac{1}{R_1} + \frac{1}{R_2} + \frac{1}{R_3} + \frac{1}{R_4}$$

and

$$G_p = G_1 + G_2 + G_3 + G_4$$

Under the definitions above, the network in Fig. 2.14(a) is equivalent to that in Fig. 2.14(b). In other words, several current sources in parallel can be replaced by one source whose value is the *algebraic* sum of the individual sources. Furthermore, *the equivalent conductance of any number of resistors in parallel is simply the sum of the individual conductances.*

EXAMPLE 2.8

Given the network in Fig. 2.15, (a) Find V, I_1, I_2, I_3, $P_{4\Omega}$, and (b) repeat part (a) if the 2 A source is changed to 20 A.

(a) Kirchhoff's current law for the network is

$$8 - I_1 - 2 - I_2 - I_3 = 0$$

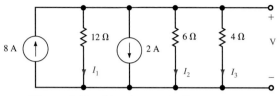

Figure 2.15 Circuit used in Example 2.8.

$$6 = I_1 + I_2 + I_3$$

$$6 = V\left[\frac{1}{12} + \frac{1}{6} + \frac{1}{4}\right]$$

or

$$V = 12 \text{ V}$$

Then using Ohm's law, $I_1 = 1$ A, $I_2 = 2$ A, and $I_3 = 3$ A. In addition

$$P_{4\Omega} = (I_3)^2\,(4) = \frac{V^2}{4} = 36 \text{ W}$$

(b) Kirchhoff's current law for the network in this case is

$$8 - I_1 - 20 - I_2 - I_3 = 0$$

or

$$-12 = V\left[\frac{1}{12} + \frac{1}{6} + \frac{1}{4}\right]$$

and hence

$$V = -24 \text{ V}$$

From Ohm's law we obtain $I_1 = -2$ A, $I_2 = -4$ A, and $I_3 = -6$ A. Finally

$$P_{4\Omega} = (I_3)^2(4) = \frac{V^2}{4} = 144 \text{ W.}$$ ∎

Consider the circuit shown in Fig. 2.16.

$$i(t) = \frac{v(t)}{R_1} + \frac{v(t)}{R_2}$$

$$= \left(\frac{1}{R_1} + \frac{1}{R_2}\right)v(t)$$

$$= \frac{v(t)}{R_p}$$

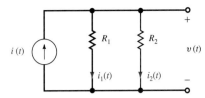

Figure 2.16 A circuit used to illustrate current division.

where

$$\frac{1}{R_p} = \frac{1}{R_1} + \frac{1}{R_2} \tag{2.12}$$

and

$$R_p = \frac{R_1 R_2}{R_1 + R_2} \tag{2.13}$$

Therefore, the equivalent resistance of two resistors connected in parallel is equal to the product of their resistances divided by their sum. Note also that this equivalent resistance R_p is always less than either R_1 or R_2. Hence, by connecting resistors in parallel we reduce the overall resistance. In the special case when $R_1 = R_2$, the equivalent resistance is equal to half of the value of the individual resistors.

The manner in which the current $i(t)$ from the source divides between the two branches is called *current division* and can be found from the expressions above. For example

$$v(t) = R_p i(t)$$

$$= \frac{R_1 R_2}{R_1 + R_2}\, i(t) \tag{2.14}$$

and

$$i_1(t) = \frac{v(t)}{R_1}$$

$$= \frac{R_2}{R_1 + R_2}\, i(t) \tag{2.15}$$

and

$$i_2(t) = \frac{v(t)}{R_2}$$

$$= \frac{R_1}{R_1 + R_2}\, i(t) \tag{2.16}$$

Equations (2.15) and (2.16) are mathematical statements of the current-division rule.

Therefore, the current divides in inverse proportion to the resistances. In other words, to determine the current in the branch containing R_1, we multiply the incoming current $i(t)$ by the opposite resistance R_2 and divide that product by the sum of the two resistors. Note that this description applies to the special case of two resistors in parallel.

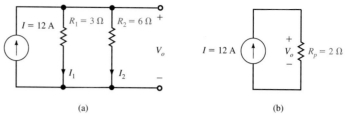

Figure 2.17   Example of a parallel circuit.

EXAMPLE 2.9

Consider the circuit shown in Fig. 2.17(a). The equivalent circuit is shown in Fig. 2.17(b). Given the information specified in the circuit, we wish to determine the currents and equivalent resistance.

We can apply current division to determine I_1 and I_2. For example

$$I_1 = \frac{R_2}{R_1 + R_2} I = \frac{6}{3 + 6} (12) \tag{12}$$

$$= 8 \text{ A}$$

and

$$I_2 = \frac{3}{3 + 6} (12) = 4 \text{ A}$$

Note that the larger current is through the smaller resistor, and vice versa. In addition, one should note that if R_1 and R_2 are equal, the current will divide equally between them.

Note that these currents satisfy Kirchhoff's current law at both the upper and lower nodes.

$$I = I_1 + I_2$$
$$12 \text{ A} = 8 \text{ A} + 4 \text{ A}$$

The equivalent resistance for the circuit is

$$R_p = \frac{R_1 R_2}{R_1 + R_2}$$

$$= \frac{(3)(6)}{3 + 6}$$

$$= 2 \ \Omega$$

Now V_o can be calculated as

$$V_o = R_p I$$
$$= (2)(12)$$
$$= 24 \text{ V}$$

Once the voltage V_o is known, Ohm's law can also be used to calculate the currents I_1 and I_2.

$$I_1 = \frac{V_o}{R_1}$$

$$= \frac{24}{3}$$

$$= 8 \text{ A}$$

and

$$I_2 = \frac{V_o}{R_2}$$

$$= \frac{24}{6}$$

$$= 4 \text{ A} \qquad \blacksquare$$

DRILL EXERCISE

D2.9. Find V, I_1, I_2, and $P_{6\Omega}$ in the network in Fig. D2.9.

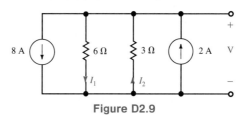

Figure D2.9

Ans: $V = -12$ V, $I_1 = -2$ A, $I_2 = 4$ A, and $P_{6\Omega} = 24$ W.

2.6
Resistor Combinations

We have already demonstrated that the equivalent resistance of N resistors in series is

$$R_s = R_1 + R_2 + \dots + R_N \qquad (2.17)$$

and the equivalent resistance of N resistors in parallel is found from

$$\frac{1}{R_p} = \frac{1}{R_1} + \frac{1}{R_2} + \dots + \frac{1}{R_N} \qquad (2.18)$$

Let us now examine a combination of these two cases.

EXAMPLE 2.10
Let us determine the resistance at terminals A–B of the network shown in Fig. 2.18(a). To determine the equivalent resistance at A–B, we begin at the opposite end of the network and combine resistors as we progress toward terminals A–B. The 8, 1, and 3–Ω resistors connected in series are equivalent to one 12 Ω resistor, which in turn is in parallel with the 6 Ω resistor. This parallel combination is equivalent to one 4 Ω resistor as shown in Fig. 2.18(b). The two 7 Ω resistors and the 4 Ω resistor are in series, and this

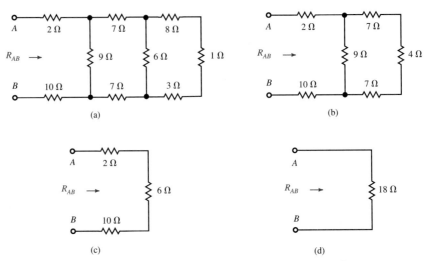

(a)   (b)

(c)   (d)

Figure 2.18 Simplification of a resistance network.

combination is in parallel with the 9 Ω resistor. Combining these resistors reduces the network to that shown in Fig. 2.18(c). Therefore, the resistance at terminals A–B is 18 Ω as shown in Fig. 2.18(d). ∎

DRILL EXERCISE

D2.10. Find the equivalent resistance at the terminals A–B in the network in Fig. D2.10.

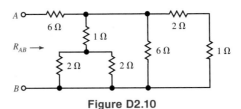

Figure D2.10

Ans: $R_{AB} = 7 \, \Omega$.

Consider now the circuit in Fig. 2.19. Note that when we attempt to reduce the circuit to an equivalent resistor R, we find that nowhere is a resistor in series or parallel with another. Therefore, we cannot attack the problem directly using the techniques we have learned thus far. We can, however, replace one portion of the network with an equivalent circuit and this conversion will permit us to reduce the combination of resistors to a single equivalent resistance with ease. This conversion is called the Y to Δ or Δ to Y transformation.

Note that the resistors in Fig. 2.20(a) form a Δ and the resistors in Fig. 2.20(b) form a Y. The transformation that relates the resistances R_1, R_2, and R_3 to the resistances R_a, R_b, and R_c is derived as follows. In order for the two networks to be equivalent at each corresponding pair of terminals, it is necessary that the resistance at the corresponding terminals be equal, for example, the resistance at terminals a and b with c open-circuited

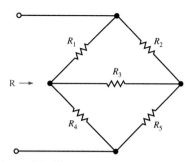

Figure 2.19 A network used to illustrate the need for the $Y \rightleftarrows \Delta$ transformation.

must be the same for both networks. Therefore, if we equate the resistances for each corresponding set of terminals, we obtain the following equations.

$$R_{ab} = R_a + R_b = \frac{R_2 (R_1 + R_3)}{R_2 + R_1 + R_3}$$

$$R_{bc} = R_b + R_c = \frac{R_3 (R_1 + R_2)}{R_3 + R_1 + R_2} \tag{2.19}$$

$$R_{ca} = R_c + R_a = \frac{R_1 (R_2 + R_3)}{R_1 + R_2 + R_3}$$

Solving this set of equations for R_a, R_b, and R_c yields

$$R_a = \frac{R_1 R_2}{R_1 + R_2 + R_3}$$

$$R_b = \frac{R_2 R_3}{R_1 + R_2 + R_3} \tag{2.20}$$

$$R_c = \frac{R_1 R_3}{R_1 + R_2 + R_3}$$

Similarly, if we solve Eq. (2.19) for R_1, R_2, and R_3, we obtain

$$R_1 = \frac{R_a R_b + R_b R_c + R_a R_c}{R_b}$$

$$R_2 = \frac{R_a R_b + R_b R_c + R_a R_c}{R_c} \tag{2.21}$$

$$R_3 = \frac{R_a R_b + R_b R_c + R_a R_c}{R_a}$$

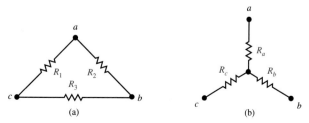

Figure 2.20 Δ and Y resistance networks.

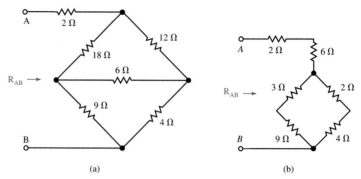

Figure 2.21 Circuits used in Example 2.11.

Equations (2.20) and (2.21) are general relationships and apply to any set of resistances connected in a Y or Δ. For the balanced case where $R_a = R_b = R_c$ and $R_1 = R_2 = R_3$, the equations above reduce to

$$R_Y = \frac{1}{3}R_\Delta \qquad (2.22)$$

and

$$R_\Delta = 3\,R_Y \qquad (2.23)$$

It is important to note that it is easy to remember the formulas since there are definite geometrical patterns associated with the conversion equations.

EXAMPLE 2.11
Let us find the equivalent resistance of the network in Fig. 2.21(a) at the terminals.

Applying the Δ to Y transformation reduces the network in Fig. 2.21(a) to that in Fig. 2.21(b). Note that the 3 Ω and 9 Ω resistors are in series and they are in parallel with the series combination of the 2 Ω and 4 Ω resistors. This total combination yields a 4 Ω resistor which is in series with the 2 Ω and 6 Ω resistors, and thus the resistance at the terminals is 12 Ω. ∎

DRILL EXERCISE

D2.11. Find the resistance at the terminals $A–B$ in the network in Fig. D2.11.

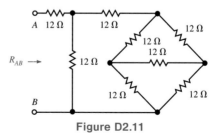

Figure D2.11

Ans: $R_{AB} = 20\ \Omega$.

2.7
Nodal Analysis

In order to perform a nodal analysis on a multiple–node circuit, we first select one node, which we refer to as the *reference node,* and measure the voltage at every other node with respect to this reference node. Thus, when we refer to the node voltage at some specific node, we mean the voltage at that node with respect to the reference node.

Quite often the reference node is the one to which the largest number of branches are connected. It is commonly called *ground* because it is said to be at ground-zero potential and it sometimes represents the chassis or ground line in a practical circuit.

A nodal analysis is based on Kirchhoff's current law (KCL) and the variables in the circuit are selected to be the node voltages.

We will select our variables as being positive with respect to the reference node. If one or more of the node voltages is actually negative with respect to the reference node, the analysis will indicate it.

Suppose for a moment that we know all the node voltages in a given network. Then with reference to Fig. 2.22, we can use Ohm's law, that is

$$i = \frac{V_m - V_n}{R} \tag{2.24}$$

to calculate the current through any resistive element. In this manner we can determine every voltage and every current in the circuit.

In a double-node circuit (i.e., one containing two nodes, one of which is the reference node), a single equation is required to solve for the unknown node voltage. In the case of an N–node circuit, $N - 1$ linearly independent simultaneous equations are required to determine the $N - 1$ unknown node voltages. These equations are written by employing KCL at $N - 1$ of the N nodes.

Consider, for example, the circuit shown in Fig. 2.23(a). This circuit has three nodes. The node at the bottom is selected as the reference node and is so labeled using the ground symbol, $\perp$. This reference node is assumed to be at zero potential and the node voltages v_1 and v_2 are defined with respect to this node.

The circuit is redrawn in Fig. 2.23(b) to indicate the nodes clearly. The branch currents are assumed to flow in the directions indicated in the figures. If one or more of the branch currents are actually flowing in a direction opposite to that assumed, the analysis will simply produce a branch current that is negative.

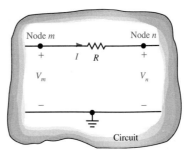

Figure 2.22 Circuit used to illustrate Ohm's law in a multiple-node network.

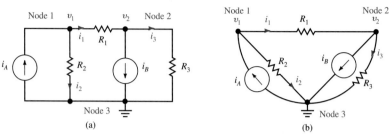

Figure 2.23 A three-node network.

Applying KCL at node 1 yields

$$i_A - i_1 - i_2 = 0$$

Using Ohm's law and noting that the reference node is at zero potential, we obtain

$$i_A - \frac{v_1 - v_2}{R_1} - \frac{v_1 - 0}{R_2} = 0$$

or

$$v_1\left(\frac{1}{R_1} + \frac{1}{R_2}\right) - v_2\left(\frac{1}{R_1}\right) = i_A$$

which could also be written in the form

$$v_1(G_1 + G_2) - v_2(G_2) = i_A$$

KCL at node 2 yields

$$i_1 - i_B - i_3 = 0$$

Using Ohm's law we obtain

$$\frac{v_1 - v_2}{R_1} - i_B - \frac{v_2 - 0}{R_3} = 0$$

or

$$-v_1\left(\frac{1}{R_1}\right) + v_2\left(\frac{1}{R_1} + \frac{1}{R_3}\right) = -i_B$$

which can also be expressed as

$$v_1(G_1) + v_2(G_2 + G_3) = -i_B$$

Therefore, the two equations, which when solved yield the node voltages, are

$$v_1\left(\frac{1}{R_1} + \frac{1}{R_2}\right) - v_2\left(\frac{1}{R_1}\right) = i_A$$

$$v_1\left(\frac{1}{R_1}\right) + v_2\left(\frac{1}{R_1} + \frac{1}{R_3}\right) = -i_B$$

Note that the analysis has produced two simultaneous equations in the unknowns v_1 and v_2. They can be solved using any convenient technique, for example, Gaussian elimination, determinants, matrices, or other methods.

Note that a nodal analysis employs KCL in conjunction with Ohm's law. Once the direction of the branch currents has been assumed, then Ohm's law, as illustrated by Fig. 2.22 and expressed by Eq. (2.24), is used to express the branch currents in terms of the unknown node voltages. We can assume the currents to be in any direction. However, once we assume a particular direction, we must be very careful to write the currents correctly in terms of the node voltages using Ohm's law.

EXAMPLE 2.12

Let us use nodal analysis to determine all the currents in the network in Fig. 2.24(a). The node voltages and branch currents are labeled as shown in Fig. 2.24(b). KCL for the two nodes with node voltages V_1 and V_2 are

$$-I_1 - 4 - I_2 - I_3 = 0$$
$$I_3 + 10 - I_4 = 0$$

Using Ohm's law these equations can be written as

$$-\frac{V_1 - 0}{2} - 4 - \frac{V_1 - 0}{2} - \frac{V_1 - V_2}{1} = 0$$

$$\frac{V_1 - V_2}{1} + 10 - \frac{V_2 - 0}{2} = 0$$

Simplifying the equations yields

$$2V_1 - V_2 = -4$$
$$-V_1 + (3/2)V_2 = 10$$

Solving the equations by Cramer's Rule, we first compute the determinant as

$$det = \begin{vmatrix} 2 & -1 \\ -1 & 3/2 \end{vmatrix}$$
$$= 2(3/2) - (-1)(-1)$$
$$= 2$$

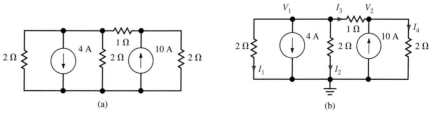

(a) (b)

Figure 2.24   Circuits used in Example 2.12.

The node voltage V_1 is then

$$V_1 = \frac{\begin{vmatrix} -4 & -1 \\ 10 & 3/2 \end{vmatrix}}{det}$$

$$= \frac{-4(3/2) - (-1)(10)}{2}$$

$$= \frac{-6 + 10}{2}$$

$$= 2 \text{ V}$$

and V_2 is

$$V_2 = \frac{\begin{vmatrix} 2 & -4 \\ -1 & 10 \end{vmatrix}}{2}$$

$$= \frac{20 - 4}{2}$$

$$= 8 \text{ V}$$

Therefore $I_1 = 1$ A, $I_2 = 1$ A, $I_3 = -6$ A, and $I_4 = 4$ A. Note that KCL is satisfied at all the nodes in the network. ■

EXAMPLE 2.13

We wish to find all the currents in the network in Fig. 2.25 using nodal analysis.
Applying KCL we obtain

$$6 - I_1 - I_2 = 0$$
$$I_2 - 3I_1 - I_3 = 0$$

Using Ohm's law the equations become

$$6 - \frac{V_1}{2} - \frac{V_1 - V_2}{2} = 0$$

$$\frac{V_1 - V_2}{2} - 3\left(\frac{V_1}{2}\right) - \frac{V_2}{1} = 0$$

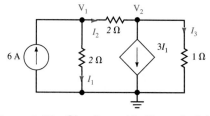

Figure 2.25 Circuit used in Example 2.13.

or

$$V_1 - \frac{1}{2} V_2 = 6$$

$$V_1 + \frac{3}{2} V_2 = 0$$

Solving these equations yields $V_1 = 4.5$ V and $V_2 = -3$ V. Therefore, $I_1 = 2.25$A, $I_2 = 3.75$ A, and $I_3 = -3$ A. Note that KCL is satisfied at every node. ∎

DRILL EXERCISE

D2.12. Use nodal analysis to find all the currents in the network in Fig. D2.12.

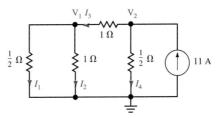

Figure D2.12

Ans: $I_1 = 2$ A, $I_2 = 1$ A, $I_3 = 3$ A, and $I_4 = 8$ A.

D2.13. Find all the currents in the network in Fig. D2.13 using nodal analysis.

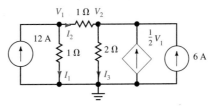

Figure D2.13

Ans: $I_1 = 16$ A, $I_2 = -4$ A, $I_3 = 10$ A.

Consider now the network in Fig. 2.26(a). If we attempt to perform a nodal analysis on this circuit to determine the unknown node voltages V_1 and V_2, we quickly find that we cannot express the branch current in the voltage source as a function of V_1 and V_2. However, we can form what is called a *supernode,* which includes the voltage source and the two nodes labeled V_1 and V_2 as shown by the dashed line in Fig. 2.26(b). KCL must hold for this supernode, that is, the algebraic sum of the currents entering or leaving the supernode must be zero. Therefore, one valid equation for the network is

$$I_1 - I_A + I_2 + I_B = 0$$

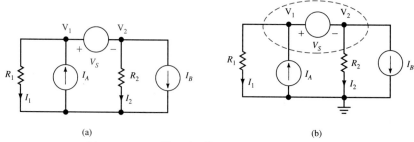

Figure 2.26 Circuits illustrating a supernode.

or

$$\frac{V_1}{R_1} + \frac{V_2}{R_2} = I_A - I_B \tag{2.25}$$

Since this is an $N = 3$ node network, we need $N - 1 = 2$ linearly independent equations to determine the node voltages. The second equation is derived from the supernode where the difference in potential between the two node voltages V_1 and V_2 is *constrained* by the voltage source, that is

$$V_1 - V_2 = V_S \tag{2.26}$$

Equations (2.25) and (2.26) will yield the node voltages which, in turn, can be used to determine all the currents.

EXAMPLE 2.14

Given the network in Fig. 2.27(a), let us find the current in each resistor.

The network is redrawn in Fig. 2.27(b). Note that a supernode exists around the 24V source and $V_1 = 12$ V. KCL for the supernode is

$$I_1 - I_2 - I_3 + 4 = 0$$

or

$$\frac{12 - V_2}{2} - \frac{V_2}{2} - \frac{V_3}{1} + 4 = 0$$

which reduces to

$$V_2 + V_3 = 10$$

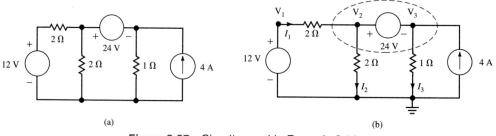

Figure 2.27 Circuits used in Example 2.14.

This equation together with the supernode constraint equation

$$V_2 - V_3 = 24$$

yields $V_2 = 17$ V and $V_3 = -7$ V. The currents in the resistors are then $I_1 = -2.5$ A, $I_2 = 8.5$ A, and $I_3 = -7$ A. ∎

DRILL EXERCISE

D2.14. Find I_1, I_2, and I_3 in the network in Fig. D2.14.

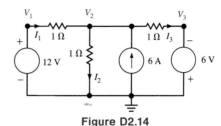

Figure D2.14

Ans: $I_1 = 8$ A, $I_2 = 4$ A, and $I_3 = 10$ A.

2.8

Loop and Mesh Analysis

In a nodal analysis the unknown parameters are the node voltages, and KCL is employed to determine them. In contrast to this approach, a loop or mesh analysis uses KVL to determine currents in the circuit. Once the currents are known, Ohm's law can be used to calculate the voltages. We have found that a single equation was sufficient to determine the current in a circuit containing a single loop. If the circuit contains N independent loops, N independent simultaneous equations will be required to describe the network.

Consider the circuit shown in Fig. 2.28. Let us identify two meshes, A–B–D–A and B–C–D–B. Let us assume that currents i_1 in the first mesh and i_2 in the second mesh flow clockwise. Then the branch current from B to D through R_2 is $i_1 - i_2$. The directions of the currents have been assumed arbitrarily. As was the case in the nodal analysis, if the actual currents are not in the direction indicated, the values calculated will be negative. Applying KVL to the first mesh yields

$$-i_1R_1 - (i_1 - i_2)R_2 + v_1 = 0$$

or

$$i_1(R_1 + R_2) - i_2R_2 = v_1$$

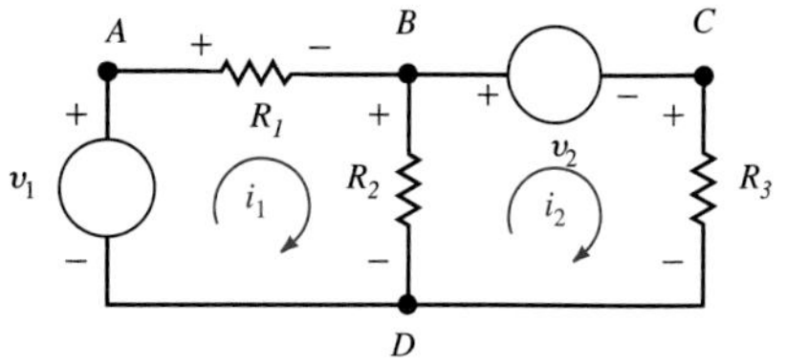

Figure 2.28 Two-mesh circuit.

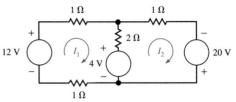

Figure 2.29 Circuit used in Example 2.15.

KVL applied to Mesh 2 yields

$$-v_2 - i_2R_3 + (i_1 - i_2)R_2 = 0$$

or

$$-i_1R_2 + i_2(R_2 + R_3) = -v_2$$

Therefore, the two simultaneous equations required to solve this two-mesh circuit are

$$i_1(R_1 + R_2) - i_2(R_2) = v_1$$
$$-i_1(R_2) + i_2(R_2 + R_3) = -v_2$$

EXAMPLE 2.15

Let us compute the mesh currents in the circuit in Fig. 2.29. The KVL equations for the two meshes are

$$-1I_1 - 2(I_1 - I_2) - 4 - 1I_1 + 12 = 0$$
$$-1I_2 + 20 + 4 - 2(I_2 - I_1) = 0$$

or

$$4I_1 - 2I_2 = 8$$
$$-2I_1 + 3I_2 = 24$$

These equations yield $I_1 = 9$ A and $I_2 = 14$ A. ∎

EXAMPLE 2.16

Let us determine V_o in the network in Fig. 2.30. Note that as we attempt to write the mesh equations using KVL, we find that we cannot express the voltage across the current source in terms of the mesh currents. The current source does, however, *constrain* the mesh currents by the equation

$$I_1 - I_2 = 4$$

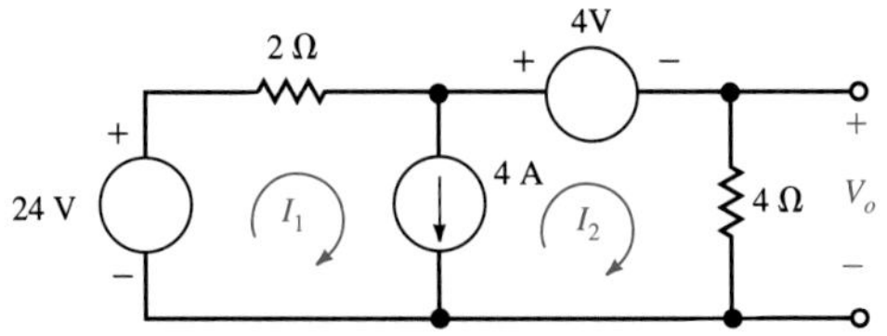

Figure 2.30 Circuit used in Example 2.16.

This constraint equation is one of the two linearly independent equations needed to find the mesh currents. The other equation can be obtained by applying KVL to the outer loop, that is

$$+ 24 - 2I_1 - 4 - 4I_2 = 0$$

Hence, the two equations for the network are

$$I_1 - I_2 = 4$$
$$2I_1 + 4I_2 = 20$$

These equations yield $I_2 = 2$ A and therefore $V_o = 8$ V.   ■

DRILL EXERCISE

D2.15. Use mesh analysis to find V_o in the network in Fig. D2.15.

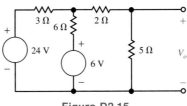

Figure D2.15

Ans: $V_o = 10$ V.

D2.16. Repeat problem D2.15 for the network in Fig. D2.16.

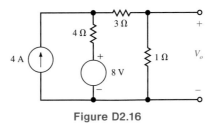

Figure D2.16

Ans: $V_o = 3$ V.

2.9
Superposition

The following example will serve to introduce this subject.

EXAMPLE 2.17
Consider the network in Fig. 2.31(a). We wish to find the voltage $v_o(t)$. Although we could use node equations or loop equations to find $v_o(t)$, only one node equation is required. Using KCL at the node labeled $v_o(t)$ yields

$$\frac{v_A(t) - v_o(t)}{2} + i_A(t) - \frac{v_o(t)}{1} = 0$$

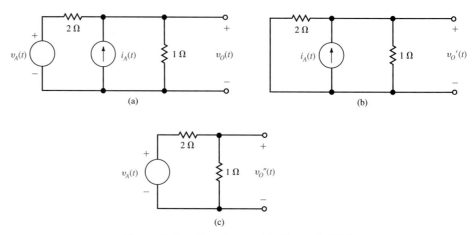

Figure 2.31   Circuits used in Example 2.17.

or

$$v_o(t) = \frac{1}{3}v_A(t) + \frac{2}{3}i_A(t)$$

In other words, the voltage $v_o(t)$ has a component due to $v_A(t)$ and a component due to $i_A(t)$. In view of the fact that $v_o(t)$ has two components, one due to each independent source, it would be interesting to examine what each source acting alone would contribute to $v_o(t)$. For $i_A(t)$ to act alone, $v_A(t)$ must be zero. As illustrated in Fig. 2.2, $v_A(t) = 0$ means that the source $v_A(t)$ is replaced with a short circuit. Therefore, to determine the value of $v_o(t)$ due to $i_A(t)$ only, we employ the circuit in Fig. 2.31(b) and refer to this value of $v_o(t)$ as $v_o'(t)$.

$$v_o'(t) = \frac{(1)\,(2)}{1+2}\,i_A(t)$$

$$= \frac{2}{3}\,i_A(t)$$

Let us now determine the value of $v_o(t)$ due to $v_A(t)$ acting alone and refer to this value as $v_o''(t)$. We employ the circuit shown in Fig. 2.31(c) and compute this value as

$$v_o''(t) = \frac{v_A(t)}{2+1}\,(1)$$

$$= \frac{v_A(t)}{3}$$

Now if we add the values $v_o'(t)$ and $v_o''(t)$, we obtain the value computed directly:

$$v_o(t) = v_o'(t) + v_o''(t)$$

$$= \frac{2}{3}A(t) + \frac{v_A(t)}{3}$$

Note that we have superposed the value of $v_o'(t)$ on $v_o''(t)$ or vice versa, to determine the total value of the unknown current.   ■

What we have demonstrated in Example 2.17 is true in general for linear circuits. Linearity requires both additivity and homogeneity (scaling), and since the voltage/current relationship for all the circuits presented in part I of this text satisfies these two conditions, the circuits are linear. *The principle of superposition states that*

> In any linear circuit containing multiple independent sources, the current or voltage at any point in the network may be calculated as the algebraic sum of the individual contributions of each source acting alone.

When determining the contribution due to an independent source, any remaining voltage sources are made zero by replacing them with short circuits, and any remaining current sources are made zero by replacing them with open circuits; however, dependent sources are not made zero and remain in the circuit.

Superposition can be applied to a circuit with any number of dependent and independent sources. In fact, superposition can be applied to such a network in a variety of ways. For example, a circuit with three independent sources can be solved using each source acting alone, as we have demonstrated above, or we could use two at a time and sum the result with that obtained from the third acting alone.

EXAMPLE 2.18

Let us use superposition to solve Example 2.16. The network for this example is redrawn in Fig. 2.32(a). As stated above, we can determine the output due to the two voltage sources and add that to the output caused by the current source. The output generated by the two voltage sources is obtained from the network in Fig. 2.32(b) where the current source has been replaced by an open circuit.

$$I = \frac{24 - 4}{2 + 4} = \frac{20}{6} \, \text{A}$$

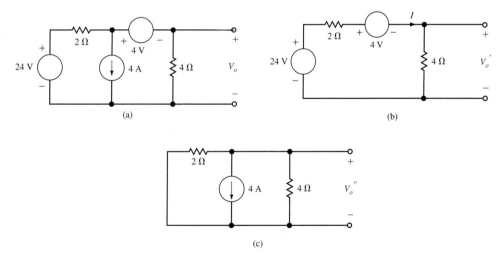

Figure 2.32 Circuits used in Example 2.18.

then

$$V'_o = \left(\frac{20}{6}\right)(4) = \frac{80}{6}\,\text{V}$$

The output generated by the current source is obtained from the network in Fig. 2.32(c) where the voltage sources have been replaced by short circuits. Since all the elements are in parallel

$$V''_o = -4\left[\frac{(2)\,(4)}{2+4}\right]$$

$$= \frac{-32}{6}\,\text{V}$$

Then using superposition

$$V_o = V'_o + V''_o$$

$$= \frac{80}{6} - \frac{32}{6}$$

$$= 8\,\text{V}$$

which is identical to that obtained in Example 2.16 ∎

DRILL EXERCISE

D2.17. Use superposition to find V_o in the network in Fig. D2.17.

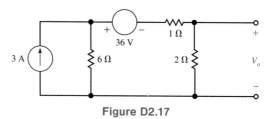

Figure D2.17

Ans: $V_o = -4$ V.

2.10

Source Exchange

We introduce this topic by considering the two circuits shown in Fig. 2.33. In one case, a voltage source in series with a resistor R_v is connected to the load R_L, and in the other case, a current source in parallel with a resistor R_i is connected to load R_L. We now ask if it is possible to *exchange* the series combination of v and R_v for the parallel combination of i and R_i, and vice versa. The answer is we can, provided that each produces exactly the same voltage and current for any load R_L.

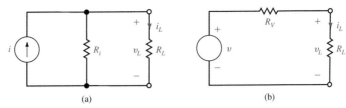

Figure 2.33 Circuits used to determine conditions for a source exchange.

In order to determine the conditions under which the series connection of v and R_v is equivalent to the parallel connection of i and R_i, let us examine the terminal conditions of each. For the network in Fig. 2.33(a)

$$i = i_L + \frac{v_L}{R_i}$$

or

$$iR_i = R_i i_L + v_L$$

For the network in Fig. 2.33(b)

$$v = i_L R_v + v_L$$

For the two networks to be equivalent, their terminal characteristics must be identical; therefore, equating like terms in the above equations,

$$v = iR_i \qquad \text{and} \qquad R_i = R_v \tag{2.27}$$

The relationships specified in Eq. (2.27) and Fig. 2.33 are extremely important and the reader should not fail to grasp their significance. What these relationships tell us is that if we have embedded within a network a current source i in parallel with a resistor R, we can replace this combination with a voltage source of value $v = iR$ in series with the resistor R. The reverse is also true; that is, a voltage source v in series with a resistor R can be replaced with a current source of value $i = v/R$ in parallel with the resistor R. Parameters within the circuit (e.g., an output voltage) are unchanged under these transformations.

The following example will demonstrate the utility of a *source exchange*. The reader is cautioned to keep the polarity of the voltage source and the direction of the current source in agreement, as shown in Fig. 2.33.

EXAMPLE 2.19

Consider the network in Fig. 2.34(a). We wish to compute the current I_o. Note that since node voltages V_1 and V_3 are known, one KCL equation at the center node will yield V_2 and therefore I_o. However, two mesh equations are required to find I_o and two separate circuits must be analyzed to determine I_o using superposition.

To solve the problem using source transformation, we first transform the 60 V source and 6 Ω resistor into a 10 A current source in parallel with the 6 Ω resistor as shown in Fig. 2.34(b). Next, the 3 Ω resistor in series with the 15 V source are transformed into a

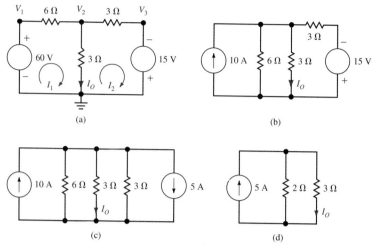

Figure 2.34   Circuits employed in Example 2.19.

5 A current source in parallel with the 3 Ω resistor, as shown in Fig. 2.34(c). Combining the resistors and current sources yields the network in Fig. 2.34(d). Now employing current division, we obtain $I_o = 2$ A.   ∎

DRILL EXERCISE

D2.18. Solve problem D2.17 using source transformation.

2.11

Thevenin's and Norton's Theorems

Thus far we have presented a number of techniques for circuit analysis. At this point we will add two theorems to our collection of tools that will prove to be extremely useful. The theorems are named after their authors, M. L. Thévenin, a French engineer, and E. L. Norton, a scientist formerly with Bell Telephone Laboratories.

Suppose that we are given a circuit and that we wish to find the current, voltage, or power that is delivered to some resistor of the network which we will call the load. *Thevenin's theorem* tells us that we can replace the entire network, exclusive of the load, by an equivalent circuit that contains only an independent voltage source in series with a resistor in such a way that the current-voltage relationship at the load is unchanged. *Norton's theorem* is identical to the statement above except that the equivalent circuit is an independent current source in parallel with a resistor.

Consider the circuit in Fig. 2.35(a). This network, exclusive of the load, R_L, can be replaced by the series combination of a voltage source V_{OC} and a resistance R_{TH} where V_{OC} is the open-circuit voltage at the terminals A–B in Fig. 2.35(b) and R_{TH} is the ratio of V_{OC} to the short-circuit current shown in Fig. 2.35(c). The *Thevenin equivalent circuit* is then shown in Fig. 2.35(d). If the network contains only independent sources, R_{TH} can be found by looking into the open-circuit terminals A–B in Fig. 2.35(b) and determining the resistance at those terminals with all voltage sources replaced by short circuits and all current sources replaced with open circuits. However, if dependent sources are present in the net-

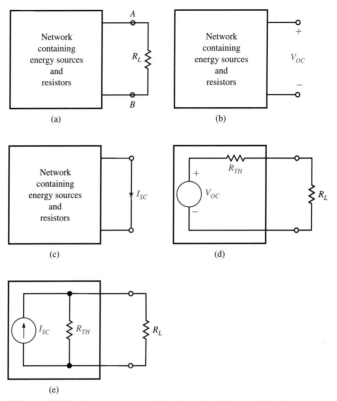

Figure 2.35   Concepts used to develop Thevenin's theorem.

work, R_{TH} must be derived from the ratio of V_{OC} to I_{SC}. The *Norton equivalent circuit* is shown in Fig. 2.35(e).

EXAMPLE 2.20

Consider the network in Fig. 2.36(a). We wish to find V_o using both Thevenin's and Norton's theorems. We can break the network at either *A–B* or *C–D*. If we break the network at *A–B*, the open circuit voltage shown in Fig. 2.36(b) is obtained as follows:

$$I_1 = \frac{36}{3 + 6} = 4 \text{ A}$$

$$V_{OC} = (4)(6) = 24 \text{ V}$$

Replacing the 36V source with a short circuit and looking into terminals *A* and *B* we find as shown in Fig. 2.36(c) that

$$R_{TH} = \frac{(3)(6)}{3 + 6} = 2 \text{ }\Omega$$

If the Thevenin equivalent circuit is connected to the terminals *A–B*, the original circuit becomes that shown in Fig. 2.36(d). Then

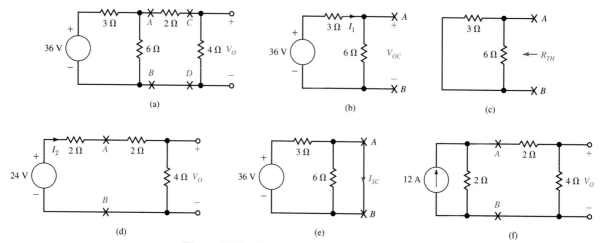

Figure 2.36   Circuits used in Example 2.20.

$$I_2 = \frac{24}{2 + 2 + 4} = 3 \text{ A}$$

$$V_o = (3)(4) = 12 \text{ V}$$

It is important to note that if we had broken the network at C–D, the open-circuit voltage would still be 24 V, since there would be no current in the 2 Ω resistor and therefore no voltage across it. R_{TH} in this case would be 4 Ω, but when this new Thevenin equivalent circuit is reconnected to the 4 Ω load the final answer would be the same.

We could also determine V_o using Norton's theorem. Once again, the network is broken at terminals A–B. The short circuit current is shown in Fig. 2.36(e). Since no current will flow in the 6 Ω resistor in parallel with the short circuit

$$I_{SC} = \frac{36}{3} = 12 \text{ A}$$

The Thevenin equivalent resistance was computed in Fig. 2.36(c) to be $R_{TH} = 2 \text{ }\Omega$. If the Norton equivalent circuit consisting of the short circuit current source in parallel with the Thevenin equivalent resistance is now attached to the remainder of the original network at terminals A–B, the resultant network is shown in Fig. 2.36(f). Using current division, we find that

$$V_o = 12 \left(\frac{2}{2 + 2 + 4} \right) (4)$$

$$= 12 \text{ V}$$

EXAMPLE 2.21

We wish to use Thevenin's theorem to solve Example 2.16. The network in Fig. 2.30 is redrawn in Fig. 2.37(a). If we break the network at the 4 Ω load, the open-circuit voltage can be computed from the network in Fig. 2.37(b). Note that $I_1 = 4$ A. Then using KVL

$$V_{CS} = 24 - 2(4) = 16 \text{ V}$$

Applying KVL again we find that

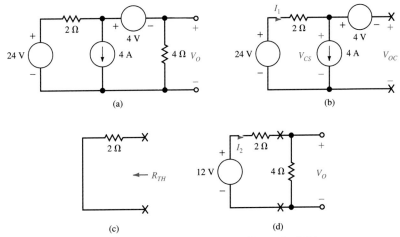

Figure 2.37   Circuits used in Example 2.21.

$$V_{OC} = V_{CS} - 4$$
$$= 12 \text{ V}$$

R_{TH} is found by replacing the voltage sources with short circuits and the current source with an open circuit and as shown in Fig. 2.37(c), $R_{TH} = 2 \Omega$. Finally, if the Thevenin equivalent circuit is reconnected to the 4 Ω load as shown in Fig. 2.37(d)

$$I_2 = \frac{12}{2 + 4} = 2 \text{ A}$$

and

$$V_o = (2)(4) = 8 \text{ V} \qquad \blacksquare$$

DRILL EXERCISE

D2.19. Solve problem D2.15 using Thevenin's theorem.
D2.20. Solve problem D2.16 using Thevenin's theorem.

Finally, let us consider one example which we will examine in a variety of ways.

EXAMPLE 2.22
Let us determine the power absorbed in the 2 Ω resistor in the circuit in Fig. 2.38(a). We will solve this problem using nodal analysis, mesh analysis, superposition, source exchange, Thevenin's theorem, and Norton's theorem.
    The node equations for the network shown in Fig. 2.38(a) are

$$V_1 = 24$$
$$V_2 = 12$$
$$\frac{V_o - V_1}{3} + \frac{V_o - V_2}{6} + \frac{V_o}{2} = 0$$

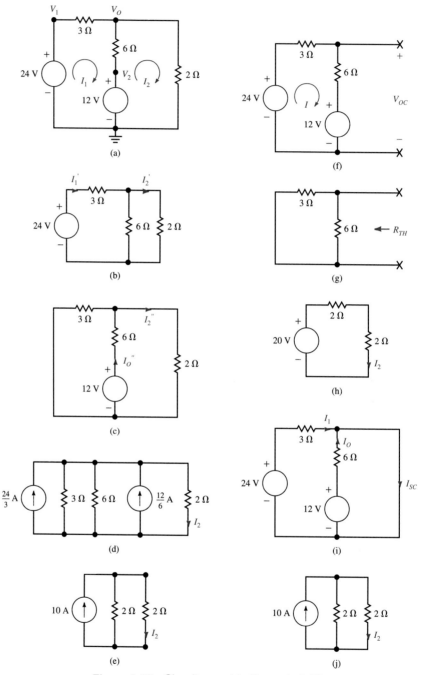

Figure 2.38 Circuits used in Example 2.22.

These equations yield

$$V_o = 10 \text{ V}$$

Therefore, the current in the $2 \, \Omega$ resistor is

$$I_{2\Omega} = \frac{10}{2} = 5 \text{ A}$$

and

$$P_{2\Omega} = (5)^2(2) = 50 \text{ W}$$

The mesh equations for the network shown in Fig. 2.38(a) are

$$+24 - 3I_1 - 6(I_1 - I_2) - 12 = 0$$
$$+12 - 6(I_2 - I_1) - 2I_2 = 0$$

or

$$9I_1 - 6I_2 = 12$$
$$-6I_1 + 8I_2 = 12$$

Solving these equations for I_2 yields

$$I_2 = 5 \text{ A}$$

Therefore, the power absorbed by the $2 \, \Omega$ resistor is 50 W.

In the application of superposition, we first consider the 24 V source acting alone as shown in Fig. 2.38(b). The equivalent resistance seen by the 24 V source is $3 \, \Omega + 6 \, \Omega$ in parallel with $2 \, \Omega$, that is

$$Req_1 = 3 + \frac{(6)\,(2)}{6 + 2}$$

and hence

$$I_1' = \frac{24}{3 + \dfrac{12}{8}}$$

I_2' can be obtained by current division as

$$I_2' = \frac{I_1'\,(6)}{6 + 2}$$

Solving for I_2' yields

$$I_2' = 4 \text{ A}$$

The component of I_2 caused by the 12V source is derived from Fig. 2.38(c). Following an identical approach to that used to determine I_1' we find that

$$Req_2 = 6 + \frac{(3)\,(2)}{3 + 2}$$

and hence

$$I_o'' = \frac{12}{6 + 6/5}$$

Once again, using current division

$$I_2'' = I_o''\left(\frac{3}{2+3}\right)$$

Solving for I_2'' we obtain

$$I_2'' = 1 \text{ A}$$

Therefore, using superposition

$$I_2 = I_2' + I_2''$$
$$= 5 \text{ A}$$

and then the power absorbed by the 2 Ω resistor is 50 W.

In applying source transformation to the network in Fig. 2.38(a), we transform the 24V source in series with the 3 Ω resistor into an 8A source in parallel with the 3 Ω resistor. The 12V source and 6 Ω resistor are transformed in a similar manner. The resulting network is shown in Fig. 2.38(d). Combining the two current sources and the 3 Ω and 6 Ω resistors yields the circuit in Fig. 2.38(e). Applying current division to this network yields $I_2 = 5$ A.

Using Thevenin's theorem, we break the network at the 2 Ω load as shown in Fig. 2.38(f). KVL around the closed path yields

$$+24 - 3I - 6I - 12 = 0$$

Hence

$$I = \frac{4}{3} \text{ A}$$

KVL applied once again to the right-hand loop including V_{OC} yields

$$+12 + 6\left(\frac{4}{3}\right) - V_{OC} = 0$$

or

$$V_{OC} = 20 \text{ V}$$

R_{TH} is derived from the network in Fig. 2.38(g) as

$$R_{TH} = \frac{(3)(6)}{3+6} = 2 \text{ } \Omega$$

Replacing the network in Fig. 2.38(f) with its Thevenin equivalent and connecting it to the 2 Ω resistor yields the circuit in Fig. 2.38(h). Obviously, the current I_2 in this network is 5 A.

The short circuit current required for the Norton equivalent is computed from the network in Fig. 2.38(i).

$$I_{SC} = I_1 + I_o$$
$$= \frac{24}{3} + \frac{12}{6} = 10 \text{ A}$$

R_{TH} has already been calculated. Connecting the Norton equivalent circuit to the 2 Ω resistor produces the network in Fig. 2.38(j). Using current division, I_2 is found to be 5 A.

As a final point, note that when we perform a source exchange we are simply replacing a Thevenin equivalent circuit with a Norton equivalent circuit, and vice versa.   ■

2.12
dc Measurements

We now describe the instruments that are employed to physically measure current, voltage, resistance and power. Often these quantities are measured with digital instruments and their value presented on a digital display. However, it is instructive to describe the analog meter movement used in many instruments for electrical measurements.

The D'Arsonval Meter Movement

The D'Arsonval Meter Movement, which is a fundamental component in many electric instruments, is shown in Fig. 2.39. This is a permanent-magnet-moving-coil (PMMC) mechanism in which a coil of fine wire is allowed to rotate between the poles of a permanent magnet. If a current is passed through the movable coil, the resulting magnetic field reacts with the magnetic field of the permanent magnet producing a torque which is counterbalanced by a restoring spring. The deflection of the pointer attached to the coil is proportional to the coil current produced by the quantity being measured. As an example, a meter may be rated at 100 mV, 30 mA, for full-scale deflection.

Current Measurement

The *ammeter* is a device employed to measure current, and a variety of meter movements are commercially available. When connected in a network it is inserted directly into the path of the current to be measured. dc ammeters should be connected such that the current enters the positive terminal for upscale deflection, or a (+) reading on a digital instrument.

In the ideal case the internal resistance of an ammeter would be zero. However, practical ammeters do have some small internal resistance, R_m, which can produce a loading

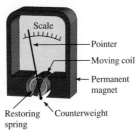

Figure 2.39 D'Arsonval meter movement.

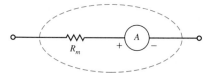

Figure 2.40   Ammeter Circuit Representation.

effect in high-current, low resistance circuits, i.e., the current measured by the meter is changed by inserting the meter. Therefore, it is convenient to represent the ammeter as shown in Fig. 2.40 in which all the resistance of the meter is lumped into R_m.

EXAMPLE 2.23

Let us determine the internal resistance of an ammeter that has a full scale rating of 30mA at 100 mV.

The internal resistance for this meter is

$$R_m = \frac{100 \times 10^{-3}}{30 \times 10^{-3}}$$

$$= 3.33 \ \Omega \qquad \blacksquare$$

Suppose now that we wish to use the meter mechanism in Example 2.23 to measure a current, as shown in Fig. 2.41a, which is larger than the meter's current rating. We can do this by shunting a large part of the current to be measured around the meter movement as shown in Fig. 2.41(b). Note that the parallel combination of R_m and R_{SH} form a current divider. By selecting the proper value of the shunt resistor we can use the meter movement to measure a current larger than the rating of the meter mechanism. The following example illustrates the selection of a shunt resistor to increase the current range.

EXAMPLE 2.24

We wish to employ a 5V, 500 mA dc adjustable current power supply to provide the power for the ringing circuit for a door bell as shown in Fig. 2.42(a). The coil for the ringing circuit has a current limit of 250 mA. The only meter movement available to us is the 30 mA, 100 mV meter described above. Therefore, we must select a shunt resistance, which when placed in parallel with the meter, will permit us to measure 250 mA full scale.

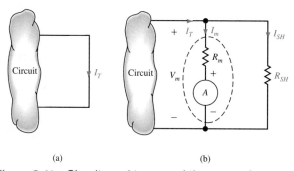

(a)   (b)

Figure 2.41   Circuit used to expand the range of a meter.

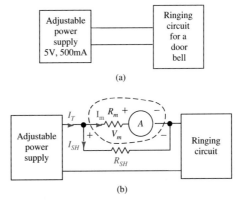

Figure 2.42 Circuits used in Example 2.24.

With reference to Fig. 2.42(b), the total current $I_T = 250$ mA. The current in the meter movement $I_m = 30$ mA. Therefore, using KCL

$$I_{SH} = I_T - I_m$$
$$= 220 \text{ mA}$$

Since the voltage $V_m = 100$ mV is across both the meter resistance and the shunt resistance, the required value of R_{SH} is

$$R_{SH} = \frac{V_m}{I_{SH}}$$
$$= \frac{0.1}{0.22}$$
$$= 0.4545 \ \Omega \qquad \blacksquare$$

DRILL EXERCISE

D2.21. A dc ammeter has a full scale rating of 200 mV, 20 mA. Find the meter resistance.

Ans: $R_m = 10 \ \Omega$

Voltage Measurement

A *voltmeter* is the basic device used to measure voltage, and is connected in parallel with the element whose voltage is desired. In order to obtain an upscale meter deflection (or a + reading on a digital instrument) the + terminal of the voltmeter should be connected to the point of higher potential.

Once again the resistance of the meter movement is R_m and is represented as shown in Fig. 2.40. A resistance R_{se}, called a multiplier, is connected in series with the meter movement. This resistance, R_{se}, and the meter movement form a voltage divider and permit the meter to be used in a manner which extends the voltage range of the meter. The following example illustrates this technique.

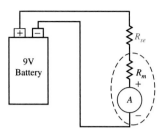

Figure 2.43   Circuit used in Example 2.25.

EXAMPLE 2.25

The remote control for our television set does not work and we are suspicious that the 9V battery in the remote is bad. We wish to measure this voltage but all we have available is a 100 mV, 30 mA D'Arsonval meter movement. Let us determine the value of the resistor, R_{se}, which must be placed in series with the meter movement to measure this voltage.

The circuit for this application is shown in Fig. 2.43 and should produce full-scale deflection of 9V using a 100 mV, 30 mA meter movement. For full-scale deflection $V_m =$ 100 mV and hence using KVL

$$9 = I_m R_{se} + V_m$$

and since $V_m = 100$ mV and $I_m = 30$mA

$$R_{se} = \frac{9 - 0.100}{0.03}$$
$$= 296.7 \ \Omega$$

Recall from Example 2.23 that the meter resistance is 3.3 Ω. Therefore, if the battery voltage has dropped to 6 V, the current read by the meter would be

$$I_m = \frac{6}{296.7 + 3.3}$$
$$= 20 \ \text{mA}$$

which is 6/9 or 2/3 of full scale.   ■

DRILL EXERCISE

D2.22. A meter has a resistance of 50 Ω and a full scale current of 1 mA. Select the multiplier resistance needed to measure 12 V full scale.

Ans: R_{se} = 11.95 kΩ

Resistance Measurement

An *ohmmeter* is an instrument used to measure resistance. An ohmmeter circuit, shown in Fig. 2.44(a), typically consists of a meter movement in series with a battery and regulating resistor. The meter scale typically reads in ohms from right to left as shown in Fig. 2.44(b). If $R_x = \infty$, i.e., the ohmmeter leads or terminals are open, there is no meter de-

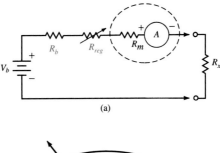

(a)

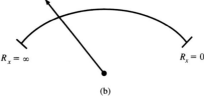

(b)

Figure 2.44   Ohmmeter circuit and scale.

flection. If however, $R_x = 0$, i.e., the meter leads are shorted, maximum current flows in the circuit and full-scale deflection occurs. The regulating resistor R_{reg} is a calibration device which allows us to zero adjust the meter movement when the leads are shorted and thus compensate for meter resistance and battery aging.

The ohmmeter differs from the ammeter and voltmeter in one important respect: It must never be connected to an energized circuit, i.e., the element to be measured must be disconnected from the original circuit when the measurement is made.

Although this type of ohmmeter is very useful, it is not a precision instrument. A much more accurate device for measuring resistance values over a wide range is what is called the *Wheatstone Bridge*. The Wheatstone Bridge circuit is shown in Fig. 2.45. In this network, the resistors R_1, R_2, and R_3 are known. R_x is the unknown resistor. The device in the center leg of the bridge is a sensitive D'Arsonval mechanism in the microamp range called a *galvanometer.*

The bridge is used in the following manner: The unknown resistance R_x is connected as shown in Fig. 2.45, and then R_3 is adjusted until there is no current in the galvanometer. At this point the bridge is said to be balanced. Under this balanced condition $I_G = 0$ and, hence, KCL applied to the center nodes of the bridge yields

$$I_1 = I_3$$

and

$$I_2 = I_x$$

Furthermore, since $I_G = 0$, there is no voltage drop across the galvanometer and, therefore, KVL requires that

$$I_1 R_1 = I_2 R_2$$

and

$$I_3 R_3 = I_x R_x$$

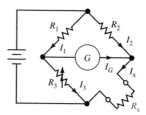

Figure 2.45 The Wheatstone Bridge circuit.

Dividing one equation by the other and using the fact that $I_1 = I_3$ and $I_2 = I_x$ yields the relationship

$$\frac{R_1}{R_3} = \frac{R_2}{R_x}$$

so that

$$R_x = \left(\frac{R_2}{R_1}\right)R_3$$

In addition to using the Wheatstone Bridge to measure an unknown resistance, it is also employed, for example, by engineers to measure strain in solid materials using a resistive strain guage.

EXAMPLE 2.26

The bridge circuit in Fig. 2.45 is used to measure an unknown resistance R_x. In this network $R_1 = 200\ \Omega$ and $R_2 = 240\ \Omega$. If R_3 is adjusted to a value of 220 Ω, which places the bridge in balance, what is the value of R_x?

The unknown resistance is

$$R_x = \left(\frac{R_2}{R_1}\right)R_3$$
$$= \left(\frac{240}{200}\right)(220) = 264\ \Omega$$ ∎

Power Measurements

In a dc circuit, power is the product of voltage and current. Therefore, power can be derived from current and voltage measurements. dc power can be measured directly using a *wattmeter*. However, since very accurate measurements can be obtained using a voltmeter in conjunction with an ammeter, the wattmeter is normally not used in dc measurements. The wattmeter remains an important device in ac measurements.

2.13
Summary

- Ohm's law states that the voltage V across a resistance R is directly proportional to the current I through it, that is, $V = IR$, where V is in volts, I is in amperes, and R is in ohms.
- Conductance G, in siemens, is the reciprocal of resistance.

- Node, loop, mesh, and branch are four common terms employed in circuit analysis and defined in section 2.2.
- Short circuit is a branch in which $R = 0$ and therefore $V = 0$ for any I.
- Open circuit is a branch in which $R = \infty$ and therefore $I = 0$ for any V.
- KCL states that the algebraic sum of the currents entering any node is zero.
- KVL states that the algebraic sum of the voltages around any loop is zero.
- Resistances in series add as $R_{\text{total}} = R_1 + R_2 + R_3 + \cdots$.
- Conductances in parallel add as $G_{\text{total}} = G_1 + G_2 + G_3 + \cdots$.
- In current division a current I is divided between parallel resistors R_1 and R_2 as

$$I_{R1} = \left(\frac{R_2}{R_1 + R_2} \right) I$$

- In voltage division a voltage V is divided between series resistors R_1 and R_2 in direct proportion to their resistance, that is,

$$V_{R1} = \left(\frac{R_1}{R_1 + R_2} \right) V$$

- The delta-wye transformation permits the conversion of a set of resistors connected in delta to a set of resistors connected in wye and vice versa.
- In a nodal analysis N-1 linearly independent simulataneous equations are required to find the N-1 unknown node voltages in a N-node circuit.
- In a loop analysis N linearly independent simultaneous equations are required to find the N unknown loop currents in a N-loop network.
- In linear network analysis we can use:
  - Superposition, which permits us to determine the effect of multiple independent sources as the sum of the effects of the individual sources acting one at a time.
  - Source exchange, which permits us to replace a voltage source V and resistance R in series with a parallel combination of a current source $I = V/R$ and the resistance R, and vice versa.
  - Thevenin's theorem permits us to replace a network of sources and resistors at a pair of terminals with a series combination of a source and resistor, that is, the open-circuit voltage and the Thevenin equivalent resistance at the terminals.
  - Norton's theorem is equivalent to Thevenin's theorem except that the network is replaced by a parallel combination of a current source and resistor, that is, the short circuit current and the Thevenin equivalent resistance at the terminals.
  - dc measurement instruments can be made to determine current, voltage, power, and resistance.

PROBLEMS

2.1. A 10 kW electric clothes dryer runs continuously for 1.5 hours while drying a load of clothes. If the cost of energy charged by the local electric utility company is $0.08 per kilowatt-hour, determine the cost of drying the load of clothes.

2.2. Many years ago a string of Christmas lights was manufactured in the form shown in Fig. P2.2(a). Today, the lights are often connected in the form shown in Fig. P2.2(b). State one good reason for the change.

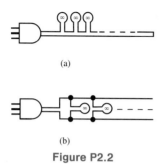

(a)

(b)

Figure P2.2

2.3. Determine the resistance of a 110-volt lightbulb that is rated at 25 watts.

2.4. Find I_1 in the network in Fig. P2.4.

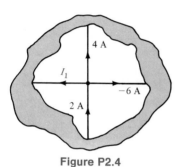

Figure P2.4

2.5. Find the currents I_1, I_2, and I_3 in the network in Fig. P2.5.

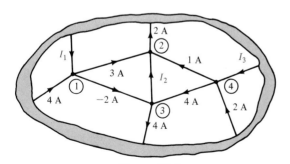

Figure P2.5

2.6. Find V_o in the network in Fig. P2.6.

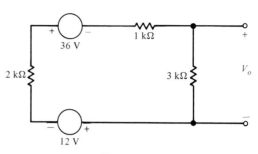

Figure P2.6

2.7. Find V_o in the network in Fig. P2.7.

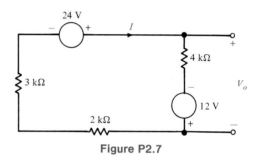

Figure P2.7

2.8. Find I and V_o in the network in Fig. P2.8.

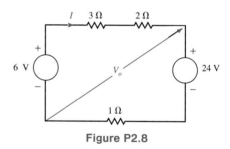

Figure P2.8

2.9. Find V_o in the network in Fig. P2.9.

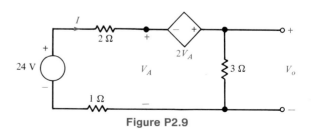

Figure P2.9

2.10. Given the network in Fig. P2.10, determine the power absorbed by the 6 Ω resistor.

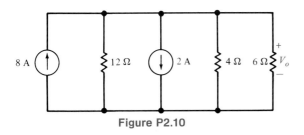

Figure P2.10

2.11. In the network in Fig. P2.11, $I_2 = 4$ mA. Find I_o.

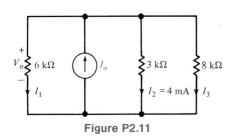

Figure P2.11

2.12. Given the network in Fig.P2.12, find V_o and I_o.

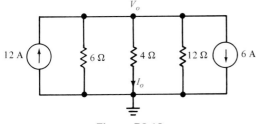

Figure P2.12

2.13 Find the equivalent resistance at the terminals of the network in Fig. P2.13.

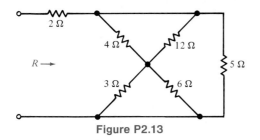

Figure P2.13

2.14. Find the equivalent resistance R_{eq} of the circuit in Fig. P2.14.

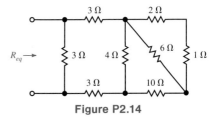

Figure P2.14

2.15. Find the equivalent resistance of the network in Fig. P2.15.

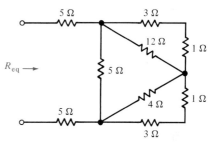

Figure P2.15

2.16. Find V_o in the network in Fig. P2.16.

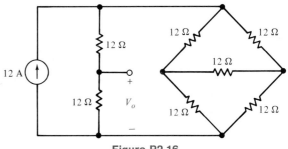

Figure P2.16

2.17. Find I_o in the network in Fig. P2.17.

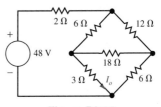

Figure P2.17

2.18. Find the power absorbed in the 1 Ω resistor in the circuit shown in Fig. P2.18.

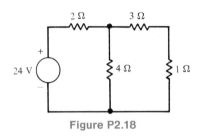

Figure P2.18

2.19. Find the power dissipated in the 2 Ω resistor in the network shown in Fig. P2.19.

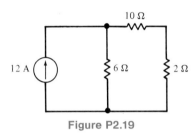

Figure P2.19

2.20. Forty lights, each with an effective resistance of 400 Ω, are connected in parallel. The power for this bank of lights is supplied by a 12 V battery with an internal resistance of 2 Ω, as shown in Fig. P2.20. Find the current in each light and the power supplied to the entire bank of lights.

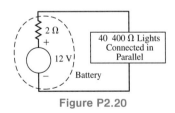

Figure P2.20

2.21. A particular automobile radio requires 9 V. The car battery is a 12 V source. A resistor must be placed between the battery and the radio to limit the voltage at the radio to 9 V. If the radio typically draws 200 mA of current, calculate the value and power rating of the required resistor.

2.22. A bank of eight 110V, 100 watt lamps are connected in parallel as shown in Fig. P2.22. The line voltage is 230 V. Compute the value of the resistor R_s, which must be connected in series with the bank to ensure that the voltage across the lamps does not exceed 110 V.

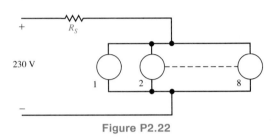

Figure P2.22

2.23. A high-voltage dc transmission line is used to transmit power from Point A to Point B. The power requirement at Point B is 1000 MW, which is delivered at 600 kV. If the resistance of the wire used in the transmission line is 0.8 milliohms per Kilometer and the distance between Point A and Point B is 1200 Kilometers, what voltage must be supplied at the sending end of the transmission line?

2.24. Find V_o in the network in Fig. P2.24.

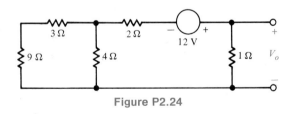

Figure P2.24

2.25. Find I_o in the network in Fig. P2.25.

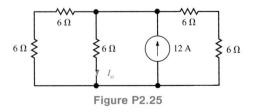

Figure P2.25

2.26. Use nodal analysis to find I_o in the network in Fig. P2.26.

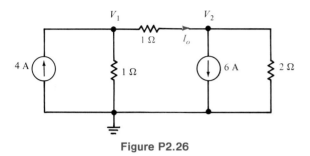

Figure P2.26

2.27. Use nodal analysis to find the node voltages in the network in Fig. P2.27.

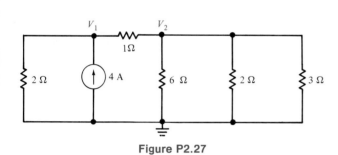

Figure P2.27

2.28. Use node equations to find V_2 in the circuit shown in Fig. P2.28.

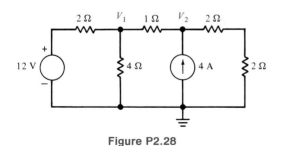

Figure P2.28

2.29. Use node equations to find V_2 in the circuit shown in Fig. P2.29.

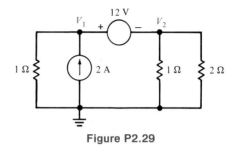

Figure P2.29

2.30. Use mesh equations to calculate the power absorbed by the 5 Ω resistor in Fig. P2.30.

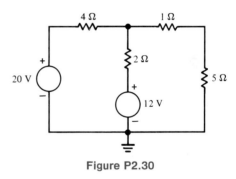

Figure P2.30

2.31. Use mesh equations to find V_o in the network shown in Fig. P2.31.

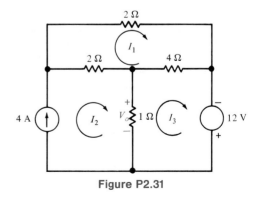

Figure P2.31

2.32. Use mesh equations to find I_o in Fig. P2.32.

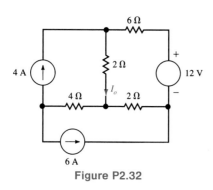

Figure P2.32

2.33. Compute V_o in the network in Fig. P2.33 using superposition.

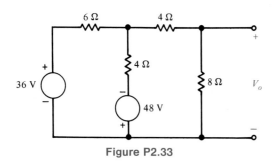

Figure P2.33

2.34. Compare V_o in the network in Fig. P2.34 using superposition.

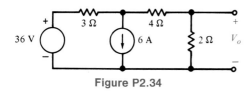

Figure P2.34

2.35. Find V_o in the network in Fig. P2.35 using superposition.

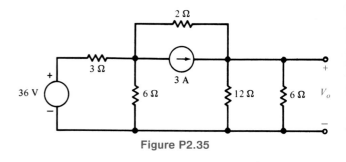

Figure P2.35

2.36. Use superposition to find V_o in Fig. P2.36.

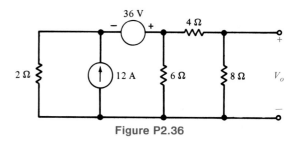

Figure P2.36

2.37. Find the voltage across the 4 Ω resistor by using superposition in the circuit in Fig. P2.37.

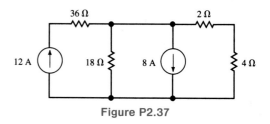

Figure P2.37

2.38. Use source transformation to find I_o in the circuit in Fig. P2.38.

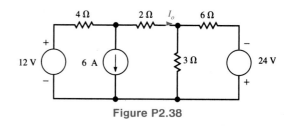

Figure P2.38

2.39. Use source transformation to find I_o in the network in Fig. P2.39.

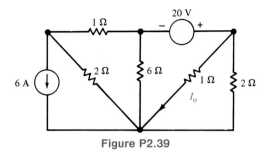

Figure P2.39

2.40. Find I_o in the network in Fig. P2.40 using source transformation.

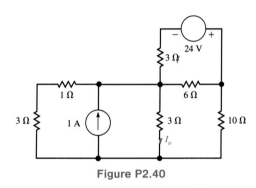

Figure P2.40

2.41. Use Thevenin's theorem to find V_o in the circuit in Fig. P2.33.

2.42. Use Thevenin's theorem to find V_o in the circuit in Fig. P2.34.

2.43. Use Thevenin's theorem to find V_o in the network in Fig. P2.35.

2.44. Use Thevenin's theorem to find V_o in the network shown in Fig. P2.44.

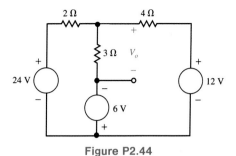

Figure P2.44

2.45. Use Thevenin's theorem to find V_o in the network shown in Fig. P2.45.

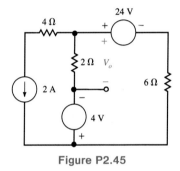

Figure P2.45

2.46. Use Thevenin's theorem to find V_o in the network in Fig. P2.46.

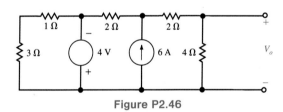

Figure P2.46

2.47. Use Norton's theorem to find I_o in the network in Fig. P2.47.

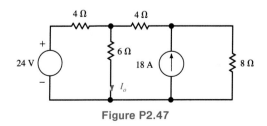

Figure P2.47

2.48 Apply Norton's theorem to find I_o in the network in Fig. P2.48.

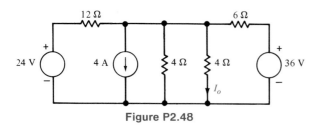

Figure P2.48

2.49. Use Norton's theorem to find I in the network in Fig. P2.49.

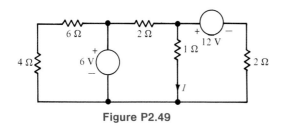

Figure P2.49

2.50. Use Norton's theorem to find V_o in the network in Fig. P2.50.

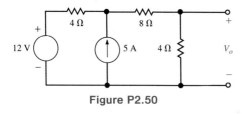

Figure P2.50

2.51. Given a meter movement rated at 50 mV, 20 mA full scale, find the meter resistance and power dissipated at full-scale deflection.

2.52. A 50 mV, 20 mA meter movement is used to measure currents between 0 and 150 mA. Find the shunt resistance required for the measurements.

2.53. A resistance of 50 Ω is connected in series with a 50 mV, 20 mA meter movement. What full scale voltage will this combination measure?

2.54. The Wheatstone Bridge in Fig. 2.45 is used to measure an unknown resistor R_x. The bridge is balanced with $R_1 = 1$ kΩ, $R_2 = 1.2$ kΩ, and $R_3 = 1.8$ kΩ. Find the value of R_x.

Transient Analysis

3.1
Introduction

In this chapter we will perform what is commonly known as a transient analysis. In this type of analysis we examine the behavior of a network as a function of time after a sudden change in the network occurs due to switches opening or closing.

Before beginning our study of transient analysis, we explore two other passive circuit elements: the capacitor and the inductor. Both these elements possess several important features: they are linear elements, their terminal characteristics are described by linear differential equations, and like a mechanical spring, they are capable of storing energy.

Techniques for the analysis of first-order networks, that is, networks that contain a single storage element, will be presented first, and the time constant of a network will be introduced. The analysis procedures will then be extended to second-order circuits, for example, those in which an inductor and capacitor are present simultaneously.

3.2
Capacitors and Inductors

Capacitors

Capacitors consist of two conducting surfaces separated by a nonconducting, or dielectric, material. A simplified capacitor and its electrical symbol are shown in Fig. 3.1. If a voltage V exists across the capacitor, positive charges will be transferred to one plate and negative charges to the other. The charge on the capacitor is proportional to the voltage across it as defined by the relationship

$$q = CV \tag{3.1}$$

where since q is in coulombs and V is in volts, the constant of proportionality C has units of coulombs per volt or "Farads" (F) after the famous English physicist Michael Faraday.

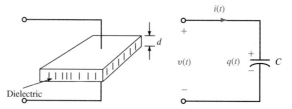

Figure 3.1   Capacitor and its electrical symbol.

Capacitors may be fixed or variable and typically range from thousands of microfarads (μF) to picofarads (pf). They find wide application in radios, TVs, high-voltage power systems, and a host of other applications.

As shown in Fig. 3.1, the parallel plates of the capacitor are separated by a dielectric, for example, air, and therefore the conduction current in the wires cannot flow internally between the plates. Therefore, the capacitor acts like an open circuit to dc—a fact we will use extensively later in this chapter. However, if the voltage across the capacitor changes with time, so will the charge, and thus

$$q(t) = Cv(t) \tag{3.2}$$

The time rate of change of charge is current. Hence

$$i(t) = \frac{dq(t)}{dt} \tag{3.3}$$

From Eqs. (3.2) and (3.3), we find that the terminal characteristics of a capacitor are defined by the linear differential equation

$$i(t) = C\frac{dv(t)}{dt} \tag{3.4}$$

If the current/voltage relationship for the capacitor defined by Eq. (3.4) is integrated, we obtain the equation

$$v(t) = \frac{1}{C} \int_{-\infty}^{t} i(x)\,dx \tag{3.5}$$

which can be expressed in the form

$$v(t) = \frac{1}{C} \int_{-\infty}^{t_o} i(x)\,dx + \frac{1}{C} \int_{t_o}^{t} i(x)\,dx$$

$$= v(t_o) + \frac{1}{C} \int_{t_o}^{t} i(x)\,dx \tag{3.6}$$

where $v(t_o)$ is the initial capacitor voltage.

The energy stored in the capacitor can be derived from the power delivered to it.

$$w_c(t) = \int p(t)\,dt$$

$$= \int_{-\infty}^{t} [v(x)] \left[C\frac{dv(x)}{dx} \right] dx \tag{3.7}$$

$$= \frac{1}{2} Cv^2(t) \; joules$$

Although we have modeled the capacitor as an ideal device, practical capacitors normally have some very large leakage resistance which provides a parallel conduction path between the plates. In addition, it is important to note that an instantaneous jump in the voltage across a capacitor is not physically realizable since such a jump would require moving a finite amount of charge in zero time, producing an infinite current. This latter point is another fact we will use extensively later in this chapter in transient analysis.

EXAMPLE 3.1

If the voltage across a 5-μF capacitor is as shown Fig. 3.2(a), let us determine the waveform for the current in the capacitor and the energy stored in the electric field of the capacitor at $t = 6$ ms.

The equations for the line segments of the waveform within the given time intervals are

$$v(t) = \frac{24}{6 \times 10^{-3}}t \qquad\qquad 0 \le t \le 6 \text{ ms}$$

$$= \frac{-24}{2 \times 10^{-3}}t + 96 \qquad 6 \le t \le 8 \text{ ms}$$

$$= 0 \qquad\qquad\qquad 8 \text{ ms} \le t$$

Using Eq. (3.4)

$$i(t) = C\frac{dv(t)}{dt}$$

$$= (5 \times 10^{-6})(4 \times 10^5) \qquad 0 \le t \le 6 \text{ ms}$$

$$= 20 \text{ mA} \qquad\qquad\qquad 0 \le t \le 6 \text{ ms}$$

$$i(t) = (5 \times 10^{-6})(-12 \times 10^3) \qquad 6 \le t \le 8 \text{ ms}$$

$$= -60 \text{ mA} \qquad\qquad\qquad 6 \le t \le 8 \text{ ms}$$

$$i(t) = 0 \qquad\qquad\qquad\qquad 8 \text{ ms} \le t$$

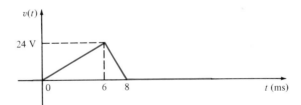

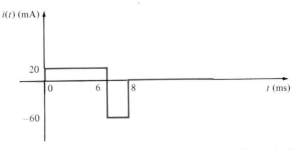

Figure 3.2   Voltage and current waveforms for Example 3.1.

The current waveform is shown in Fig. 3.2(b).

The energy stored in the electric field of the capacitor at $t = 6$ ms is

$$w_c(t) = \frac{1}{2}Cv^2(t)$$

$$w_c(6 \text{ ms}) = \frac{1}{2}(5 \times 10^{-6})(24)^2$$

$$= 1440 \ \mu\text{J} \qquad \blacksquare$$

Just as resistors can be combined in series and parallel combinations, interconnected capacitors can be combined to yield a single equivalent capacitance. Using Kirchhoff's voltage law, we can show that if N capacitors are connected in series, their equivalent capacitance is

$$\frac{1}{C_s} = \frac{1}{C_1} + \frac{1}{C_2} + \cdots + \frac{1}{C_N} \qquad (3.8)$$

Using Kirchhoff's current law we can show that if N capacitors are connected in parallel, their equivalent capacitance is

$$C_p = C_1 + C_2 + \cdots + C_N \qquad (3.9)$$

EXAMPLE 3.2

Let us find the total capacitance C_T at the terminals A-B of the network shown in Fig. 3.3(a). C_3 and C_4 are in series and therefore

$$\frac{1}{C_{34}} = \frac{1}{C_3} + \frac{1}{C_4} = \frac{1}{6 \ \mu\text{F}} + \frac{1}{3 \ \mu\text{F}}$$

$$C_{34} = 2 \ \mu\text{F}$$

Then C_2 and C_{34} are in parallel as shown in Fig. 3.3(b) and hence

$$C_{234} = C_2 + C_{34}$$

$$= 4 \ \mu\text{F}$$

Finally, C_1 and C_{234} are in series as shown in Fig. 3.3(c) and hence the total capacitance is

$$\frac{1}{C_T} = \frac{1}{C_1} + \frac{1}{C_{234}}$$

$$C_T = 3 \ \mu\text{F} \qquad \blacksquare$$

DRILL EXERCISE

D3.1. Find the total capacitance of the network in Fig. D3.1.

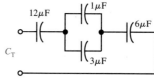

Figure D3.1

Ans: 2 μF.

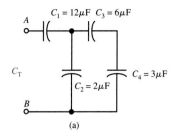

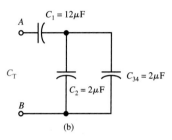

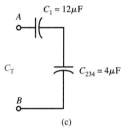

Figure 3.3   Circuits used in Example 3.2

Inductors

The ideal *inductor* is a circuit element that consists of a conducting wire that is wound around a core. The core material may range from some nonmagnetic material to a ferromagnetic material. These elements are employed in a variety of electric equipment and form the basis for large power transformers.

Two typical inductors and their electrical symbol are shown in Fig. 3.4. The flux lines for nonmagnetic core inductors, shown in Fig. 3.4(a), extend beyond the inductor itself. In contrast, a magnetic core confines the flux as shown in Fig. 3.4(b).

The American inventor Joseph Henry discovered that the voltage/current relationship for an inductor is

$$v(t) = L \frac{di(t)}{dt} \tag{3.10}$$

where the constant of proportionality L is called the inductance, and is measured in the unit "Henry." Practical inductors typically range from a few microhenrys to tens of henry. Equation (3.10) indicates that 1 henry is dimensionally equal to 1 volt-second per ampere. The relationship described by Eq. (3.10) indicates that a voltage which is produced by a changing magnetic field is proportional to the time rate of change of the current

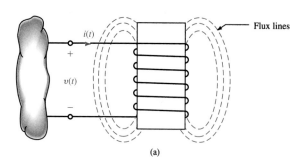

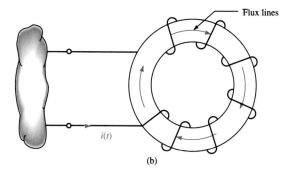

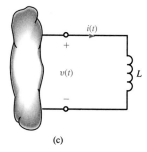

Figure 3.4   Inductors and their electrical symbol.

which generates the magnetic field. Note, however, if the current is constant, the voltage across the inductor is zero, that is, $di/dt = 0$. Hence, the inductor looks like a short circuit to dc—a fact we will use extensively later in the chapter.

Following an identical development of the equations for a capacitor, the current in an inductor can be written as

$$i(t) = \frac{1}{L}\int_{-\infty}^{t} v(x)\, dx \tag{3.11}$$

or

$$i(t) = i(t_o) + \frac{1}{L}\int_{t_o}^{t} v(x)\, dx \tag{3.12}$$

The power delivered to an inductor is

$$p(t) = v(t)i(t)$$
$$= \left[L\frac{di(t)}{dt}\right]i(t) \tag{3.13}$$

By integrating the power we find the energy stored in the magnetic field is

$$w_L(t) = \int_{-\infty}^{t} \left[L\frac{di(x)}{dx}\right]i(x)\, dx$$

or

$$w_L(t) = \frac{1}{2} L\, i^2(t) \qquad joules \qquad\qquad (3.14)$$

A practical inductor cannot store energy as well as a practical capacitor because there is always some winding resistance in the coil which quickly dissipates energy. In addition, just as the voltage across a capacitor cannot change instantaneously, the current in an inductor cannot change instantaneously. This point is very useful in transient analysis and we will apply it there.

EXAMPLE 3.3

If the current in a 10mH inductor has the waveform shown in Fig. 3.5(a), let us determine the waveform for the voltage.

The equations which describe the waveform within the time intervals are

$$i(t) = \frac{(20 \times 10^{-3})t}{2 \times 10^{-3}} \qquad\qquad 0 \le t \le 2 \text{ ms}$$

$$= \frac{-(20 \times 10^{-3})t}{2 \times 10^{-3}} + 40 \times 10^{-3} \qquad 2 \le t \le 4 \text{ ms}$$

$$= 0 \qquad\qquad 4 \text{ ms} \le t$$

Using Eq. (3.10) we obtain

$$v(t) = (10 \times 10^{-3}) \frac{20 \times 10^{-3}}{2 \times 10^{-3}} \qquad 0 \le t \le 2 \text{ ms}$$

$$= 100 \text{ mV}$$

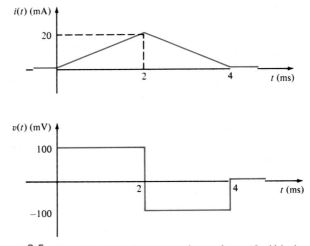

Figure 3.5 Current and voltage waveforms for a 10mH inductor.

and

$$v(t) = (10 \times 10^{-3}) \left(\frac{-20 \times 10^{-3}}{2 \times 10^{-3}} \right) \qquad 2 \le t \le 4 \text{ ms}$$

$$= -100 \text{ mV}$$

and

$$v(t) = 0 \qquad 4 \text{ ms} \le t$$

Hence, the voltage waveform is shown in Fig. 3.5(b).　　　　　　　■

Inductors combine in exactly the same manner as resistors. For example, using Kirchhoff's voltage law we can show that if N inductors are connected in series, their equivalent inductance is

$$L_s = L_1 + L_2 + \cdots + L_N \tag{3.15}$$

Furthermore, using Kirchhoff's current law we can show that if N inductors are connected in parallel, their equivalent inductance is

$$\frac{1}{L_p} = \frac{1}{L_1} + \frac{1}{L_2} + \cdots + \frac{1}{L_N} \tag{3.16}$$

EXAMPLE 3.4

We wish to determine the total inductance L_T at the terminals A-B of the circuit shown in Fig. 3.6(a).

Note that L_5 cannot be combined with L_3 or L_4. However, L_2 and L_3 are in parallel, and hence

$$\frac{1}{L_{23}} = \frac{1}{L_2} + \frac{1}{L_3} = \frac{1}{4 \text{ mH}} + \frac{1}{12 \text{ mH}}$$

$$L_{23} = 3 \text{ mH}$$

As shown in Fig. 3.6(b) L_{23} and L_4 are in series, and hence

$$L_{234} = L_{23} + L_4$$

$$= 12 \text{ mH}$$

L_{234} and L_5 are in parallel and therefore

$$\frac{1}{L_{2345}} = \frac{1}{L_{234}} + \frac{1}{L_5}$$

$$L_{2345} = 4 \text{ mH}$$

Then, from Fig. 3.6(c)

$$L_T = L_1 + L_{2345}$$

$$= 10 \text{ mH}$$

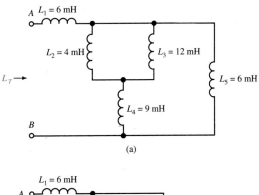

(a)

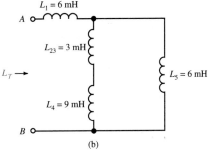

(b)

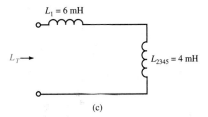

(c)

Figure 3.6 Circuits used in Example 3.4.

DRILL EXERCISE

D3.2. Find the total inductance in the network in Fig. D3.2.

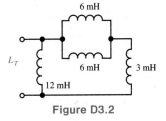

Figure D3.2

Ans: $L_T = 4$ mH.

3.3

First-order Networks **Mathematical Development of the Response Equations**

To begin our discussion, we consider the two networks in Figure 3.7. In each case, at time $t = 0$, a switch is closed and we wish to determine the equation which describes the operation of the network for time $t > 0$. KVL applied to the RC circuit yields

$$\frac{1}{C} \int_{-\infty}^{t} i(x) \, dx + Ri(t) = V_s$$

Taking the derivative of this equation with respect to t yields

$$\frac{i(t)}{C} + R \frac{di(t)}{dt} = 0$$

or

$$\frac{di(t)}{dt} + \frac{1}{RC} i(t) = 0 \tag{3.17}$$

In a similar manner, for $t > 0$, KVL for the RL network yields

$$\frac{Ldi(t)}{dt} + Ri(t) = V_s$$

or

$$\frac{di(t)}{dt} + \frac{R}{L} i(t) = \frac{V_s}{L} \tag{3.18}$$

Thus, we find that a network with only a single storage element can be described by a differential equation of the form

$$\frac{dx(t)}{dt} + ax(t) = f(t) \tag{3.19}$$

where $x(t)$ represents a current or voltage somewhere in the network.

In order to solve Eq. (3.19) we recall that a fundamental theorem of differential equations states that if $x(t) = x_p(t)$ is any solution to the equation

$$\frac{dx_p(t)}{dt} + ax_p(t) = f(t) \tag{3.20}$$

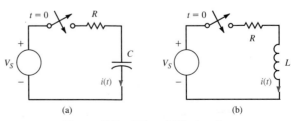

(a)   (b)

Figure 3.7 RC and RL circuits.

and if $x(t) = x_c(t)$ is any solution to the homogeneous equation

$$\frac{dx_c(t)}{dt} + ax_c(t) = 0 \tag{3.21}$$

then

$$x(t) = x_p(t) + x_c(t) \tag{3.22}$$

is a solution to Eq. (3.19). The term $x_p(t)$ is called the *particular integral solution*, or *forced response*, and $x_c(t)$ is called the *complementary solution*, or *natural response*. We note from Eq. (3.20) that, in general, $x_p(t)$ will consist of functional forms such as $f(t)$ and its first derivative—a notable exception is the case in which $f(t) = Ae^{-at}$ where the a in $f(t)$ is the same as the a in Eq. (3.20). Deductive reasoning suggests that the form of the solution of Eq. (3.21) must be such that the derivative of the function has the same time varying form as the function itself. Otherwise $dx_c(t)/dt$ and $ax_c(t)$ cannot add identically to zero for every t.

In the case where $f(t)$ is equal to some constant, for example, A, then the solution of Eq. (3.20) is $x_p(t) = K_1$. A solution to Eq. (3.21) is $x_c(t) = K_2e^{-at}$ and hence the general solution to the equation

$$\frac{dx(t)}{dt} + ax(t) = A \tag{3.23}$$

is

$$x(t) = K_1 + K_2e^{-at} \tag{3.24}$$

Note that as $t \to \infty$, $K_2e^{-at} \to 0$ and therefore the remaining term K_1 is referred to as the *steady state solution*. If Eq. (3.24) is written in the form

$$x(t) = K_1 + K_2e^{-t/\tau} \tag{3.25}$$

the term τ is called the *time constant* of the network. Note that $a = \frac{1}{\tau}$ where $\tau = RC$ for an RC network and $\frac{L}{R}$ for an RL network.

Some properties of the time constant are illustrated in Fig. 3.8. Figure 3.8(a) illustrates that the value of $x_c(t) = K_2e^{-t/\tau}$ is $0.368K_2$ at $t = \tau$, $0.135K_2$ at $t = 2\tau$, etc. In other words, in one time constant the value of the function drops by 63.2%. In five time constants $x_c(t) = .0067K_2$, which is less than 1% of its original value.

Figure 3.8(b) illustrates the difference between a large time constant, that is, slow response, and a small time constant, that is, fast response. If the time constant is large, a long time is required for the network to settle down to its steady state value. However, if the time constant is small, steady state is reached quickly. Air conditioning systems have relatively slow time constants. If the thermostat in a room is moved from 75°F to 70°F, several minutes will be required to cool the area to the desired temperature, and thus on a plot of temperature vs. time, the curve would be relatively flat. In contrast, the switching time for transistors on a very large scale integrated circuit is typically measured in nanoseconds to picoseconds.

Finally, it is very important to note that, although the two networks in Figure 3.7—upon which our analysis thus far has been based—are quite simple, by using Thevenin's theorem we can reduce more complicated networks to these two forms.

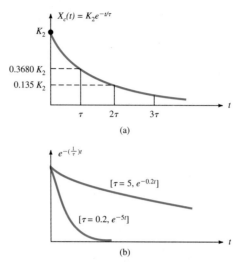

Figure 3.8   Time constant properties.

The Analysis Procedure

The following examples illustrate one approach to a transient analysis.

EXAMPLE 3.5

For the circuit in Fig. 3.9(a), the switch opens at $t = 0$. We wish to find $v_o(t)$ for $t > 0$.

We begin our analysis by assuming that the network in Fig. 3.9(a) is in steady state prior to opening the switch. Under this condition, the capacitor can be replaced by an open circuit because the source voltage is constant and therefore the current in the capacitor is $C\dfrac{dv_c(t)}{dt} = 0$. The resulting network in Fig. 3.9(b) is a simple voltage divider and thus $v_c(0)$, the initial capacitor voltage, is 8 V. Now that the initial condition on the capacitor is known, we examine the circuit which results from opening the switch.

The network for $t > 0$ is shown in Fig. 3.9(c). At this point we write an equation in terms of the voltage across the capacitor. Applying KCL to this circuit yields

$$C\frac{dv_c(t)}{dt} + \frac{v_c(t)}{R_2} = 0$$

or

$$\frac{dv_c(t)}{dt} + \frac{v_c(t)}{R_2C} = 0$$

As indicated earlier, if we assume the solution is of the form

$$v_c(t) = K_2 e^{-t/\tau}$$

then substituting this expression into the differential equation yields

$$\frac{-K_2 e^{-t/\tau}}{\tau} + \frac{K_2 e^{-t/\tau}}{R_2C} = 0$$

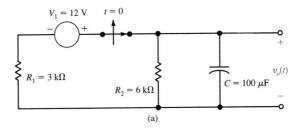

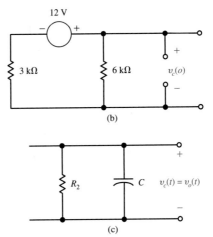

Figure 3.9   Circuits used in Example 3.5.

and the time constant for the circuit is

$$\tau = R_2 C$$

Therefore

$$v_c(t) = K_2 e^{-t/R_2 C}$$

However, from the initial condition we know that

$$v_c(0) = 8$$

Hence, the complete solution is

$$v_c(t) = 8e^{-t/R_2 C}$$
$$= 8e^{-t/0.6} \text{ volts}$$

where, as shown in Fig. 3.9(a), $v_o(t) = v_c(t)$. A plot of this function is shown in Fig. 3.10. ■

EXAMPLE 3.6

In the network in Fig. 3.11(a), the switch closes at $t = 0$. We wish to find $v_o(t)$ for $t > 0$.

Assuming the network in Fig. 3.11(a) is in steady state prior to closing the switch, the inductor can be replaced by a short circuit because the voltage source is constant, therefore the current in the inductor is constant, and hence $v_L(t) = L di_2(t)/dt = 0$. The resulting network in Fig. 3.11(b) illustrates that $i_L(0) = 2A$. Now that the initial condition on the inductor is known, we examine the circuits which result from closing the switch.

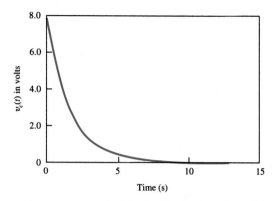

Figure 3.10 A plot of $v_c(t)$ as a function of t.

The network for $t > 0$ is shown in Fig. 3.11(c). At this point we write an equation in terms of the current through the inductor. Applying KVL to this circuit yields

$$L\frac{di_L(t)}{dt} + R_2 i_L(t) = 0$$

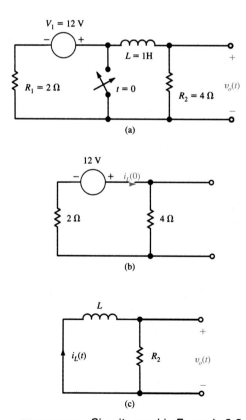

Figure 3.11 Circuits used in Example 3.6.

or

$$\frac{di_L(t)}{dt} + \frac{R_2}{L} i_L(t) = 0$$

Again, we assume a solution of the form

$$i_L(t) = K_2 e^{-t/\tau}$$

Following a development identical to that illustrated in Example 3.5 we find that

$$\tau = \frac{L}{R_2}$$

and hence

$$i_L(t) = K_2 e^{-\frac{R_2}{L} t}$$

Employing the initial condition

$$i_L(0) = 2 = K_2$$

Therefore

$$i_L(t) = 2e^{-4t} \text{ A}$$

Then

$$v_o(t) = R_2 i_L(t)$$
$$= 8e^{-4t} \text{ V}$$

A plot of this function is shown in Fig. 3.12.   ∎

EXAMPLE 3.7

We wish to find both $v_c(t)$ and $v_o(t)$ for $t > 0$ in the network in Fig. 3.13(a).

Assuming the network is in steady state prior to closing the switch, we find that the initial voltage on the capacitor is 0 since the voltage source is not yet connected to the remainder of the circuit.

The network for $t > 0$ is shown in Fig. 3.13(b). This network can be simplified by replacing V_1, R_1, and R_2 with its Thevenin equivalent. The open-circuit voltage derived from Fig. 3.13(c) is $V_{oc} = 4$ V and the Thevenin equivalent resistance determined from the cir-

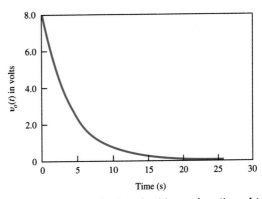

Figure 3.12   A plot of $v_o(t)$ as a function of t.

cuit in Fig. 3.13(d) is $R_{TH} = 2$ kΩ. Hence, the circuit in Fig. 3.13(b) reduces to that shown in Fig. 3.13(e). The voltage across the capacitor in Fig. 3.13(e) can be expressed as

$$C \frac{dv_c(t)}{dt} = i_c(t)$$

Applying KVL to the circuit yields the equation

$$V_{oc} = R_{TH}i_c(t) + v_c(t) + R_2i_c(t)$$

Combining the last two equations yields the expression

$$\frac{dv_c(t)}{dt} + \frac{v_c(t)}{(R_{TH} + R_2)C} = \frac{V_{oc}}{(R_{TH} + R_1)C}$$

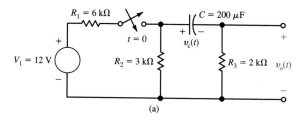

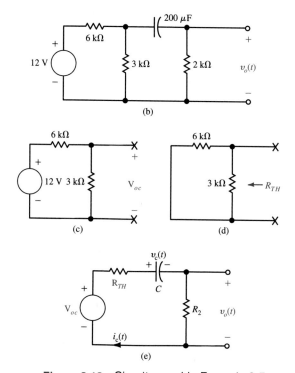

Figure 3.13 Circuits used in Example 3.7.

Since there is a constant forcing function present in the equation, the solution of this first-order differential equation will be of the form

$$v_c(t) = K_1 + K_2 e^{-t/\tau}$$

where as indicated earlier

$$\tau = (R_{TH} + R_2)C$$

Since in steady state the capacitor looks like an open circuit, $v_c(t = \infty) = V_{oc}$. Therefore

$$v_c(\infty) = V_{oc} = K_1 + 0$$

since the exponential approaches 0 as $t \to \infty$. Then

$$v_c(t) = V_{oc} + K_2 e^{-t/\tau}$$

However, employing the initial condition

$$v_c(0) = 0 = V_{oc} + K_2$$

or

$$K_2 = -V_{oc}$$

Therefore, using the circuit parameters the expression for $v_c(t)$ becomes

$$v_c(t) = 4 - 4e^{-t/0.8} \text{ V}$$

Note that this equation satisfies both the initial and final values of the function $v_c(t)$. A plot of this function is shown in Fig. 3.14. Recall, however, that we also wish to find the output voltage, $v_o(t)$, for $t > 0$. Knowing the voltage across the capacitor, we can compute the current via the equation

$$i_c(t) = C \frac{dv_c(t)}{dt}$$

and using this result we can determine the voltage $v_o(t)$ from the expression

$$v_o(t) = R_2 i_c(t)$$

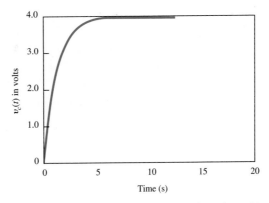

Figure 3.14   A plot of $v_c(t)$ as a function of t.

Performing the operation indicated by the two equations above yields

$$v_o(t) = 2e^{-t/0.8} \text{ V}$$

■

Before proceeding further, it is instructive to summarize the approach that is being employed to perform a transient analysis. In a systematic fashion we have first used the original circuit prior to switch action to determine the initial capacitor voltage $v_c(0)$ or inductor current $i_L(0)$. Second, for the network which exists after switch action we have derived an expression for the capacitor voltage $v_c(t)$ or inductor current $i_L(t)$. This expression yields the time constant of the network, τ. Third, the initial and final conditions are used to determine the constraints in the equation for either the capacitor voltage, $v_c(t)$, or inductor current $i_L(t)$. Finally, if the unknown is neither the capacitor voltage nor inductor current, equations which describe the network are used in conjunction with $v_c(t)$ or $i_L(t)$ to determine the desired quantity.

Let us consider now an example in which the forcing function is not a constant. In this case, as stated earlier, the particular integral solution will consist of functional forms of the forcing function and its first derivative.

EXAMPLE 3.8

We wish to determine $v_o(t)$ for $t > 0$ in the network in Fig. 3.15(a).

In the steady state prior to closing the switch the current in the inductor is zero and hence $i(0) = 0$.

For $t > 0$, KVL for the network yields

$$L \frac{di(t)}{dt} + Ri(t) = v_s(t)$$

or

$$\frac{di(t)}{dt} + \frac{R}{L} i(t) = \frac{v_s(t)}{L}$$

Using the known parameters the equation becomes

$$\frac{di(t)}{dt} + 2i(t) = 6e^{-4t} \qquad t > 0$$

The complementary solution is derived from the equation

$$\frac{di_c(t)}{dt} + 2i_c(t) = 0 \qquad t > 0$$

and hence

$$i_c(t) = K_2 e^{-2t}$$

The particular solution is obtained from the equation

$$\frac{di_p(t)}{dt} + 2i_p(t) = 6e^{-4t} \qquad t > 0$$

Assuming that $i_p(t)$ is of the form of the input and its derivatives, we select $i_p(t) = K_1 e^{-4t}$. Then

$$\frac{d}{dt} (K_1 e^{-4t}) + 2K_1 e^{-4t} = 6e^{-4t}$$

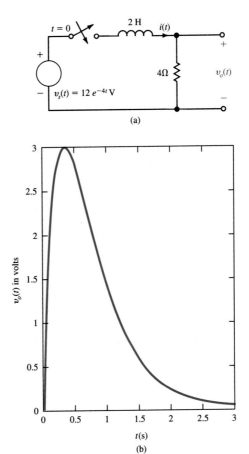

Figure 3.15 A circuit and its transient response.

or

$$-4K_1 + 2K_1 = 6$$
$$K_1 = -3$$

Hence

$$i(t) = i_c(t) + i_p(t)$$
$$= K_2 e^{-2t} - 3e^{-4t}$$

However, $i(0) = 0$, and therefore

$$0 = K_2 - 3$$

or

$$K_2 = 3$$

then

$$i(t) = 3[e^{-2t} - e^{-4t}] \text{ A}$$

and

$$v_o(t) = 12[e^{-2t} - e^{-4t}] \text{ V}$$

This voltage is shown plotted in Fig. 3.15(b).

DRILL EXERCISES

D3.3. Find $i_o(t)$ and $v_o(t)$ for $t > 0$ in the network in Fig. D3.3.

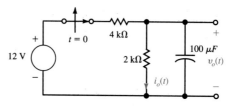

Figure D3.3

Ans: $v_o(t) = 4e^{-t/0.2}$ V and $i_o(t) = 2e^{-t/0.2}$ mA.

D3.4. Find $v_o(t)$ for $t > 0$ in the network in Fig. D3.4.

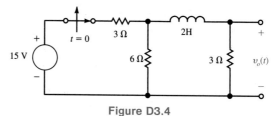

Figure D3.4

Ans: $v_o(t) = 6e^{-4.5t}$ V.

D3.5. Find $i(t)$ for $t > 0$ in the network in Fig. D3.5.

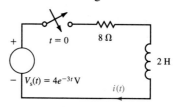

Figure D3.5

Ans: $i(t) = 2[e^{-3t} - e^{-4t}]$ A.

3.4
Second-order
Circuits

Mathematical Development of the Response Equations

We now extend our analysis to the case where an inductor and a capacitor are present simultaneously. Consider, for example, the two basic *RLC* circuits shown in Fig. 3.16. As-

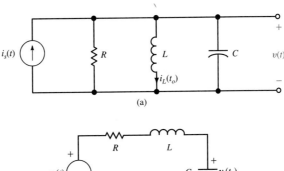

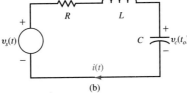

Figure 3.16 Parallel and series *RLC* circuits.

suming that energy may be initially stored in the inductor and capacitor, the node equation for the network in Fig. 3.16(a) is

$$\frac{v(t)}{R} + \frac{1}{L}\int_{t_o}^{t} v(x)\,dx + i_L(t_o) + C\,\frac{dv(t)}{dt} = i_s(t) \tag{3.26}$$

and the mesh equation for the network in Fig. 3.16(b) is

$$Ri(t) + \frac{1}{C}\int_{t_o}^{t} i(x)\,dx + v_c(t_o) + L\,\frac{di(t)}{dt} = v_s(t) \tag{3.27}$$

If both equations are differentiated with respect to time, the first equation is divided by *C* and the second by *L*, the equations become

$$\frac{d^2v(t)}{dt^2} + \frac{1}{RC}\frac{dv(t)}{dt} + \frac{1}{LC}v(t) = \frac{1}{C}\frac{di_s(t)}{dt}$$

$$\frac{d^2i(t)}{dt^2} + \frac{R}{L}\frac{di(t)}{dt} + \frac{1}{LC}i(t) = \frac{1}{L}\frac{dv_s(t)}{dt} \tag{3.28}$$

Note that both circuits lead to second-order differential equations with constant coefficients. Therefore, in general we are confronted with an equation of the form

$$\frac{d^2x(t)}{dt^2} + a_1\,\frac{dx(t)}{dt} + a_2x(t) = f(t) \tag{3.29}$$

Once again, if $x_p(t)$ is any solution to Eq. (3.29) and $x_c(t)$ is a solution to the homogeneous equation

$$\frac{d^2x_c(t)}{dt^2} + a_1\,\frac{dx_c(t)}{dt} + a_2x_c(t) = 0 \tag{3.30}$$

then

$$x(t) = x_p(t) + x_c(t)$$

is a solution to Eq. (3.29). If the forcing function is constant, that is, $f(t) = A$, then $x_p(t) = A/a_2$.

For simplicity we will now express the homogeneous equation in the form

$$\frac{d^2x(t)}{dt^2} + 2\alpha \frac{dx(t)}{dt} + \omega_0^2 x(t) = 0 \tag{3.31}$$

Assuming that $x(t) = Ke^{st} \neq 0$, Eq. (3.31) yields

$$s^2 Ke^{st} + 2\alpha s Ke^{st} + \omega_0^2 Ke^{st} = 0$$

or

$$s^2 + 2\alpha s + \omega_0^2 = 0 \tag{3.32}$$

This equation is called the *characteristic equation,* α is the exponential *damping coefficient,* and ω_o is the *undamped resonant frequency.* The quadratic formula indicates that

$$s = \frac{-2\alpha \pm \sqrt{4\alpha^2 - 4\omega_o^2}}{2}$$
$$= -\alpha \pm \sqrt{\alpha^2 - \omega_o^2}$$

The two values of s that satisfy this equation are

$$s_1 = -\alpha + \sqrt{\alpha^2 - \omega_0^2}$$
$$s_2 = -\alpha - \sqrt{\alpha^2 - \omega_0^2} \tag{3.33}$$

Hence, $x_1(t) = K_1 e^{s_1 t}$ and $x_2(t) = K_2 e^{s_2 t}$ are solutions to Eq. (3.31). In addition, the sum of the solutions is also a solution. Therefore

$$x_c(t) = K_1 e^{s_1 t} + K_2 e^{s_2 t} \tag{3.34}$$

s_1 and s_2 are called *natural frequencies* because they determine the natural, that is, unforced, response of the network. K_1 and K_2 are constants that can be evaluated via the initial conditions $x(0)$ and $dx(0)/dt$. For example, since

$$x(t) = K_1 e^{s_1 t} + K_2 e^{s_2 t}$$

then

$$x(0) = K_1 + K_2$$

and

$$\left.\frac{dx(t)}{dt}\right|_{t=0} = \frac{dx(0)}{dt} = s_1 K_1 + s_2 K_2$$

Hence, $x(0)$ and $dx(0)/dt$ produce two simultaneous equations, which when solved yield the constants K_1 and K_2. Equations (3.33) and (3.34) indicate that the form of the solution of the homogeneous equation is dependent upon the relative magnitude of the values of α and ω_o. There are three possible cases.

Case 1, $\alpha > \omega_o$. This is commonly called *overdamped.* The natural frequencies s_1 and s_2 are real and unequal, and therefore the natural response is of the form

$$x_c(t) = K_1 e^{-(\alpha - \sqrt{\alpha^2 - \omega_o^2})t} + K_2 e^{-(\alpha + \sqrt{\alpha^2 - \omega_o^2})t} \tag{3.35}$$

where K_1 and K_2 are found from the initial conditions. This indicates that the natural response is the sum of two decaying exponentials.

Case 2, $\alpha < \omega_o$. This case is called *underdamped.* Since $\omega_o > \alpha$, the roots of the characteristic equation given in Eq. (3.33) can be written as

$$s_1 = -\alpha + \sqrt{-(\omega_o^2 - \alpha^2)} = -\alpha + j\omega_n$$
$$s_2 = -\alpha - \sqrt{-(\omega_o^2 - \alpha^2)} = -\alpha - j\omega_n$$

where $j = \sqrt{-1}$ and $\omega_n = \sqrt{\omega_o^2 - \alpha^2}$. Thus, the natural frequencies are complex numbers. The natural response is then

$$x_c(t) = K_1 e^{-(\alpha - j\omega_n)t} + K_2 e^{-(\alpha + j\omega_n)t}$$

Using Euler's identities, this equation can be simplified to the following form.

$$x_c(t) = e^{-\alpha t}(A_1 \cos\omega_n t + A_2 \sin\omega_n t) \tag{3.36}$$

where A_1 and A_2, like K_1 and K_2, are constants, which are evaluated using the initial conditions $x(0)$ and $dx(0)/dt$. The natural response in this case is an exponentially damped oscillation.

Case 3, $\alpha = \omega_o$. This case, called *critically damped,* results in

$$s_1 = s_2 = -\alpha$$

as shown in Eq. (3.33). Therefore, Eq. (3.34) reduces to

$$x_c(t) = K_3 e^{-\alpha t}$$

where $K_3 = K_1 + K_2$. However, this cannot be a solution to the second-order homogeneous differential equation because in general it is not possible to satisfy the two initial conditions $x(0)$ and $dx(0)/dt$ with the single constant K_3.

In the case where the characteristic equation has repeated roots, we can show that the general solution can be written as

$$X_c(t) = B_1 e^{-\alpha t} + B_2 t e^{-\alpha t} \tag{3.37}$$

where B_1 and B_2 are constants derived from the initial conditions.

The results we have just presented are summarized in Table 3.1.

Table 3.1 Homogeneous Network Equation $\dfrac{d^2x(t)}{dt^2} + 2\alpha\dfrac{dx(t)}{dt} + \omega_o^2 x(t) = 0$

Damping Condition	Natural (Unforced) Response Equation	Type of Response
$\alpha > \omega_o$	$x(t) = K_1 e^{-(\alpha - \sqrt{\alpha^2 - \omega_o^2})t} + K_2 e^{-(\alpha + \sqrt{\alpha^2 - \omega_o^2})t}$	Overdamped
$\alpha < \omega_o$	$x(t) = e^{-\alpha t}[K_1 \cos\sqrt{\omega_o^2 - \alpha^2}\, t + K_2 \cos\sqrt{\omega_o^2 - \alpha^2}\, t]$	Underdamped
$\alpha = \omega_o$	$x(t) = K_1 e^{-\alpha t} + K_2 t e^{-\alpha t}$	Critically Damped

D3.6. A parallel *RLC* circuit has the following circuit parameters: $R = 1\,\Omega$, $L = 2$ H, and $C = 2$ F. Using Eq. (3.28) and (3.32), compute the damping ratio and the undamped natural frequency of this network.

    Ans: $\alpha = 0.25$, $\omega_o = 0.5$ rad/s.

D3.7. A series *RLC* circuit consists of $R = 2\,\Omega$, $L = 1$ H, and a capacitor. Determine the type of response exhibited by the network if (a) $C = 1/2$ F, (b) $C = 1$ F, and (c) $C = 2$ F.

    Ans: (a) underdamped, (b) critically damped, (c) overdamped.

The Analysis Procedure

We will now analyze some simple *RLC* circuits that contain both nonzero initial conditions and constant forcing functions.

EXAMPLE 3.9

Consider the network in Fig. 3.17. The circuit parameters are $R = 2\,\Omega$, $L = 5$ H, $C = 1/5$ F, $i_L(0) = -1$ A, and $v_c(0) = 4$ V. We wish to determine the equation for the voltage $v(t)$.

We know from Eq. (3.28) that the general form of the second-order differential equation that describes the voltage $v(t)$ is

$$\frac{d^2v(t)}{dt^2} + \frac{1}{RC}\frac{dv(t)}{dt} + \frac{1}{LC}v(t) = 0$$

With the given parameter values the equation becomes

$$\frac{d^2v(t)}{dt^2} + 2.5\frac{dv(t)}{dt} + v(t) = 0$$

The characteristic equation for the network is

$$s^2 + 2.5s + 1 = 0$$

and the roots are

$$s_1 = -2$$
$$s_2 = -0.5$$

Since the roots are real and unequal, the circuit is overdamped, and $v(t)$ is of the form

$$v(t) = K_1 e^{-2t} + K_2 e^{-0.5t}$$

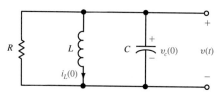

Figure 3.17   A parallel *RLC* circuit with initial conditions.

The initial conditions are now employed to determine the constants K_1 and K_2. Since $v(t) = v_c(t)$

$$v_c(0) = v(0) = 4 = K_1 + K_2$$

The second equation needed to determine K_1 and K_2 is normally obtained from the expression

$$\frac{dv(t)}{dt} = -2K_1e^{-2t} - 0.5K_2e^{-0.5t}$$

However, the second initial condition is not given as $dv(0)/dt$; we were given $i_L(0)$. We can, however, circumvent this problem by noting that the node equation for the circuit can be written as

$$C\frac{dv(t)}{dt} + \frac{v(t)}{R} + i_L(t) = 0$$

or

$$\frac{dv(t)}{dt} = \frac{-1}{RC}v(t) - \frac{i_L(t)}{C}$$

At $t = 0$

$$\frac{dv(0)}{dt} = \frac{-1}{RC}v(0) - \frac{1}{C}i_L(0)$$
$$= -2.5(4) - 5(-1)$$
$$= -5$$

But since

$$\frac{dv(t)}{dt} = -2K_1e^{-2t} - 0.5K_2e^{-0.5t}$$

then when $t = 0$

$$-5 = -2K_1 - 0.5K_2$$

This equation, together with the equation

$$4 = K_1 + K_2$$

produces the constants $K_1 = 2$ and $K_2 = 2$. Therefore, the final equation for the voltage is

$$v(t) = 2e^{-2t} + 2e^{-0.5t} \text{ V}$$

In addition, note that in comparison with the RL and RC circuits analyzed earlier, the response of this circuit is controlled by two time constants.   ■

EXAMPLE 3.10
Let us determine the equation for $v(t)$ in the network in Fig. 3.18 if the network parameters are as follows: $R_1 = 10 \, \Omega$, $L = 2$ H, $C = 1/8$ F, $R_2 = 8 \, \Omega$, $i_L(0) = 1/2$ A, and $v_c(0) = 1$ V.

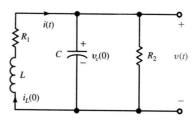

Figure 3.18   Series/parallel *RLC* network.

The two equations that describe the network are a Kirchhoff's voltage law equation for the outer loop

$$L \frac{di(t)}{dt} + R_1 i(t) + v(t) = 0$$

and a Kirchhoff's current law equation at the upper node

$$i(t) = C \frac{dv(t)}{dt} + \frac{v(t)}{R_2}$$

Substituting the second equation into the first yields

$$\frac{d^2 v}{dt^2} + \left(\frac{1}{R_2 C} + \frac{R_1}{L} \right) \frac{dv}{dt} + \frac{R_1 + R_2}{R_2 L C} v = 0$$

Using the circuit parameter values, the equation becomes

$$\frac{d^2 v}{dt^2} + 6 \frac{dv}{dt} + 9v = 0$$

The characteristic equation is then

$$s^2 + 6s + 9 = 0$$

and hence the roots are

$$s_1 = -3$$
$$s_2 = -3$$

Since the roots are real and equal, the circuit is critically damped. The term $v(t)$ is then given by the expression

$$v(t) = K_1 e^{-3t} + K_2 t e^{-3t}$$

Since $v(t) = v_c(t)$

$$v(0) = v_c(0) = 1 = K_1$$

In addition

$$\frac{dv(t)}{dt} = -3K_1 e^{-3t} + K_2 e^{-3t} - 3K_2 t e^{-3t}$$

However, our second defining equation for the network is

$$\frac{dv(t)}{dt} = \frac{i(t)}{C} - \frac{v(t)}{R_2 C}$$

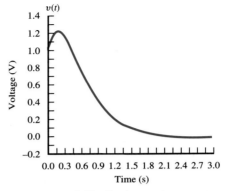

Figure 3.19   Critically damped response.

Setting these two expressions equal to one another and evaluating the resultant equation at $t = 0$ yields

$$\frac{1/2}{1/8} - \frac{1}{1} = -3K_1 + K_2$$

$$3 = -3K_1 + K_2$$

Since $K_1 = 1$, $K_2 = 6$ and the expression for $v(t)$ is

$$v(t) = e^{-3t} + 6te^{-3t} \text{ V}$$

Note that the expression satisfies the initial condition $v(0) = 1$. The function $v(t)$ is shown graphically in Fig. 3.19. ∎

EXAMPLE 3.11

Consider the network in Fig. 3.20. We wish to find an expression for $v_c(t)$ for $t > 0$ given the following circuit parameters: $R = 6\ \Omega$, $L = 1$ H, $C = 0.04$ F, $i_L(0) = 4$ A, and $v_c(0) = -4$ V.

Recall from Fig. 3.16 and Eq. 3.28 that the homogenous equation for the network current $i(t)$ is, in general, of the form

$$\frac{d^2i(t)}{dt^2} + \frac{R}{L}\frac{di(t)}{dt} + \frac{1}{LC}i(t) = 0$$

Note that for this series *RLC* circuit the damping coefficient is $R/2L$ and the resonant frequency is $1/\sqrt{LC}$. Using the circuit element values, the differential equation becomes

$$\frac{d^2i(t)}{dt^2} + 6\frac{di(t)}{dt} + 25i(t) = 0$$

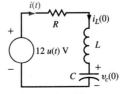

Figure 3.20   A series *RLC* circuit with a constant forcing function.

The characteristic equation is then

$$s^2 + 6s + 25 = 0$$

and the roots are

$$s_1 = -3 + j4$$
$$s_2 = -3 - j4$$

Since the roots are complex, the circuit is underdamped, and the complementary solution is of the form $K_1 e^{-3t}\cos 4t + K_2 e^{-3t}\sin 4t$. The forcing function is constant, and therefore the particular solution will be a constant. The general solution of the voltage $v_c(t)$ will then be of the form

$$v_c(t) = K_3 e^{-3t}\cos 4t + K_4 e^{-3t}\sin 4t + K_5$$

Now recall that in steady state the inductor acts like a short circuit and the capacitor acts like an open circuit, therefore $v_c(\infty) = 12$ V and hence $K_5 = 12$. The general solution is then

$$v_c(t) = K_3 e^{-3t}\cos 4t + K_4 e^{-3t}\sin 4t + 12$$

The initial conditions can now be used to evaluate the constants K_3 and K_4. For example

$$v_c(0) = -4 = K_3 + 12$$
$$K_3 = -16$$

In addition

$$\frac{dv_c(t)}{dt} = -3K_3 e^{-3t}\cos 4t - 4K_3 e^{-3t}\sin 4t - 3K_4 e^{-3t}\sin 4t + 4K_4 e^{-3t}\cos 4t$$

This expression can be related to the known initial condition using the expression for the current in the capacitor

$$i(t) = C\,\frac{dv_c(t)}{dt}$$

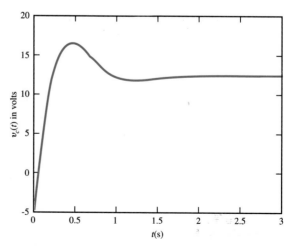

Figure 3.21   An underdamped response.

or

$$\frac{dv_c(t)}{dt} = \frac{i(t)}{C}$$

Setting the two expressions above equal to one another and evaluating the resultant equation at $t = 0$ yields

$$\frac{i(0)}{C} = -3K_3 + 4K_4$$

$$100 = -3(-16) + 4K_4$$

$$K_4 = 13$$

Therefore, the general solution for $v_c(t)$ is

$$v_c(t) = 12 - 16e^{-3t}\cos4t + 13e^{-3t}\sin4t \ V.$$

Note that this equation satisfies the initial and steady state conditions. The voltage is plotted in Fig. 3.21. ∎

DRILL EXERCISES

D3.8. The switch in the network in Fig. D3.8 moves from position 1 to position 2 at $t = 0$. Find $v(t)$ for $t > 0$.

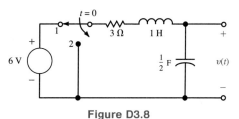

Figure D3.8

Ans: $v(t) = 12e^{-t} - 6e^{-2t} \ V.$

D3.9. The switch in the network in Fig. D3.9 moves from position 1 to position 2 at $t = 0$. Find $v(t)$ for $t > 0$.

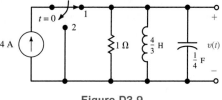

Figure D3.9

Ans: $v(t) = -8e^{-t} + 8e^{-3t} \ V.$

3.5

Summary

- The defining equations for a capacitor are $i = C\dfrac{dv}{dt}$, $v = 1/C \int i\, dt$,

$$\text{and } w_c = \left(\frac{1}{2}\right) Cv^2.$$

- The equivalent capacitance of N capacitors in series is

$$\frac{1}{C_s} = \frac{1}{C_1} + \frac{1}{C_2} + \text{-----} \frac{1}{C_N}$$

- The equivalent capacitance of N capacitors in parallel is

$$C_p = C_1 + C_2 + \text{-----} C_N$$

- The defining equations for an inductor are $v = L\dfrac{di}{dt}$, $i = 1/L \int v\, dt$,

$$\text{and } w_L = \left(\frac{1}{2}\right) Li^2.$$

- The equivalent inductance of N inductors in series is

$$L_s = L_1 + L_2 + \text{-----} L_N$$

- The equivalent inductance of N inductors in parallel is

$$\frac{1}{L_p} = \frac{1}{L_1} + \frac{1}{L_2} \text{-----} \frac{1}{L_N}$$

- The forced response of a circuit is that due to external energy sources.
- The natural response of a circuit is that due to energy storage elements within the network, that is, inductors and capacitors.
- The transient response of a first-order network is of the form

$$x(t) = K_1 + K_2 e^{-t/\tau}$$

where $x(t)$ is any current or voltage in the network, K_1 is the steady state response, and τ is the time constant of the network.

- The procedure for determining the transient behavior of a first-order network is
  - For the network which exists prior to switch action, determine the initial capacitor voltage or inductor current.
  - For the network which exists after switch action, derive an expression for the capacitor voltage or inductor current which will yield the time constant of the network.
  - Use the initial and final conditions to determine the constants in the equation for the capacitor voltage or inductor current.
  - If the desired quantity is not the capacitor voltage or inductor current, use equations which describe the network to find the desired quantity.

- The transient response of a second-order network is determined by the roots of the network's characteristic equation.
  - If the roots are real and unequal, the response is overdamped.
  - If the roots are complex conjugates, the response is underdamped.
  - If the roots are real and unequal, the response is critically damped.
- The response equations for each case are shown in Table 3.1.

PROBLEMS

3.1. The voltage across a $2\mu F$ capacitor is given by the waveform in Fig. P3.1. Compute the current waveform.

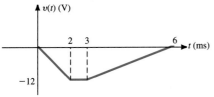

Figure P3.1

3.2. Find the equivalent capacitance at terminals $A-B$ in Fig. P3.2.

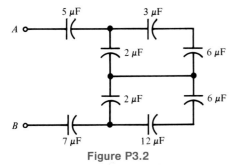

Figure P3.2

3.3. Compute the equivalent capacitance of the network in Fig. P3.3.

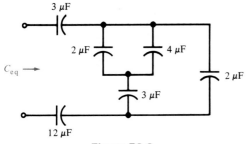

Figure P3.3

3.4. The current in a 5mH inductor has the waveform shown in Fig. P3.4. Compute the waveform for the inductor voltage.

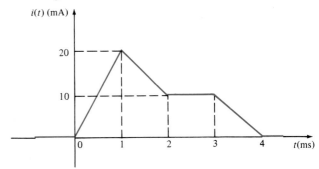

Figure P3.4

3.5. Determine the inductance at terminals $A-B$ in the network in Fig. P3.5.

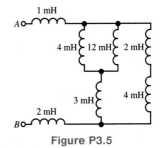

Figure P3.5

3.6. In the network in Fig. P3.6, the total inductance is $L_T = 4$ H. Find the value of the inductor L_1.

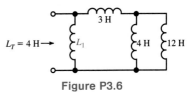

Figure P3.6

3.7. A flash bulb circuit for a camera is shown in Fig. P3.7. Initially both switches are open. Switch 1 closes to charge the capacitor. Once the capacitor is charged, switch 1 opens. When switch 2 closes, the energy stored in the capacitor is discharged through the light bulb represented by the 0.2 Ω resistor. Find the equation of the current which causes the bulb to flash.

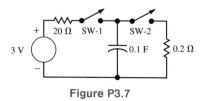

Figure P3.7

3.8. Consider the circuit shown in Fig. P3.8. Assuming that the switch has been in position 1 for a long time, at time $t = 0$ the switch is moved to position 2. Calculate the current $i(t)$ for $t > 0$.

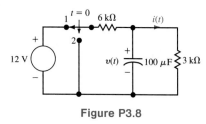

Figure P3.8

3.9. Find $v_c(t)$ for $t > 0$ in the circuit shown in Fig. P3.9.

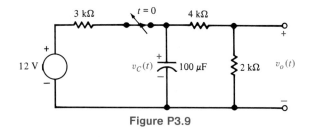

Figure P3.9

3.10. Find $v_o(t)$ for $t > 0$ in Fig. P3.9.

3.11. Consider the network in Fig. P3.11. At $t = 0$, the switch opens. Find $v_o(t)$ for $t > 0$.

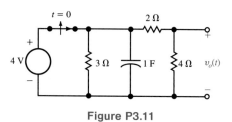

Figure P3.11

3.12. Consider the network in Fig. P3.12. The switch closes at $t = 0$. Find $v_o(t)$ for $t > 0$.

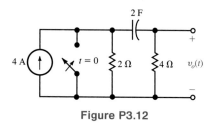

Figure P3.12

3.13. Find $i_o(t)$ for $t > 0$ in Fig. P3.13.

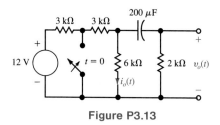

Figure P3.13

3.14. Find $v_o(t)$ for $t > 0$ in Fig. P3.13.

3.15. Given the circuit shown in Fig. P3.15, at $t = 0$ the switch is opened. Calculate the current $i(t)$ for $t > 0$.

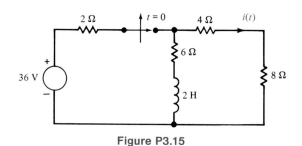

Figure P3.15

3.16. In the circuit shown in Fig. P3.16 the switch opens at $t = 0$. Find $i_1(t)$ for $t > 0$.

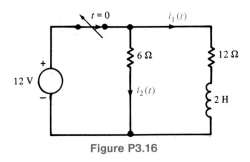

Figure P3.16

3.17. Determine the expression for $i_2(t)$, $t > 0$, in Fig. P3.16.

3.18. In Fig. P3.16, if the switch was open for $t < 0$ and then closes at $t = 0$, find $i_1(t)$ for $t > 0$.

3.19. Find $i_o(t)$ for $t > 0$ in Fig. P3.19.

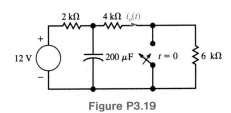

Figure P3.19

3.20. Consider the circuit shown in Fig. P3.20. The circuit is in steady state prior to time $t = 0$, when the switch is closed. Calculate the current $i(t)$ for $t > 0$.

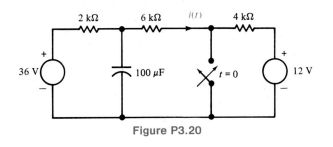

Figure P3.20

3.21. Consider the network shown in Fig. P3.21. If the switch opens at $t = 0$, find the output voltage $v_o(t)$ for $t > 0$.

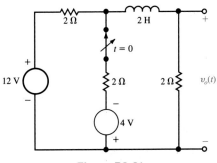

Figure P3.21

3.22 The switch in the network in Fig. P3.22 closes at $t = 0$. Find $i_o(t)$ for $t > 0$.

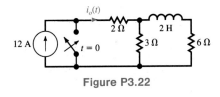

Figure P3.22

3.23. The switch in the network in Fig. P3.23 opens at $t = 0$. Find $v_o(t)$ for $t > 0$.

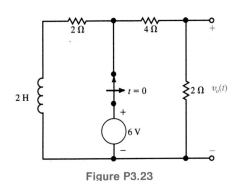

Figure P3.23

3.24. Determine $i_o(t)$ for $t > 0$ in Fig. P3.24.

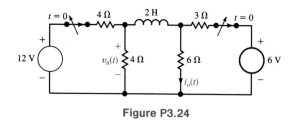

Figure P3.24

3.25. Find $v_A(t)$ for $t > 0$ in Fig. P3.24.

3.26. Consider the circuit in Fig. P3.26. The capacitor has an initial charge and an exponentially decaying source is applied at $t = 0$. We wish to determine both $v_o(t)$ and $i(t)$ for $t > 0$.

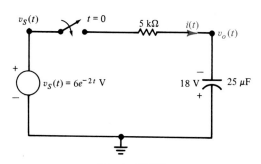

Figure P3.26

3.27. Find $v_o(t)$ for $t > 0$ in the network in Fig. P3.27.

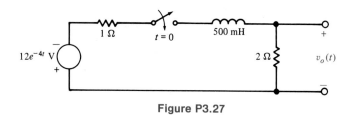

Figure P3.27

3.28. The parameters for a parallel RLC circuit are $R = 1\ \Omega$, $L = 1/5$ H, and $C = 1/4$ F. Determine the type of damping exhibited by the circuit.

3.29. If the inductor in problem 3.28 is changed from $L = 1/5$ H to $L = 4/3$ H, what effect does this have on the circuit response?

3.30. A series RLC circuit contains a resistor $R = 2\ \Omega$ and a capacitor $C = 1/8$ F. Select the value of the inductor so that the circuit is critically damped.

3.31. The parallel RLC circuit shown in Fig. P3.31 has the following parameters: $R = 1\ \Omega$, $L = 1/8$ H, and $C = 1/12$ F. Is it possible to select a positive value of R_1 that will produce a voltage response that is critically damped?

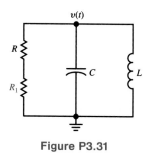

Figure P3.31

3.32. The series RLC circuit shown in Fig. P3.32 has the following parameters: $C = 0.04$ F, $L = 1$ H, $R = 6\ \Omega$, $i_L(0) = 4$ A, and $v_c(0) = -4$ V. Find the equation for the current $i(t)$.

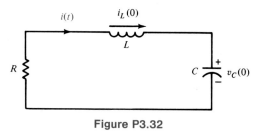

Figure P3.32

3.33. The switch in the network in Fig. P3.33 opens at $t = 0$. Find $i(t)$ for $t > 0$.

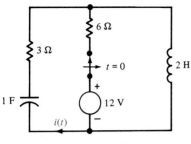

Figure P3.33

3.34. For the underdamped circuit shown in Fig. P3.34, determine the voltage $v(t)$ if the initial conditions on the storage elements are $i_L(0) = 1$ A and $v_c(0) = 10$ V.

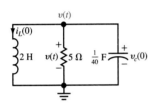

Figure P3.34

3.35. Given the circuit and the initial conditions of problem 3.34 determine the current through the inductor.

3.36. In the critically damped circuit shown in Fig. P3.36, the initial conditions on the storage elements are $i_L(0) = 2$ A and $v_c(0) = 5$ V. Determine the voltage $v(t)$.

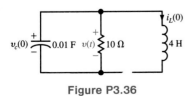

Figure P3.36

3.37. Given the circuit and the initial conditions from problem 3.36 determine the current $i_L(t)$ that is flowing through the inductor.

3.38. Determine the equation for the current $i(t)$, $t > 0$, in the circuit shown in Fig. P3.38.

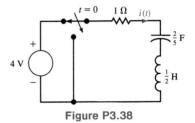

Figure P3.38

3.39. Given the circuit in Fig. P3.39, find the equation for $i(t)$, $t > 0$.

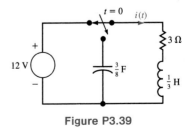

Figure P3.39

3.40. Find the equation for $v(t)$, $t > 0$, in Fig. P3.40.

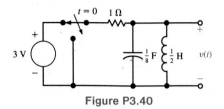

Figure P3.40

3.41. In the circuit shown in Fig. P3.41 find $v(t)$, $t > 0$.

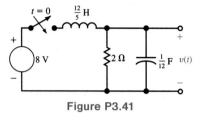

Figure P3.41

AC Steady State Analysis

4.1
Introduction

At this point in our analysis of linear circuits we have learned how to determine both their natural and forced responses. We found that the natural response is simply a characteristic of the network.

In this chapter we will concentrate on the steady state forced response of circuits with *sinusoidal forcing functions*. In this *steady state analysis* we ignore the transient nature of the response since it will vanish prior to the steady state. Our interest in the network response to sinusoidal forcing functions stems from the fact that this forcing function is the dominant waveform in the worldwide electric power industry and therefore is present, for example, at the ac outlets in our home, office, and laboratory. Furthermore, in Fourier series analysis we can demonstrate that *any* periodic signal can be represented by a sum of sinusoids.

4.2
Sinusoidal Functions

We begin our discussion of these functions by considering the sine wave

$$x(\omega t) = X_M \sin\omega t \qquad (4.1)$$

where $x(t)$ could represent a voltage $v(t)$ or current $i(t)$. In Eq. (4.1), X_M is the *amplitude* or *maximum value*, ω is the *radian* or *angular frequency*, and ωt is the *argument*. Two plots of Eq. (4.1) as a function of its argument are shown in Fig. 4.1. Note that the function in Fig 4.1(a) is periodic and repeats after 2π radians. The function in Fig. 4.1(b) is periodic with a *period* of T seconds. At the end of one period the function repeats itself.

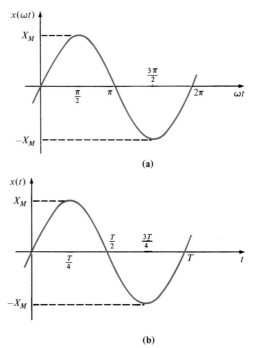

Figure 4.1   Plots of a sine wave as a function of both ωt and t.

Periodicity can be expressed in general for a function $x(t)$ with period T as

$$x(t + nT) = x(t) \qquad n = 1,2,3, \cdots \qquad (4.2)$$

The *frequency* in Hertz (cycles per second) is related to the period by the equation

$$f = \frac{1}{T} \qquad (4.3)$$

For the function shown in Fig. 4.1(a)

$$\omega T = 2\pi$$

and therefore

$$\omega = \frac{2\pi}{T} = 2\pi f \qquad (4.4)$$

Equation (4.4) is the general relationship among period in seconds, frequency in Hertz, and radian frequency.

A more general expression for a sinusoidal function is

$$x(t) = X_M \sin (\omega t + \theta) \qquad (4.5)$$

where θ is called the *phase angle.* For comparison, Eqs. (4.1) and (4.5) are plotted in Fig. 4.2. Since any point on the waveform $X_M \sin(\omega t + \theta)$ occurs θ radians earlier in time than the corresponding point on the waveform $X_M \sin \omega t$, $X_M \sin \omega t$ lags $X_M \sin (\omega t + \theta)$ by θ

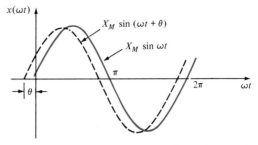

Figure 4.2 Graphical illustration of $X_M\sin(\omega t + \theta)$ leading $X_M\sin\omega t$ by θ radians.

radians. In general, if

$$x_1(t) = X_{M1}\sin(\omega t + \theta)$$

and

$$x_2(t) = X_{M2}\sin(\omega t + \phi)$$

Then $x_1(t)$ *leads* $x_2(t)$ by $\theta - \phi$ radians and $x_2(t)$ *lags* $x_1(t)$ by $\theta - \phi$ radians. If $\theta = \phi$, the waveforms are *in phase,* and if $\theta \neq \phi$, the waveforms are *out of phase.* It is important to note that when comparing sinusoidal functions *of the same frequency* to determine the phase difference, both functions should be expressed as sines or cosines with positive amplitudes.

In the equations above, ωt, ϕ, and θ are expressed in radians, and since they are added to form the total argument of the trigonometric function, they all must be in the same units. However, in practice the phase angle is normally expressed in degrees. Therefore, throughout this text we will express phase angles in degrees and convert ωt from radians to degrees using the fact that 2π radians corresponds to $360°$. The following equations illustrate that the sine and cosine function differ only by a phase angle

$$\sin\omega t = \cos(\omega t - 90°)$$
$$\cos\omega t = \sin(\omega t + 90°)$$

(4.6)

$$-A\sin(\omega t + \theta) = A\sin(\omega t + \theta \pm 180°)$$
$$-A\cos(\omega t + \theta) = A\cos(\omega t + \theta \pm 180°)$$

(4.7)

EXAMPLE 4.1
Let us compute the phase angle between the two voltages $v_1(t) = 12\sin(\omega t + 60°)$V and $v_2(t) = -6\cos(\omega t + 30°)$V.

Since the minus sign corresponds to a phase angle of $\pm 180°$, $v_2(t)$ can be expressed as

$$v_2(t) = -6\cos(\omega t + 30°) = 6\cos(\omega t + 210°) \text{ V}$$

Then using Eq. (4.7), $v_2(t)$ can be written as

$$v_2(t) = 6\sin(\omega t + 300°) \text{ V}$$

Now that both voltages of the same frequency are expressed as sine waves with positive amplitudes, the phase difference can be determined. $v_1(t)$ leads $v_2(t)$ by $60° - 300° = -240° = +120°$ and $v_2(t)$ lags $v_1(t)$ by $+120°$. ∎

D4.1. Given the voltage $v(t) = 120 \cos (314t + \frac{\pi}{4})$ V, determine the frequency of the voltage in Hertz and the phase angle in degrees.

Ans: $f = 50$ Hz and $\theta = 45°$.

D4.2. Given the two branch currents $i_1(t) = 2\sin (377t + 45°)$ and $i_2(t) = -4\sin (377t + 60°)$ A, determine the phase angle by which $i_1(t)$ leads $i_2(t)$.

Ans: $i_1(t)$ leads $i_2(t)$ by $-195°$.

4.3

The Sinusoidal Function/Complex Number Connection

If we apply a sinusoidal forcing function to a linear network, the steady state voltages and currents in the network will also be sinusoidal. For example, KVL would dictate that if one branch voltage is a sinusoid of some frequency, the other branch voltages must be sinusoids of the same frequency.

Consider for a moment a series circuit consisting of a resistor, inductor, and a voltage source of value $V_M \cos \omega t$. The current $i(t)$ leaving the positive terminal of the source is described by the KVL equation

$$L \frac{di(t)}{dt} + Ri(t) = V_M \cos \omega t \qquad (4.8)$$

Since the input voltage is $V_M \cos \omega t$, the current must be of the form

$$i(t) = I_M \cos(\omega t + \phi) \qquad (4.9)$$

If we substitute Eq. (4.9) into Eq. (4.8) and solve for the two unknowns I_M and ϕ, it is straightforward but tedious to show that

$$I_M = \frac{V_M}{\sqrt{R^2 + \omega^2 L^2}}$$

and

$$\phi = -\tan^{-1} \frac{\omega L}{R}$$

therefore

$$i(t) = \frac{V_M}{\sqrt{R^2 + \omega^2 L^2}} \cos \left(\omega t - \tan^{-1} \frac{\omega L}{R} \right) \qquad (4.10)$$

Since the solution to this very simple problem is quite laborious, as we imagine trying to attack a more complicated network we are prompted to ask if there isn't a better way! There is indeed a better way, and it involves establishing a correspondence between sinusoidal time functions and complex numbers. This correspondence will lead to a set of algebraic equations for the loop currents or node voltages in which the coefficients of these variables are complex numbers.

The vehicle we will employ to establish a relationship between time varying sinusoidal functions and complex numbers is Euler's equation, which for our purposes is written as

$$e^{j\omega t} = \cos\omega t + j\sin\omega t \tag{4.11}$$

This complex function has a real part and an imaginary part:

$$\text{Re}(e^{j\omega t}) = \cos\omega t$$
$$\text{Im}(e^{j\omega t}) = \sin\omega t \tag{4.12}$$

where Re $(-)$ and Im $(-)$ represent the real part and the imaginary part, respectively, of the function in the parentheses.

Now suppose that we select as our forcing function the nonrealizable voltage

$$v(t) = V_M e^{j\omega t} \tag{4.13}$$

which because of Euler's identity can be written as

$$v(t) = V_M\cos\omega t + jV_M\sin\omega t \tag{4.14}$$

The real and imaginary parts of this function are each realizable. We think of this complex forcing function as two forcing functions, a real one and an imaginary one, and as a consequence of linearity, the principle of superposition applies and thus the current response can be written as

$$i(t) = I_M\cos(\omega t + \phi) + jI_M\sin(\omega t + \phi) \tag{4.15}$$

Where $I_M\cos(\omega t + \phi)$ is the response due to $V_M\cos\omega t$ and $jI_M\sin(\omega t + \phi)$ is the response due to $jV_M\sin\omega t$. This expression for the current containing both a real term and an imaginary term can be written via Euler's equation as

$$i(t) = I_M e^{j(\omega t + \phi)} \tag{4.16}$$

Because of the relationships above we find that rather than applying the forcing function $V_M\cos\omega t$ and calculating the response $I_M\cos(\omega t + \phi)$, we can apply the complex forcing function $V_M e^{j\omega t}$ and calculate the response $I_M e^{j(\omega t + \phi)}$, the real part of which is the desired response $I_M\cos(\omega t + \phi)$. Although this procedure may initially appear to be more complicated, it is not. It is via this technique that we will convert the differential equation to an algebraic equation which is much easier to solve.

Once again, let us determine the current in the RL circuit described by Eq. (4.8). However, rather than applying $V_M\cos\omega t$, we will apply $V_M e^{j\omega t}$. The forced response will be of the form

$$i(t) = I_M e^{j(\omega t + \phi)}$$

where only I_M and ϕ are unknown. Substituting $v(t)$ and $i(t)$ into the differential equation for the circuit, we obtain

$$RI_M e^{j(\omega t + \phi)} + L\frac{d}{dt}(I_M e^{j(\omega t + \phi)}) = V_M e^{j\omega t}$$

Taking the indicated derivative, we obtain

$$RI_M e^{j(\omega t + \phi)} + j\omega L I_M e^{j(\omega t + \phi)} = V_M e^{j\omega t}$$

Dividing each term of the equation by the common factor $e^{j\omega t}$ yields

$$RI_M e^{j\phi} + j\omega L I_M e^{j\phi} = V_M$$

which is an algebraic equation with complex coefficients. This equation can be written as

$$I_M e^{j\phi} = \frac{V_M}{R + j\omega L}$$

Converting the right-hand side of the equation to exponential or polar form produces the equation

$$I_M e^{j\phi} = \frac{V_M}{\sqrt{R^2 + \omega^2 L^2}} \; e^{j\left(-\tan^{-1}\left(\frac{\omega L}{R}\right)\right)}$$

(A quick refresher on complex numbers is given in Appendix A for readers who need to sharpen their skills in this area.) The form above clearly indicates that the magnitude and phase of the resulting current are

$$I_M = \frac{V_M}{\sqrt{R^2 + \omega^2 L^2}}$$

and

$$\phi = -\tan^{-1} \frac{\omega L}{R}$$

However, since our actual forcing function was $V_M \cos^{\omega t}$ rather than $V_M e^{j\omega t}$, our actual response is the real part of the complex response:

$$i(t) = I_M \cos(\omega t + \phi)$$
$$= \frac{V_M}{\sqrt{R^2 + \omega^2 L^2}} \cos\left(\omega t - \tan^{-1} \frac{\omega L}{R}\right)$$

Note that this is identical to the response obtained by solving the differential equation for the current $i(t)$.

In summary, our correspondence between sinusoidal time functions and complex numbers is based upon Euler's equation, which permits us to express a voltage or current in the form

$$x(t) = X_M \cos(\omega t + \theta)$$
$$= R_e[X_M e^{j(\omega t + \theta)}]$$

where $x(t)$ is a current $i(t)$ or a voltage $v(t)$. This latter equation can be expressed as

$$x(t) = R_e[(X_M e^{j\theta})e^{j\omega t}]$$

As the previous analysis indicates, the term $e^{j\omega t}$ is a common factor in the defining equations for a network and therefore is simply implicit in the analysis. It is the remaining terms, that is, X_M and θ that completely describe the magnitude and phase angle of the unknown current or voltage. We define the complex representation, $X_M e^{j\theta}$, as a *phasor*. Thus, a phasor is a complex number, expressed in polar form, consisting of two terms: X_M, which represents the *magnitude* of the sinusoidal signal, and θ, which represents the *phase angle* of the sinusoidal signal, which is measured with respect to a cosine signal. As a distinguishing feature, phasors will be written in boldface type.

If we now employ phasors in the analysis of the *RL* circuit, Eq. (4.8) becomes

$$L \frac{d}{dt} (\mathbf{I}e^{j\omega t}) + R\mathbf{I}e^{j\omega t} = \mathbf{V}e^{j\omega t}$$

where $\mathbf{I} = I_M \angle \phi$ and $\mathbf{V} = V_M \angle 0°$. Performing the indicated derivative and eliminating the common factor $e^{j\omega t}$ yields the phasor equation

$$j\omega L\mathbf{I} + R\mathbf{I} = \mathbf{V}$$

Therefore

$$\mathbf{I} = \frac{\mathbf{V}}{R + j\omega L} = I_M \angle \phi = \frac{V_M}{\sqrt{R^2 + \omega^2 L^2}} \Big/ \! -\tan^{-1} \frac{\omega L}{R}$$

Thus

$$i(t) = \frac{V_M}{\sqrt{R^2 + \omega^2 L^2}} \cos\!\left(\omega t - \tan^{-1} \frac{\omega L}{R}\right)$$

which, once again, is the function we obtained earlier.

We refer to a phasor analysis as a *frequency domain* analysis. Thus, we have transformed a set of differential equations with sinusoidal forcing functions in the time domain to a set of algebraic equations containing complex numbers in the frequency domain. In effect, we are now faced with solving a set of algebraic equations for the unknown phasors. The phasors are then simply transformed back to the time domain to yield the solution of the original set of differential equations. In addition, as indicated earlier, we have tacitly assumed that sinusoidal functions would be represented as phasors with a phase angle based on a cosine function. Therefore, in both the transformation between the time domain and the frequency domain, as well as the reverse transformation

$$A \cos(\omega t \pm \theta) \leftrightarrow A \angle \pm \theta$$
$$A \sin(\omega t \pm \theta) \leftrightarrow A \angle \pm \theta - 90° \tag{4.17}$$

EXAMPLE 4.2
Convert the time functions $v(t) = 24 \cos(377t - 45°)$ and $i(t) = 12 \sin(377t + 120°)$ to phasors.

Using the phasor transformation shown above, we have

$$\mathbf{V} = 24 \angle -45°$$
$$\mathbf{I} = 12 \angle 120° - 90° = 12 \angle 30°$$ ∎

EXAMPLE 4.3
Convert the phasors $\mathbf{V} = 16 \angle 20°$ and $\mathbf{I} = 10 \angle -75°$ from the frequency domain to the time domain if the frequency is 60 Hz.

Employing the reverse transformation for phasors, we find that

$$v(t) = 16 \cos(377t + 20°)$$
$$i(t) = 10 \cos(377t - 75°)$$ ∎

D4.3. Convert the following voltage functions to phasors.

$$v_1(t) = 12 \cos(377t - 425°) \text{ V}$$
$$v_2(t) = 18 \sin(2513t + 4.2°) \text{ V}$$

Ans: $\mathbf{V}_1 = 12\angle{-425°}\text{V}, \mathbf{V}_2 = 18\angle{-85.8°}\text{V}.$

D4.4. Convert the following phasors to the time domain if the frequency is 400 Hz.

$$\mathbf{V}_1 = 10\angle{20°}$$
$$\mathbf{V}_2 = 12\angle{-60°}$$

Ans: $v_1(t) = 10 \cos(800\pi t + 20°)\text{V}, v_2(t) = 12 \cos(800\pi t - 60°) \text{ V}.$

4.4

Phasor Relationships for Circuit Elements

The next step in our development of the phasor analysis technique is the establishment of the phasor relationships between voltage and current for the passive elements R, L, and C.

For a resistor, the voltage–current relationship is

$$v(t) = R\,i(t) \tag{4.18}$$

Applying $v(t) = V_M e^{j(\omega t + \theta_v)}$ yields $i(t) = I_M e^{j(\omega t + \theta_i)}$ and

therefore

$$V_M e^{j(\omega t + \theta_v)} = R I_M e^{j(\omega t + \theta_i)}$$

which can be written in phasor form as

$$\mathbf{V} = R\mathbf{I} \tag{4.19}$$

where $\mathbf{V} = V_M e^{j\theta_v} = V_M\angle{\theta_v}$ and $\mathbf{I} = I_M e^{j\theta_i} = I_M\angle{\theta_i}$. In this case $\theta_v = \theta_i$ and the current and voltage are *in phase*.

The voltage–current relationship for an inductor is

$$v(t) = L\frac{di(t)}{dt} \tag{4.20}$$

The same arguments which lead to the development of Eq. (4.19) yield the relationship

$$\mathbf{V} = j\omega L\mathbf{I} \tag{4.21}$$

The imaginary operator $j = 1e^{j90°} = 1\angle{90°} = \sqrt{-1}$. Therefore, in an inductor, the voltage leads the current by 90° or the current lags the voltage by 90°.

For a capacitor the voltage–current relationship is

$$i(t) = C\frac{dv(t)}{dt} \tag{4.22}$$

which can be written in phasor form as

$$\mathbf{I} = j\omega C\mathbf{V} \tag{4.23}$$

Hence, in a capacitor, the current leads the voltage by 90° or the voltage lags the current by 90°.

Since phasors are complex numbers, they can be graphically represented as line segments on a *phasor diagram*. This diagram illustrates the relative magnitude of one phasor with another, the angle between them, and their relative position with respect to one another. Figure 4.3 illustrates the voltage–current relationship, the sinusoidal waveforms, and the phasor diagrams for the three passive elements.

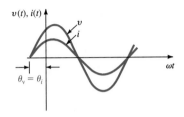

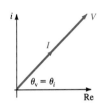

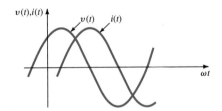

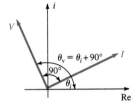

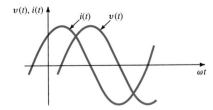

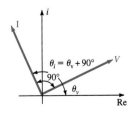

Figure 4.3   Characteristics of the passive elements *R*, *L*, and *C*.

4.5

Impedance and Admittance

Having examined each of the passive circuit elements in the frequency domain, we now define the *Impedance* **Z** in exactly the same manner in which we defined resistance earlier. Impedance is defined as the ratio of the phasor voltage **V** to the phasor current **I**:

$$\mathbf{Z} = \frac{\mathbf{V}}{\mathbf{I}} \tag{4.24}$$

at the two terminals of the element related to one another by the passive sign convention, as illustrated in Fig. 4.4. Equation (4.24) can be written in the form

$$\mathbf{Z} = \frac{V_M \angle \theta_v}{I_M \angle \theta_i} = \frac{V_M}{I_M} \angle \theta_v - \theta_i = Z \angle \theta_z \tag{4.25}$$

Since **Z** is the ratio of **V** to **I,** the units of **Z** are ohms. Thus, impedance in an ac circuit is analogous to resistance in a dc circuit. In rectangular form, impedance is expressed as

$$\mathbf{Z}(j\omega) = R(\omega) + jX(\omega) \tag{4.26}$$

where $R(\omega)$ is the *real, or resistive, component* and $X(\omega)$ is the *imaginary, or reactive, component.* Equation (4.26) clearly indicates that **Z**, which is frequency dependent, is a complex number; however, it is not a phasor. Recall that the phasors **V** and **I** correspond to time domain, steady state signals. However, the impedance, **Z**, has no meaning in the time domain

Equations (4.25) and (4.26) indicate that

$$Z \angle \theta_z = R + jX$$

Therefore

$$Z = \sqrt{R^2 + X^2} \tag{4.27}$$

$$\theta_z = \tan^{-1} \frac{X}{R} \tag{4.28}$$

where

$$R = Z \cos \theta_z$$
$$X = Z \sin \theta_z$$

Just as it was advantageous to know how to determine the equivalent resistance of a series and/or parallel combination of resistors in a dc circuit, we now want to learn how to determine the equivalent impedance in an ac circuit when the passive elements are interconnected. The determination of equivalent impedance is based on Kirchhoff's current

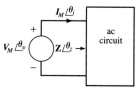

Figure 4.4   General impedance relationship.

law (KCL) and Kirchhoff's voltage law (KVL). Therefore, we must see if these laws are valid in the frequency domain. Suppose, for example, that a circuit is driven by voltage sources of the form $V_{M_m}\cos(\omega t + \theta_m)$. Then, since the network is linear, every steady state current in the network will be of the form $I_{M_k}\cos(\omega t + \phi_k)$. At any node in the circuit, KCL, written in the time domain, is

$$i_1(t) + i_2(t) + \cdots i_n(t) = 0$$
$$I_{M_1}\cos(\omega t + \phi_1) + I_{M_2}\cos(\omega t + \phi_2) + \cdots + I_{M_n}\cos(\omega t + \phi_n) = 0$$

From our previous work we can immediately apply the phasor transformation to the equation above to obtain

$$\mathbf{I}_1 + \mathbf{I}_2 + \cdots \mathbf{I}_n = 0$$

However, this equation is simply KCL in the frequency domain. In a similar manner we can show that KVL applies in the frequency domain. Using the fact that KCL and KVL are valid in the frequency domain, we can show, as was done in Chapter 2 for resistors, that impedances can be combined using the same rules that we established for resistor combinations, that is, if $\mathbf{Z}_1, \mathbf{Z}_2, \mathbf{Z}_3, \ldots, \mathbf{Z}_n$ are connected in series, the equivalent impedance $\mathbf{Z}_s$ is

$$\mathbf{Z}_s = \mathbf{Z}_1 + \mathbf{Z}_2 + \mathbf{Z}_3 + \cdots + \mathbf{Z}_n \tag{4.29}$$

and if $\mathbf{Z}_1, \mathbf{Z}_2, \mathbf{Z}_3, \ldots, \mathbf{Z}_n$ are connected in parallel, the equivalent impedance is

$$\frac{1}{\mathbf{Z}_p} = \frac{1}{\mathbf{Z}_1} + \frac{1}{\mathbf{Z}_2} + \frac{1}{\mathbf{Z}_3} + \cdots + \frac{1}{\mathbf{Z}_n} \tag{4.30}$$

EXAMPLE 4.4

In a farmhouse, the wiring circuit for the basement area uses a 25 A circuit breaker. The circuit breaker will open if the current in it exceeds 25 A. The basement circuit is modeled as shown in Fig. 4.5. Given this network, we wish to know (a) if all the lamps and the dryer can be turned on at the same time, and (b) if the space heater can be used if everything else in the circuit is on.

(a) If all lamps and the dryer are on, that is, switches $sw1$, $sw2$, $sw3$, and $sw5$ are closed, then

$$\mathbf{I}_1 = \mathbf{I}_2 = \mathbf{I}_5 = \frac{120 \angle 0°}{50}$$
$$= 2.4 \angle 0° \text{ A}$$

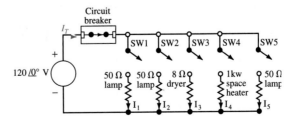

Figure 4.5   Circuit used in Example 4.4.

and

$$I_3 = \frac{120 \angle 0°}{8}$$

$$= 15 \angle 0° \text{ A}$$

Therefore, the total current

$$I_T = I_1 + I_2 + I_3 + I_5$$
$$= 22.2 \angle 0° \text{ A}$$

Since this has a magnitude less than 25 A, the circuit breaker will permit this current in the circuit.

(b) The 1 kW space heater will draw a current of

$$I_4 = \frac{1000}{120} \angle 0°$$

$$= 8.33 \angle 0° \text{ A}$$

If the space heater is turned on while everything else is on, the total current will be

$$I_T = I_1 + I_2 + I_3 + I_4 + I_5$$
$$= 30.53 \angle 0° \text{ A}$$

and the circuit breaker will open, interrupting the circuit. ∎

EXAMPLE 4.5

Determine the equivalent impedance of the network shown in Fig. 4.6 if the frequency is $f = 60$ Hz. Then compute the current $i(t)$ if the voltage source is $v(t) = 50 \cos (\omega t + 30°)$ V. Finally, calculate the equivalent impedance if the frequency is that used in aircraft, that is, $f = 400$ Hz.

The impedances of the individual elements at 60 Hz are

$$Z_R = 25 \ \Omega$$
$$Z_L = j\omega L = j(2\pi \times 60)(20 \times 10^{-3}) = j7.54 \ \Omega$$
$$Z_C = \frac{-j}{\omega C} = \frac{-j}{(2\pi \times 60)(50 \times 10^{-6})} = -j53.05 \ \Omega$$

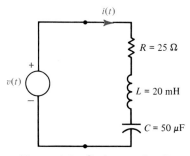

Figure 4.6 Series ac circuit.

Since the elements are in series

$$\mathbf{Z} = \mathbf{Z}_R + \mathbf{Z}_L + \mathbf{Z}_C$$
$$= 25 - j45.51 \ \Omega$$

the current in the circuit is given by

$$\mathbf{I} = \frac{\mathbf{V}}{\mathbf{Z}} = \frac{50 \angle 30°}{25 - j45.51} = \frac{50 \angle 30°}{51.92 \angle -61.22°} = 0.96 \angle 91.22° \mathrm{A}$$

or in the time domain, $i(t) = 0.96 \cos (377t + 91.22°)$ A.

If the frequency is 400 Hz, the impedance of each element is

$$\mathbf{Z}_R = 25 \ \Omega$$
$$\mathbf{Z}_L = j\omega L = j50.27 \ \Omega$$
$$\mathbf{Z}_c = \frac{-j}{\omega C} = -j7.96 \ \Omega$$

The total impedance is then

$$\mathbf{Z} = 25 + j42.31 = 49.14 \angle 59.42° \ \Omega$$

It is important to note that at the frequency $f = 60$ Hz, the reactance of the circuit is capacitive, that is, if the impedance is written as $R + jX$, $X < 0$; however, at $f = 400$ Hz the reactance is inductive since $X > 0$. ∎

Another quantity that is very useful in the analysis of ac circuits is *admittance,* which is the reciprocal of impedance, that is

$$\mathbf{Y} = \frac{1}{\mathbf{Z}} = \frac{\mathbf{I}}{\mathbf{V}} \tag{4.31}$$

The units of $\mathbf{Y}$ are Siemens, and this quantity is analogous to conductance in resistive dc circuits. Since $\mathbf{Z}$ is a complex number, $\mathbf{Y}$ is also a complex number

$$\mathbf{Y} = Y_M \angle \theta_Y \tag{4.32}$$

which is written in rectangular form as

$$\mathbf{Y} = G + jB \tag{4.33}$$

where G and B are called *conductance* and *susceptance,* respectively. Because of the relationship between $\mathbf{Y}$ and $\mathbf{Z},$ we can express the components of one quantity as a function of the components of the other

$$G + jB = \frac{1}{R + jX} \tag{4.34}$$

Rationalizing the right-hand side of this equation yields

$$G + jB = \frac{R - jX}{R^2 + X^2}$$

and therefore

$$G = \frac{R}{R^2 + X^2} \qquad B = \frac{-X}{R^2 + X^2} \tag{4.35}$$

In a similar manner we can show that

$$R = \frac{G}{G^2 + B^2} \tag{4.36}$$

$$X = \frac{-B}{G^2 + B^2} \tag{4.37}$$

It is very important to note that in general R and G are not reciprocals of one another. The same is true for X and B. The purely resistive case is an exception. In the purely reactive case the quantities are negative reciprocals of one another.

The admittance of the individual passive elements is

$$\mathbf{Y}_R = \frac{1}{R} = G$$

$$\mathbf{Y}_L = \frac{1}{j\omega L} = \frac{1}{\omega L} \underline{/-90^\circ} \tag{4.38}$$

$$\mathbf{Y}_C = j\omega C = \omega C \underline{/90^\circ}$$

Once again, since KCL and KVL are valid in the frequency domain, we can show, using the same approach outlined in Chapter 2 for conductance in resistive circuits, that the rules for combining admittances are the same as those for combining conductances; that is, if $\mathbf{Y}_1, \mathbf{Y}_2, \mathbf{Y}_3, \ldots, \mathbf{Y}_n$ are connected in parallel, the equivalent admittance is

$$\mathbf{Y}_p = \mathbf{Y}_1 + \mathbf{Y}_2 + \cdots + \mathbf{Y}_n \tag{4.39}$$

and if $\mathbf{Y}_1, \mathbf{Y}_2, \ldots, \mathbf{Y}_n$ are connected in series, the equivalent admittance is

$$\frac{1}{\mathbf{Y}_s} = \frac{1}{\mathbf{Y}_1} + \frac{1}{\mathbf{Y}_2} + \cdots + \frac{1}{\mathbf{Y}_n} \tag{4.40}$$

EXAMPLE 4.6

Let us compute the impedance $\mathbf{Z}_T$ at the terminals of the network shown in Fig. 4.7(a). We will first determine $\mathbf{Z}_T$ using only impedances and then we will solve the problem using a combination of Impedance and Admittance.

The circuit is redrawn as shown in Fig. 4.7(b). From that network we find that

$$\mathbf{Z}_{23} = \frac{\mathbf{Z}_2 \mathbf{Z}_3}{\mathbf{Z}_2 + \mathbf{Z}_3}$$

$$= \frac{(2 + j4)(4 - j2)}{2 + j4 + 4 - j2}$$

$$= 3 + j1 \ \Omega$$

Furthermore

$$\mathbf{Z}_1 = \frac{(2)(-j2)}{2 - j2}$$

$$= 1 - j1 \ \Omega$$

Therefore

$$\mathbf{Z}_T = \mathbf{Z}_1 + \mathbf{Z}_{23}$$

$$= 4 \ \Omega$$

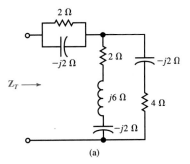

(a)

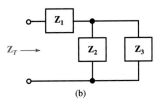

(b)

Figure 4.7   Circuits used in Example 4.6.

Using admittances we find that

$$\mathbf{Y}_3 = \frac{1}{\mathbf{Z}_3} = \frac{1}{4 - j2} = 0.20 + j0.10 \ S$$

And

$$\mathbf{Y}_2 = \frac{1}{\mathbf{Z}_2} = \frac{1}{2 + j4} = 0.10 - j0.20 \ S$$

Hence

$$\mathbf{Y}_{23} = \mathbf{Y}_2 + \mathbf{Y}_3 = 0.30 - j0.10 \ S$$

And

$$\mathbf{Z}_{23} = \frac{1}{\mathbf{Y}_{23}} = \frac{1}{0.30 - j0.10} = 3 + j1 \ \Omega$$

Now since

$$\mathbf{Y}_1 = \frac{1}{2} + \frac{1}{-j2} = 0.5 + j0.5 \ S$$

Then

$$\mathbf{Z}_1 = \frac{1}{\mathbf{Y}_1} = \frac{1}{0.5 + j0.5} = 1 - j1 \ \Omega$$

And therefore

$$\mathbf{Z}_T = \mathbf{Z}_1 + \mathbf{Z}_{23} = 4 \ \Omega$$

which, of course, is identical to that obtained above.   ■

D4.5. Compute the impedance $\mathbf{Z}_T$ in the network in Fig. D4.5.

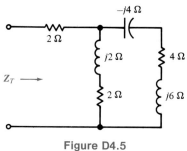

Figure D4.5

Ans: $\mathbf{Z}_T = 3.38 + j1.08 \; \Omega.$

4.6

Y ⇆ Δ Transformations The wye-to-delta and delta-to-wye transformations presented earlier for resistance are also valid for impedance in the frequency domain. Therefore, the impedances shown in Fig. 4.8 are related by the following equations:

$$\mathbf{Z}_a = \frac{\mathbf{Z}_1 \mathbf{Z}_2}{\mathbf{Z}_1 + \mathbf{Z}_2 + \mathbf{Z}_3}$$

$$\mathbf{Z}_b = \frac{\mathbf{Z}_1 \mathbf{Z}_3}{\mathbf{Z}_1 + \mathbf{Z}_2 + \mathbf{Z}_3} \qquad (4.41)$$

$$\mathbf{Z}_c = \frac{\mathbf{Z}_2 \mathbf{Z}_3}{\mathbf{Z}_1 + \mathbf{Z}_2 + \mathbf{Z}_3}$$

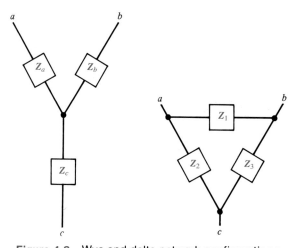

Figure 4.8 Wye and delta network configurations.

and

$$\mathbf{Z}_1 = \frac{\mathbf{Z}_a\mathbf{Z}_b + \mathbf{Z}_b\mathbf{Z}_c + \mathbf{Z}_c\mathbf{Z}_a}{\mathbf{Z}_c}$$

$$\mathbf{Z}_2 = \frac{\mathbf{Z}_a\mathbf{Z}_b + \mathbf{Z}_b\mathbf{Z}_c + \mathbf{Z}_c\mathbf{Z}_a}{\mathbf{Z}_b} \qquad (4.42)$$

$$\mathbf{Z}_3 = \frac{\mathbf{Z}_a\mathbf{Z}_b + \mathbf{Z}_b\mathbf{Z}_c + \mathbf{Z}_c\mathbf{Z}_a}{\mathbf{Z}_a}$$

These equations are general relationships and therefore apply to any set of impedances connected in a wye or delta configuration.

EXAMPLE 4.7

Let us determine the impedance Z_{eq} at the terminals $A-B$ of the network in Fig. 4.9(a). To simplify the network we must convert one of the back-to-back deltas into a wye.

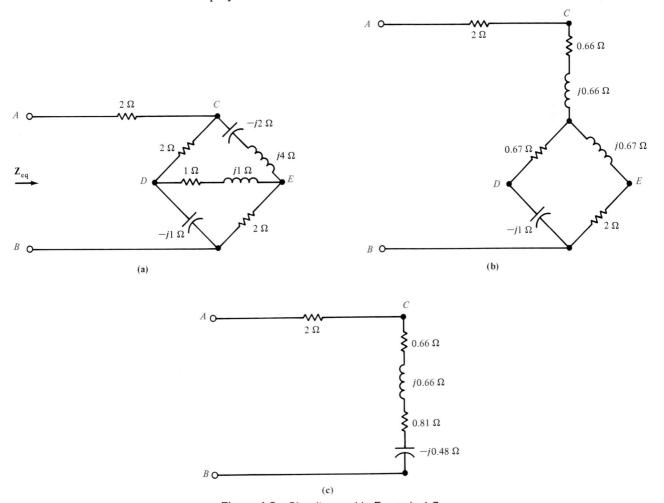

Figure 4.9   Circuits used in Example 4.7.

If we select the delta defined by the points C, D, and E, the impedances of the corresponding wye are calculated from Eq. (4.41) as

$$\frac{(2)(-j2 + j4)}{2 - j2 + j4 + 1 + j1} = \frac{2(j2)}{3 + j3} = 0.66 + j0.66 \ \Omega$$

$$\frac{(2)(1 + j1)}{3 + j3} = 0.67 \ \Omega$$

$$\frac{(-j2 + j4)(1 + j1)}{3 + j3} = j0.67 \ \Omega$$

and shown in Fig. 4.9(b). If we now combine the two impedances in parallel, we obtain

$$\frac{(0.67 - j1)(2 + j0.67)}{0.67 - j1 + 2 + j0.67} = 0.81 - j0.48 \ \Omega$$

which reduces the network to that shown in Fig. 4.9(c). Hence, the equivalent impedance is

$$\mathbf{Z}_{eq} = 2 + 0.66 + j0.66 + 0.81 - j0.48$$
$$= 3.47 + j0.18 \ \Omega$$ ■

DRILL EXERCISE

D4.6. Determine $\mathbf{Z}_{eq}$ at the terminals $A-B$ of the network shown in Fig. D4.6.

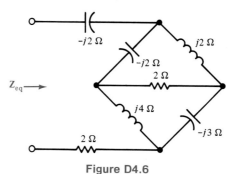

Figure D4.6

Ans: $\mathbf{Z}_{eq} = 4 - j4 \ \Omega$.

4.7

Basic Analysis Using Kirchhoff's Laws

We have shown that Kirchhoff's laws apply in the frequency domain, and therefore they can be used to compute steady state voltages and currents in ac circuits. This approach involves expressing the voltages and currents as phasors, and once this is done, the ac steady state analysis employing phasor equations is performed in an identical fashion to that used in the dc analysis of resistive circuits—the only difference being that in steady state ac circuit analysis the algebraic phasor equations have complex coefficients.

EXAMPLE 4.8

We wish to calculate the voltages and currents in the circuit shown in Fig. 4.10. Our approach will be as follows. We will calculate the total impedance seen by the source $\mathbf{V}_s$.

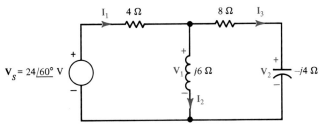

Figure 4.10   Example of an ac circuit.

Then we will use this to determine $\mathbf{I}_1$. Knowing $\mathbf{I}_1$, we can compute $\mathbf{V}_1$ using KVL. Knowing $\mathbf{V}_1$, we can compute $\mathbf{I}_2$ and $\mathbf{I}_3$, and so on.

$$\mathbf{Z}_{eq} = 4 + \frac{(j6)(8 - j4)}{j6 + 8 - j4}$$

$$= 4 + \frac{24 + j48}{8 + j2}$$

$$= 4 + 4.24 + j4.94$$

$$= 9.61 \, \underline{/\, 30.94°} \; \Omega$$

Then

$$\mathbf{I}_1 = \frac{\mathbf{V}_s}{\mathbf{Z}_{eq}} = \frac{24 \, \underline{/\, 60°}}{9.61 \, \underline{/\, 30.94°}}$$

$$= 2.5 \, \underline{/\, 29.06°} \, \text{A}$$

$\mathbf{V}_1$ can be determined using KVL:

$$\mathbf{V}_1 = \mathbf{V}_s - 4\mathbf{I}_1$$

$$= 24 \, \underline{/\, 60°} - 10 \, \underline{/\, 29.06°}$$

$$= 3.26 + j15.92 = 16.25 \, \underline{/\, 78.43°} \; \text{V}$$

Note that $\mathbf{V}_1$ could also be computed via voltage division:

$$\mathbf{V}_1 = \frac{\mathbf{V}_s \dfrac{(j6)(8 - j4)}{j6 + 8 - j4}}{4 + \dfrac{(j6)(8 - j4)}{j6 + 8 - j4}} \; \text{V}$$

which from our previous calculation is

$$\mathbf{V}_1 = \frac{(24 \, \underline{/\, 60°})(6.5 \, \underline{/\, 49.30°})}{9.61 \, \underline{/\, 30.49°}}$$

$$= 16.25 \, \underline{/\, 78.43°} \; \text{V}$$

Knowing $\mathbf{V}_1$, we can calculate both $\mathbf{I}_2$ and $\mathbf{I}_3$:

$$\mathbf{I}_2 = \frac{\mathbf{V}_1}{j6} = \frac{16.25 \, \underline{/\, 78.43°}}{6 \, \underline{/\, 90°}}$$

$$= 2.71 \, \underline{/\, -11.56°} \; \text{A}$$

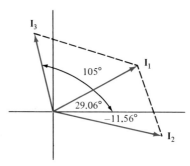

Figure 4.11   Phasor diagram for the currents in Example 4.8.

and

$$I_3 = \frac{V_1}{8 - j4}$$
$$= 1.82\underline{/105°}\ A$$

I_2 and I_3 could have been calculated by current division. For example, I_2 could be determined by

$$I_2 = \frac{I_1(8 - j4)}{8 - j4 + j6}$$
$$= \frac{(2.5\underline{/29.06°})(8.94\underline{/-26.57°})}{8 + j2}$$
$$= 2.71\ \underline{/-11.56°}\ A$$

Finally, V_2 can be computed as

$$V_2 = I_3(-j4)$$
$$= 7.28\underline{/15°}\ V$$

This value could also have been computed by voltage division. The phasor diagram for the currents I_1, I_2, and I_3 is shown in Fig. 4.11 and is an illustration of KCL.   ∎

We have demonstrated the use of KCL, KVL, and the relationship $V = IZ$ in the solution of ac steady state circuit problems. Since the passive circuit elements R, L, and C are linear components, the circuit analysis tools used in dc circuit analysis are applicable in ac steady state analysis also. The primary difference between these two analyses is that in the ac case, the use of phasors, impedance, and admittance lead to equations involving complex numbers.

The following example illustrates that *nodal analysis* is performed in exactly the same

DRILL EXERCISE

D4.7. Given the network in Fig. D4.7, find the voltages V_L, V_R, and V_C and plot them on a phasor diagram.

Figure D4.7

Ans: $\mathbf{V}_L = 4\sqrt{2}\underline{/45°}$ V, $\mathbf{V}_R = 4\underline{/90°}$ V, $\mathbf{V}_C = 4\underline{/0°}$ V, and the phasor diagram is as shown below.

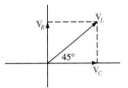

4.8

Nodal Analysis

manner as it was in dc resistive circuits. In this section nodal analysis will be employed in the frequency domain using the phasor technique.

EXAMPLE 4.9

Let us compute the node voltages and use them to find $\mathbf{I}_0$ in the circuit shown in Fig. 4.12. There are four nodes and therefore three equations are needed to solve the circuit.

Note that $\mathbf{V}_0 = 12\underline{/0°}$ V and therefore is known. KCL applied to the remaining nodes yields the equations

$$\frac{\mathbf{V}_1 - \mathbf{V}_0}{1} + \frac{\mathbf{V}_1}{j1} + \frac{\mathbf{V}_1 - \mathbf{V}_2}{-j1} = 0$$

$$\frac{\mathbf{V}_2 - \mathbf{V}_1}{-j1} + \frac{\mathbf{V}_2}{1} + 2\underline{/0°} = 0$$

which when simplified become

$$1\mathbf{V}_1 - j1\mathbf{V}_2 = 12\underline{/0°}$$
$$-j1\mathbf{V}_1 + (1 + j)\mathbf{V}_2 = -2\underline{/0°}$$

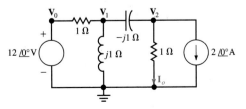

Figure 4.12   Circuit analyzed in Example 4.9.

Solving the equations by Cramer's rule, we first compute the determinant as

$$det = \begin{vmatrix} 1 & -j1 \\ -j1 & 1+j \end{vmatrix}$$
$$= 1(1+j) - (-j)(-j)$$
$$= 2+j$$

The node voltage $\mathbf{V}_1$ is then

$$\mathbf{V}_1 = \frac{\begin{vmatrix} 12\angle 0° & -j1 \\ -2\angle 0° & 1+j \end{vmatrix}}{2+j}$$

$$= \frac{12(1+j) - (-2)(-j1)}{2+j}$$

$$= \frac{12 + j10}{2+j}$$

$$= \frac{34}{5} + j\frac{8}{5} \ \text{V}$$

and the node voltage $\mathbf{V}_2$ is

$$\mathbf{V}_2 = \frac{\begin{vmatrix} 1 & 12\angle 0° \\ -j & -2\angle 0° \end{vmatrix}}{2+j}$$

$$= \frac{-2 + j12}{2+j}$$

$$= \frac{8}{5} + j\frac{26}{5} \ \text{V}$$

Then

$$\mathbf{I}_0 = \frac{\mathbf{V}_2}{1}$$

$$= \frac{8}{5} + j\frac{26}{5} \ \text{A}$$

DRILL EXERCISE

D4.8. Use nodal analysis to find $\mathbf{I}_0$ in the circuit in Fig. D4.8.

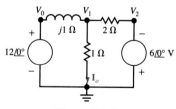

Figure D4.8

Ans: $\mathbf{I}_0 = \left(\dfrac{30}{13} - j\dfrac{84}{13} \right)$ A.

4.9
Mesh and Loop Analysis

The following example will illustrate the use of *mesh* and *loop analysis* in ac steady state circuits. For comparison we will use the circuit in Example 4.9.

EXAMPLE 4.10

Consider the network in Fig. 4.13 in which the mesh currents are labeled. Note that since the mesh current $\mathbf{I}_3$ goes through the current source, $\mathbf{I}_3 = 2\angle 0°$A. The KVL equations for the two remaining meshes are

$$(1 + j)\mathbf{I}_1 - j1\mathbf{I}_2 = 12\angle 0°$$
$$-j1\mathbf{I}_1 + (j1 - j1 + 1)\mathbf{I}_2 - 1\mathbf{I}_3 = 0$$

Simplifying the equations yields

$$(1 + j)\mathbf{I}_1 - j\mathbf{I}_2 = 12\angle 0°$$
$$-j\mathbf{I}_1 + \mathbf{I}_2 = 2\angle 0°$$

The determinant for the equation is

$$det = \begin{vmatrix} 1 + j & -j \\ -j & 1 \end{vmatrix}$$
$$= (1 + j)(1) - (-j)(-j)$$
$$= 2 + j$$

The current $\mathbf{I}_1$ is then

$$\mathbf{I}_1 = \frac{\begin{vmatrix} 12\angle 0° & -j \\ 2\angle 0° & 1 \end{vmatrix}}{2 + j}$$
$$= \frac{12 + 2j}{2 + j}$$
$$= \left(\frac{26}{5} - j\frac{8}{5}\right) \text{A}$$

and the current $\mathbf{I}_2$ is

$$\mathbf{I}_2 = \frac{\begin{vmatrix} 1 + j & 12\angle 0° \\ -j & 2\angle 0° \end{vmatrix}}{2 + j}$$
$$= \frac{2(1 + j) + 12j}{2 + j}$$
$$= \left(\frac{18}{5} + j\frac{26}{5}\right) \text{A}$$

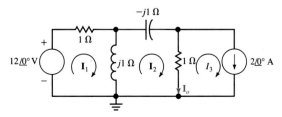

Figure 4.13   Circuit used in Example 4.10.

Then as shown in Fig. 4.13, I_0 is

$$I_0 = I_2 - I_3$$

$$= \frac{18}{5} + j\frac{26}{5} - 2$$

$$= \left(\frac{8}{5} + j\frac{26}{5}\right) \text{ A}$$

which is of course identical to that obtained in Example 4.9. ■

DRILL EXERCISE

D4.9. Use mesh equations to find V_0 in the network in Fig. D4.9.

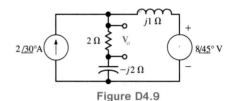

Figure D4.9

Ans: $V_0 = 0.76 + j7.76$ V.

4.10

Superposition

The *principle of superposition* will be illustrated by considering, once again, the circuit analyzed in Example 4.9.

EXAMPLE 4.11

We wish to find I_0 in the network in Fig. 4.14(a). First, we begin making the current source zero by replacing it with an open circuit and deriving the component of I_0 caused by the voltage source acting alone. We call this component I_0' and obtain it from the network in Fig. 4.14(b). The current I_v is equal to the voltage source value divided by the total impedance seen by the source or

$$I_v = \frac{12 \angle 0°}{1 + \frac{(1-j)(j)}{1-j+j}}$$

Then, using current division

$$I_0' = I_v\left(\frac{j}{j+1-j}\right)$$

$$= \frac{12}{1+j+1}(j)$$

$$= \frac{12j}{2+j} \text{ A}$$

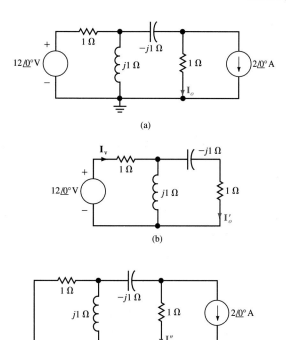

Figure 4.14   Circuits used in Example 4.11.

Next, the component of I_0 caused by the current source acting alone, that is, I_0'', is obtained from the network in Fig. 4.14(c), where the voltage source has been replaced with a short circuit.

Using current division we find that

$$I_0'' = -2\angle 0° \; \frac{-j + \dfrac{(1)(j)}{1+j}}{1 - j + \dfrac{(1)(j)}{1+j}}$$

$$= \frac{-2}{1+j+1}$$

$$= \frac{-2}{2+j} \text{ A}$$

Therefore

$$I_0 = I_0' + I_0''$$

$$= \frac{12j}{2+j} + \frac{-2}{2+j}$$

$$= \frac{-2 + j12}{2+j}$$

$$= \left(\frac{8}{5} + j\frac{26}{5}\right) \text{ A}$$

which is identical to that obtained earlier. ■

D4.10. Using superposition, find $\mathbf{V}_0$ in the network in Fig. D4.10.

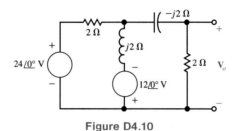

Figure D4.10

Ans: $\mathbf{V}_0 = 12\,\underline{/\,90°}$ V.

4.11

Source Transformation

Source transformation will also be illustrated by solving the network examined in Examples 4.9, 4.10, and 4.11.

EXAMPLE 4.12

The original network examined in the previous three examples is shown in Fig. 4.15(a). Let us employ source transformation to find $\mathbf{I}_0$ in this network.

We begin the network simplification by transforming the $12\,\underline{/\,0°}$ V source in series with the 1-ohm resistor into a $12\,\underline{/\,0°}$ A current source in parallel with the 1-ohm resistor as shown in Fig. 4.15(b). The parallel combination of the 1-ohm resistor and the $j1$-ohm inductor is

$$\mathbf{Z}_p = \frac{(1)\,(j)}{1+j}$$

$$= \frac{1}{2} + j\frac{1}{2}\,\Omega$$

If we now transform this impedance, $\mathbf{Z}_p$, which is in parallel with the $12\,\underline{/\,0°}$ A current source, into a voltage source in series with the impedance $\mathbf{Z}_p$, we obtain the circuit shown in Fig. 4.15(c). Note carefully that with each successive transformation we are absorbing circuit elements, and thus simplifying the network, as we reduce the network around the unknown $\mathbf{I}_0$. The impedance in series with the $6(1+j)$ V source is

$$\mathbf{Z}_s = \frac{1}{2} + j\frac{1}{2} - j$$

$$= \frac{1}{2} - j\frac{1}{2}\,\Omega$$

Thus, transforming the $6(1+j)$ V source in series with the impedance $\mathbf{Z}_s$ into a current source in parallel with this impedance yields the circuit in Fig. 4.15(d). At this point we note that the two current sources in the network in Fig. 4.15(d) are in parallel and there-

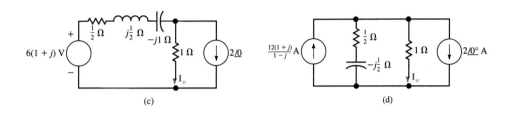

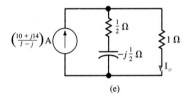

Figure 4.15 Circuits used in Example 4.12.

fore can be added to produce one resultant source which is

$$\mathbf{I}_T = \frac{12(1 + j)}{1 - j} - 2$$

$$= \left(\frac{10 + j14}{1 - j}\right) \text{ A}$$

Thus, the network in Fig. 4.15(d) has been simplified to that in Fig. 4.15(e). Employing current division yields the current $\mathbf{I}_0$

$$\mathbf{I}_0 = \left(\frac{10 + j14}{1 - j}\right)\left(\frac{\frac{1}{2} - j\frac{1}{2}}{\frac{1}{2} - j\frac{1}{2} + 1}\right)$$

$$= \left(\frac{8}{5} + j\frac{26}{5}\right) \text{ A}$$ ∎

DRILL EXERCISE

D4.11 Find $\mathbf{V}_0$ in problem D4.10 using source transformation.

4.12

Thevenin's and Norton's Theorems

We will now demonstrate the use of *Thevenin's and Norton's theorems* by using them to solve the network we have examined in the last several examples.

EXAMPLE 4.13

Once again, let us find the current $\mathbf{I}_0$ in the circuit in Fig. 4.16(a).

We will analyze this circuit by removing the 1 Ω resistor, and then developing an equivalent circuit for the remainder of the network. Finally, we will recombine the 1 Ω resistor with this equivalent circuit.

The open circuit voltage at the resistor containing $\mathbf{I}_0$ is shown in Fig. 4.16(b). Note that the voltage $\mathbf{V}_{oc}$ is the voltage across the current source. However, recall that although we know the voltage across a voltage source, we have to use the rest of the circuit to compute the current in the voltage source. In a similar manner, although we know the current in a current source, we have to use the balance of the network to find the voltage across it. Therefore, we have shown in Fig. 4.16(b) two mesh currents $\mathbf{I}_A$ and $\mathbf{I}_B$. The equation for the first mesh is

$$12\underline{/0°} - 1\mathbf{I}_A - j(\mathbf{I}_A - \mathbf{I}_B) = 0$$

and since $\mathbf{I}_B$ passes directly through the current source

$$\mathbf{I}_B = 2\underline{/0°}$$

Solving the two equations for $\mathbf{I}_A$ yields

$$\mathbf{I}_A = \left(\frac{12 + j2}{1 + j}\right)\text{ A}$$

Now KVL for the closed path containing the voltage source and the open circuit voltage is

$$12\underline{/0°} - 1\mathbf{I}_A - (-j)\mathbf{I}_B - \mathbf{V}_{oc} = 0$$

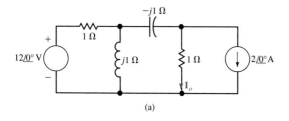

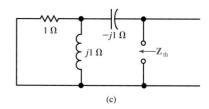

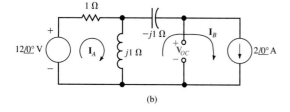

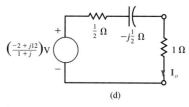

Figure 4.16   Circuits used to illustrate Thevenin's theorem.

which yields

$$V_{oc} = \left(\frac{-2 + j12}{1 + j} \right) V$$

The Thevenin equivalent impedance, Z_{TH}, is derived from the network in Fig. 4.16(c) where the voltage source has been replaced with a short circuit and the current source replaced with an open circuit. This impedance is

$$Z_{TH} = -j + \frac{(1)(j)}{1 + j}$$

$$= \frac{1}{2} - j\frac{1}{2}, \Omega$$

Now the Thevenin equivalent circuit consisting of the open-circuit voltage in series with the Thevenin equivalent impedance is used to replace the entire network exclusive of the 1-ohm resistor containing the current I_0. This network is shown in Fig. 4.16(d). The current I_0 is then

$$I_0 = \frac{\dfrac{-2 + j12}{1 + j}}{\dfrac{1}{2} - j\dfrac{1}{2} + 1}$$

$$= \left(\frac{8}{5} + j\frac{26}{5} \right) A$$

Now if we were to solve the same problem using Norton's theorem, the 1-ohm resistor containing I_0 is replaced with a short circuit as shown in Fig. 4.17(a). The current in this short circuit, I_{sc}, will consist of two components: one due to the $12 \angle 0°$ V source and one due to the $2 \angle 0°$ A source. It is important to note at this point that current emanating from the current source will simply return through the short circuit since this is the path of least resistance. Furthermore, the current generated by the voltage source that flows in the capacitor will all flow through the short circuit. Sometimes it is easier to understand these points if we redraw the circuit in Fig. 4.17(a) by collapsing the short circuit to a single point.

The network in Fig. 4.17(a) is just another network and therefore all the techniques we have learned for finding voltages and currents apply to it also. Therefore, we will demonstrate the use of Thevenin's theorem to provide some simplification. Let us find the Thevenin equivalent circuit for the portion of the network to the left of the capacitor. Thus, we break the network at that point and find the open-circuit voltage shown in Fig. 4.17(b). Using voltage division

$$V_{oc} = 12 \angle 0° \left(\frac{j}{1 + j} \right)$$

$$= \frac{12j}{1 + j}$$

$$= 6(1 + j) V$$

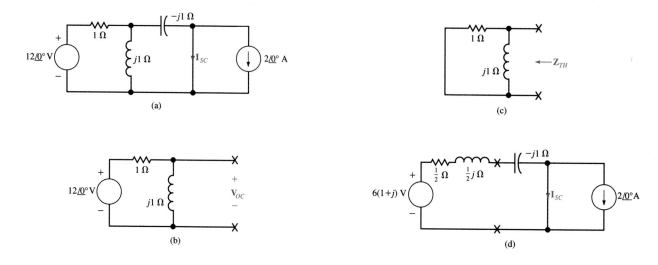

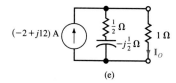

Figure 4.17   Circuits used to illustrate Norton's theorem.

The Thevenin equivalent impedance derived from Fig. 4.17(c) is

$$\mathbf{Z}_{TH} = \frac{(1)(j)}{1+j}$$

$$= \frac{1}{2} + j\frac{1}{2} \; \Omega$$

If we connect the Thevenin equivalent circuit to the original network, we obtain the circuit shown in Fig. 4.17(d). Compare this network to that shown in Fig. 4.15(c). From this latter network we find that

$$\mathbf{I}_{sc} = \frac{6(1+j)}{\frac{1}{2} - j\frac{1}{2}} - 2\angle 0°$$

$$= (-2 + j12) \; \text{A}$$

The Thevenin equivalent impedance was found earlier to be

$$\mathbf{Z}_{TH} = \frac{1}{2} - j\frac{1}{2} \; \Omega$$

Now, if the Norton equivalent, consisting of the short-circuit current source in parallel with the Thevenin equivalent impedance, is connected to the 1-ohm resistor as shown in

Fig. 4.17(e), we find that the current I_0 is obtained by current division as

$$I_0 = (-2 + j12)\left[\frac{\left(\frac{1}{2} - j\frac{1}{2}\right)}{\frac{1}{2} - j\frac{1}{2} + 1}\right]$$

$$= \left(\frac{8}{5} + j\frac{26}{5}\right) \text{ A}$$

■

DRILL EXERCISE

D4.12. Use Thevenin's theorem to find V_0 in problem D4.10.

4.13

Nonsinusoidal Steady State Response

Thus far the signals we have considered have either been a dc or an ac sinusoidal signal with a specific frequency. However, in electrical engineering we encounter a variety of periodic signals, many of which are nonsinusoidal.

In general, a periodic signal is one that satisfies the relationship

$$f(t) = f(t + nT) \quad n = \pm1, \pm2, \pm3, \cdots$$

for every value of t where T is the period. For example, the pulse train shown in Fig. 4.18(a) is used as a timing marker in digital systems, and the saw tooth waveform shown in Fig. 4.18(b) is used to drive the beam across the cathode ray tube in an oscilloscope.

At this point, if we could somehow decompose a nonsinusoidal periodic signal into a set of sinusoidal periodic signals of different frequencies, our background would permit us to determine the steady state response. The vehicle we will employ to accomplish this is the *Fourier Series,* which is based upon the work of Jean Baptiste Joseph Fourier (1768–1830). Although our analysis will be confined to electric circuits, it is important to point out that the techniques are applicable to a wide range of engineering problems. In fact, it was Fourier's work in heat flow that led to the techniques which will be presented here.

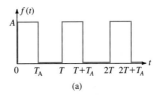

(a)

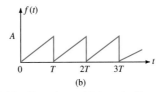

(b)

Figure 4.18 Nonsinusoidal periodic waveforms.

In his work, Fourier demonstrated that a periodic function $f(t)$ could be expressed as a sum of sinusoidal functions. Therefore, given this result and the fact that if a periodic function is expressed as a sum of linearly independent functions, and each function in the sum is periodic with the same period, then the function $f(t)$ can be expressed in the form

$$f(t) = a_0 + \sum_{n=1}^{\infty} D_n \cos(n\omega_0 t + \theta_n) \tag{4.43}$$

where $\omega_0 = 2\pi/T$ and a_0 is the average value of the waveform. An examination of this expression illustrates that all sinusoidal waveforms that are periodic with period T have been included. For example, for $n = 1$, one cycle covers T seconds and $D_1 \cos(\omega_0 t + \theta_1)$ is called the fundamental. For $n = 2$, two cycles fall within T seconds and the term $D_2 \cos(2\omega_0 t + \theta_2)$ is called the second harmonic. In general, for $n = k$, k cycles fall within T seconds, $D_k \cos(k\omega_0 t + \theta_k)$ is the kth harmonic term, and $D_k \angle \theta_k$ is the *phasor* for this term.

Since the function $\cos(n\omega_0 t + \theta_n)$ can be written in exponential form using Euler's identity or as a sum of cosine and sine terms of the form $\cos n\omega_0 t$ and $\sin n\omega_0 t$, the series in Eq. (4.43) can be written as

$$f(t) = a_0 + \sum_{\substack{n=-\infty \\ n \neq 0}}^{\infty} \mathbf{c}_n e^{jn\omega_0 t} = \sum_{n=-\infty}^{\infty} \mathbf{c}_n e^{jn\omega_0 t} \tag{4.44}$$

$$= a_0 + \sum_{n=1}^{\infty} (a_n \cos n\omega_0 t + b_n \sin n\omega_0 t) \tag{4.45}$$

The approach we will take will be to represent a nonsinusoidal periodic input by a sum of complex exponential functions, which because of Euler's identity is equivalent to a sum of sines and cosines. We will then use (1) the superposition property of linear systems and (2) our knowledge that the steady state response of a time-invariant linear system to a sinusoidal input of frequency ω_0 is a sinusoidal function of the same frequency to determine the response of such a system.

In order to illustrate the manner in which a nonsinusoidal periodic signal can be represented by a Fourier series, consider the periodic function shown in Fig. 4.19(a). In Fig. 4.19 we can see the impact of using a specific number of terms in the series to represent the original function. Note that the series more closely represents the original function as we employ more and more terms.

Exponential Fourier Series

Any physically realizable periodic signal may be represented over the interval $t_1 < t < t_1 + T$ by the *exponential Fourier series*

$$f(t) = \sum_{n=-\infty}^{\infty} \mathbf{c}_n e^{jn\omega_0 t} \tag{4.46}$$

where the $\mathbf{c}_n$ are the complex (phasor) Fourier coefficients. These coefficients are derived as follows. Multiplying both sides of Eq. (4.46) by $e^{-jk\omega_0 t}$ and integrating over the inter-

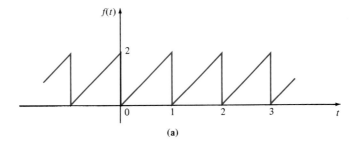

(a)

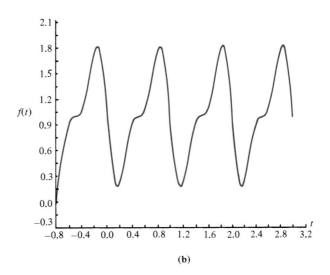

(b)

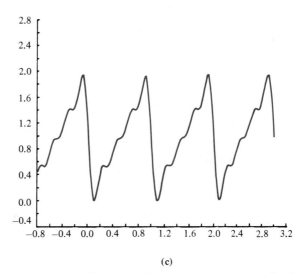

(c)

Figure 4.19 Periodic function (a) and its representation by a fixed number of Fourier series terms: (b) 2 terms; (c) 4 terms; (d) 100 terms. (continues)

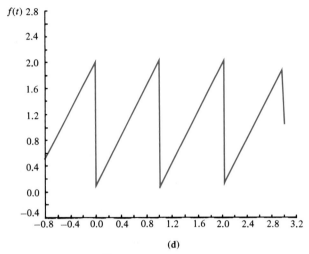

(d)

Figure 4.19   (cont)

val t_1 to $t_1 + T$, we obtain

$$\int_{t_1}^{t_1+T} f(t)e^{-jk\omega_0 t}\, dt = \int_{t_1}^{t_1+T} \left(\sum_{n=-\infty}^{\infty} c_n e^{jn\omega_0 t} \right) e^{-jk\omega_0 t}\, dt$$

since

$$\int_{t_1}^{t_1-T} e^{j(n-k)\omega_0 t}\, dt = \begin{cases} 0 & \text{for } n \neq k \\ T & \text{for } n = k \end{cases}$$

Therefore, the Fourier coefficients are defined by the equation

$$c_n = \frac{1}{T_0} \int_{t_1}^{t_1+T} f(t)e^{-jn\omega_0 t}\, dt \qquad (4.47)$$

The following example illustrates the manner in which we can represent a periodic signal by an exponential Fourier series.

EXAMPLE 4.14

We wish to determine the exponential Fourier series for the periodic voltage waveform shown in Fig. 4.20.

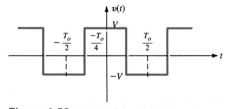

Figure 4.20   Periodic voltage waveform.

The Fourier coefficients are determined using Eq. (4.47) by integrating over one complete period of the waveform.

$$
\begin{aligned}
\mathbf{c_n} &= \frac{1}{T_0} \int_{-T/2}^{T/2} f(t)e^{-jn\omega_0 t}\, dt \\[2mm]
&= \frac{1}{T_0}\int_{-T/2}^{-T/4} -Ve^{-jn\omega_0 t}\, dt + \int_{-T/4}^{T/4} Ve^{-jn\omega_0 t}\, dt + \int_{T/4}^{T/2} -Ve^{-jn\omega_0 t} \\[2mm]
&= \frac{V}{jn\omega_0 T}\left[+e^{-jn\omega_0 t}\Big|_{-T/2}^{-T/4} - e^{-jn\omega_0 t}\Big|_{-T/4}^{T/4} + e^{jn\omega_0 t}\Big|_{T/4}^{T/2} \right] \\[2mm]
&= \frac{V}{jn\omega_0 T}\left(2e^{jn\pi/2} - 2e^{-jn\pi/2} + e^{-jn\pi} - e^{+jn\pi} \right) \\[2mm]
&= \frac{V}{n\omega_0 T}\left[4\sin\frac{n\pi}{2} - 2\sin(n\pi) \right] \\[2mm]
&= 0 \qquad\qquad \text{for } n \text{ even} \\[2mm]
&= \frac{2V}{n\pi}\sin\frac{n\pi}{2} \qquad \text{for } n \text{ odd}
\end{aligned}
$$

Since $\mathbf{c}_0$ corresponds to the average value of the waveform $\mathbf{c}_0 = 0$, this term can also be evaluated using the original equation for $\mathbf{c}_n$. Therefore

$$
v(t) = \sum_{\substack{n=-\infty \\ n\neq 0 \\ n \text{ odd}}}^{\infty} \frac{2V}{n\pi}\sin\frac{n\pi}{2}\, e^{jn\omega_0 t}
$$

This equation can be written as

$$
v(t) = \sum_{\substack{n=1 \\ n \text{ odd}}}^{\infty} \frac{2V}{n\pi}\sin\frac{n\pi}{2}\, e^{jn\omega_0 t} + \sum_{\substack{n=-1 \\ n \text{ odd}}}^{-\infty} \frac{2V}{n\pi}\sin\frac{n\pi}{2}\, e^{jn\omega_0 t}
$$

$$
= \sum_{\substack{n=1 \\ n \text{ odd}}}^{\infty} \left(\frac{2V}{n\pi}\sin\frac{n\pi}{2} \right)e^{jn\omega_0 t} + \left(\frac{2V}{n\pi}\sin\frac{n\pi}{2} \right)^{*} e^{-jn\omega_0 t}
$$

Since a number plus its complex conjugate is equal to two times the real part of the number, $v(t)$ can be written as

$$
v(t) = \sum_{\substack{n=1 \\ n \text{ odd}}}^{\infty} 2Re\left(\frac{2V}{n\pi}\sin\frac{n\pi}{2}\, e^{jn\omega_0 t} \right)
$$

or

$$
v(t) = \sum_{\substack{n=1 \\ n \text{ odd}}}^{\infty} \frac{4V}{n\pi}\sin\frac{n\pi}{2}\cos n\omega_0 t
$$

Note that this same result could have been obtained by integrating over the interval $-T/4$ to $3T/4$ or any other one period interval.   ■

D4.13. Find the Fourier coefficients for the waveform in Fig. D4.13.

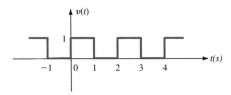

Figure D4.13

Ans: $c_n = \dfrac{1 - e^{-jn\pi}}{j2\pi n}$

A short listing of the Fourier series for a number of common waveforms is shown in Table 4.1. These series can be derived using the approach illustrated in Example 4.14.

Network Response

If a periodic signal is applied to a network, the steady state voltage or current response at some point in the circuit can be found in the following manner. First, we represent the periodic forcing function by a Fourier series. If the input forcing function for a network is a voltage, the input can be expressed in the form

$$v(t) = v_0 + v_1(t) + v_2(t) + \cdots$$

and therefore represented in the time domain as shown in Fig. 4.21. Each source has its

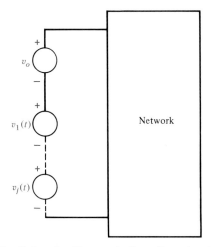

Figure 4.21   Network with a periodic voltage forcing function.

Table 4.1. *Fourier Series for Some Common Waveforms*

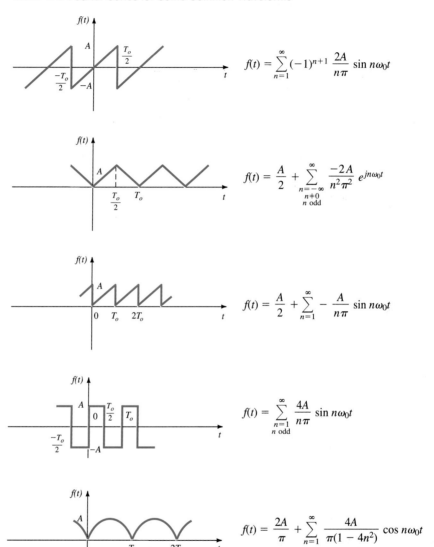

$$f(t) = \sum_{n=1}^{\infty} (-1)^{n+1} \frac{2A}{n\pi} \sin n\omega_0 t$$

$$f(t) = \frac{A}{2} + \sum_{\substack{n=-\infty \\ n \neq 0 \\ n \text{ odd}}}^{\infty} \frac{-2A}{n^2 \pi^2} e^{jn\omega_0 t}$$

$$f(t) = \frac{A}{2} + \sum_{n=1}^{\infty} - \frac{A}{n\pi} \sin n\omega_0 t$$

$$f(t) = \sum_{\substack{n=1 \\ n \text{ odd}}}^{\infty} \frac{4A}{n\pi} \sin n\omega_0 t$$

$$f(t) = \frac{2A}{\pi} + \sum_{n=1}^{\infty} \frac{4A}{\pi(1 - 4n^2)} \cos n\omega_0 t$$

own amplitude and frequency. Next we determine the response due to each component of the input Fourier series; that is, we use phasor analysis in the frequency domain to determine the network response due to each source. The network response due to each source in the frequency domain is then transformed to the time domain. Finally, we add the time domain solutions due to each source using the principle of superposition to obtain the Fourier series for the total **steady state** network response.

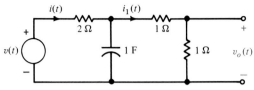

Figure 4.22   *RC* circuit employed in Example 4.15.

EXAMPLE 4.15

We wish to determine the steady state voltage $v_0(t)$ in Fig. 4.22 if the input voltage $v(t)$ is given by the expression

$$v(t) = 7.5 \cos(2t - 122°) + 2.2 \cos(6t - 102°) + 1.3 \cos(10t - 97°)$$
$$0.91 \cos(14t - 95°) + \cdots$$

Note that this source has no constant term, and therefore its dc value is zero.

From the network we find that

$$\mathbf{I} = \frac{\mathbf{V}}{2 + \dfrac{2/j\omega}{2 + 1/j\omega}} = \frac{\mathbf{V}(1 + 2j\omega)}{4 + 4j\omega}$$

$$\mathbf{I}_1 = \frac{\mathbf{I}(1/j\omega)}{2 + 1/j\omega} = \frac{\mathbf{I}}{1 + 2j\omega}$$

$$\mathbf{V}_0 = (1)\mathbf{I}_1 = 1 \cdot \frac{\mathbf{V}(1 + 2j\omega)}{4 + 4j\omega} \frac{1}{1 + 2j\omega} = \frac{\mathbf{V}}{4 + 4j\omega}$$

Therefore, since $\omega_0 = 2$

$$\mathbf{V}_0(n) = \frac{\mathbf{V}(n)}{4 + j8n}$$

The individual components of the output due to the components of the input source are then

$$\mathbf{V}_0(\omega_0) = \frac{7.5 \angle -122°}{4 + j8} = 0.84 \angle -185.4° \text{ V}$$

$$\mathbf{V}_0(3\omega_0) = \frac{2.2 \angle -102°}{4 + j24} = 0.09 \angle -182.5° \text{ V}$$

$$\mathbf{V}_0(5\omega_0) = \frac{1.3 \angle -97°}{4 + j40} = 0.03 \angle -181.3° \text{ V}$$

$$\mathbf{V}_0(7\omega_0) = \frac{0.91 \angle -95°}{4 + j56} = 0.016 \angle -181° \text{ V}$$

Hence, the steady state output voltage $v_0(t)$ can be written as

$$v_0(t) = 0.84 \cos(2t - 185.4°) + 0.09 \cos(6t - 182.5°)$$
$$+ 0.03 \cos(10t - 181.3°) + 0.016 \cos(14t - 181°) + \cdots \text{ V} \quad \blacksquare$$

DRILL EXERCISE

D4.14. Determine the expression for the steady state current $i(t)$ in Fig. D4.14 if the input voltage $V_s(t)$ is given by the expression

$$V_s(t) = 10 + 8\cos(2t) + 6\cos(4t - 60°) + 4\cos(6t - 45°) \text{ V}$$

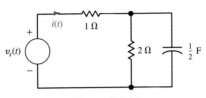

Figure D4.14

Ans: $i(t) = 3.33 + 4.96\cos(2t + 29.74°) + 4.96\cos(4t - 37.17°) +$
$3.63\cos(6t - 27.89°) + \cdots \text{ A}$

4.14
Summary

- The frequency of the U.S. and Canadian power grid is 60 Hz.
- Two sinusoidal functions $x(t) = A\sin(\omega t + \theta)$ and $y(t) = B\sin(\omega t + \phi)$ are said to be in phase if $\theta = \phi$, and out of phase if $\theta \neq \phi$.
- A sinusoidal time function $x(t) = A\cos(\omega t + \theta)$ can be represented as a phasor, $\mathbf{X} = A\angle\theta$, which consists of a magnitude A and a phase angle θ. The phasor is measured with respect to a cosine signal and written in boldface type.
- Phasor manipulation is performed in complex algebra, and phasor analysis is called a frequency domain analysis.
- The phasor relationship for a resistor, capacitor, and inductor are $\mathbf{V} = R\mathbf{I}$, $\mathbf{I} = j\omega C\mathbf{V}$, and $\mathbf{V} = j\omega L\mathbf{I}$.
- Impedance is defined as the ratio of the phasor voltage to the phasor current. Admittance is the reciprocal of impedance.
- Kirchhoff's laws hold in the frequency domain.
- The wye-delta transformation is applicable in the frequency domain.
- All the analysis techniques, such as nodal analysis, and the network theorems that are used in dc analysis, are valid in the frequency domain.
- A periodic nonsinusoidal function can be expressed as a sum of sinusoidal functions called a Fourier series.
- Superposition can be used to determine the response of a network to a nonsinusoidal input function.

PROBLEMS

4.1. Given the voltages $v_1(t)$ and $v_2(t)$, determine their frequency and the phase angle between them.

$$v_1(t) = 12\cos(2513t - 70°) \text{ V}$$
$$v_2(t) = 4\cos(2513t - 30°) \text{ V}$$

4.2. Determine the relative position of the two sine waves.

$$v_1(t) = 12\sin(377t - 45°)$$
$$v_2(t) = 6\sin(377t + 675°)$$

4.3. The current in a 4 Ω resistor is known to be $\mathbf{I} = 12\underline{/60°}$ A. Express the voltage across the resistor as a time function if the frequency of the current is 60 Hz.

4.4. The current in a 150 μF capacitor is $\mathbf{I} = 3.6\underline{/-145°}$ A. If the frequency of the current is 60 Hz, determine the voltage across the capacitor.

4.5. The voltage $v(t) = 100\cos(377t + 15°)$ V is applied to a 100 μF capacitor. Find the resultant phasor current.

4.6. Determine the impedance, in polar form, for the following elements.
(a) A 0.2 H inductor and a 300 Ω resistor in parallel at 377 rad/s.
(b) A 125 μF capacitor and a 40 Ω resistor in parallel at 60 Hz.
(c) A 1 H inductor and a 50 μF capacitor in parallel at 100 rad/s.

4.7. If the equivalent impedance of a network is $\mathbf{Z} = 10\underline{/30°}$ Ω, compute the equivalent admittance and draw the equivalent circuits.

4.8. Given the following values of $\mathbf{Z}_1$, $\mathbf{Z}_2$, and $\mathbf{Z}_3$, determine the equivalent impedance of the circuit in Fig. P4.8.

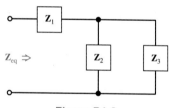

Figure P4.8

$\mathbf{Z}_1 = 5\underline{/40°}$Ω, $\mathbf{Z}_2 = 3\underline{/-35°}$Ω, $\mathbf{Z}_3 = 7\underline{/135°}$Ω.

4.9. Calculate the equivalent impedance at terminals A-B in the circuit shown in Fig. P4.9.

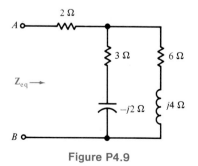

Figure P4.9

4.10. Find $\mathbf{Z}_T$ in the network in Fig. P4.10.

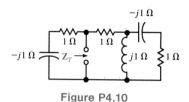

Figure P4.10

4.11. Calculate $\mathbf{Y}_{eq}$ as shown in Fig. P4.11.

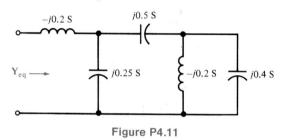

Figure P4.11

4.12. Calculate the equivalent impedance for the circuit in Fig. P4.12 and use this value to determine the voltage $\mathbf{V}_S$ if the current is $\mathbf{I} = 10\underline{/30°}$ A.

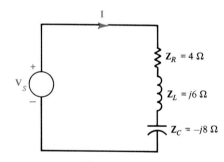

Figure P4.12

4.13. Calculate the equivalent admittance $\mathbf{Y}_P$ for the network in Fig. P4.13 and use it to determine the current $\mathbf{I}$ if $\mathbf{V}_S = 60\underline{/45°}$ V.

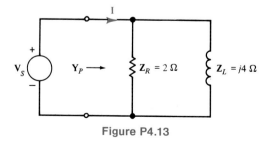

Figure P4.13

4.14. Determine the equivalent impedance of the circuit in Fig. P4.14 at 60 Hz. Compute the voltage $\mathbf{V}_s$ if the current $\mathbf{I}$ is known to be $\mathbf{I} = 0.5\,\angle{-22.98°}$ A.

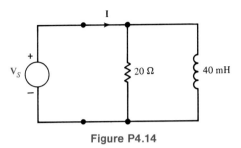

Figure P4.14

4.15. Compute the voltage $v(t)$ in the network in Fig. P4.15.

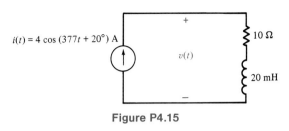

Figure P4.15

4.16. Determine the voltage across the current source in the network in Fig. P4.16.

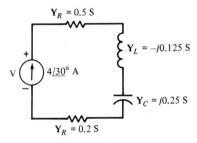

Figure P4.16

4.17. Find the current i(t) in the network in Fig. P4.17.

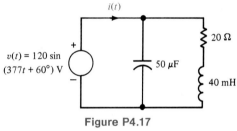

Figure P4.17

4.18. Find the current $\mathbf{I}$ in the network in Fig. P4.18.

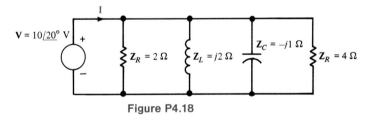

Figure P4.18

4.19. Compute the voltage $\mathbf{V}_0$ in the circuit shown in Fig. P4.19.

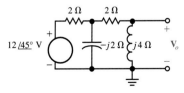

Figure P4.19

4.20. Find the currents I_1, I_2, and I_3 in the network in Fig. P4.20.

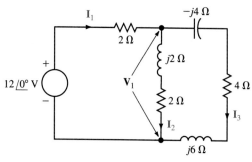

Figure P4.20

4.21. Calculate all the currents in the circuit shown in Fig. P4.21.

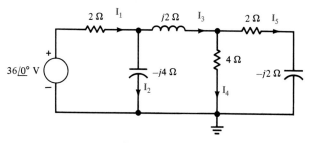

Figure P4.21

4.22. Find I_0 in the network in Fig. P4.22.

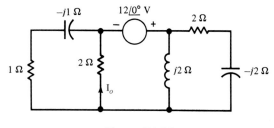

Figure P4.22

4.23. We wish to compute the voltage V_0 in the network in Fig. P4.23.

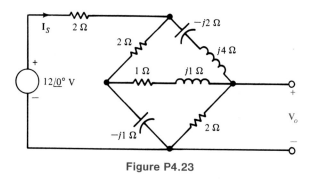

Figure P4.23

4.24. Draw a phasor diagram illustrating all currents and voltages for the network in Fig. P4.24.

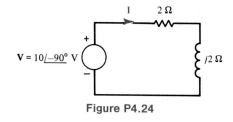

Figure P4.24

4.25. Draw a phasor diagram illustrating all currents and voltages for the network in Fig. P4.25.

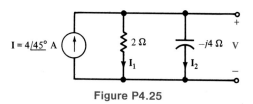

Figure P4.25

4.26. If $I_0 = 4 \angle 0°$ A, find V_S in Fig. P4.26.

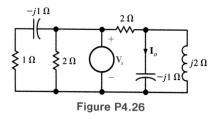

Figure P4.26

4.27. In the network in Fig. P4.27, V_0 is known to be $8 \angle 45°$ V. Compute V_S.

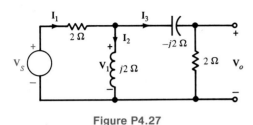

Figure P4.27

4.28. The current in the 2 Ω resistor in the circuit shown in Fig. P4.28 is $I_0 = 4 \angle 30°$A. Compute V_S.

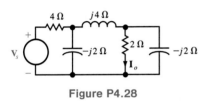

Figure P4.28

4.29. Use nodal equations to find the current in the inductor in the circuit shown in Fig. P4.29.

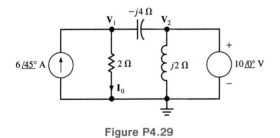

Figure P4.29

4.30. Use nodal analysis to find I_0 in the circuit shown in Fig. P4.30.

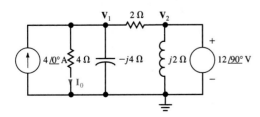

Figure P4.30

4.31. Determine I_0 in the circuit shown in Fig. P4.31 using nodal analysis.

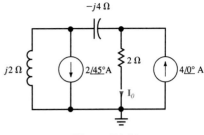

Figure P4.31

4.32. Use nodal analysis to find V_0 in the network in Fig. P4.32.

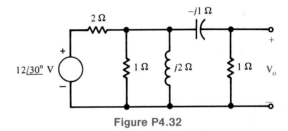

Figure P4.32

4.33. Find the current in the inductor in the circuit shown in Fig. P4.33, using mesh analysis.

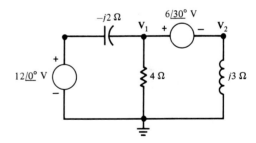

Figure P4.33

4.34. Use mesh equations to find V_0 in the network in Fig. P4.34.

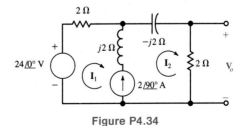

Figure P4.34

4.35. Use mesh analysis to find V_0 in the circuit shown in Fig. P4.35.

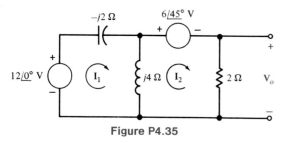

Figure P4.35

4.36. Use mesh analysis to find V_0 in the circuit shown in Fig. P4.36.

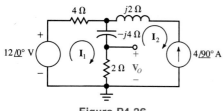

Figure P4.36

4.37. Given the network in Fig. P4.37, write the mesh equations for I_1 and I_2.

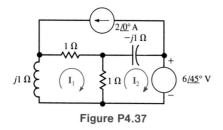

Figure P4.37

4.38. Use superposition to find V_0 in the network in Fig. P4.38.

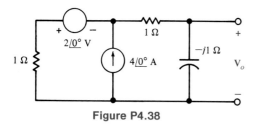

Figure P4.38

4.39. Solve problem 4.31 using superposition.

4.40. Solve problem 4.35 using superposition.

4.41. Solve problem 4.36 using superposition.

4.42. Determine V_x in Fig. P4.42 using superposition.

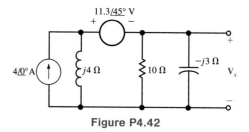

Figure P4.42

4.43. Solve problem 4.33 using Thevenin's theorem.

4.44. Solve problem 4.35 using Thevenin's theorem.

4.45. Solve problem 4.36 using Thevenin's theorem.

4.46. Solve problem 4.38 using Thevenin's theorem.

4.47. Solve problem 4.31 using Thevenin's theorem.

4.48. Find the exponential Fourier series for the periodic function shown in Fig. P4.48.

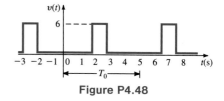

Figure P4.48

4.49. Find the Fourier coefficients for the waveform in Fig. P4.49.

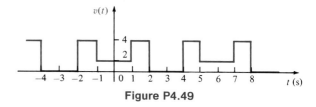

Figure P4.49

4.50. Derive the exponential Fourier series for the periodic signal shown in Fig. P4.50.

Figure P4.50

4.51. If the input voltage in Fig. P4.51 is

$$v_s(t) = 1 - \frac{2}{\pi} \sum_{n=1}^{\infty} \frac{1}{n} \sin 0.2 \, \pi n t \text{ V}$$

find the expression for the steady state current $i_0(t)$.

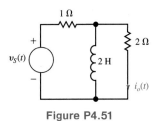

Figure P4.51

4.52. Determine the first three terms of the steady state voltage $v_0(t)$ in Fig. P4.52 if the input voltage is a periodic signal of the form

$$v(t) = \frac{1}{2} + \sum_{n=1}^{\infty} \frac{1}{n\pi} (\cos n\pi - 1) \sin nt \text{ V}$$

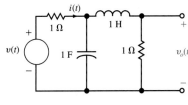

Figure P4.52

Steady State Power Analysis

5.1
Introduction

In the preceding chapters, we were primarily concerned with determining the voltage or current at some point in a network. Of equal importance to us in many situations is the power that is supplied or absorbed by some element. Therefore, in this chapter we explore the many ramifications of power in ac circuits. We will first treat the single phase case, that is, the types of networks we have examined thus far. Then we will investigate three-phase circuits, which are typically employed in high-power applications.

5.2
Instantaneous Power

By employing the sign convention adopted in the earlier chapters, we can compute the *instantaneous power* supplied or absorbed by any device as the product of the instantaneous voltage across the device and the instantaneous current through it. Consider the circuit shown in Fig. 5.1. In general, the steady state voltage and current for the network can be written as

$$v(t) = V_M \cos(\omega t + \theta_v) \tag{5.1}$$
$$i(t) = I_M \cos(\omega t + \theta_i) \tag{5.2}$$

The instantaneous power is then

$$p(t) = v(t)i(t) \tag{5.3}$$
$$= V_M I_M \cos(\omega t + \theta_v) \cos(\omega t + \theta_i)$$

Employing the following trigonometric identity

$$\cos\phi_1 \cos\phi_2 = \tfrac{1}{2}[\cos(\phi_1 - \phi_2) + \cos(\phi_1 + \phi_2)] \tag{5.4}$$

the instantaneous power can be written as

$$p(t) = \frac{V_M I_M}{2} [\cos(\theta_v - \theta_i) + \cos(2\omega t + \theta_v + \theta_i)] \qquad (5.5)$$

Note that the instantaneous power consists of two terms. The first term is a constant (i.e., it is time independent); and the second term is a cosine wave of twice the excitation frequency. We will examine this equation in more detail in the next section.

EXAMPLE 5.1

The circuit in Fig. 5.1 has the following parameters: $v(t) = 4\cos(\omega t + 60°)$ V and $\mathbf{Z} = 2\angle 30°$ Ω. We wish to determine equations for the current and the instantaneous power as a function of time, and plot these functions with the voltage on a single graph for comparison.

Since

$$\mathbf{I} = \frac{4\angle 60°}{2\angle 30°}$$

$$= 2\angle 30° \text{ A}$$

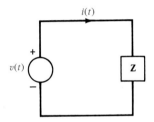

Figure 5.1   Simple ac network.

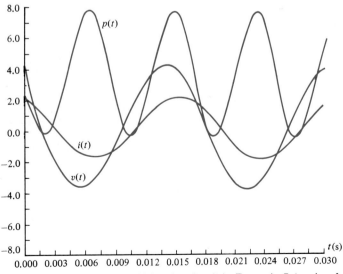

Figure 5.2   Plots of $v(t)$, $i(t)$, and $p(t)$ for the circuit in Example 5.1 using $f = 60$ Hz.

then

$$i(t) = 2 \cos(\omega t + 30°) \text{ A}$$

From Eq. (5.5)

$$p(t) = 4[\cos(30°) + \cos(2\omega t + 90°)]$$
$$= 3.46 + 4\cos(2\omega t + 90°) \text{ W}$$

A plot of this function, together with plots of the voltage and current, are shown in Fig. 5.2. As can be seen in the figure, the instantaneous power has an average value, and the frequency is twice that of the voltage or current. ∎

5.3

Average Power

The average value of any periodic waveform (e.g., a sinusoidal function) can be computed by integrating the function over a complete period and dividing this result by the period. Therefore, if the voltage and current are given by Eqs. (5.1) and (5.2), respectively, the *average power* is

$$P = \frac{1}{T} \int_{t_0}^{t_0+T} p(t) \, dt$$
$$= \frac{1}{T} \int_{t_0}^{t_0+T} V_M I_M \cos(\omega t + \theta_v) \cos(\omega t + \theta_i) \, dt \qquad (5.6)$$

where t_0 is arbitrary, $T = 2\pi/\omega$ is the period of the voltage or current, and P is measured in watts.

Employing Eq. (5.5) in the expression (5.6), we obtain

$$P = \frac{1}{T} \int_{t_0}^{t_0+T} \frac{V_M I_M}{2} [\cos(\theta_v - \theta_i) + \cos(2\omega t + \theta_v + \theta_i)] \, dt \qquad (5.7)$$

The first term is independent of t, and therefore a constant in the integration. Integrating the constant over the period and dividing by the period simply results in the original constant. The second term is a cosine wave. It is well known that the average value of a cosine wave over one complete period or an integral number of periods is zero. Therefore, Eq. (5.7) reduces to

$$P = \tfrac{1}{2} V_M I_M \cos(\theta_v - \theta_i) \qquad (5.8)$$

Note that since $\cos(-\theta) = \cos(\theta)$, the argument for the cosine function can be either $\theta_v - \theta_i$ or $\theta_i - \theta_v$. In addition, note that $\theta_v - \theta_i$ is the angle of the circuit impedance as shown in Fig. 5.1. Therefore, for a purely resistive circuit

$$P = \tfrac{1}{2} V_M I_M \qquad (5.9)$$

and for a purely reactive circuit

$$P = \tfrac{1}{2} V_M I_M \cos(90°)$$
$$= 0$$

Because purely reactive impedances absorb no average power, they are often called lossless elements. The purely reactive network operates in a mode in which it stores energy over one part of the period and releases it over another.

EXAMPLE 5.2

For the circuit shown in Fig. 5.3 we wish to determine both the total average power absorbed and the total average power supplied. From the figure we note that

$$\mathbf{I}_1 = \frac{12\underline{/45°}}{4} = 3\underline{/45°}\,A$$

$$\mathbf{I}_2 = \frac{12\underline{/45°}}{2 - j1} = \frac{12\underline{/45°}}{2.24\underline{/-26.57°}} = 5.37\underline{/71.57°}\,A$$

and therefore

$$\mathbf{I} = \mathbf{I}_1 + \mathbf{I}_2$$
$$= 3\underline{/45°} + 5.37\underline{/71.57°}$$
$$= 8.16\underline{/62.08°}\,A$$

The average power absorbed in the 4 Ω resistor is

$$P_{4\Omega} = \tfrac{1}{2}V_M I_M = \tfrac{1}{2}(12)(3) = 18\,W$$

The average power absorbed in the 2 Ω resistor is

$$P_{2\Omega} = \tfrac{1}{2}I_M^2 R = \tfrac{1}{2}(5.37)^2(2) = 28.8\,W$$

Therefore, the total average power absorbed is

$$P_A = 18 + 28.8 = 46.8\,W$$

Note that we could have calculated the power absorbed in the 2 Ω resistor using $\tfrac{1}{2}V_M^2/R$ if we had first calculated the voltage across the 2 Ω resistor.

The total average power supplied by the source is

$$P_s = + \tfrac{1}{2}V_M I_M \cos(\theta_v - \theta_i)$$
$$= + \tfrac{1}{2}(12)\,(8.16)\cos(45° - 62.08°)$$
$$= + 46.8\,W$$

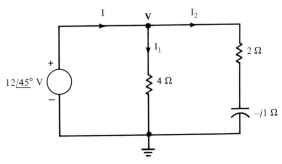

Figure 5.3 Example circuit for illustrating a power balance.

Thus, the total average power supplied is, of course, equal to the total average power absorbed. ∎

D5.1. Find the average power absorbed by each resistor in the network in Fig. D5.1.

Ans: $P_{2\,\Omega} = 7.18$ W, $P_{4\Omega} = 7.14$ W.

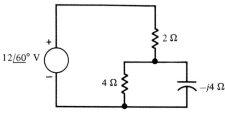

Figure D5.1

D5.2. Given the network in Fig. D5.2, find the average power absorbed by each passive circuit element and the total average power supplied by the current source.

Ans: $P_{3\Omega} = 56.62$ W, $P_{4\Omega} = 33.95$ W, $P_L = 0$, $P_{cs} = -90.50$ W.

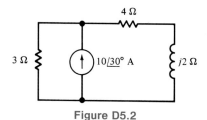

Figure D5.2

5.4
Effective or RMS Values

In the preceding sections of this chapter we have shown that the average power absorbed by a resistive load is directly dependent on the type, or types, of sources that are delivering power to the load. For example, if the source was dc, the average power absorbed was I^2R; and if the source was sinusoidal, the average power was $\frac{1}{2}I_M^2R$. Although these two types of waveforms are extremely important, they are by no means the only waveforms we will encounter in circuit analysis. Therefore, a technique by which we can compare the effectiveness of different sources in delivering power to a resistive load would be quite useful.

In order to accomplish this comparison, we define what is called the *effective value of a periodic waveform.* We define the effective value of a periodic current as a constant, or dc value, which delivers the same average power to a resistor R. Let us call the constant current I_{eff}. Then the average power delivered to a resistor as a result of this current is

$$P = I_{eff}^2 R$$

Similarly, the average power delivered to a resistor by a periodic current $i(t)$ is

$$P = \frac{1}{T} \int_{t_0}^{t_0 + T} i^2(t)R \, dt \tag{5.10}$$

Equating these two expressions, we find that

$$I_{eff} = \sqrt{\frac{1}{T} \int_{t_0}^{t_0+T} i^2(t) \, dt} \tag{5.11}$$

Note that this effective value is found by first determining the square of the current, then computing the average or mean value, and finally taking the square root. Thus, in "reading" the mathematical Eq. (5.11), we are determining the *root-mean-square* which we abbreviate as *rms*, and therefore I_{eff} is called I_{rms}.

EXAMPLE 5.3

The waveform for the current in a 5 Ω resistor is shown in Fig. 5.4. We wish to find the average power absorbed by the resistor.

The equations for the current within the time interval $0 \le t \le 4$ seconds are

$$i(t) = \begin{cases} 2t & 0 \le t \le 2 \text{ sec} \\ -4 & 2 \le t \le 4 \text{ sec} \end{cases}$$

Therefore, the rms value of the current is

$$\begin{aligned}
I_{rms} &= \left[\frac{1}{4} \left(\int_0^2 (2t)^2 dt + \int_2^4 (-4)^2 dt \right) \right]^{1/2} \\
&= \left[\frac{1}{4} \left(\frac{4t^3}{3} \Big|_0^2 + 16t \Big|_2^4 \right) \right]^{1/2} \\
&= 3.27 \text{ A rms}
\end{aligned}$$

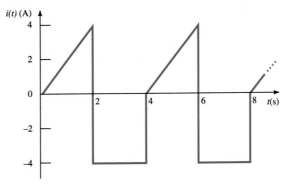

Figure 5.4   Waveform used in Example 5.3.

Then the average power absorbed by the 5 Ω resistor is

$$P = I_{rms}^2(R)$$
$$= (3.27)^2(5)$$
$$= 53.3 \text{ W}$$

■

EXAMPLE 5.4

We wish to compute the rms value of the waveform $i(t) = I_M\cos(\omega t - \theta)$, which has a period of $T = 2\pi/\omega$.

Substituting these expressions into Eq. (5.11) yields

$$I_{rms} = \left[\frac{1}{T}\int_0^T I_M^2\cos^2(\omega t - \theta)dt\right]^{\frac{1}{2}}$$

Using the trigonometric identity

$$\cos^2\phi = \tfrac{1}{2} + \tfrac{1}{2}\cos 2\phi$$

the equation above can be expressed as

$$I_{rms} = I_M\left[\frac{\omega}{2\pi}\int_0^{2\pi/\omega}[\tfrac{1}{2} + \tfrac{1}{2}\cos(2\omega t - 2\theta)]dt\right]^{\frac{1}{2}}$$

Since we know that the average or mean value of a cosine wave is zero

$$I_{rms} = I_M\left[\frac{\omega}{2\pi}\int_0^{2\pi/\omega}\frac{1}{2}\,dt\right]^{\frac{1}{2}}$$

$$= I_M\left[\frac{\omega}{2\pi}\left(\frac{t}{2}\right)\Big|_0^{2\pi/\omega}\right]^{\frac{1}{2}} = \frac{I_M}{\sqrt{2}}$$

Therefore, the rms value of a sinusoid is equal to the maximum value divided by the $\sqrt{2}$. Hence, a sinusoidal current with a maximum value of I_M delivers the same average power to a resistor R as a dc current with a value of $I_M/\sqrt{2}$. ■

Using the rms values for voltage and current, the average power can be written in general as

$$P = V_{rms}I_{rms}\cos(\theta_v - \theta_i) \tag{5.12}$$

The power absorbed by a resistor R is

$$P = I_{rms}^2R = \frac{V_{rms}^2}{R} \tag{5.13}$$

In dealing with voltages and currents in numerous electrical applications, it is important to know whether the values quoted are the maximum, average, or rms. For example, the normal 120 V ac electrical outlets have an rms value of 120 V, an average value of 0 V, and a maximum value, or amplitude, of $120\sqrt{2}$ V.

DRILL EXERCISES

D5.3 Compute the rms value of the voltage waveform shown in Fig. D5.3.

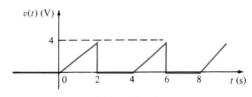

Figure D5.3

Ans: V_{rms} = 1.633 V rms.

D5.4 The current waveform in Fig. D5.4 is flowing through a 10 Ω resistor. Determine the average power delivered to the resistor.

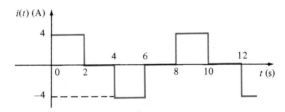

Figure D5.4

Ans: P = 80 W.

5.5

The Power Factor

The power factor is a very important quantity. Its importance stems in part from the economic impact it has on industrial users of large amounts of power. In this section we carefully define this term and then illustrate its significance via some practical examples. Earlier we showed that a load operating in the ac steady state is delivering an average power of

$$P = V_{rms}I_{rms} \cos(\theta_v - \theta_i)$$

We will now further define the terms in this important equation. The product $V_{rms}I_{rms}$ is referred to as the *apparent power*. Although the term $\cos(\theta_v - \theta_i)$ is a dimensionless quantity, and the units of P are watts, apparent power is normally stated in volt-amperes (VA) or kilovolt-amperes (kVA) in order to distinguish it from average power. We now define the *power factor* (PF) as the ratio of the average power to the apparent power, that is

$$PF = \frac{P}{V_{rms}I_{rms}} = \cos(\theta_v - \theta_i) \tag{5.14}$$

where

$$\cos(\theta_v - \theta_i) = \cos\theta_{ZL} \tag{5.15}$$

The angle $\theta_v - \theta_i = \theta_{ZL}$ is the phase angle of the load impedance and is often referred to as the *power factor angle*. The extreme positions for this angle correspond to a purely resistive load where $\theta_{ZL} = 0$, the PF is 1, and the purely reactive load where $\theta_{ZL} = \pm 90°$ and the PF is 0. It is, of course, possible to have a unity PF for a load containing R, L, and C elements if the values of the circuit elements are such that a zero phase angle is obtained at the particular operating frequency.

There is, of course, a whole range of power factor angles between $\pm 90°$ and $0°$. If the load is an equivalent RC combination, then the PF angle lies between the limits $-90° < \theta_{ZL} < 0°$. On the other hand, if the load is an equivalent RL combination, then the PF angle lies between the limits $0 < \theta_{ZL} < 90°$. Obviously, confusion in identifying the type of load could result, due to the fact that $\cos(+\theta_{ZL}) = \cos(-\theta_{ZL})$. To circumvent this problem, the PF is said to be either *leading* or *lagging,* where these two terms refer to the phase of the current with respect to the voltage. Since the current leads the voltage in an RC load, the load has a *leading* PF. In a similar manner, an RL load has a *lagging* PF; therefore, load impedances of $\mathbf{Z}_L = 1 - j1$ and $\mathbf{Z}_L = 2 + j1$ have power factors of $\cos(-45°) = 0.707$ leading and $\cos(26.59°) = 0.894$ lagging, respectively.

EXAMPLE 5.5

An industrial load consumes 88 kW at a PF of 0.707 lagging from a 480 V rms line. The transmission line resistance from the power company's transformer to the plant is 0.08 Ω. Let us determine the power that must be supplied by the power company (a) under present conditions, and (b) if the PF is somehow changed to 0.90 lagging.

(a) The equivalent circuit for these conditions is shown in Fig. 5.5. Using Eq. (5.14), the magnitude of the rms current into the plant is

$$I_{rms} = \frac{P_L}{(PF)(V_{rms})}$$

$$= \frac{(88)(10^3)}{(0.707)(480)}$$

$$= 259.3 \text{ A rms}$$

The power that must be supplied by the power company is

$$P_s = P_L + (0.08)I_{rms}^2$$

$$= 88{,}000 + (0.08)(259.3)^2$$

$$= 93.38 \text{ kW}$$

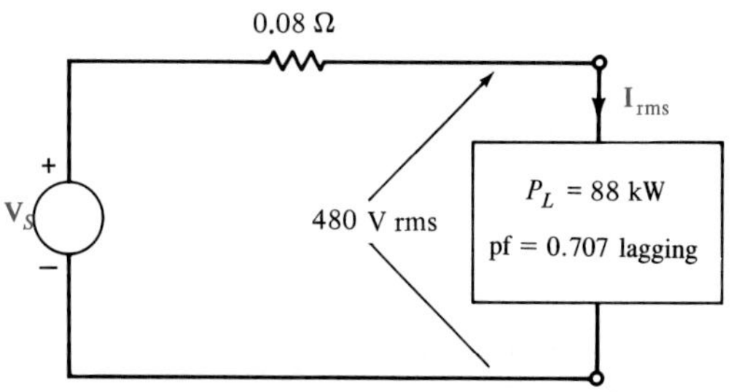

Figure 5.5 Example circuit for examining

(b) Suppose now that the PF is somehow changed to 0.90 lagging but the voltage remains constant at 480 V. The rms load current for this condition is

$$I_{rms} = \frac{P_L}{(\text{PF})(V_{rms})}$$

$$= \frac{(88)(10^3)}{(0.90)(480)}$$

$$= 203.7 \text{ A rms}$$

Under these conditions, the power company must generate

$$P_s = P_L + (0.08)I^2_{rms}$$

$$= 88{,}000 + (0.08)(203.7)^2$$

$$= 91.32 \text{ kW}$$

Note carefully the difference between the two cases. A simple change in the PF of the load from 0.707 lagging to 0.90 lagging has had an interesting effect. Note that in the first case the power company must generate 93.38 kW in order to supply the plant with 88 kW of power because the low power factor means that the line losses will be high—5.38 kW. However, in the second case the power company need only generate 91.32 kW in order to supply the plant with its required power, and the corresponding line losses are only 3.32 kW. ■

The example clearly indicates the economic impact of the load's power factor. A low power factor at the load means that the utility generators must be capable of carrying more current at constant voltage, and they must also supply power for higher $I^2_{rms}R$ line losses than would be required if the load's power factor were high. Since line losses represent energy expended in heat and benefit no one, the utility will insist that a plant maintain a high PF, typically 0.90 lagging or better, and adjust their rate schedule to penalize plants that do not conform to this requirement. In the next section we will demonstrate a simple and economical technique for achieving this power factor correction.

DRILL EXERCISE

D5.5 An industrial load consumes 100 kW at 0.707 PF lagging. The 60 Hz line voltage at the load is $480\underline{/0°}\text{V}$ rms. The transmission line resistance between the power company's transformer and the load is 0.1 Ω. Determine the power savings that could be obtained if the PF is changed to 0.94 lagging.

Ans: Power saved is 3.771 kW.

5.6
Complex Power

In our study of ac steady state power, it is convenient to introduce another quantity, which is commonly called *complex power*. To develop the relationship between this quantity and others we have presented in the preceding sections, consider the circuit shown in Fig. 5.6.

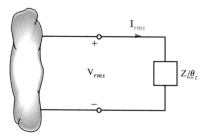

Figure 5.6   Circuit used to explain power relationships.

The complex power is defined to be

$$\mathbf{S} = \mathbf{V}_{rms}\mathbf{I}_{rms}^{*} \tag{5.16}$$

where $\mathbf{I}_{rms}^{*}$ refers to the complex conjugate of $\mathbf{I}_{rms}$; that is, if $\mathbf{I}_{rms} = I_{rms}\angle\,\theta_i = I_r + jI_i$, then $\mathbf{I}_{rms}^{*} = I_{rms}\angle -\theta_i = I_r - jI_i$. Complex power is then

$$\mathbf{S} = V_{rms}\angle\,\theta_v\, I_{rms}\angle -\theta_i = V_{rms}I_{rms}\angle\,\theta_v - \theta_i \tag{5.17}$$

or

$$\mathbf{S} = V_{rms}I_{rms}\cos(\theta_v - \theta_i) + jV_{rms}I_{rms}\sin(\theta_v - \theta_i) \tag{5.18}$$

where, of course, $\theta_v - \theta_i = \theta_z$. We note from Eq. (5.18) that the real part of the complex power is simply the *real* or *average power.* The imaginary part of **S** we call the *reactive* or *quadrature power.* Therefore, complex power can be expressed in the form

$$\mathbf{S} = P + jQ \tag{5.19}$$

where

$$P = Re(\mathbf{S}) = V_{rms}I_{rms}\cos(\theta_v - \theta_i) \tag{5.20}$$
$$Q = Im(\mathbf{S}) = V_{rms}I_{rms}\sin(\theta_v - \theta_i) \tag{5.21}$$

and as shown in Eq. (5.18), the magnitude of the complex power is what we have called the *apparent power,* and the phase angle for complex power is simply the *power factor angle.* Complex power, like apparent power, is measured in volt-amperes, real power is measured in watts, and in order to distinguish Q from other quantities, which in fact have the same dimensions, it is measured in volt-amperes reactive, or var.

Equations (5.20) and (5.21) can be written as

$$P = I_{rms}^2 Re\,(\mathbf{Z}) \tag{5.22}$$
$$Q = I_{rms}^2 Im\,(\mathbf{Z}) \tag{5.23}$$

and therefore Eq. (5.19) can be expressed as

$$\mathbf{S} = I_{rms}^2\mathbf{Z} \tag{5.24}$$

The relationships among **S**, P, and Q can be expressed via the diagrams shown in Figs. 5.7(a) and (b). In Fig. 5.7(a) we note the following conditions. If Q is positive, the load is inductive, the power factor is lagging, and the complex number **S** lies in the first quadrant. If Q is negative, the load is capacitive, the power factor is leading, and the complex number **S** lies in the fourth quadrant. If Q is zero, the load is resistive, the power factor

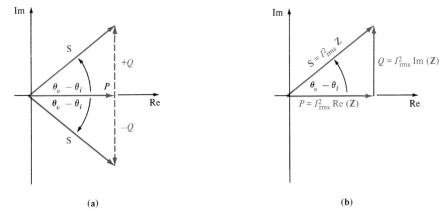

(a)   **(b)**

Figure 5.7   Diagram for illustrating power relationships.

is unity, and the complex number **S** lies along the positive real axis. Figure 5.7(b) illustrates the relationships expressed by Eqs.(5.22) to (5.24) for an inductive load.

EXAMPLE 5.6

A small manufacturing plant is located one mile down a transmission line which has a resistance of 0.1 Ω/mile. The line reactance is negligible. At the plant the line voltage is $480\underline{/0°}$ V rms, and the plant consumes 120 kW at 0.85 PF lagging. We wish to determine the voltage and power factor at the input of the transmission line using (a) a complex power approach, and (b) a circuit analysis approach.

(a) The problem is modeled as shown in Fig. 5.8(a). At the load, P_L = 120 kW and the power factor angle is

$$\theta_{ZL} = \cos^{-1}0.85$$
$$= 31.79°$$

The power triangle for the load is shown in Fig. 5.8(b). The trigonometric relationship among the parameters permits us to determine the unknowns S_L and Q_L, for example

$$\tan 31.79° = \frac{Q_L}{P_L}$$

and hence

$$Q_L = 74.364 \text{ kvar}$$

then

$$\mathbf{S}_L = P_L + jQ_L$$
$$= 141.180\underline{/31.79°} \text{ kVA}$$

The magnitude of the current is

$$I_{rms} = \frac{S_L}{V_{rms}}$$
$$= \frac{141,180}{480}$$
$$= 294.13 \text{ A rms}$$

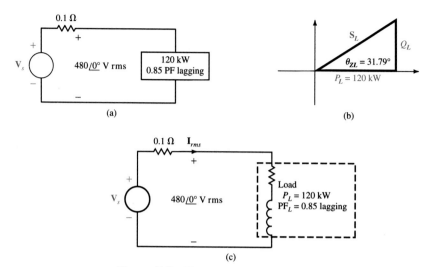

Figure 5.8   Diagrams used in Example 5.6.

The real power losses in the line are

$$P_{Line} = (294.13)^2(0.1)$$
$$= 8651.24 \text{ W}$$

Then the real power that must be supplied is

$$P_S = P_L + P_{Line}$$
$$= 128{,}651.24 \text{ W}$$

Therefore

$$\mathbf{S}_s = P_s + jQ_s$$
$$= 128{,}651.24 + j74{,}364$$
$$= 148{,}597.26 \underline{/30.03°} \text{ VA}$$

Hence, the input voltage is

$$V_s = \frac{S_s}{I_{rms}}$$
$$= 505.22 \text{ V rms}$$

and the power factor at the input is

$$PF_s = \cos 30.03°$$
$$= 0.866 \text{ lagging}$$

(b) From the previous analysis we know that

$$\mathbf{S}_L = 141.180 \underline{/31.79°} \text{ kVA}$$

and since

$$\mathbf{S}_L = \mathbf{V}_{rms}\mathbf{I}_{rms}^*$$

the line current is

$$\mathbf{I}_{rms} = 294.13 \angle -31.79° \text{ A rms}$$

Although the information is not needed for the solution, it is interesting to note that the equivalent impedance at the load is

$$\mathbf{Z}_L = \frac{480 \angle 0°}{294.13 \angle -31.79°}$$
$$= 1.39 + j0.086 \ \Omega$$

and the relationships

$$P_L = I_{rms}^2 Re(\mathbf{Z}_L)$$
$$Q_L = I_{rms}^2 Im(\mathbf{Z}_L)$$

are satisfied. The circuit diagram is shown in Fig. 5.8(c).
Applying KVL

$$\mathbf{V}_s = R_{Line}\mathbf{I}_{rms} + \mathbf{V}_L$$
$$= (0.1)(294.11 \angle -31.79°) + 480 \angle 0°$$
$$= 505.22 \angle -1.76° \text{ V rms}$$

and

$$PF_s = \cos 30.03°$$
$$= 0.866 \text{ lagging}$$

∎

DRILL EXERCISE

D5.6. An industrial load operates at 20 kW, 0.8 PF lagging with a line voltage of 220 $\angle 0°$ V rms. Construct the power triangle for the load.

Ans:

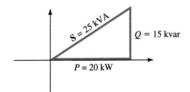

$S = 25$ kVA

$Q = 15$ kvar

$P = 20$ kW

Figure D5.6

5.7
Power Factor Correction

Industrial plants that require large amounts of power have a wide variety of loads. However, by nature the loads normally have a lagging power factor. In view of the results obtained in Example 5.5, we are naturally led to ask if there is any convenient technique for raising the power factor of a load. Since a typical load may be a bank of induction mo-

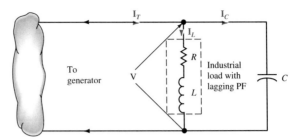

Figure 5.9   Circuit for power factor correction.

tors or other expensive machinery, the technique for raising the PF should be an economical one in order to be feasible.

To answer the question we pose, let us examine the circuit shown in Fig. 5.9. The circuit illustrates a typical industrial load. In parallel with this load we have placed a capacitor. The original complex power for the load $\mathbf{Z}_L$, which we will denote as $\mathbf{S}_{old}$ is

$$\mathbf{S}_{old} = P_{old} + jQ_{old} = |\mathbf{S}_{old}| \angle \theta_{old}$$

The new complex power that results from adding a capacitor is

$$\mathbf{S}_{new} = P_{old} + jQ_{new} = |\mathbf{S}_{new}| \angle \theta_{new}$$

where θ_{new} is specified by the required power factor. The difference between the new and old complex powers is caused by the addition of the capacitor. Hence

$$\mathbf{S}_{cap} = \mathbf{S}_{new} - \mathbf{S}_{old} \tag{5.25}$$

and since the capacitor is purely reactive

$$\mathbf{S}_{cap} = +jQ_{cap} = -j\omega C V_{rms}^2 \tag{5.26}$$

Equation (5.26) can be used to find the required value of C in order to achieve the new specified power factor. In general, we want the power factor to be large, and therefore the power factor angle must be small (i.e., the larger the desired power factor, the smaller the angle $\theta_{new} = \theta_{vT} - \theta_{iT}$.

The following example illustrates the simplicity of the *power factor correction scheme*, which is a favorite type of question on the fundamentals of engineering (FE) examination taken in preparation for obtaining a professional engineers license.

EXAMPLE 5.7

An industrial load consisting of a bank of induction motors consumes 50 kW at a PF of 0.8 lagging from a 220 $\angle 0°$ V rms, 60 Hz line. We wish to raise the PF to 0.95 lagging by placing a bank of capacitors in parallel with the load.

The circuit diagram for this problem is shown in Fig. 5.10. $P = 50$ kW and since $\cos^{-1}0.8 = 36.87°$, $\theta_{old} = 36.87°$. Therefore,

$$Q_{old} = P_{old}\tan\theta_{old} = (50)(10^3)(0.75) = 37.5 \text{ kvar}$$

Hence

$$\mathbf{S}_{old} = 50,000 + j37,500$$

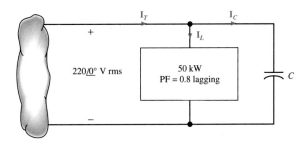

Figure 5.10   Example circuit for power factor correction.

Since the required power factor is 0.95, $\theta_{new} = 18.19°$. Then

$$\mathbf{S}_{new} = 50,000 + j50,000 \tan 18.19°$$
$$= 50,000 + j16,430$$

and therefore

$$\mathbf{S}_{cap} = \mathbf{S}_{new} - \mathbf{S}_{old}$$
$$= -j21,070 \text{ VA}$$

Hence

$$C = \frac{21,070}{(377)(220)^2}$$
$$= 1155 \ \mu\text{F}$$

By using a capacitor of this magnitude in parallel with the industrial load, we create, from the utility's perspective, a load with a PF of 0.95 lagging. However, the parameters of the actual load remain unchanged.   ■

DRILL EXERCISE

D5.7. Compute the value of the capacitor necessary to change the power factor in Drill Exercise D5.5 to 0.95 lagging.

Ans: C = 773 μF.

5.8
Power Measurements

In the preceding sections we have illustrated techniques for computing power. We will now show how power is actually measured in an electrical network. An instrument used to measure average power is the *wattmeter*. This instrument contains a low-impedance current coil (which ideally has zero impedance) that is connected in series with the load, and a high-impedance voltage coil (which ideally has infinite impedance) that is connected across the load. If the voltage and current are periodic and the wattmeter is connected as shown in Fig. 5.11, it will read

$$P = \frac{1}{T} \int_0^T v(t)i(t) \, dt$$

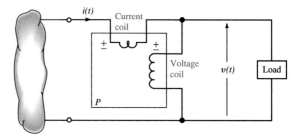

Figure 5.11 Wattmeter connection for power measurement.

where $v(t)$ and $i(t)$ are defined on the figure. Note that $i(t)$ is referenced entering the $\pm$ terminal of the current coil and $v(t)$ is referenced positive at the $\pm$ terminal of the voltage coil. (The $\pm$ designation for one end of a coil is common practice for wattmeters.) In the frequency domain the wattmeter reading will be

$$P = Re(\mathbf{VI}^*) = |\mathbf{V}||\mathbf{I}|\cos(\theta_v - \theta_i)$$

which is average power.

EXAMPLE 5.8

Given the network shown in Fig. 5.12, we wish to determine the wattmeter reading.

The impedance of the load is

$$\mathbf{Z}_L = \frac{(1)(1 + j)}{1 + 1 + j}$$
$$= 0.6 + j0.2 \ \Omega$$

The current through the wattmeter is then

$$\mathbf{I} = \frac{120\angle 0°}{1 - j1 + 0.6 + j0.2}$$
$$= 67.08\angle 26.56° \ A \ rms$$

The load voltage measured by the wattmeter is

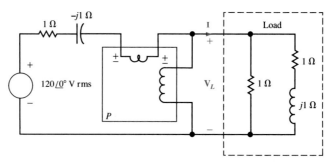

Figure 5.12 Use of the wattmeter in power measurement.

$$\mathbf{V}_L = \mathbf{IZ}_L$$
$$= (67.08 \underline{/26.56°})(0.6325 \underline{/18.43°})$$
$$= 42.425 \underline{/44.99°} \text{ V rms}$$

The wattmeter reading, which is the power absorbed by the load, is

$$P = |\mathbf{V}||\mathbf{I}| \cos(\theta_v - \theta_i)$$
$$= (42.425)(67.08)\cos(44.99° - 26.56°)$$
$$= 2700 \text{ W}$$ ■

DRILL EXERCISE

D5.8. Determine the wattmeter reading in the network in Fig. D5.8.

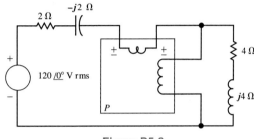

Figure D5.8

Ans: P = 1440 W.

5.9
Typical Residential ac Power Circuits

Consider as an application of ac circuit analysis the standard American residential electrical power service. A comprehensive discussion of the subject is beyond the scope of this book; it is treated here only to acquaint the reader with some basic principles. The reader is referred to *The National Electric Code* and *The National Electric Safety Code* for an authoritative discussion of the issues.

Furthermore, the reader is strongly cautioned that working with residential power circuits can be fatal, and such work may be done *only* by persons properly licensed to do so.

Consider now the typical residential electric power service illustrated in Fig. 5.13(a).

The single-phase two-wire utility primary distribution line serves as a 7200 V rms voltage source. Note that one, typically the bottom, conductor is grounded via a bare conductor running down the pole and connected to a ground rod, which is a copperclad steel rod driven into the earth close to the base of the pole. This voltage is applied to the primary winding of a distribution transformer—a device which will be described in some detail in Chapter 18. The transformer steps the input voltage down to a 3-wire 240/120 V rms level represented by the circuit diagram shown in Fig. 5.13(b). Typically, lights or small appliances are connected from one line to neutral *n* and operate at 120 V rms;

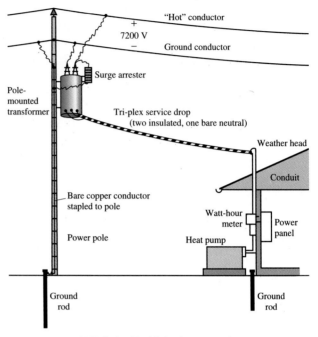

(a) Typical residential electric power service

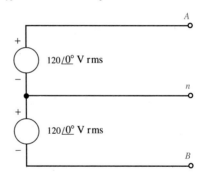

(b) Circuit diagram of transformer output

Figure 5.13   Diagrams for residential electric power service.

large appliances, for example, hot water heaters, are connected line to line and operate at approximately 240 V rms.

As indicated in Fig. 5.13(a), a "surge arrester" is typically connected at the high-voltage transformer terminals. The surge arrester is basically a nonlinear resistor, offering high (ideally infinite) resistance at normal operating voltage, and low (ideally zero) resistance at abnormally high system voltage, typically caused by lightning.

"Grounding" means to connect a part, or parts, of an electrical circuit to the local earth, its purpose being primarily for safety (i.e., to minimize shock hazards), and to suppress lightning and switching-induced transient overvoltages. The "service drop" to the residence typically consists of triplex cable, made up of three conductors: one bare aluminum neutral, which is grounded at both ends, and two insulated, spiraled about the neutral. The cable enters "conduit," a metallic pipe used for mechanical support and protection, through a "weather head," and enters a "watthour meter." The watthour meter is basically an integrating wattmeter, which records the total electrical energy flow from the service drop into the residence. If it is read at time t_1, and subsequently at time t_2, the difference in the readings represents the electrical energy supplied in the interval $t_2 - t_1$. Typical units used are "kW-hrs" (1 kW-hr = 3.6 MJ).

The "electrical panel" is basically a metal box containing circuit breakers (switches, that automatically open if the current through them is excessive), thermal devices (fuses, which melt when the current in them is excessive thus breaking a closed circuit), and terminals for all electrical conductors. Figure 5.14(a) illustrates three typical branch circuits that might be found in a residence, and Fig. 5.14(b) illustrates the corresponding equivalent circuit diagram.

Consider the following example which describes some of the practical aspects of residential power circuits.

EXAMPLE 5.9
Suppose that three residential power circuits are composed of a lighting branch, appliance branch, and an electric dryer circuit as shown in Fig. 5.14(b). The lighting branch circuit must illuminate an area of 640 sq. ft., the appliance circuit must serve 10 electrical receptacles, and the dryer is rated at 5 kVA. We wish to determine (a) the size of circuit breaker required for each individual circuit, (b) the currents in all parts of the power net-

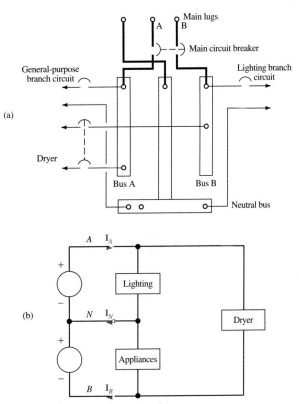

Figure 5.14   Diagrams used to explain residential power circuit.

work in Fig. 5.14(b), and (c) the monthly cost, at $0.075 per kW-hr, for operating the lights, appliances, and dryer 40%, 8%, and 3% of the time, respectively.

(a) For general residential illumination, electrical distribution designers use a general "rule of thumb" of 3 watts per sq. ft. Therefore, the power required for lighting is $P = (3)(640) = 1920$ W. Assuming unity power factor, $S = 1920$ VA and since the lighting voltage is 120 V rms, the current is $I = 1920/120 = 16$ A rms. For this current, an AWG (American Wire Gauge) #12 copper conductor with a typical polyethelene insulation such as RH or RHW (these are standard wiring insulation designations found in the National Electric Code) would typically be used. Since circuit breakers come in 15 A, 20 A, 25 A, and 30 A rms sizes, a 20 A rms circuit breaker would be used to protect the lighting branch circuit.

For the appliance branch circuit, when the actual load is unknown, the figure frequently used is 180 VA per receptacle. For 10 receptacles the apparent power is $S = (10)(180) = 1800$ VA. At 120 V rms, the current would be $1800/120 = 15$ A rms, and therefore once again #12 conductors protected by a 20 A rms breaker would serve this circuit.

The 5 kVA electric dryer operates at 240 V rms, and therefore the current is $5000/240 = 20.83$ A rms. However, for this higher power circuit the National Electric Code specifies #10 copper conductors, protected by a 2-pole (breaks both lines as shown in Fig. 5.14(a) 30 A rms breaker.

(b) Rigorously speaking, the load should be summed as complex power. However, power factors (PF) are frequently unknown; also they are typically between 0.9 and 1.0. Conservatively, then, we may assume unity PF for loads, that is, currents computed by this method will be slightly higher than is actually the case. The magnitude of the current $\mathbf{I}_A$ shown in Fig. 5.14(b) is the lighting current plus the dryer current or

$$I_A = 20.83 + 16$$
$$= 36.83 \text{ A rms}$$

The magnitude of the current $\mathbf{I}_B$ shown in Fig. 5.14(b) is the sum of the appliance current and the dryer current or

$$I_B = 20.83 + 15$$
$$= 35.83 \text{ A rms}$$

Assuming that the phase angle for the two voltage sources in Fig. 5.14(b) is 0°, the magnitude of the current $\mathbf{I}_N$ is the difference between I_A and I_B or 1 A.

(c) Assuming a 30-day month, the total number of hours in the month is $(24)(30) = 720$ hours. Then at $0.075 per kW-hr, the operation of the lights, appliances, and dryer 40%, 8%, and 3% of the time, respectively, would cost

$$\text{Cost} = 0.075[(0.4)(1.92) + (0.08)(1.8) + (0.03)(5.0)] = \$57.35 \qquad \blacksquare$$

In the basic single-phase 120 V rms branch circuit, three conductors are normally used: one with black insulation, that is, connected to Bus A or B in Example 5.9; one with white insulation that is connected to the neutral bus which serves as the return path for the load current; and one bare conductor, also connected to the neutral bus, which is not intended to carry load current. These conductors are illustrated in Fig. 5.15. The bare conductor serves to maintain the potential of the receptacle metal enclosure, and exposed metal parts of connected appliances, at ground level, for safety purposes. Carefully examine a com-

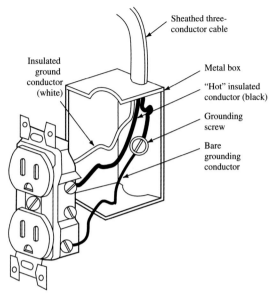

Figure 5.15 Typical general-purpose appliance receptacle.

mon three-wire plug. Note the three blades: one small-ended spade connector—the black wire; one large-ended spade connector—the white wire; and one round connector—the bare wire.

5.10
Three-Phase Circuits

Up to this point we have dealt with what we refer to as single-phase circuits. Now we extend our analysis techniques to *three-phase circuits,* that is, circuits containing three voltage sources that are one-third of a cycle apart in time.

The study of three phase circuits is important because it is more advantageous and economical to transmit electric power in the three-phase mode. That is why we find that all the very high voltage transmission lines have three conductors, that is, three phases.

As the name implies, three-phase circuits are those in which the forcing function is a three-phase system of voltages. If the three sinusoidal voltages have the same magnitude and frequency and each voltage is 120° out of phase with the other two, the voltages are said to be *balanced.* If the loads are such that the currents produced by the voltages are also balanced, the entire circuit is referred to as a *balanced three-phase circuit.*

A balanced set of three-phase voltages can be represented in the frequency domain as shown in Fig. 5.16(a) where we have assumed that their magnitudes are 120 V rms. From the figure we note that

$$\mathbf{V}_{an} = 120 \angle 0° \text{ V rms}$$
$$\mathbf{V}_{bn} = 120 \angle -120° \text{ V rms}$$
$$\mathbf{V}_{cn} = 120 \angle -240° \text{ V rms}$$

(5.27)

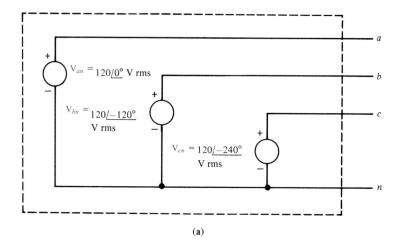

(a)

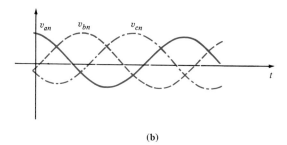

(b)

Figure 5.16   Balanced three-phase voltages.

Note that our double-subscript notation is exactly the same as that employed in the earlier chapters.

The phasor voltages, which can be expressed in the time domain as

$$v_{an}(t) = 120\sqrt{2} \cos\omega t \text{ V}$$
$$v_{bn}(t) = 120\sqrt{2} \cos(\omega t - 120°) \text{ V} \qquad (5.28)$$
$$v_{cn}(t) = 120\sqrt{2} \cos(\omega t - 240°) \text{ V}$$

are shown in Fig. 5.16(b).

Let us examine the instantaneous power generated by a three-phase system. Assume that the voltages are given by Eq. (5.28).

If the load is balanced, the currents produced by the sources are

$$i_a(t) = I_M\cos(\omega t - \theta) \text{ A}$$
$$i_b(t) = I_M\cos(\omega t - \theta - 120°) \text{ A} \qquad (5.29)$$
$$i_c(t) = I_M\cos(\omega t - \theta - 240°) \text{ A}$$

where θ is the angle between the voltage and the current. The instantaneous power produced by the system is

$$p(t) = p_a(t) + p_b(t) + p_c(t)$$
$$= V_M I_M[\cos\omega t \cos(\omega t - \theta) + \cos(\omega t - 120°)\cos(\omega t - \theta - 120°) \quad (5.30)$$
$$+ \cos(\omega t - 240°)\cos(\omega t - \theta - 240°)]$$

Using the two trigonometric identities

$$\cos\alpha\cos\beta = \frac{1}{2}[\cos(\alpha - \beta) + \cos(\alpha + \beta)] \quad (5.31)$$

and

$$\cos\phi + \cos(\phi - 120°) + \cos(\phi + 120°) = 0 \quad (5.32)$$

We can show that the expression for the power reduces to

$$p(t) = 3\,\frac{V_M I_M}{2}\,\cos\theta \text{ W} \quad (5.33)$$

This equation is very interesting. It states that the *instantaneous* power is *independent of time.* Recall that in the single-phase case, described by Eq. (5.5) and illustrated in Fig. 5.2, the power is pulsating at twice the excitation frequency. However, in the three-phase case the power is always *constant,* which reduces wear on the generators and is therefore a good reason to generate power in three-phase form.

The phasor diagram for the three-phase voltages is shown in Fig. 5.17. The phase sequence of this set is said to be *abc,* meaning that V_{bn} lags V_{an} by 120°.

We will standardize our notation so that we always label the voltages $\mathbf{V}_{an}$, $\mathbf{V}_{bn}$, and $\mathbf{V}_{cn}$ and observe them in the order *abc,* which will be referred to as a *positive phase sequence.* Furthermore, we will normally assume with no loss of generality that $\angle\mathbf{V}_{an} = 0°$.

An important property of the balanced voltage set is that

$$\mathbf{V}_{an} + \mathbf{V}_{bn} + \mathbf{V}_{cn} = 0 \quad (5.34)$$

This property can easily be seen by resolving the voltage phasors into components along the real and imaginary axes. It can also be demonstrated via Eq. (5.32).

From the standpoint of the user who connects a load to the balanced three-phase voltage source, it is not important how the voltages are generated. It is important to note, however, that if the load currents generated by connecting a load to the power source are also

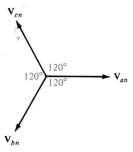

Figure 5.17   Phasor diagram for a balanced three-phase voltage source.

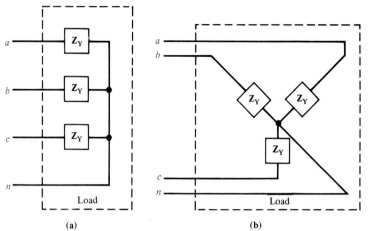

Figure 5.18   Wye (Y)-connected loads.

balanced, there are two possible equivalent configurations for the load. The equivalent load can be considered as being connected in either a *wye* (Y) or a *delta* (Δ) configuration. The balanced wye configuration is shown in Fig. 5.18(a) and equivalently in Fig. 5.18(b). The delta configuration is shown in Fig. 5.19(a) and equivalently in Fig. 5.19(b). Note that in the case of the delta connection, there is no neutral line. The actual function of the neutral line in the wye connection will be examined and it will be shown that in a balanced system the neutral line carries no current and, for purposes of analysis, may be omitted.

Balanced Wye-Wye Connection

Suppose now that the source and load are both connected in a wye, as shown in Fig. 5.20. The phase voltages with positive phase sequence are

$$\mathbf{V}_{an} = V_P \angle 0°$$
$$\mathbf{V}_{bn} = V_P \angle -120° \tag{5.35}$$
$$\mathbf{V}_{cn} = V_P \angle +120°$$

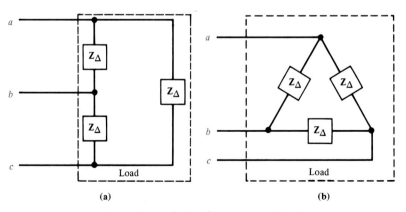

Figure 5.19   Delta (Δ)-connected loads.

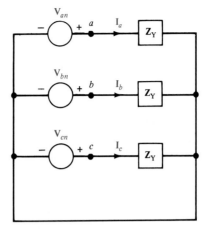

Figure 5.20 Balanced three-phase wye-wye connection.

where V_P, the *phase voltage,* is the magnitude of the phasor voltage from the neutral to any line. The *line-to-line* or simply, *line voltages* V_L, can be calculated using KVL; for example

$$
\begin{aligned}
\mathbf{V}_{ab} &= \mathbf{V}_{an} - \mathbf{V}_{bn} \\
&= V_P \angle 0° - V_P \angle -120° \\
&= V_P - V_P \left[-\frac{1}{2} - j\frac{\sqrt{3}}{2} \right] \\
&= V_P \left[\frac{3}{2} + j\frac{\sqrt{3}}{2} \right] \\
&= \sqrt{3}\, V_P \angle 30°
\end{aligned}
\tag{5.36}
$$

The phasor addition is shown graphically in Fig. 5.21(a). In a similar manner, we obtain

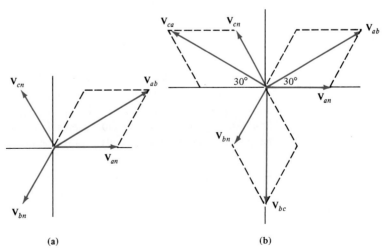

(a)   (b)

Figure 5.21 Phasor representation of phase and line voltages in a balanced wye-wye system.

the set of line-to-line voltages as

$$\mathbf{V}_{ab} = \sqrt{3}\, V_P \angle 30°$$
$$\mathbf{V}_{bc} = \sqrt{3}\, V_P \angle -90°$$
$$\mathbf{V}_{ca} = \sqrt{3}\, V_P \angle -210°$$

All the line voltages together with the phase voltages are shown in Fig. 5.21(b).

Note carefully that the phase voltages are multiplied by $\sqrt{3}$ and shifted by 30° to obtain the line voltages. Furthermore, if V_L, I_L, V_P, and I_P represent the line voltage, line current, phase voltage, and phase current, respectively, then for a balanced wye-wye system

$$V_L = \sqrt{3}\, V_P$$
$$I_L = I_P \tag{5.37}$$

As shown in Fig. 5.20, the line currents are

$$\mathbf{I}_a = \frac{\mathbf{V}_{an}}{\mathbf{Z}_Y} = \frac{V_P\angle 0°}{\mathbf{Z}_Y}$$

$$\mathbf{I}_b = \frac{\mathbf{V}_{bn}}{\mathbf{Z}_Y} = \frac{V_P\angle -120°}{\mathbf{Z}_Y} \tag{5.38}$$

$$\mathbf{I}_c = \frac{\mathbf{V}_{cn}}{\mathbf{Z}_Y} = \frac{V_P\angle +120°}{\mathbf{Z}_Y}$$

The neutral current $\mathbf{I}_n$ is then

$$\mathbf{I}_n = (\mathbf{I}_a + \mathbf{I}_b + \mathbf{I}_c) = 0 \tag{5.39}$$

Since there is no current in the neutral, this conductor could contain any impedance or it could be an open or a short circuit, without changing the results found above. Therefore, for simplicity, we will assume that the neutral line is not present, that is, an open circuit.

It is important to note that although we have a three-phase system composed of three sources and three loads, Eq. (5.38) illustrates that we can compute the currents and voltages in each phase independent of the other phases. Therefore, in a balanced three-phase system, we can analyze only one phase and use the phase sequence to obtain the voltages and currents in the other phases. This is, of course, a direct result of the balanced condition. We may even have impedances present in the lines; however, as long as the system remains balanced, we need analyze only one phase.

EXAMPLE 5.10

A three-phase wye-connected load is supplied by an *abc* sequence balanced three-phase wye-connected source as shown in Fig. 5.22(a). Note that we employ lowercase letters to represent the generator end of the network and uppercase letters to represent the load end. If the phase voltage of the source is 120 V rms, the line impedance and load impedance per phase are $1 + j1\ \Omega$ and $20 + j10\ \Omega$, respectively, we wish to determine the value of the line currents and the load voltages.

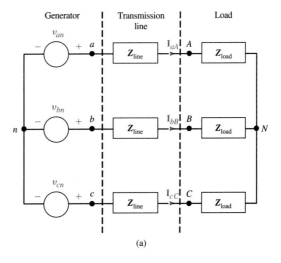

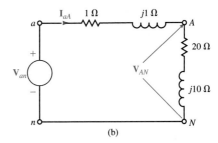

Figure 5.22   Circuit diagrams for the problem in Example 5.10.

The phase voltages are

$$\mathbf{V}_{an} = 120 \underline{/\,0°} \text{ V rms}$$
$$\mathbf{V}_{bn} = 120 \underline{/-120°} \text{ V rms}$$
$$\mathbf{V}_{cn} = 120 \underline{/+120°} \text{ V rms}$$

For a balanced system, we need only analyze the *a* phase, since the results for the *b* and *c* phases are displaced by 120° and 240° respectively. The per phase circuit diagram is shown in Fig. 5.22(b). The line current for the *a* phase is

$$\mathbf{I}_{aA} = \frac{120 \underline{/\,0°}}{21 + j11}$$
$$= 5.06 \underline{/-27.65°} \text{ A rms}$$

where $\mathbf{I}_{aA}$ represents the current from point *a* to point *A*. The load voltage for the *a* phase, which we call $\mathbf{V}_{AN}$, is

$$\mathbf{V}_{AN} = (5.06 \underline{/-27.65°})(20 + j10)$$
$$= 113.15 \underline{/-1.08°} \text{ V rms}$$

The corresponding line currents and load voltages for the b and c phases are

$$\mathbf{I}_{bB} = 5.06\angle{-147.65°} \text{ A rms} \qquad \mathbf{V}_{BN} = 113.15\angle{-121.08°} \text{ V rms}$$
$$\mathbf{I}_{cC} = 5.06\angle{-267.65°} \text{ A rms} \qquad \mathbf{V}_{CN} = 113.15\angle{-241.08°} \text{ V rms}$$

■

To reemphasize and clarify our terminology, phase voltage, V_P, is the magnitude of the phasor voltage from the neutral to any line, while line voltage, V_L, is the magnitude of the phasor voltage between any two lines. Thus, the values of V_L and V_P will depend on the point at which they are calculated in the system.

D5.9. The voltage for the a phase of an abc-phase-sequence balanced wye-connected source is $\mathbf{V}_{an} = 120\angle{90°}$ V rms. Determine the line voltages for this source.

Ans: $\mathbf{V}_{ab} = 208\angle{120°}$ V rms, $\mathbf{V}_{bc} = 208\angle{0°}$ V rms, $\mathbf{V}_{ca} = 208\angle{-120°}$ V rms.

D5.10. A three-phase wye load is supplied by an abc-sequence-balanced three-phase wye-connected source through a transmission line with an impedance of $1 + j1$ ohms per phase. The load impedance is $8 + j3$ ohms per phase. If the load voltage for the a phase is $104.02\angle{26.6°}$ V rms (i.e., $V_P = 104.02$ V rms at the load end), determine the phase voltages of the source.

Ans: $\mathbf{V}_{an} = 120\angle{30°}$ V rms, $\mathbf{V}_{bn} = 120\angle{-90°}$ V rms, $\mathbf{V}_{cn} = 120\angle{-210°}$ V rms.

Balanced Wye-Delta Connection

Another important three-phase circuit is the balanced wye-delta system; that is, a wye-connected source and a delta-connected load, as shown in Fig. 5.23. From Fig. 5.23 note that the line-to-line voltage is the voltage across each load impedance in the delta-connected load. Therefore, if the phase voltages of the source are

$$\mathbf{V}_{an} = V_P\angle{0°}$$
$$\mathbf{V}_{bn} = V_P\angle{-120°} \qquad (5.40)$$
$$\mathbf{V}_{cn} = V_P\angle{+120°}$$

then the line voltages are

$$\mathbf{V}_{ab} = \sqrt{3}\, V_P\angle{30°}$$
$$\mathbf{V}_{bc} = \sqrt{3}\, V_P\angle{-90°} \qquad (5.41)$$
$$\mathbf{V}_{ca} = \sqrt{3}\, V_P\angle{+150°}$$

Knowing the voltage across each phase of the Δ, we can easily compute the phase currents in the Δ. KCL can then be employed in conjunction with the phase currents to determine the line currents. For example

$$\mathbf{I}_{aA} = \mathbf{I}_{AB} + \mathbf{I}_{AC}$$
$$= \mathbf{I}_{AB} - \mathbf{I}_{CA}$$

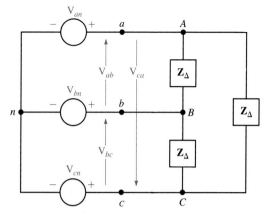

Figure 5.23 Balanced three-phase wye-delta system.

Following a development similar to that which led to Eq. (5.36), we can show that the magnitudes of the line and phase voltages and currents for the delta-connected load are related by the equations

$$V_L = V_P$$
$$I_L = \sqrt{3}\,I_P \qquad (5.42)$$

EXAMPLE 5.11
The balanced three-phase wye-delta system shown in Fig. 5.23 has the following parameters: the voltage source has an *abc* sequence and $\mathbf{V}_{an} = 120\underline{/30°}$ V rms, and the impedance per phase in the delta load is $\mathbf{Z}_\Delta = 6 + j6$ Ω. We wish to determine the currents in the delta and the line currents.

At the generator, $\mathbf{V}_{an} = 120\underline{/30°}$ V rms, and therefore the line voltage is $\mathbf{V}_{ab} = 120\sqrt{3}\underline{/60°}$ V rms. Since there is no line impedance, the line voltage at the delta load is

$$\mathbf{V}_{AB} = \mathbf{V}_{ab} = 120\sqrt{3}\underline{/60°}\ \text{V rms}$$

and hence

$$\mathbf{I}_{AB} = \frac{120\sqrt{3}\underline{/60°}}{6 + j6}$$
$$= 24.50\underline{/15°}\ \text{A rms}$$

Recall that the line voltages at the load are 120° out of phase with one another and the load impedance is balanced. Therefore, the phase currents at the load differ by 120° and hence

$$\mathbf{I}_{BC} = 24.50\underline{/-105°}\ \text{A rms}$$
$$\mathbf{I}_{CA} = 24.50\underline{/135°}\ \text{A rms}$$

The line current can be computed using KCL at the node labeled A.

$$\mathbf{I}_{aA} = \mathbf{I}_{AB} - \mathbf{I}_{cA}$$
$$= 24.50\underline{/15°} - 24.50\underline{/135°}$$
$$= 42.44\underline{/-15°} \text{ A rms}$$

and hence

$$\mathbf{I}_{bB} = 42.44\underline{/-135°} \text{ A rms}$$
$$\mathbf{I}_{cC} = 42.44\underline{/105°} \text{ A rms}$$

It is interesting to note that from the standpoint of determining the line currents we could simply employ the $Y \rightleftarrows \Delta$ transformation to the delta-connected load and compute the line currents directly. For the balanced impedance case, the transformation Eqs. (4.41) and (4.42) reduce to

$$\mathbf{Z}_Y = \frac{1}{3}\mathbf{Z}_\Delta$$

Therefore, the delta load can be replaced by the equivalent wye load which would then produce an equivalent balanced wye-wye system in which the phase voltage at the generator is $\mathbf{V}_{an} = 120\underline{/30°}$ V rms and the impedance in one branch of the wye load is $2 + j2 \ \Omega$. Hence, the line current

$$\mathbf{I}_{aA} = \frac{120\underline{/30°}}{2 + j2}$$

$$= 42.44\underline{/-15°} \text{ A rms}$$

which is, of course, identical to that obtained by applying KCL at the delta load. ∎

The wye/delta transformation relationships are also useful in solving three-phase systems with multiple loads. The following example, which for simplicity has only resistive loads, illustrates their use.

EXAMPLE 5.12

Consider the network shown in Fig. 5.24(a). We wish to determine all the line currents. Note that the 30 Ω resistors are connected in wye and the 60 Ω resistors are connected in delta. If we convert the delta to an equivalent wye, the impedance of each branch of the equivalent wye would contain a 20 Ω resistor. The two wye loads are now in parallel and the equivalent circuit for the a phase is shown in Fig. 5.24(b). The line current $\mathbf{I}_{aA}$ is then

$$\mathbf{I}_{aA} = \frac{120\underline{/0°}}{4 + 30\|20}$$

$$= \frac{120\underline{/0°}}{16}$$

$$= 7.5\underline{/0°} \text{ A rms}$$

and hence $\mathbf{I}_{bB} = 7.5\underline{/-120°}$ A rms and $\mathbf{I}_{cC} = 7.5\underline{/+120°}$ A rms. ∎

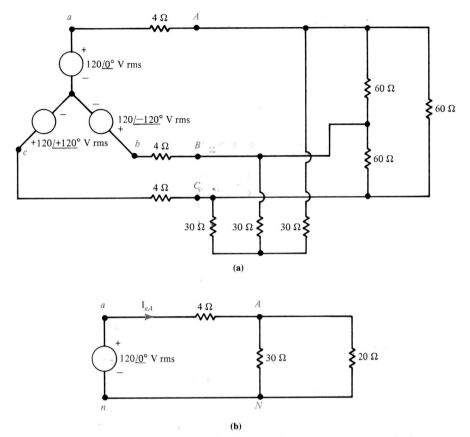

Figure 5.24 Networks used in Example 5.12; (a) original network; (b) *a*-phase equivalent network.

D5.11. A balanced three-phase wye-connected source has a line voltage of $\mathbf{V}_{ab} = 208 \angle 50°$ V rms. Compute the phase voltages of the source.

Ans: $\mathbf{V}_{an} = 120 \angle 20°$ V rms, $\mathbf{V}_{bn} = 120 \angle -100°$ V rms, $\mathbf{V}_{cn} = 120 \angle +140°$ V rms.

D5.12. In a balanced three-phase wye-delta system, the delta-connected load consists of a 20 Ω resistor in series with a 20 mH inductor. If the system frequency is 60 Hz, find the impedance per phase of an equivalent wye load.

Ans: $\mathbf{Z}_Y = 6.67 + j2.51$ Ω.

Delta-Connected Source

Up to this point we have concentrated our discussion on circuits that have wye-connected sources. However, our analysis of the wye-wye and wye-delta connections provides us with the information necessary to handle a delta-connected source.

Consider the delta-connected source shown in Fig. 5.25(a). Note that the sources are connected line to line. We found earlier that the relationship between line-to-line and line-to-neutral voltages was given by Eq. (5.36) and illustrated in Fig. 5.21 for an *abc* phase sequence of voltages. Therefore, if the delta sources are

$$
\begin{aligned}
\mathbf{V}_{ab} &= V_L \angle 0° \\
\mathbf{V}_{bc} &= V_L \angle -120° \\
\mathbf{V}_{ca} &= V_L \angle +120°
\end{aligned}
\tag{5.43}
$$

where V_L is the magnitude of the voltage sources connected in delta, the equivalent wye sources shown in Fig. 5.25(b) are

$$
\mathbf{V}_{an} = \frac{V_L}{\sqrt{3}} \angle -30°
$$

$$
\mathbf{V}_{bn} = \frac{V_L}{\sqrt{3}} \angle -150°
\tag{5.44}
$$

$$
\mathbf{V}_{cn} = \frac{V_L}{\sqrt{3}} \angle -270°
$$

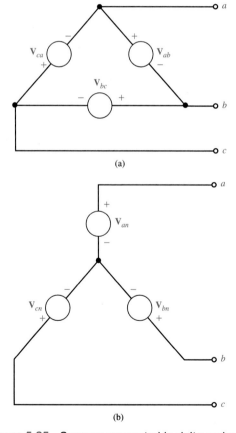

(a)

(b)

Figure 5.25   Sources connected in delta and wye.

Therefore, if we encounter a network containing a delta-connected source, we can easily convert the source from delta to wye so that all the techniques we have discussed previously can be appied in an analysis.

D5.13. A delta-connected source has the following values: $\mathbf{V}_{ab} = 440 \angle 60°$ V rms, $\mathbf{V}_{bc} = 440 \angle -60°$ V rms, and $\mathbf{V}_{ca} = 440 \angle -180°$ V rms. Find the values of the equivalent wye-connected source.

Ans: $\mathbf{V}_{an} = 254 \angle 30°$ V rms, $\mathbf{V}_{bn} = 254 \angle -90°$ V rms, and $\mathbf{V}_{cn} = 254 \angle -210°$ V rms.

Power Relationships

It is interesting to note that the complex power for a single phase of a three-phase system can be expressed using Eq. (5.17) as

$$\mathbf{S}_{1\phi} = V_P I_P \angle \theta_z \tag{5.45}$$

where θ_z is the impedance angle of the load. Since the line current and line voltage are the quantities most easily measured, we wish to express the total complex power $\mathbf{S}_T$ in terms of these quantities. For a wye-connected load, Eqs. (5.37) indicate that the complex power for a single phase can be expressed as

$$\mathbf{S}_{1\phi} = \frac{V_L I_L}{\sqrt{3}} \angle \theta_z \tag{5.46}$$

The same results are obtained from Eqs. (5.42) if the load is delta-connected.

The total complex power for all three phases is then

$$\mathbf{S}_T = \sqrt{3} \, V_L I_L \angle \theta_z \tag{5.47}$$

where

$$\mathbf{S}_T = P_T + jQ_T$$
$$P_T = \sqrt{3} \, V_L I_L \cos\theta_z \tag{5.48}$$
$$Q_T = \sqrt{3} \, V_L I_L \sin\theta_z$$

and

$$|\mathbf{S}_T| = \sqrt{P_T^2 + Q_T^2}$$
$$= \sqrt{3} \, V_L I_L \tag{5.49}$$

EXAMPLE 5.13

A balanced three-phase source serves two small industrial plants which are located at the intersection of two major country roads. Each plant represents a load, and the characteristics of the loads are as follows.

Load 1: 28 kW at 0.85 lagging power factor
Load 2: 18 kW at unity power factor

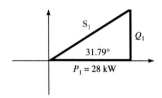

Figure 5.26　Power triangle used in Example 5.13.

If the line voltage at the loads is 208 V rms at 60 Hz, we wish to determine the total line current and the combined power factor for the total load.

The power factor angle for load 1 is

$$\theta_1 = \cos^{-1}0.85 = 31.79°$$

The power triangle for this load is shown in Fig. 5.26. The magnitude of the complex power for load 1 is then

$$|\mathbf{S}_1| = \frac{P_1}{\cos\theta_1} = \frac{28,000}{0.85}$$
$$= 32.941 \text{ kVA}$$

and

$$\mathbf{S}_1 = 32,941 \angle 31.79° \text{ VA}$$
$$= 28,000 + j17,354 \text{ VA}$$

The power angle for load 2 is 0° and hence

$$\mathbf{S}_2 = 18,000 \text{ VA}$$

The total load is

$$\mathbf{S}_T = \mathbf{S}_1 + \mathbf{S}_2$$
$$= 46,000 + j17,354 \text{ VA}$$
$$= 49,165 \angle 20.67° \text{ VA}$$

The line current which serves both loads is

$$I_L = \frac{|\mathbf{S}_T|}{\sqrt{3}\ V_L}$$
$$I_L = 136.47 \text{ A rms}$$

and the combined power factor is

$$\mathrm{PF}_T = \cos 20.67°$$
$$= 0.936 \text{ lagging}$$ ∎

5.11
Summary

- If the voltage and current are both sinusoidal functions, that is, $v(t) = V_M\cos(\omega t + \theta_v)$ and $i(t) = I_M\cos(\omega t + \theta_i)$, then the average power is $P = (1/2)V_M I_M\cos(\theta_v - \theta_i)$, which for a purely resistive load is $P = (1/2)V_M I_M$ and for a purely reactive load is $P = 0$.

- The root-mean-square (rms) or effective value of a periodic current is a constant, that is, dc value, which delivers the same average power to a resistor.
- The rms value of a waveform $X_M\cos(\omega t + \theta)$ is $X_M/\sqrt{2}$ and therefore the power absorbed by a resistor R is $I_{rms}^2 R = V_{rms}^2/R$.
- Real power is measured in watts. Apparent power is measured in volt-amperes and the power factor is

$$PF = \cos(\theta_v - \theta_i) = \frac{P}{V_{rms}I_{rms}}$$

where $\theta_v - \theta_i$ is the power factor angle.
- In a network, if the current leads the voltage, for example, in an *RC* load, the power factor is leading, and if the current lags the voltage, for example, in an *RL* load, the power factor is lagging.
- Apparent power or complex power consists of two terms: real or average power and reactive or quadrature power. Power factor correction can be accomplished, typically by adding a bank of capacitors.
- Average power can be measured using a device called a wattmeter.
- In a balanced three-phase network the *instantaneous* power is *constant*.
- Balanced three-phase systems are connected in either delta or wye.
- Balanced three-phase networks can be analyzed by considering only a single phase.
- The total complex power for a three-phase network is the same whether the circuit is connected in delta or wye.

PROBLEMS

5.1. The voltage and current at the input of a network are given by the expressions

$$v(t) = 6 \cos \omega t \text{ V}$$
$$i(t) = 4 \sin \omega t \text{ A}$$

Determine the average power absorbed by the network.

5.2. Compute the power generated and the power absorbed in the network in Fig. P5.2.

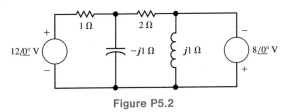

Figure P5.2

5.3. Show that the conservation of power holds for the network shown in Fig. P5.3.

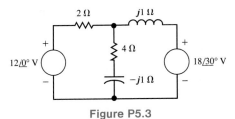

Figure P5.3

5.4. Determine the average power absorbed by the impedance shown in Fig. P5.4.

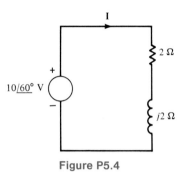

Figure P5.4

5.5. Find the average power absorbed by the network shown in Fig. P5.5.

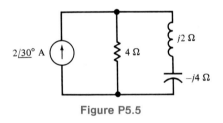

Figure P5.5

5.6. Consider the network shown in Fig. P5.6. Determine the total average power absorbed and supplied by each element.

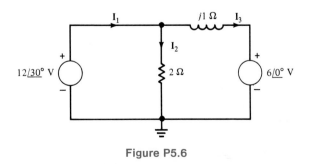

Figure P5.6

5.7. Given the network in Fig. P5.7, determine the total average power absorbed or supplied by each element.

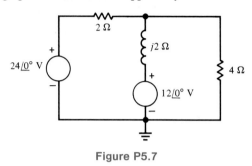

Figure P5.7

5.8. Determine the total average power absorbed and supplied by each element in the network in Fig. P5.8.

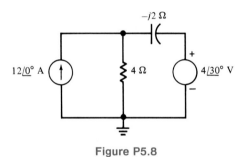

Figure P5.8

5.9. The current waveform in Fig. P5.9 exists in a 4 Ω resistor. Compute the average power delivered to the resistor.

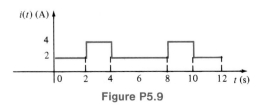

Figure P5.9

5.10. Calculate the rms value of the waveform shown in Fig. P5.10.

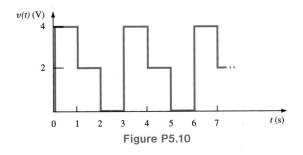

Figure P5.10

5.11. The current waveform in Fig. P5.11 exists in a 10 Ω resistor. Determine the average power delivered to the resistor.

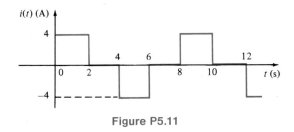

Figure P5.11

5.12. Determine the rms value of the current waveform in Fig. P5.12 and use this value to compute the average power delivered to a 2 Ω resistor in which this current exists.

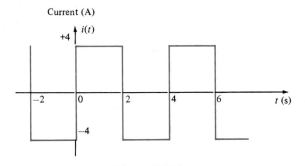

Figure P5.12

5.13. Determine the rms value of the waveform shown in Fig. P5.13.

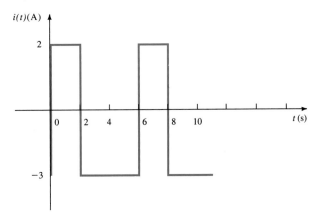

Figure P5.13

5.14. Compute the rms value of the voltage waveform shown in Fig. P5.14.

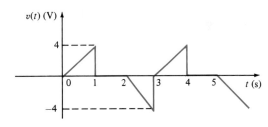

Figure P5.14

5.15. Calculate the rms value of the waveform in Fig. P5.15.

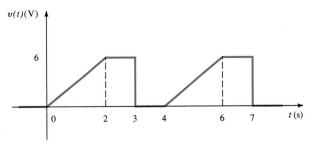

Figure P5.15

5.16. A load operates at 20 kW, 0.8 PF lagging. The load voltage is 220 $\angle 0°$ V rms at 60 Hz. The impedance of the line is $0.09 + j0.3$ Ω. Determine the voltage and power factor at the input to the line.

5.17. A particular load has a PF of 0.8 lagging. The power delivered to the load is 40 kW from a 220 V rms 60 Hz line. If the transmission line resistance is 0.085 Ω, determine the real power that must be generated at the supply.

5.18. As shown in Fig. P5.18, a plant consumes 50 kW at a lagging power factor of 0.85 from a 440 $\angle 0°$ V rms line. If the transmission line resistance is 0.09 Ω, what is the required supply voltage?

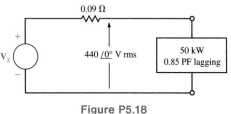

Figure P5.18

5.19. Calculate the voltage V_s that must be supplied to obtain 2 kW, 240 $\angle 0°$ V rms, and a power factor of 0.8 leading at the load Z_L in the network in Fig. P5.19.

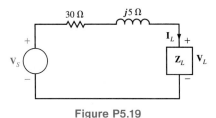

Figure P5.19

5.20. A plant draws 250 A rms from a 440 V rms line to supply a load with 100 kW. What is the power factor of the load?

5.21. A small plant has a bank of induction motors that consume 64 kW at a PF of 0.68 lagging. The 60 Hz line voltage across the motors is 220 $\angle 0°$ V rms. The local power company has told the plant to raise the PF to 0.92 lagging. What value of capacitance is required?

5.22. What value of capacitance, placed in parallel with the load in problem 5.17, will raise the PF to 0.9 lagging?

5.23. Given the network in Fig. P5.23, determine the wattmeter reading.

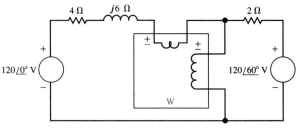

Figure P5.23

5.24. Given the network in Fig. P5.24, determine the wattmeter reading.

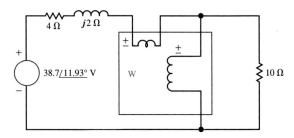

Figure P5.24

5.25. An *abc*-phase-sequence three-phase voltage source connected in a balanced wye has a line voltage of $V_{ab} = 208 \angle 0°$ V rms. Determine the phase voltages of the source.

5.26. An *abc*-sequence three-phase voltage source connected in a balanced wye has a line voltage of $V_{ab} = 208 \angle -30°$ V rms. Determine the phase voltages.

5.27. In a balanced three-phase wye-wye system the load impedance is $20 + j12$ Ω. The source has an *abc* phase sequence and $V_{an} = 120 \angle 0°$ V rms. If the load voltage is $V_{AN} = 111.49 \angle -0.2°$ V rms, determine the magnitude of the line current if the load is suddenly short circuited.

5.28. An *abc*-sequence balanced three-phase wye-connected source with a phase voltage of 100 V rms supplies power to a balanced wye-connected load. The per phase load impedance is $40 + j10$ Ω. Determine the line currents in the circuit if $\angle V_{an} = 0°$.

5.29. An *abc* sequence balanced three-phase wye-connected source supplies power to a balanced wye-connected load. The line impedance per phase is $1 + j0\ \Omega$, and the load impedance per phase is $20 + j20\ \Omega$. If the source line voltage $\mathbf{V}_{ab}$ is $100\ \underline{/0°}$ V rms, find the line currents.

5.30. In a balanced wye-delta system the load per phase is $50 + j20\ \Omega$. Also, $\mathbf{V}_{an}$ for the abc-sequence balanced wye source is $120\ \underline{/-20°}$ V rms. Determine all the delta currents and line currents in the circuit.

5.31. In a balanced three-phase wye-delta system the source has an abc phase sequence. The load impedance is $12 + j8\ \Omega$. If the phase voltage at the load is $\mathbf{V}_{AB} = 260\ \underline{/45°}$ V rms, find the line currents and the phase voltages of the source.

5.32. An abc-phase-sequence three-phase voltage source connected in a balanced wye supplies power to a balanced delta-connected load. The load current $\mathbf{I}_{AB} = 4\ \underline{/20°}$ A rms. Determine the line currents.

5.33. An abc-sequence balanced three-phase wye-connected source supplies power to a balanced delta-connected load. The load impedance per phase is $12 + j8\ \Omega$. If the current $\mathbf{I}_{AB}$ in one phase of the delta is $14.42\ \underline{/86.31°}$ A rms, determine the line currents and phase voltages at the source.

5.34. In a balanced three-phase delta-wye system the source has an abc phase sequence. The line and load impedances are $0.4 + j0.3\ \Omega$ and $10 + j4\ \Omega$, respectively. Calculate the phase voltages and delta currents of the source if the load voltage $\mathbf{V}_{AN} = 116\ \underline{/-30°}$ V rms.

5.35. Consider the network shown in Fig. P5.35. Compute the magnitude of the line voltages at the load and the magnitude of the phase currents in the delta-connected source.

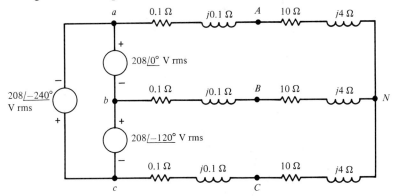

Figure P5.35

5.36. A three-phase load impedance consists of a balanced wye in parallel with a balanced delta, as shown in Fig. P5.36. Determine the equivalent delta load.

5.37. An abc-sequence wye-connected source having a phase voltage of $120\ \underline{/0°}$ V rms is attached to a wye-connected load having an impedance of $80\ \underline{/70°}\ \Omega$. If the line impedance is $4\ \underline{/20°}\ \Omega$, determine the total complex power produced by the voltage sources and the real and reactive power dissipated by the load.

5.38. A balanced three-phase system has a load consisting of a balanced wye in parallel with a balanced delta.

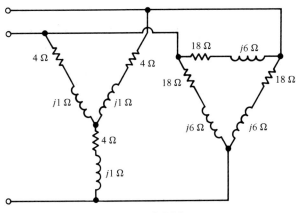

Figure P5.36

The impedance per phase for the wye is $10 + j6 \ \Omega$ and for the delta is $24 + j9 \ \Omega$. The source is a balanced wye with an *abc* phase sequence, and the line voltage $\mathbf{V}_{ab} = 208$ $\underline{/30°}$ V rms. If the line impedance per phase is $1 + j0.5 \ \Omega$, determine the line currents and the load phase voltages when the load is converted to an equivalent wye.

5.39. A 480 V rms line feeds two balanced three-phase loads. If the two loads are rated as follows:

Load 1: 5 kVA at 0.8 PF lagging

Load 2: 10 kVA at 0.9 PF lagging

determine the magnitude of the line current from the 480 V rms source.

5.40. A balanced three-phase source serves the following loads:

Load 1: 48 kVA at 0.9 PF lagging

Load 2: 24 kVA at 0.75 PF lagging

The line voltage at the load is 208 V rms at 60 Hz. Determine the line currents and the combined power factor at the load.

5.41. A balanced three-phase wye-wye system has two parallel loads. Load 1 is rated at 3000 VA, 0.7 PF lagging, and load 2 is rated at 2000 VA, 0.75 PF leading. If the line voltage is 208 V rms, find the magnitude of the line current.

5.42. Two industrial plants represent balanced three-phase loads. The plants receive their power from a balanced three-phase source with a line voltage of 4.6 kV rms. Plant 1 is rated at 300 kVA, 0.8 PF lagging, and plant 2 is rated at 350 kVA, 0.84 PF lagging. Determine the power line current.

5.43. A balanced three-phase source serves three loads:

Load 1: 88 kVA at 0.9 PF lagging

Load 2: 24 kVA at 0.7 PF leading

Load 3: 18 kW at unity power factor

The load voltage is 480 V rms at 60 Hz. If the line impedance is $0.06 + j0.04 \ \Omega$, find the line voltage and power factor at the source.

Network Frequency Characteristics

6.1
Introduction

In this chapter we examine the frequency characteristics of electrical networks. Of particular interest to us will be the effect of the source frequency on a network. Techniques for determining the frequency response of a network are presented, and standard plots that display the network's performance as a function of frequency are discussed.

The concept of resonance is examined. The various parameters used to define the frequency selectivity of a network, such as bandwidth, cutoff frequency, and quality factor, are discussed.

Networks with special filtering properties are examined. Specifically, low-pass, high-pass, band-pass, and band-elimination filters are discussed.

Networks that exhibit particular filtering characteristics have wide application in numerous types of communication and control systems where it is necessary to pass certain frequencies and reject others.

6.2
Sinusoidal Frequency Analysis

Many dynamic systems display phenomena that are frequency dependent. An example is a tuning fork; regardless of how the fork is pulsed, it will vibrate at only one frequency. In addition, electrical networks are designed to operate at a single frequency, for example, the power grid that includes the ac outlets in our homes and offices. However, as a general rule, circuits will respond differently to sinusoidal signals of different frequency. Recall from Example 4.5 that the network characteristics were quite different at the frequency used in aircraft power systems than at the U.S. power grid frequency. In this section we add a new dimension to our analysis of electric circuits by examining the performance of a network as a function of frequency. We are typically interested in the *sinusoidal frequency response,* or what we simply call the *frequency response,* of a circuit. The frequency response defines the behavior of the network as a function of frequency. Consider the following example, which demonstrates that networks respond differently at different frequencies.

EXAMPLE 6.1

Let us determine the transfer function, that is, the ratio of output voltage to input voltage, for the circuit in Fig. 6.1(a), and sketch the magnitude and phase of this function over the frequency range 0 to 8 rad/sec.

Using voltage division, the output voltage can be expressed as

$$\mathbf{V}_0 = \left(\frac{R}{j\omega L + \dfrac{1}{j\omega C} + R} \right) \mathbf{V}_1$$

and therefore the transfer function is

$$\frac{\mathbf{V}_0}{\mathbf{V}_1}(j\omega) = \frac{j\omega \dfrac{R}{L}}{(j\omega)^2 + j\omega \dfrac{R}{L} + \dfrac{1}{LC}}$$

which with the parameter values is

$$\frac{\mathbf{V}_0}{\mathbf{V}_1}(j\omega) = \frac{2j\omega}{(j\omega)^2 + 2j\omega + 17}$$

If we now evaluate this function at $\omega = 0, 1, 2, \ldots 8$, we obtain the following values of the magnitude and phase as a function of frequency.

ω	0	1	2	3	4	$\sqrt{17}$	5	6	7	8
Magnitude	0	0.12	0.29	0.60	0.99	1.0	0.78	0.53	0.40	0.32
Phase	90°	83°	73°	53°	7°	0°	−39°	−58°	−66°	−71°

A sketch of the magnitude and phase characteristics for the network transfer function is shown in Fig. 6.1(b). Thus, the network response is quite different depending upon the frequency of the input voltage source. ■

The frequency response of certain types of circuits is extremely important, and enormous effort is often expended in the design of these networks in order to achieve certain frequency response characteristics. For example, in a stereo system the amplifiers are often designed so that each frequency in the audio range is amplified the same amount, that is, the magnitude-vs.-frequency characteristic is flat in the audio range. If each frequency in the audio range is amplified the same amount, high-fidelity sound reproduction is achieved. However, if the magnitude-vs.-frequency characteristic is not flat in the audio range, some frequencies are amplified more than others, distortion occurs, and the output is not an exact duplicate of the input. An "equalizer" is a component in an audio system that allows the user to adjust the frequency response of the system to the individual taste of the listener.

The previous example indicates something that is true in general, that is, transfer functions can be expressed as the ratio of two polynomials in $j\omega$. In addition, we note that since the values of our circuit elements, or controlled sources, are real numbers, the coefficients of the two polynomials will be real. Therefore, we will express transfer function, which in general we will call $\mathbf{G}(j\omega)$, in the form

$$\mathbf{G}(j\omega) = \frac{N(j\omega)}{D(j\omega)} = \frac{a_m(j\omega)^m + a_{m-1}(j\omega)^{m-1} + \ldots + a_1 j\omega + a_0}{b_n(j\omega)^n + b_{n-1}(j\omega)^{n-1} + \ldots + b_1 j\omega + b_0} \tag{6.1}$$

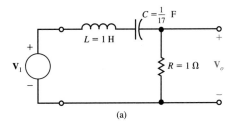

(a)

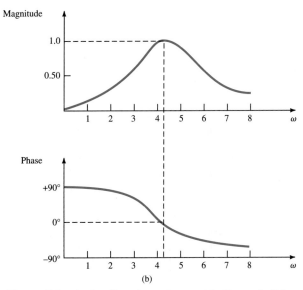

(b)

Figure 6.1   A circuit and graphs used in Example 6.1.

where $N(j\omega)$ is the numerator polynomial of degree m and $D(j\omega)$ is the denominator polynomial of degree n. Equation (6.1) can also be written in the form

$$\mathbf{G}(j\omega) = \frac{K_0(j\omega - z_1)(j\omega - z_2)\dots(j\omega - z_m)}{(j\omega - p_1)(j\omega - p_2)\dots(j\omega - p_n)} \qquad (6.2)$$

where K_0 is a constant, $z_1 \dots z_m$ are the roots of $N(j\omega)$, and $p_1, \dots p_n$ are the roots of $D(j\omega)$. Note that if $j\omega = z_1$ or $z_2 \dots z_m$, then $\mathbf{G}(j\omega)$ becomes zero and hence $z_1 \dots z_m$ are called *zeros* of the function. Similarly, if $j\omega = p_1$ or $p_2 \dots p_n$, then $\mathbf{G}(j\omega)$ becomes infinite and therefore $p_1 \dots p_n$ are called *poles* of the function. The zeros or poles may actually be complex. However, if they are complex, they must occur in conjugate pairs since the coefficients of the polynomial are real. The importance of this form (Eq. 6.2) stems from the fact that the dynamic properties of a system can be gleaned from an examination of the system poles. For example, recall that the transient performance of circuits with a second-order characteristic equation could be overdamped, underdamped, or critically damped depending upon the values of the circuit's poles.

DRILL EXERCISE

D6.1. Determine the transfer function $\dfrac{\mathbf{V}_0}{\mathbf{V}_1}(j\omega)$ for the circuit in Fig. D6.1 and locate the poles and zeros of the function.

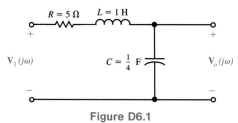

Figure D6.1

Ans: $\dfrac{\mathbf{V}_0}{\mathbf{V}_1}(j\omega) = \dfrac{4}{(j\omega)^2 + 5jw + 4}$, there are no zeros and the pole locations are $j\omega = -1$ and $j\omega = -4$.

6.3
Passive Filter Networks

A filter network is generally designed to pass signals with a specific frequency range and reject or attenuate signals whose frequency spectrum is outside this passband. The most common filters are *low-pass filters,* which pass low frequencies and reject high frequencies; *high-pass filters,* which pass high frequencies and block low frequencies; *band-pass filters,* which pass some particular band of frequencies and reject all frequencies outside the range; and *band-rejection filters,* which are specifically designed to reject a particular band of frequencies and pass all other frequencies.

The ideal frequency characteristic for a low-pass filter is shown in Fig. 6.2(a). Also shown is a typical or physically realizable characteristic. Ideally, we would like the low-pass filter to pass all frequencies up to some frequency ω_0 and pass no frequency above that value; however, it is not possible to design such a filter with linear circuit elements. Hence, we must be content to employ filters that we can actually build in the laboratory, and these filters have frequency characteristics that are simply not ideal.

A simple low-pass filter network is shown in Fig. 6.2(b). The voltage gain for the network is

$$\frac{\mathbf{V}_0}{\mathbf{V}_1}(j\omega) = \frac{\dfrac{1}{j\omega C}}{R + \dfrac{1}{j\omega C}}$$

This transfer function can be expressed as $\mathbf{G}_v(j\omega)$ in the form

$$\mathbf{G}_v(j\omega) = \frac{1}{1 + j\omega RC} \tag{6.3}$$

which can be written as

$$\mathbf{G}_v(j\omega) = \frac{1}{1 + j\omega \tau} \tag{6.4}$$

where $\tau = RC$, the *time constant.*

In general, in a sinusoidal steady state analysis the transfer function can be expressed as

$$\mathbf{G}(j\omega) = M(\omega)e^{j\phi(\omega)}$$

where $M(\omega) = |\mathbf{G}(j\omega)|$ is the magnitude function and $\phi(\omega)$ is the phase. For the low-pass filter, the magnitude function is

$$\mathbf{M}(\omega) = \frac{1}{[1 + (\omega\tau)^2]^{1/2}} \tag{6.5}$$

and the phase characteristic is

$$\phi(\omega) = -\tan^{-1}\omega\tau \tag{6.6}$$

The magnitude and phase curves for this simple low-pass circuit are shown in Fig. 6.2(c). Note that the magnitude curve is flat for low frequencies and rolls off at high frequencies. The phase shifts from 0° at low frequencies to $-90°$ at high frequencies. Note that at the frequency $\omega = 1/\tau$, which we call the *break frequency*, the amplitude is

$$\mathbf{M}(\omega = 1/\tau) = \frac{1}{\sqrt{2}} \tag{6.7}$$

The break frequency is also commonly called the *half-power frequency.* This name is derived from the fact that if the voltage or current is $1/\sqrt{2}$ of its maximum value, then the power, which is proportional to the square of the voltage or current, is one-half its maximum value. The phase angle at the break frequency is $-45°$.

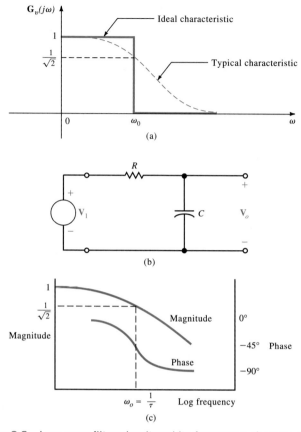

Figure 6.2   Low-pass filter circuit and its frequency characteristics.

The ideal frequency characteristic for a high-pass filter is shown in Fig. 6.3(a) together with a typical characteristic that we could achieve with linear circuit components. Ideally, the high-pass filter passes all frequencies above some frequency ω_0 and no frequencies below that value.

A simple high-pass filter network is shown in Fig. 6.3(b). This is a network similar to that shown in Fig. 6.2(b) with the output voltage taken across the resistor. The voltage gain for this network is

$$\mathbf{G}_v(j\omega) = \frac{j\omega\tau}{1 + j\omega\tau} \tag{6.8}$$

where once again $\tau = RC$. The magnitude of this function is

$$M(\omega) = \frac{\omega\tau}{[1 + (\omega\tau)^2]^{1/2}} \tag{6.9}$$

and the phase is

$$\phi(\omega) = \frac{\pi}{2} - \tan^{-1}\omega\tau \tag{6.10}$$

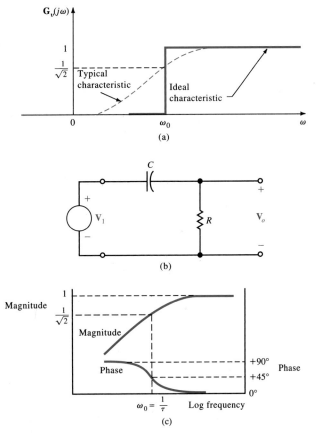

Figure 6.3 High-pass filter circuit and its frequency characteristics.

The half-power frequency is $\omega = 1/\tau$, and the phase at this frequency is 45°.

The magnitude and phase curves for this high-pass filter are shown in Fig. 6.3(c). At low frequencies the magnitude curve has a positive slope due to the term $\omega\tau$ in the numerator of Eq. (6.9). Then at the break frequency the curve begins to flatten out. The phase curve is derived from Eq. (6.10).

Note that since we know the basic form of the low-pass and high-pass filters shown in Figs. 6.2 and 6.3, we can select parameters to *design* a filter with certain characteristics.

EXAMPLE 6.2

A source generates a low-frequency signal which contains noise. It appears that the noise could be eliminated with a simple low-pass filter with a half-power frequency of 1 kHz. If a 1.6 μF capacitor is available, select a resistor which when used with the capacitor in the network in Fig. 6.2 will eliminate the noise.

At the half-power frequency, $\omega\tau = 1$. Therefore

$$\omega RC = 1$$

or

$$R = \frac{1}{2\pi(10)^3(1.6)(10)^{-6}}$$
$$\cong 100\ \Omega$$ ∎

DRILL EXERCISE

D6.2. Design a simple RC high-pass filter with a break frequency of 20 kHz. Use a resistor value of 1 kΩ.

Ans: $C = 8\ nF$.

Ideal and typical amplitude characteristics for simple band-pass and band-rejection filters are shown in Figs. 6.4(a) and (b), respectively. Simple networks that are capable of realizing the band-pass and band-rejection filters are shown in Figs. 6.4(c) and (d), respectively. ω_0 is the *center frequency* of the pass or rejection band and the frequency at which the maximum or minimum amplitude occurs. ω_{LO} and ω_{HI} are the *lower* and *upper break frequencies* or *cutoff frequencies,* where the amplitude is $1/\sqrt{2}$ of the maximum value. The width of the pass or rejection band is called the *bandwidth,* and hence

$$BW = \omega_{HI} - \omega_{LO} \tag{6.11}$$

To illustrate these points, let us consider the band-pass filter in Fig. 6.4(c). Note carefully that this is the same network examined in Example 6.1. The voltage transfer function is

$$\mathbf{G}_v(j\omega) = \frac{R}{R + j(\omega L - 1/\omega C)}$$

and therefore the amplitude characteristic is

$$M(\omega) = \frac{RC\omega}{\sqrt{(RC\omega)^2 + (\omega^2 LC - 1)^2}} \tag{6.12}$$

It is tedious but straightforward to show that the center frequency obtained from the expression

$$\frac{dM(\omega)}{d\omega} = 0$$

yields

$$\omega_0 = \frac{1}{\sqrt{LC}}$$

At the upper and lower cutoff frequencies, the magnitude characteristic is

$$M(\omega) = \frac{1}{\sqrt{2}} \tag{6.13}$$

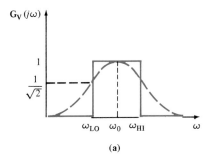

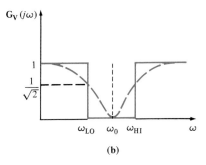

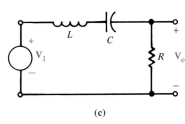

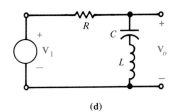

Figure 6.4   Band-pass and band-rejection filters and characteristics.

Substituting this value into Eq. (6.12) and simplifying the resulting expression yields the equation

$$\omega^2 LC - 1 = \pm RC\omega \tag{6.14}$$

At the lower cutoff frequency

$$\omega^2 LC - 1 = -RC\omega \tag{6.15}$$

and using the fact that $\omega_0 = 1/\sqrt{LC}$, Eq. (6.15) can be expressed in the form

$$\omega^2 + \frac{R}{L}\omega - \omega_0^2 = 0 \tag{6.16}$$

Solving this expression for ω_{LO}, we obtain

$$\omega_{LO} = \frac{-(R/L) + \sqrt{(R/L)^2 + 4\omega_0^2}}{2}$$

At the upper cutoff frequency

$$\omega^2 LC - 1 = +RC\omega$$

or

$$\omega^2 - \frac{R}{L}\omega - \omega_0^2 = 0$$

Solving this expression for ω_{HI}, we obtain

$$\omega_{HI} = +(R/L) + \frac{\sqrt{(R/L)^2 + 4\omega_0^2}}{2}$$

Therefore, the bandwidth of the filter is

$$BW = \omega_{HI} - \omega_{LO} = \frac{R}{L} \tag{6.17}$$

In addition, we note that

$$\begin{aligned}
\omega_{LO}\omega_{HI} &= \frac{-R^2}{4L^2} + \frac{(R/L)^2 + \dfrac{4}{LC}}{4} \\[2mm]
&= \frac{1}{LC} \\[2mm]
&= \omega_0^2
\end{aligned} \tag{6.18}$$

which illustrates that *the center frequency is the geometric mean of the two half-power frequencies.*

EXAMPLE 6.3

Consider the band-pass filter network shown in Fig. 6.5. Let us determine the filter's center frequency, half-power frequencies, and bandwidth.

The center frequency is

$$\omega_0 = \frac{1}{\sqrt{LC}} = \frac{1}{\sqrt{(1)(10)^{-6}}}$$

$$= 1 \text{ k rad/s}$$

The half-power frequencies are

$$\begin{aligned}
\omega_{LO} &= \frac{-\left(\dfrac{R}{L}\right) + \sqrt{\left(\dfrac{R}{L}\right)^2 + 4\omega_0^2}}{2} \\[3mm]
&= \frac{-1000 + \sqrt{10^6 + 4(10)^6}}{2} \\[3mm]
&= \frac{-1000 + 10^3\sqrt{5}}{2} \\[3mm]
&= 618 \text{ rad/s}
\end{aligned}$$

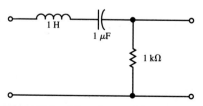

Figure 6.5   A band-pass filter network.

and

$$\omega_{HI} = \frac{1000 + 10^3\sqrt{5}}{2}$$

$$= 1618 \text{ rad/s}$$

The bandwidth, obtained from either Eq. (6.17) or (6.18), is $BW = 1000$ rad/s. ∎

DRILL EXERCISE

D6.3. Determine the half-power frequency for each of the filters in Fig. D6.3 and identify the type of filter.

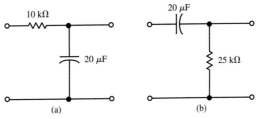

Figure D6.3

Ans: (a) Low-pass filter, $\omega_{HI} = 5$ rad/s (b) High-pass filter $\omega_{LO} = 2$ rad/s.

6.4
Resonant Circuits

Resonance is a fascinating phenomenon which exists in a wide variety of engineering systems. For example, consider an old automobile with weak shock absorbers such that the suspension system can be approximated by a second-order underdamped system. If we press down on the front end of the car, we find that it is difficult to do so. However, if we push the front end down in a repetitive mode with the proper rhythm, that is, at the proper frequency, we find that the front end of the automobile will begin to oscillate and we will not have to use as much force to achieve fairly wide deflections. What is happening in this case is that we are forcing the mechanical system at its resonant frequency.

As another example, it is interesting to note that the engineers designing the attitude control system for the Saturn moon rocket had to ensure that the control system frequency did not excite the body bending (resonant) frequencies of the rocket. Excitation of the bending frequencies would cause oscillation which, if continued unchecked, would result in a buildup of stress until the vehicle would finally break apart.

Having briefly mentioned some practical aspects of resonance, let us now examine the phenomenon in some basic circuits.

Two fundamental circuits with extremely important frequency characteristics are shown in Fig. 6.6. The input impedance for the series *RLC* circuit is

$$\mathbf{Z}(j\omega) = R + j\omega L + \frac{1}{j\omega C} \tag{6.19}$$

and the input admittance for the parallel *RLC* circuit is

$$\mathbf{Y}(j\omega) = G + j\omega C + \frac{1}{j\omega L} \qquad (6.20)$$

Note that these two equations have the same general form. The imaginary terms in both of the equations above will be zero if

$$\omega L = \frac{1}{\omega C}$$

The value of ω that satisfies this equation is

$$\omega_0 = \frac{1}{\sqrt{LC}} \qquad (6.21)$$

and at this value of ω the impedance of the series circuit becomes

$$\mathbf{Z}(j\omega_0) = R \qquad (6.22)$$

and the admittance of the parallel circuit is

$$\mathbf{Y}(j\omega_0) = G \qquad (6.23)$$

This frequency ω_0, at which the impedance of the series circuit or the admittance of the parallel circuit is purely real, is called the *resonant frequency,* and the circuits themselves, at this frequency, are said to be *in resonance.*

At resonance the voltage and current are in phase, and therefore the phase angle is zero and the power factor is unity. In the series case, at resonance the impedance is a minimum, and therefore the current is maximum for a given voltage. At low frequencies, the impedance of the series circuit is dominated by the capacitive term and the admittance of the parallel circuit is dominated by the inductive term. At high frequencies, the impedance of the series circuit is dominated by the inductive term and the admittance of the parallel circuit is dominated by the capacitive term.

Resonance can be viewed from another perspective—that of the phasor diagram. Once again we will consider the series and parallel cases together in order to illustrate the similarities between them. In the series case the current is common to every element, and in the parallel case the voltage is a common variable. Therefore, the current in the series circuit and the voltage in the parallel circuit are employed as references. Phasor diagrams for both circuits are shown in Fig. 6.7 for the three frequency values $\omega < \omega_0$, $\omega = \omega_0$, and $\omega > \omega_0$.

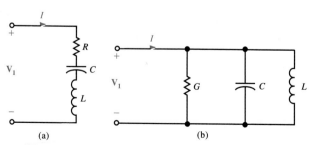

Figure 6.6 Series and parallel *RLC* circuits.

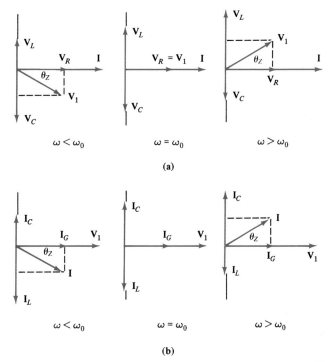

Figure 6.7   Phasor diagrams for (a) a series *RLC* circuit, and (b) a parallel *GLC* circuit.

In the series case when $\omega < \omega_0$, $\mathbf{V}_C > \mathbf{V}_L$, θ_Z is negative, and the voltage $\mathbf{V}_1$ lags the current. If $\omega = \omega_0$, $\mathbf{V}_L = \mathbf{V}_C$, θ_Z is zero, and the voltage $\mathbf{V}_1$ is in phase with the current. If $\omega > \omega_0$, $\mathbf{V}_L > \mathbf{V}_C$, θ_Z is positive, and the voltage $\mathbf{V}_1$ leads the current. Similar statements can be made for the parallel case in Fig. 6.7(b). Because of the close relationship between series and parallel resonance, as illustrated by the preceding material, we will concentrate most of our discussion on the series case in the following developments.

For the series circuit we define what is commonly called the *quality factor Q* as

$$Q = \frac{\omega_0 L}{R} \tag{6.24}$$

Using Eq. (6.21), the quality factor can be expressed in the alternate forms.

$$Q = \frac{1}{\omega_0 CR} = \frac{1}{R}\frac{\sqrt{L}}{C} \tag{6.25}$$

Q is a very important factor in resonant circuits, and its ramifications will be illustrated throughout the remainder of this section.

EXAMPLE 6.4

Consider the network shown in Fig. 6.8. Let us determine the resonant frequency, the voltage across each element at resonance, and the value of the quality factor.

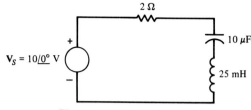

Figure 6.8   Series circuit.

The resonant frequency is obtained from the expression

$$\omega_0 = \frac{1}{\sqrt{LC}}$$

$$= \frac{1}{\sqrt{(25)(10^{-3})(10)(10^{-6})}}$$

$$= 2000 \text{ rad/s}$$

At this resonant frequency

$$\mathbf{I} = \frac{\mathbf{V}}{\mathbf{Z}} = \frac{\mathbf{V}}{R} = 5\angle 0° \text{ A}$$

Therefore

$$\mathbf{V}_R = (5\angle 0°)(2) = 10\angle 0° \text{ V}$$

$$\mathbf{V}_L = j\omega_0 L\mathbf{I} = 250\angle 90° \text{ V}$$

$$\mathbf{V}_C = \frac{\mathbf{I}}{j\omega_0 C} = 250\angle -90° \text{ V}$$

Note the magnitude of the voltages across the inductor and capacitor with respect to the input voltage. Note also that these voltages are equal and 180° out of phase with one another. Therefore, the phasor diagram for this condition is shown in Fig. 6.7 for $\omega = \omega_0$. The quality factor Q derived from Eq. (6.24) is

$$Q = \frac{\omega_0 L}{R} = \frac{(2)(10^3)(25)(10^{-3})}{2} = 25 \qquad \blacksquare$$

It is interesting to note that the voltages across the inductor and capacitor can be written in terms of Q as

$$|\mathbf{V}_L| = \omega_0 L|\mathbf{I}| = \frac{\omega_0 L}{R}|\mathbf{V}_s| = Q|\mathbf{V}_s|$$

and

$$|\mathbf{V}_c| = \frac{|\mathbf{I}|}{\omega_0 C} = \frac{1}{\omega_0 CR}|\mathbf{V}_s| = Q|\mathbf{V}_s|$$

This analysis indicates that for a given current there is a resonant voltage rise across the inductor and capacitor which is equal to the product of Q and the applied voltage.

D6.4. Given the network in Fig. D6.4, find the value of C that will place the circuit in resonance at 1800 rad/s.

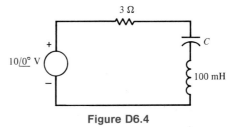

Figure D6.4

Ans: $C = 3.09\ \mu F.$

D6.5. Given the network in D6.4, determine the Q of the network and the magnitude of the voltage across the capacitor.

Ans: $Q = 60, |\mathbf{V}_c| = 600$ V.

Equation (6.24) indicates the dependence of Q on R. A high-Q series circuit has a small value of R, and as we will illustrate later, a high-Q parallel circuit has a relatively large value of R.

In the previous section we showed that the circuit bandwidth could be expressed as

$$BW = \omega_{HI} - \omega_{LO} = \frac{R}{L}$$

which can be written as

$$BW = \frac{R\omega_0}{L\omega_0}$$

$$= \frac{\omega_0}{Q}$$

(6.26)

This equation illustrates that the bandwidth is inversely proportional to Q. Therefore, the *frequency selectivity* of the circuit is determined by the value of Q. A high-Q circuit has a small bandwidth, and therefore the circuit is very selective. The manner in which Q affects the frequency selectivity of the network is graphically illustrated in Fig. 6.9. Hence, if we pass a signal with a wide frequency range through a high-Q circuit, only the frequency components within the bandwidth of the network will not be attenuated; that is, the network acts like a band-pass filter.

From an energy standpoint, recall that an inductor stores energy in its magnetic field and a capacitor stores energy in its electric field. When a network is in resonance, there is a continuous exchange of energy between the magnetic field of the inductor and the electric field of the capacitor. During each half-cycle, the energy stored in the inductor's magnetic field will vary from zero to a maximum value and back to zero again. The

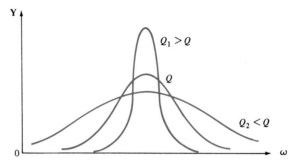

Figure 6.9   Network frequency response as a function of Q.

capacitor operates in a similar manner. The energy exchange takes place in the following way. During one quarter-cycle, the capacitor absorbs energy as quickly as the inductor gives it up, and during the following one quarter-cycle, the inductor absorbs energy as fast as it is released by the capacitor. Energy is exchanged back and forth between the two reactive elements.

EXAMPLE 6.5

Given a series *RLC* circuit with $R = 2\ \Omega$, $L = 2$ mH, and $C = 5\ \mu$F, we wish to determine the resonant frequency, the quality factor, and the bandwidth for the circuit. Then we wish to determine the change in Q and the *BW* if R is changed from 2 to 0.2 Ω.

Using Eq. (6.21), we have

$$\omega_0 = \frac{1}{\sqrt{LC}} = \frac{1}{[(2)(10^{-3})(5)(10^{-6})]^{1/2}}$$
$$= 10^4 \text{ rad/s}$$

and therefore the resonant frequency is $10^4/2\pi = 1592$ Hz. The quality factor is

$$Q = \frac{\omega_0 L}{R} = \frac{(10^4)(2)(10^{-3})}{2}$$
$$= 10$$

and the bandwidth is

$$BW = \frac{\omega_0}{Q} = \frac{10^4}{10}$$
$$= 10^3 \text{ rad/s}$$

If R is changed to $R = 0.2\ \Omega$, the new value of Q is 100, and therefore the new *BW* is 10^2 rad/s.   ∎

EXAMPLE 6.6

We wish to determine the parameters R, L, and C so that the circuit shown in Fig. 6.10 operates as a band-pass filter with an ω_0 of 1000 rad/s and a bandwidth of 100 rad/s.

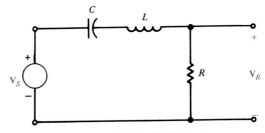

Figure 6.10 Series *RLC* circuit.

The resonant frequency for the circuit is

$$\omega_0 = \frac{1}{\sqrt{LC}}$$

and since $\omega_0 = 10^3$

$$\frac{1}{\sqrt{LC}} = 10^6$$

The bandwidth is

$$BW = \frac{\omega_0}{Q}$$

and hence

$$Q = \frac{\omega_0}{BW} = \frac{1000}{100}$$
$$= 10$$

However

$$Q = \frac{\omega_0 L}{R}$$

and therefore

$$\frac{1000L}{R} = 10$$

This equation, together with the equation for the resonant frequency, represents *two* equations in the *three* unknown circuit parameters R, L, and C. Hence, if we select $C = 1$ μF, then

$$L = \frac{1}{10^6 C} = 1 \text{ H}$$

and

$$\frac{1000(1)}{R} = 10$$

yields

$$R = 100 \ \Omega$$

Therefore, the parameters $R = 100 \ \Omega$, $L = 1$ H, and $C = 1 \ \mu F$ are one set of parameters that will produce the proper filter characteristics. ∎

DRILL EXERCISES

D6.6. A series RLC circuit is composed of $R = 2 \ \Omega$, $L = 40$ mH, and $C = 100 \ \mu F$. Determine the bandwidth of this circuit about its resonant frequency.

Ans: BW = 50 rad/s, ω_0 = 500 rad/s.

D6.7. A series RLC circuit has the following properties: $R = 4 \ \Omega$, $\omega_0 = 4000$ rad/s, and the $BW = 100$ rad/s. Determine the values of L and C.

Ans: L = 40 mH, C = 1.56 μF.

In our presentation of resonance thus far, we have focused most of our discussion on the series resonant circuit. We should recall, however, that the equations for the impedance of the series circuit and the admittance of the parallel circuit are similar. Therefore, the networks possess similar properties.

Consider the network shown in Fig. 6.11. The source current $\mathbf{I}_s$ can be expressed as

$$\mathbf{I}_s = \mathbf{I}_G + \mathbf{I}_C + \mathbf{I}_L$$

$$= \mathbf{V}_s G + j\omega C \mathbf{V}_s + \frac{\mathbf{V}_s}{j\omega L}$$

$$= \mathbf{V}_s \left[G + j\left(\omega C - \frac{1}{\omega L} \right) \right]$$

When the network is in resonance

$$\mathbf{I}_s = G \mathbf{V}_s$$

that is, all the source current flows through the conductance G. Does this mean that there is no current in L or C? Definitely not! $\mathbf{I}_c$ and $\mathbf{I}_L$ are equal in magnitude but 180° out of phase with one another and there is an energy exchange between the electric field of the capacitor and the magnetic field of the inductor. As one increases, the other decreases, and vice versa.

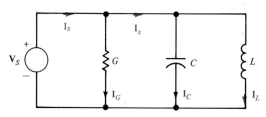

Figure 6.11   Parallel *RLC* circuit.

EXAMPLE 6.7

The network in Fig. 6.11 has the following parameters:

$$\mathbf{V}_s = 120 \; \underline{/0°} \; \text{V}, \; G = 0.01 \; \text{S}, \; C = 600 \; \mu\text{F}, \; \text{and} \; L = 120 \; \text{mH}.$$

If the source operates at the resonant frequency of the network, compute all the branch currents.

The resonant frequency for the network is

$$\omega_0 = \frac{1}{\sqrt{LC}}$$

$$= \frac{1}{\sqrt{(120)(10^{-3})(600)(10^{-6})}}$$

$$= 117.85 \; \text{rad/s}$$

At this frequency

$$\mathbf{Y}_C = j\omega_0 C = j \, 7.07 \times 10^{-2} \; \text{S}$$

and

$$\mathbf{Y}_L = -j\left(\frac{1}{\omega_0 L}\right) = -j \, 7.07 \times 10^{-2} \; \text{S}$$

The branch currents are then

$$\mathbf{I}_G = G\mathbf{V}_s = 1.2 \underline{/0°} \; \text{A}$$
$$\mathbf{I}_C = Y_C \mathbf{V}_s = 8.49 \underline{/90°} \; \text{A}$$
$$\mathbf{I}_L = Y_L \mathbf{V}_s = 8.49 \underline{/-90°} \; \text{A}$$

and

$$\mathbf{I}_s = \mathbf{I}_G + \mathbf{I}_C + \mathbf{I}_L = 1.2 \underline{/0°} \; \text{A}$$

As the analysis indicates, the source supplies only the losses in the resistive element. In addition, the source voltage and current are in phase and therefore the power factor is unity.   ∎

Consider the network in Fig. 6.12. The output voltage can be written as

$$\mathbf{V}_{out} = \frac{\mathbf{I}_{in}}{\mathbf{Y}_t}$$

and therefore the magnitude of the transfer characteristic can be expressed as

$$\left|\frac{\mathbf{V}_{out}}{\mathbf{I}_{in}}\right| = \frac{1}{\sqrt{(1/R^2) + (\omega C - 1/\omega L)^2}} \tag{6.27}$$

$$\omega_0 = \frac{1}{\sqrt{LC}} \tag{6.28}$$

and at this frequency

$$\left|\frac{\mathbf{V}_{out}}{\mathbf{I}_{in}}\right|_{max} = R \tag{6.29}$$

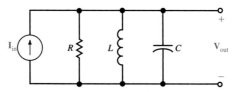

Figure 6.12   A parallel *RLC* circuit.

As demonstrated earlier, at the half-power frequencies the magnitude is equal to $1/\sqrt{2}$ of its maximum value, and hence the half-power frequencies can be obtained from the expression

$$\frac{1}{\sqrt{(1/R^2) + (\omega C - 1/\omega L)^2}} = \frac{R}{\sqrt{2}} \tag{6.30}$$

Solving this equation and taking only the positive values of ω yields

$$\omega_{LO} = -\frac{1}{2RC} + \sqrt{\frac{1}{(2RC)^2} + \frac{1}{LC}} \tag{6.31}$$

and

$$\omega_{HI} = \frac{1}{2RC} + \sqrt{\frac{1}{(2RC)^2} + \frac{1}{LC}} \tag{6.32}$$

Subtracting these two half-power frequencies yields the bandwidth

$$BW = \omega_{HI} - \omega_{LO}$$
$$= \frac{1}{RC} \tag{6.33}$$

Therefore, the quality factor is

$$Q = \frac{\omega_0}{BW}$$
$$= \frac{RC}{\sqrt{LC}} \tag{6.34}$$
$$= R\sqrt{\frac{C}{L}}$$

EXAMPLE 6.8

A stereo receiver is tuned to 98 MHz on the FM band. The tuning knob controls a variable capacitor in a parallel resonant circuit. If the inductance of the circuit is $0.1\,\mu$H and the Q is 120, determine the values of C and G.

Using the expression for the resonant frequency, we obtain

$$C = \frac{1}{\omega_0^2 L}$$

$$= \frac{1}{(2\pi \times 98 \times 10^6)^2 (0.1 \times 10^{-6})}$$

$$= 26.4 \text{ PF}$$

The conductance is obtained from Eq. (6.34) where

$$Q = \frac{RC}{\sqrt{LC}}$$

which can be expressed using the equation for the resonant frequency as

$$Q = RC\omega_0$$

If C is expressed in terms of the resonant frequency, we obtain

$$Q = \frac{R}{\omega_0 L}$$

and since $G = 1/R$, the conductance is

$$\mathbf{G} = \frac{1}{\omega_0 L Q}$$

$$= \frac{1}{(2\pi \times 98 \times 10^6)(10^{-7})(120)}$$

$$= 135 \ \mu S \qquad \blacksquare$$

DRILL EXERCISES

D6.8. A parallel *RLC* circuit has the following parameters: $R = 2 \text{ k}\Omega$, $L = 20 \text{ mH}$, and $C = 150 \ \mu F$. Determine the resonant frequency, the Q, and the bandwidth of the circuit.

Ans: $\omega_0 = 577$ rad/s, $BW = 3.33$ rad/s, $Q = 173$.

D6.9. A parallel *RLC* circuit has the following parameters: $R = 6 \text{ k}\Omega$, $BW = 1000$ rad/s, and $Q = 120$. Determine the values of L, C, and ω_0.

Ans: $C = 0.167 \ \mu F$, $L = 416.7 \ \mu H$, $\omega_0 = 119760$ rad/s.

6.5
Summary

- Network transfer functions can be expressed as a ratio of two polynomials in $j\omega$. The roots of the numerator polynomial are called zeros and the roots of the denominator polynomial are called poles.
- Four important types of passive filter networks are low-pass, high-pass, band-pass, and band-rejection.

- The half-power frequency for filter networks is the frequency at which the voltage or current is $1/\sqrt{2}$ of its maximum value.
- The frequency range between a band-pass or band-elimination filter's upper and lower half-power frequencies is called the filter bandwidth.
- The resonant frequency is the frequency at which the impedance of a series RLC circuit or the admittance of a parallel RLC circuit is purely real.
- Q is a factor which specifies a network's frequency selectivity, for example, high-Q circuits are very selective and low-Q circuits are not selective.

PROBLEMS

6.1. Find the transfer function $\dfrac{V_0}{V_s}(j\omega)$ for the network in Fig. P6.1.

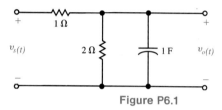

Figure P6.1

6.2. Find the transfer function $\dfrac{V_0}{V_s}(j\omega)$ for the circuit in Fig. P6.2.

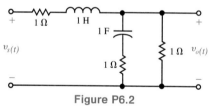

Figure P6.2

6.3. Find the transfer function $\dfrac{I_0}{V_s}(j\omega)$ for the network in Fig. P6.3.

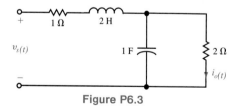

Figure P6.3

6.4. Given the following transfer function $\mathbf{G}(j\omega)$, tabulate the values of $M(\omega)$ and $\phi(\omega)$ versus ω for $0 \le \omega \le 10$ rad/s in increments of 1 rad/s.

$$\mathbf{G}(j\omega) = \frac{j\omega + 2}{(j\omega + 6)(j\omega + 8)}$$

6.5. Find the poles and zeros of the transfer function $\dfrac{V_0}{V_s}(j\omega)$ for the circuit in Fig. P6.5.

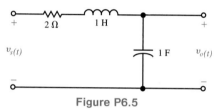

Figure P6.5

6.6. Find the poles and zeros of the transfer function $\dfrac{V_0}{V_s}(j\omega)$ for the network in Fig. P6.6.

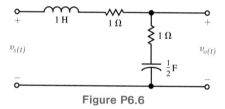

Figure P6.6

6.7. An RC low-pass filter has the following parameters: $R = 1$ kΩ and $C = 50$ μF. Determine the half-power frequency of the filter.

6.8. An RC high-pass filter has the following parameters: $R = 10$ kΩ and $C = 200$ μF. Determine the half-power frequency of the filter and the phase angle of the magnitude transfer function at this frequency.

6.9. A simple low-pass RC filter employs a 20 Ω resistor. Find the capacitor value which will produce a half-power frequency of 18 kHz.

6.10. A band-pass filter has a center frequency of 924 rad/s. If the upper half-power frequency is 1132 rad/s, determine the bandwidth of the filter.

6.11. Compute the voltage transfer function for the network shown in Fig. P6.11 and tell what type of filter the network represents.

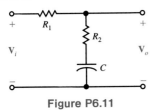

Figure P6.11

6.12. Determine what type of filter the network in Fig. P6.12 represents by determining the voltage transfer function.

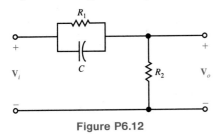

Figure P6.12

6.13. Given the lattice network shown in Fig. P6.13, determine what type of filter this network represents by determining the voltage transfer function.

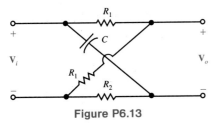

Figure P6.13

6.14. Determine the type of filter that is represented by the network in Fig. P6.14.

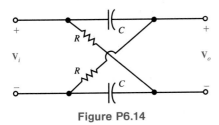

Figure P6.14

6.15. Given the network in Fig. P6.15, find the value of C that will place the circuit in resonance at 1800 rad/s.

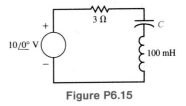

Figure P6.15

6.16. Given the network in P6.15, determine the Q of the network and the magnitude of the voltage across the capacitor.

6.17. For the network in Fig. P6.15, compute the two half-power frequencies and the bandwidth of the network.

6.18. Given the series RLC circuit in Fig. P6.18, if $R = 10\ \Omega$, find the values of L and C such that the network will have a resonant frequency of 100 kHz and a bandwidth of 1 kHz.

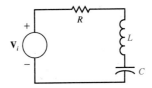

Figure P6.18

6.19. A series circuit is composed of $R = 2\ \Omega$, $L = 40$ mH, and $C = 100\ \mu$F. Determine the bandwidth of this circuit about its resonant frequency.

6.20. A series RLC circuit has the following properties: $R = 4\ \Omega$, $\omega_0 = 4000$ rad/s, and the $BW = 100$ rad/s. Determine the values of L and C.

6.21. A series RLC circuit resonates at 2000 rad/s. If $C = 20\ \mu$F and it is known that the impedance at resonance is $2.4\ \Omega$, compute the value of L, the Q of the circuit, and the bandwidth.

6.22. A series resonant circuit has a Q of 120 and a resonant frequency of 60,000 rad/s. Determine the half-power frequencies and the bandwidth of the circuit.

6.23. Given the network in Fig. P6.23, find ω_0 and Q.

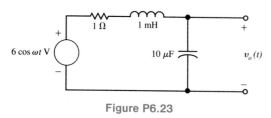

Figure P6.23

6.24. Repeat problem 6.23 if the value of R is changed to 0.1 Ω.

6.25. Given the series RLC circuit in Fig. P6.25,
(a) Derive the expression for the half-power frequencies, the resonant frequency, the bandwidth, and the quality factor for the transfer characteristic I/V_{in} in terms of R, L, C.
(b) Compute the quantities in part (a) if $R = 10$ Ω, $L = 100$ mH, and $C = 10$ μF.

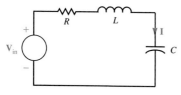

Figure P6.25

6.26. A parallel RLC resonant circuit has a resistance of 200 Ω. If it is known that the bandwidth is 80 rad/s and the lower half-power frequency is 800 rad/s, find the values of the parameters L and C.

6.27. A parallel RLC circuit has the following parameters: $R = 2$ kΩ, $L = 20$ mH, and $C = 150$ μF. Determine the resonant frequency, the Q, and the bandwidth of the circuit.

6.28. A parallel RLC circuit has the following parameters: $R = 6$ kΩ, $BW = 1000$ rad/s, and $Q = 120$. Determine the values of L, C, and ω_0.

6.29. Given the parallel RLC circuit in Fig. P6.29,
(a) Derive the expression for the resonant frequency, the

half-power frequencies, the bandwidth, and the quality factor for the transfer characteristic V_{out}/I_{in} in terms of the circuit parameters R, L, and C.
(b) Compute the quantities in part (a) if $R = 1$ kΩ, $L = 10$ mH, and $C = 100$ μF.

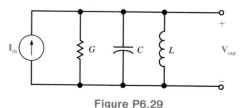

Figure P6.29

6.30. A stereo receiver is tuned to 98 MHz on the FM band. The tuning knob controls a variable capacitor in a parallel resonant circuit. If the inductance of the circuit is 2 μH and the Q is 100, determine the values of C and G.

6.31. Given the data in problem 6.30, suppose that another FM station in the vicinity is broadcasting at 98.1 MHz. Let us determine the relative value of the voltage across the resonant circuit at this frequency compared with that at 98 MHz, assuming that the current produced by both signals has the same amplitude.

6.32. A parallel RLC resonant circuit with a resonant frequency of 20,000 rad/s has an admittance at resonance of 1 mS. If the capacitance of the network is 5 μF, find the values of R and L.

6.33. A parallel RLC circuit, which is driven by a variable frequency 2A current source, has the following values: $R = 1$ kΩ, $L = 100$ mH, and $C = 10$ μF. Find the bandwidth of the network, the half-power frequencies, and the voltage across the network at the half-power frequencies.

6.34. Determine the value of C in the network shown in Fig. P6.34 in order for the circuit to be in resonance.

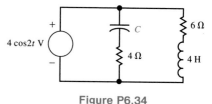

Figure P6.34

BASIC ELECTRONICS

Overview of Basic Concepts in Electronics

7.1
Introduction

As we described in Chapter 1, electrical engineering involves the transfer and use of electrical power or information. The utilization of electronics is essential for both of these applications. In general, the field of *electronics* involves the use of devices and circuits which specifically control the flow of electric current to achieve some purpose. All electronic systems involve circuits, and therefore this section of the text builds heavily on the first. Electronic circuits often contain resistors, capacitors, and inductors; however, they also utilize "electronic devices" such as transistors or diodes. These latter nonlinear elements are used to control or direct the current into desired paths in the circuit. Most modern electronic devices are termed *solid state*, because the current flow is achieved through solid materials called *semiconductors*.

This chapter will provide an overview of some of the concepts to be presented in the following chapters and set the stage for a closer look, with some initial understanding of the concepts already in hand.

7.2
Energy and Power

Power can be transferred from its point of generation to its point of use as either ac or dc, or first one and then the other. Alternating current (ac) is almost universally used for distributing power to households throughout the world. There are a number of reasons for this choice. However, the most important consideration is that transformers can only be used with ac, and they allow the voltage and current to be stepped up or down as needed with very little energy loss.

An automobile typically utilizes a dc power distribution system at approximately 12 volts. The electrical power, however, is created at the alternator as ac and converted to dc by a process called rectification using electronic devices called diodes. Many electronic circuits require dc to operate. Therefore, they receive their operating power from either batteries, which universally produce dc, or from ac (such as a wall socket), which has been rectified and thus converted to dc. The solar cell is being used more frequently as a power source, particularly in remote locations. When illuminated with a light of constant intensity, such as the sun, it produces dc.

In Chapter 8 we will examine the diode and several electronic circuits using this device that provide electrical power conversion, that is, the changing of electrical energy from one condition to another.

7.3
Signals and Pulses; Analog and Digital

A voltage or current which, in some manner, is varied over time in order to encode and transmit information is called a *signal*. Typically, signals are termed either *analog* if they vary continuously with time, or *digital* if they switch between discrete levels. In a digital signal there are typically two such levels, a "high" level and a "low" level, which are arbitrarily termed a "1" and a "0," respectively. The information conveyed by the single presentation of a high or low is called a "*bit*." In the analog signal in Fig. 7.1(a), the amplitude of the signal represents the information at every point in time. For the digital signal in Fig. 7.1(b), the information is conveyed by the presence or absence of a pulse.

A *positive pulse* is a single rectangular-shaped waveform representing a change of voltage or current which goes from a low level to a high level and returns. The waveform for both a positive and a negative pulse is shown in Fig. 7.2. A rectangular pulse always has two edges—one positive going and the other negative going. The digital signal of Fig.

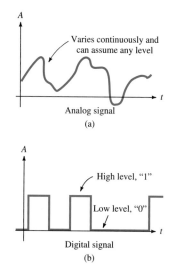

Figure 7.1   Analog and digital signals.

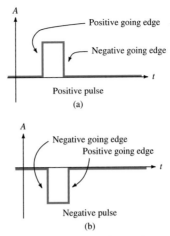

Figure 7.2   Positive and negative pulses.

7.1(b) conveys information as a result of the signal being high or low at sequential points in time. Such a digital signal is termed *serial* because on a single conductor, the pulses must all follow one after the other in time. This approach is viable; however, it is also time consuming.

In order to increase the rate of data transmission, many digital systems convey information by simultaneously presenting 1s or 0s on a number of parallel conductors. Such an arrangement is termed a *parallel* digital data bus, and the input or output (I/O) of such a bus is termed a parallel input or output port.

By operating at the same pulse transmission speed, an 8-bit parallel data bus can deliver information 8 times faster than a single conductor in a serial mode. Most modern personal computers have a serial port and a parallel port for connecting the computer to different types of peripheral equipment.

Since digital systems operate using 1s and 0s, the data in these signals is encoded in the *binary number system*. Therefore, prior to a detailed study of digital systems, a review of the binary system and basic logic will be presented in Chapter 11.

Each electronic circuit can generally be categorized as either *analog* (linear as it's

Table 7.1 Analog parameters measurable by electronic sensors

Pressure	Temperature
Light Intensity	Magnetic Fields
Electric Fields	Proximity
Position	Acoustic Waves
Seismic Vibrations	Humidity/Dew Point
Acceleration/Vibration	Force
Rotation	Radiation
Gas Concentrations	Gas Flow
Acidity (pH)	Imaging (visible/IR)
Biopotentials	Ion Concentrations
Gas Composition	Torque
Tactile/Touch	

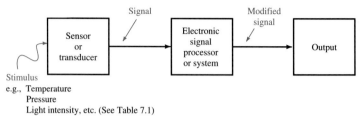

Figure 7.3   A typical electronic measurement system.

sometimes called) or *digital.* Until the 1960s, virtually all electronic circuitry was analog. Analog signals are very common today in certain applications. For example, in a stereo, an audio signal, which is actually a time-varying current, moves the cone of a speaker back and forth, reproducing the appropriate sound. Almost all of observable nature is analog, with continuously varying parameters such as temperature, pressure, light intensity, etc. Therefore, many systems receive as their input an analog signal representing an analog parameter. The electronic signal generally originates from a sensor or transducer which provides an analog signal proportional to the parameter it's designed to measure. Table 7.1 provides a partial list of analog parameters that various types of electronic sensors can measure. Many of the outputs from electronic systems are also analog, such as audio and video images. Therefore, many electronic systems contain: (1) one or more sensors which convert a measurement parameter to an electronic signal, (2) a signal processor which may use analog or digital electronics to modify the signal, and (3) an output device as illustrated in Fig. 7.3.

In the last several decades there has been an increasing trend toward digital systems, primarily as a result of the advances in digital electronic circuitry, for example the microprocessor, which is a computer on a single silicon chip (see Chapter 12). Digital circuitry has become increasingly less expensive and more powerful. Digital circuitry now dominates most electronic systems, in particular those which perform calculations. However, analog circuits remain essential in certain applications, especially those which interface electronics with the real world.

7.4

Amplifiers and Analog Systems

An analog signal can be a complex waveform such as that shown in Fig. 7.4(a), or it can be repetitive or periodic, like the sinusoidal signal shown in Fig. 7.4(b). The sinusoidal signal response is fundamental to understanding the performance of many electronic circuits and systems. First, as shown in Chapter 4, a complex waveform can be shown to be composed of the sum of a number of sinusoidal signals using Fourier analysis techniques. In addition, the waveform normally produced by an ac generator is sinusoidal.

Recall that a voltage that varies sinusoidally with time can be represented by the function

$$v(t) = A \sin(\omega t + \phi) \tag{7.1}$$

In such a sinusoidally varying signal there are three parameters that can be varied: the amplitude A, the angular frequency ω, and the phase angle ϕ. The value of each of these parameters determines certain characteristics of the sinusoidal signal.

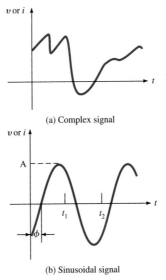

(a) Complex signal

(b) Sinusoidal signal

Figure 7.4 Complex and sinusoidal signals.

Recall the angular frequency, ω, in radians per second, is related to f, the frequency in hertz by the relation

$$\omega = 2\pi f \tag{7.2}$$

The standard frequency for the transmission of electric power in the United States and Canada is 60 hertz. The audio frequency range, or the range of frequencies the human ear can detect, is from approximately 30 hertz to 15,000 hertz, which is the frequency range of the electronic signal observed from the output of a high-quality microphone. The entire frequency spectrum was presented in Fig. 1.2.

The amplitude of a signal is an extremely important parameter and can vary over many orders of magnitude. The voltage received on the antenna of a television set may be in the order of microvolts (10^{-6} volts), whereas the voltage applied to the picture tube in the same set may be 20,000 volts (2×10^4 volts). This represents a voltage range of some ten orders of magnitude.

In many applications, an exact replica of a particular signal is needed, but with a larger amplitude. Consider the following example.

EXAMPLE 7.1.

Assume the voltage signal from an electric guitar is 20 millivolts in amplitude, and must be amplified in order to provide 100 watts of power to an 8-ohm speaker. Let us determine the required *rms* sinusoidal voltage across the speaker.

For this case, the voltage and power are related by the expression

$$P = \frac{V_{rms}^2}{R}$$

and therefore

$$V_{rms} = \sqrt{PR} = \sqrt{(100)(8)} = 28 \text{ volts}$$

Assuming this signal is sinusoidal, the amplitude is

$$V_m = \sqrt{2}\,V_{rms} = (1.41)(28) = 40 \text{ volts}$$

Since the guitar's output amplitude is 20 millivolts, the voltage must be increased by a factor of

$$\frac{40}{20 \times 10^{-3}} = 2000 \qquad\blacksquare$$

An *amplifier*, shown symbolically in Fig. 7.5, is an electronic circuit which is used to increase the amplitude of a signal. The magnitude of the amplifier's *voltage gain*, A_v is defined as the ratio of the amplitude of the output voltage, v_o to the amplitude of the input voltage, v_i or

$$A_v = \frac{|v_o|}{|v_i|} \qquad (7.3)$$

In the example just described, the voltage gain, A_v, of the required amplifier is 2000.

Thus far we have discussed only the magnitude of the gain; however, there is also a phase relation between the input and output signal. Thus, the voltage gain is more accurately represented as a complex quantity.

The complex voltage gain, $\mathbf{A}_v$, is the ratio of the phasor transform of the output voltage to that of the input voltage, assuming the signals are sinusoidal,

$$\mathbf{A}_v = \frac{\mathbf{V}_o}{\mathbf{V}_i} \qquad (7.4)$$

Most electronic circuits are designed to operate in what is called the *midband frequency region*, and in this range the phase shift of amplifiers is either zero or 180°, that is, noninverting or inverting. For this reason, in the basic analysis of electronic amplifiers we most often speak only of the magnitude of the voltage gain and include a plus sign if the amplifier is noninverting and a minus sign if it is inverting. In Chapter 17 the effects of various *RC* networks that provide additional phase shifts will be considered.

There are also current amplifiers, which produce an output current that is larger than its input current by a constant factor called the *current gain*.

An amplifier requires dc power in order to operate. The supply voltage to the amplifier must be at least slightly larger than the maximum voltage swing possible at the output of the amplifier. For example, the amplifier just described in Example 7.1 delivers a peak voltage to the speaker of 40 volts. Therefore, the amplifier would require a dc volt-

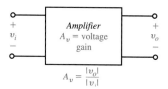

Figure 7.5   A voltage amplifier.

age source of more than 40 volts for proper operation. These concepts will become clear as amplifiers are described in more detail in Chapter 15.

EXAMPLE 7.2

A current of 3×10^{-5} amperes, dc, is derived from a solar cell illuminated by the sun. Let us determine the current gain required in order to activate a small electric motor requiring 5×10^{-2} amperes. The current gain, A_i, needed is

$$A_i = \frac{5.00 \times 10^{-2}}{3.00 \times 10^{-5}} = 1670 \qquad \blacksquare$$

DRILL EXERCISE

D7.1. Determine the voltage gain required in Example 7.1 if a 4 Ω speaker is used.

Ans: 1410.

Another important characteristic of an amplifier is its *frequency response,* that is, the way in which the gain varies as the frequency of a sinusoidal input signal is changed. No real amplifier can amplify an input signal over an infinite range of frequencies. At some high frequency the amplifier will no longer be able to produce the same amplitude of output signal as it could at a lower frequency, and hence the gain is said to "roll off." Many amplifiers also roll off as frequency is decreased below some particular low-frequency value. The high frequency at which the gain is reduced by a factor of 0.707 from the midband value is termed the *upper* or *high-corner frequency,* f_{HI}, and the low-frequency at which the gain has dropped by a factor of 0.707 is termed the *low-corner frequency,* f_{LO}. These corner frequencies are also called *half-power frequencies* as outlined in Chapter 6.

The difference in frequency between the high-corner frequency and the low-corner frequency is called the *bandwidth, BW.* Some amplifiers do not have a lower corner frequency, and will amplify signals all the way down to dc or zero frequency. These amplifiers are called *dc amplifiers.* A plot of such an amplifier's frequency response is shown in Fig. 7.6. In most plots of frequency response, the frequency, f, and the magnitude of the gain are plotted on logarithmic scales. Figure 7.7 shows a similar plot for an ac amplifier. These concepts will be described in more detail in Chapter 17.

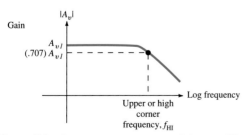

Figure 7.6   Frequency response of dc amplifier.

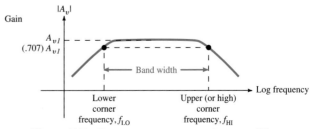

Figure 7.7   Frequency response of ac amplifier.

EXAMPLE 7.3

The mid-band gain of a voltage amplifier is 20. If the gain drops to 14.1 at a high frequency of 11 MHz, and the amplifier has a bandwidth of 3 MHz, let us determine the low-frequency half-power point.

Since the voltage gain drops to 0.707 of its mid-band value at 11 MHz, this frequency is the high-frequency half-power point. The bandwidth is the difference between the high- and the low-corner frequencies, and therefore, the low-frequency half-power point is

$$11 \text{ MHz} - 3 \text{ MHz} = 8 \text{ MHz} \qquad \blacksquare$$

DRILL EXERCISE

D7.2. An ac amplifier has a lower corner frequency of 0.5 MHz and a bandwidth of 1.0 MHz. (a) Determine the amplifier's upper corner frequency. (b) Determine the upper corner frequency of a dc amplifier with the same bandwidth.

Ans: (a) 1.5 MHz (b) 1.0 MHz.

7.5

Modulation and Demodulation; Encoding and Decoding

Modulation is the process used to encode, or "attach and carry," information on an analog signal. *Demodulation* is the reverse process in which the information is extracted from an analog signal. The various processes by which this is accomplished are fundamental topics in *communications*, which are discussed in Part VI of this text.

Recall that there are three basic ways of carrying information via a sinusoidal signal: by varying the amplitude, the frequency, or the phase. Assume, for example, the information we wish to send is represented by the "voltage ramp" shown in Fig. 7.8(a); this is called the *baseband signal*.

The sinusoidal wave, called the carrier wave, which is amplitude modulated with this information, would appear as shown in Fig. 7.8(b). Note that the amplitude of the sinusoidal signal varies in relation to the information desired to be transmitted. *Amplitude modulation* (AM) is commonly used for transmission of radio signals. For example, in the United States the Federal Communications Commission has established a range of frequencies for the commercial broadcast of AM signals. The AM band is available on most home and automobile radios.

Frequency Modulation, used, for example, on the commercial FM radio band, employs changes in frequency to represent the desired information. The same wave transmitted by FM would appear as shown in Fig. 7.8(c). Note that the frequency of the sinusoidal signal changes in proportion to the information to be transmitted.

Phase modulation is similar to frequency modulation; however, rather than varying the frequency, the phase of the sinusoidal signal is varied relative to some fixed reference.

When television was first developed and introduced in the United States, it utilized black and white transmission. It was decided early that the picture would be transmitted via AM and the sound via FM. Therefore, if you have an FM radio receiver with a little

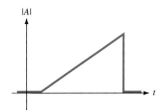

Figure 7.8(a)   Baseband signal.

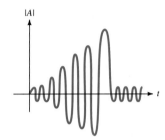

Figure 7.8(b)   AM modulated signal.

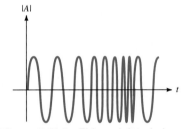

Figure 7.8(c)   FM modulated signal.

extra range, you can receive the audio for television channel 6 at the very bottom of the FM dial. Later when color television was developed, one requirement imposed by the government was that the color transmission be compatible with the old black and white sets. An ingenious system was devised whereby the color information was added using phase modulation, and did not affect black and white reception on the older sets.

The telephone was designed as an instrument to transmit analog signals, primarily the tones of the human voice. With the advent of the computer, a need developed to transmit digital signals via phone lines. To accomplish this, a modulation system was designed that translated the digital information into analog tones which could be sent over the phone lines and reconstructed at the other end. Such a system is called a *MODEM,* an abbreviation for "*mo*dulator-*dem*odulator."

The process of modifying information so that it is suitable for transmission by a digital signal and constructing the digital signal to represent the information is called simply *encoding.* The opposite process, in which the encoded information is extracted from a digital signal, is termed *decoding.* Examples of modulation/demodulation and encoding/decoding will be presented in later chapters.

7.6

Digital Systems, Information, and Gates

Almost all digital systems utilize circuits with *binary logic.* In binary digital logic, data is represented by a voltage (or occasionally a current) switching between one of two possible levels, a low level called a logic 0, and a high level called a logic 1. This process is termed "asserted positive logic." The system could be implemented using the opposite convention where the high level is considered a logic 0 and the low level a logic 1, and such a system is said to use "asserted negative logic." Positive logic is by far the most common.

The single unit of digital information, the *bit,* is represented by the transmission of a 1 or a 0. The bit has been found to be a very useful means of quantizing information. Information theory, a subspecialty within the field of communications, has developed expressions for calculating the number of equivalent bits of information contained in various signals.

Each bit conveys to us whether the information or value is in the top half or lower half of its possible range. For example, if we consider a signal, v, which we know has values in the range $0 < v < 2$, one bit of information would tell us the following: if the voltage is in the range $0 < v < 1$, then the bit status could be 0, and if the voltage is in the range $1 < v < 2$, the bit status would be 1.

This is very coarse information about this signal, v. We originally knew its value was between zero and two, and now we have started to "pin it down" by breaking the total range into two zones. We use this first bit, the *most significant bit* (MSB), to tell us whether we are in the upper 50% or lower 50% of the entire range.

We can add a second bit to tell us whether we are in the top half or lower half of the smaller range just selected by the previous bit. We are now capable of quantifying the signal to within four possible "zones" using the following correspondence:

Range	Bit status	
	MSB	LSB
$0.0 < v < 0.5$	0	0
$0.5 < v < 1.0$	0	1
$1.0 < v < 1.5$	1	0
$1.5 < v < 2.0$	1	1

Where MSB = most significant bit and LSB = least significant bit.

We could now add a third bit which would give us a resolution of eight possible ranges, or zones. Note that in the binary system, the number of zones of resolution, Z, is related to the number of bits, n, by the expression

$$Z = 2^n \tag{7.5}$$

The series of bits that represent a particular analog level is termed a digital *word*. If this series of bits is presented at the same time via parallel conductors, that is, a parallel bus, we receive a parallel word; if they appear one after the other via a single conductor, that is, a serial data bus, then we receive a serial word. Eight-bit word segments are normally referred to as a *byte.*

EXAMPLE 7.4

Let us determine the voltage resolution that can be achieved if a signal with a maximum value of 12 volts and minimum value of 0 volts is represented by 10 bits.

The number of zones defined by 10 bits is:

$$Z = 2^n = 2^{10} = 1024 \text{ zones}$$

$$Resolution = \frac{12.0}{1024} = 0.0117 \text{ volts/zone}$$

∎

D7.3. A 4-bit parallel data bus can transmit 8 bits of information in 40 microseconds. If the same technology were used to transmit data serially over a single conductor, how many bits could be transmitted in the same 40 microseconds?

Ans: 2 bits.

D7.4. A signal that can have values between -10 to $+10$ volts is represented by a 12-bit digital word; what is the voltage resolution obtainable with this system?

Ans: 4.9 mV.

A/D and D/A Converters

Electronic circuits, which are designed to automatically convert an analog voltage into a digital or binary representation of that voltage, are called *Analog to Digital Converters,* or A/D Converters. Circuits that convert a digital word into an analog voltage level are called *Digital to Analog Converters,* or D/A converters. Both converters are shown symbolically in Fig. 7.9. A timing signal is used to specify the point in time at which a sample is taken.

An 8-bit A/D converter, for example, converts a continuously variable voltage within a specified range at its input, into an 8-bit digital word at its output which represents the

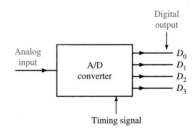

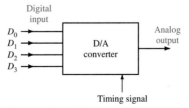

Figure 7.9   A/D and D/A converters.

quantized value of the input voltage at a specific point in time. Therefore, we are told which of 256 zones the input voltage is contained within ($2^8 = 256$) at the time of the sample. A/D and D/A converters are available from many manufacturers integrated on a single IC chip.

Logic Gates

Information represented as 1s and 0s in a digital system is processed primarily by elements called *logic gates.*

A practical *logic gate* is an electronic circuit which has a digital output that can assume voltage levels which correspond to a 1 or a 0 state, and one or more digital inputs, which determine the state of the output.

Three simple logic gates will be described here: the *inverter,* the *AND-gate,* and the *OR-gate.* Later, in Chapters 10 through 12, further details of these and more complex logical functions will be described.

The simplest logic gate is the inverter. The inverter has only one input and one output, and its output is always the opposite or *complement* of its input. *Truth tables* are often used in order to describe logic gate operation. A truth table is a chart showing the logical state of the output for every possible input combination. The truth table for the *inverter* is relatively simple since it has only one input. If A is the input and B is the output, then the truth table is

A (input)	B (output)
0	1
1	0

There are sets of standard symbols for the basic logic gates. The symbol for the inverter is shown in Fig. 7.10(a).

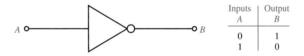

Inputs A	Output B
0	1
1	0

Figure 7.10(a) Inverter or NOT-gate.

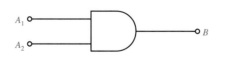

Inputs A_1 A_2		Output B
0	0	0
0	1	0
1	0	0
1	1	1

Figure 7.10(b) AND-gate.

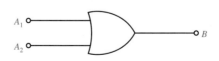

Inputs A_1 A_2		Output B
0	0	0
0	1	1
1	0	1
1	1	1

Figure 7.10(c) OR-gate.

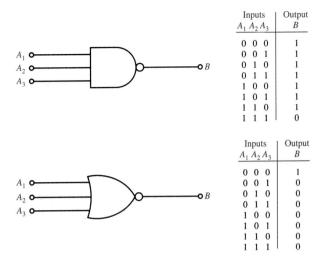

Inputs $A_1\ A_2\ A_3$	Output B
0 0 0	1
0 0 1	1
0 1 0	1
0 1 1	1
1 0 0	1
1 0 1	1
1 1 0	1
1 1 1	0

Inputs $A_1\ A_2\ A_3$	Output B
0 0 0	1
0 0 1	0
0 1 0	0
0 1 1	0
1 0 0	0
1 0 1	0
1 1 0	0
1 1 1	0

Figure 7.11   NAND and NOR logic gates.

The *AND-gate* has a single output and can have any number of inputs greater than 1. The output of this gate is high only when all the inputs are high, and for any other combination of inputs, the output is low. The logic symbol and truth table for a two-input AND-gate are shown in Fig. 7.10(b).

The *OR-gate* has a single output and can have any number of inputs greater than 1. This gate produces a high level at its output if *any* of its inputs are high, and the output is low only if none of the inputs are high. The truth table and logic symbol for a two-input OR-gate are shown in Fig. 7.10(c).

In order to implement certain logic functions, it often becomes necessary to use an AND-gate followed by an inverter. This combination produces an output which is the exact complement of the AND-gate, and the combination is defined as a *NAND-gate*. Similarly, a gate which produces an output which is the complement of the OR-gate is called the *NOR-gate*. The logic symbols and truth tables for three input NAND and NOR gates are given in Fig. 7.11. Many electronic circuits utilized for implementing logic inherently invert the signal while performing the logic function, and therefore NAND/NOR logic is quite common in practical electronic systems.

7.7
Summary

- Electronic circuits contain linear and nonlinear elements.
- Electric power can be transmitted by ac or dc.
- Rectification is the process of converting ac to dc.
- A signal is a voltage or current which in some manner is varied over time in order to encode and transmit information.
- Signals are either analog if they vary continuously or digital if they switch between discrete levels.
- Digital signals typically have 2 levels, a high called a 1, and a low called a 0. The single presentation of a 1 or 0 is called a bit.
- An amplifier produces an output which is a factor of *A* larger than its input where *A* is called the gain.

- Modulation is the process of attaching information to an analog signal; demodulation is the reverse process of extracting information from an analog signal.
- An analog range can be quantized into Z zones where n is the number of bits by the expression $Z = 2^n$.
- An A/D converter samples an analog signal and creates a digital output. A D/A converter performs the opposite task of converting a digital word into an analog level.
- Practical logic gates are electronic circuits which have a digital output and one or more digital inputs; the logic state of the output is unequally determined by the logic values of the inputs.

PROBLEMS

7.1. For each of the following, decide if it is most analogous to digital information or analog information.
(a) the result of a coin toss
(b) the sound of a slide trombone
(c) the beat of a drum
(d) a human voice
(e) the ticking of a clock
(f) the sound of a siren
(g) the motion of a cat
(h) the motion of a frog

7.2. Construct digital voltage signals by drawing the pulse waveforms based on the following descriptions: high level = 5 V; low level = 0.3 V.
(a) low at $t = 0$, positive going edge at $t = 1$, negative going edge at $t = 1.5$, positive going edge at $t = 2$, negative going edge at $t = 3$.
(b) low at $t = 0$, positive going edge at $t = .5$, negative going edge at $t = 1.5$, positive going edge at $t = 1.75$, negative going edge at $t = 2$.
(c) low at $t = 0$, followed by 5 seconds of alternating positive and negative going edges equally spaced by 1 second.

7.3. Construct serial digital signals representing the following sequences of bits using pulses of width 0.5 seconds.
(a) 0 0 1 0 1 0 0 0 1 1 1 0 1
(b) 1 0 1 1 0 0 1 0 1 1 0 0 1

7.4. An 8-bit parallel data line is used to transmit 16 bits of information in 10 μs. If the same technology were used to transmit data serially over a conductor, how many bits could be transmitted in 80 μs?

7.5. Assume a computer's output port sends out data serially to a printer, and that each character can be represented by an 8-bit word. If a single printed line contains 72 characters, how many lines can be sent out by the computer in 1 minute? The data transmission rate is 9600 bits/second.

7.6. The Central Processing Unit (CPU) of a microprocessor operates at a clock frequency of 12 MHz. Assuming that it takes 4 clock cycles to perform one memory fetch operation, how many 8-bit words can be fetched from memory in 1 ms if:
(a) the bits are fetched serially (1 bit per fetch operation)?
(b) the bits are fetched in parallel (8 bits per fetch operation)?

7.7. A temperature-sensitive resistor is an element often used to measure temperature in electronic systems. Assume that the resistance of such an element is given by $R(T) = R_0[1 + \alpha(T - T_0)]$ where T is temperature, R_0 is the resistance value at a temperature of T_0 (°C), and α is a constant called the temperature coefficient. If $\alpha = 0.01$, $R_0 = 1$ kΩ, what constant dc current must be established in this element so that the voltage change across it will be 0.05 V when the temperature change is 10 degrees?

7.8. A thermistor is a resistor with a strong negative temperature coefficient (see equation in problem 7.7). Assume such a device shows a decrease in resistance of 7% per degree increase in temperature. If the room temperature current is 2 mA when connected across a 3 V source, what is the expected current when the temperature is lowered 5 degrees?

7.9. A solar cell is placed outdoors and has an output of 7 mA at peak sun intensity (noon), and 0.1 mA when exposed to moonlight. Plot the approximate current output expected versus time over a 24-hour period, assuming
(a) a cat does *not* stand over the cell.
(b) a cat stands over the cell for 1 hour beginning at noon. (Assume an opaque cat.)

7.10. The amplitude of the signal received by the antenna of an AM-radio receiver is 5 μV. The signal is then amplified to supply 3 V to a speaker. What is the voltage gain of this system?

7.11. The input and the output signals of four different amplifiers are given below. Obtain the voltage gain and the phase shift for each of the amplifiers.
  (a) $v_{in}(t) = 3\cos(\omega t)$ $v_{out}(t) = 27\cos(\omega t)$
  (b) $v_{in}(t) = 5\sin(10t)$ $v_{out}(t) = 15\cos(10t)$
  (c) $v_{in}(t) = 4\sin(\omega t)$ $v_{out}(t) = 20\sin(\omega t + \pi/2)$
  (d) $v_{in}(t) = 7\,e^{20jt}$ $v_{out}(t) = j5\,e^{20jt}$

7.12. An electronic thermometer generates a voltage signal of 6 mV for every 1°C rise in temperature. This signal is used to activate a switch which requires an input signal of 5 V to operate. What is the voltage gain of the amplifier which will allow one to operate the switch at the boiling point of water (100°C) if the thermometer generates 0 V at 0°C?

7.13. Calculate the upper and lower corner frequencies and the bandwidth of the amplifier whose frequency response is given in Fig. P7.13.

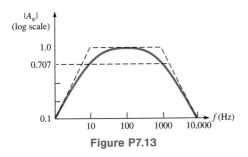

Figure P7.13

7.14. An audio amplifier is designed for signals in the frequency range of 10^2 Hz to 10^4 Hz. The amplifier has an input of 30 mV and its output is to provide 10 W to an 8 Ω speaker. Plot the frequency response of this amplifier showing the mid-band gain, the two corner frequencies, and the bandwidth.

7.15. The signal in Fig. P7.15(a) modulates the high-frequency sinusoidal carrier signal shown in Fig. P7.15(b). Sketch the resulting waveform for (a) amplitude modulation and (b) frequency modulation. (Note different time scales on the figures.)

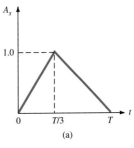

(a)

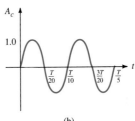

(b)

Figure P7.15

7.16. Find the minimum number of bits necessary to represent a 4 V analog signal with a resolution greater than 1 mV.

7.17. An analog signal having values from −4 V to 6 V is represented using 8 bits. (a) Calculate the voltage resolution obtained. (b) How many additional bits are required to double the voltage resolution?

7.18. An analog signal which has a maximum value of 5 V is represented digitally using a 7-bit word. Due to a fault in the digital circuit, the middle bit is incorrect. Find the magnitude of the error in the analog representation of the signal.

7.19. Obtain the digital words corresponding to the indicated analog signal levels, a through j, as shown in Fig. P7.19, that would be assigned by an A/D system with:
  (a) 2 bits
  (b) 3 bits

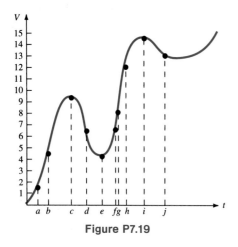

Figure P7.19

7.20. An analog signal with a maximum amplitude of 1 volt is sampled 15 times at one-second intervals. The samples are converted to 4-bit digital words as shown below. Reconstruct the analog signal as accurately as possible from this list and comment on the accuracy of your result. Identify two ways of obtaining a better digital representation of the analog signal.

Sample time(s)	4-bit digital word
1	0000
2	0010
3	0011
4	0001
5	0100
6	0101
7	0110
8	1010
9	1101
10	1111
11	1000
12	0110
13	0111
14	1011
15	1111

7.21. An Exclusive OR gate (XOR) is a logic gate with an output which is high only when all its inputs are at the same logic value. Construct the truth table for a 2-input XOR gate.

7.22. Construct the truth table for the logic circuit shown in Fig. P7.22. (Assume A_1 is the MSB.)

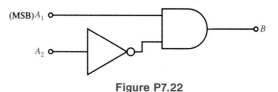

Figure P7.22

7.23. Construct the truth table for the logic circuit shown in Fig. P7.23. (Assume A_1 is the MSB.)

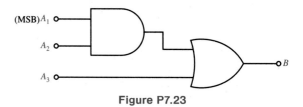

Figure P7.23

Semiconductors, Diodes, and Applications

8.1

Introduction

One of the most basic and yet remarkable effects in solid state science is the electronic behavior of the solid state *diode*. In a conductor such as copper, or in any other homogeneous conducting material, current flows in either direction equally well. If a voltage source is applied across such material, a certain current can be measured and if the polarity of the applied source is reversed, the same current flows in the opposite direction. This is not true of current flow through a semiconductor diode; it is a nonlinear device. The network analysis techniques that require the circuit elements to be linear cannot be generally applied; however, they can be used over limited regions where the diode's characteristics are assumed to be linear.

A *diode* is a two-terminal electronic device which conducts current if a voltage source is applied in one direction, and refuses to conduct significant current when the voltage is applied with the opposite polarity.

A diode with a voltage applied so that the diode is conducting a current is said to be *forward biased;* a diode with a voltage applied in the opposite direction such that no current flows is said to be *reverse biased.*

The schematic symbol for the diode and a photograph of a common semiconductor diode installed on a circuit board is shown in Fig. 8.1. The direction of the "arrow" contained in the symbol indicates the direction of positive current flow that is allowed, in this case top to bottom. Current will not flow in the opposite direction.

The fundamental characteristics of a diode can be illustrated by an analogous fluid system component—a valve that allows fluid to flow only in one direction. Usually these

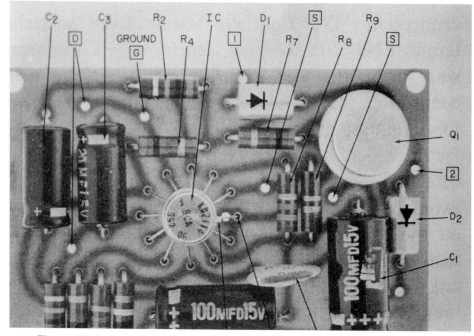

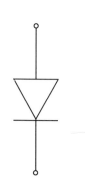

Figure 8.1 (a)
Diode symbol

Figure 8.1 (b) Photograph of Diodes (D_1 and D_2) on Circuit Board with Other Components.

valves are called "check" valves or "flapper" valves. Such valves allow flow in one direction, but close and restrict fluid flow if the fluid attempts to move in the other direction.

An illustration of such a valve is shown in Fig. 8.2. In this analogy remember that fluid flow is analogous to current, and pressure is analogous to voltage. Part (a) of this figure illustrates that if a pressure (voltage) difference is applied so the left side is at a higher pressure than the right, then this pressure attempts to move fluid to the right and the valve opens and fluid (current) flows. The application of a pressure difference of opposite polarity, as shown in part (b) of the figure, attempts to force fluid to the left; the valve closes and no fluid flows.

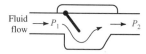

Figure 8.2 (a) Fluid passes through from left to right—valve forced open
$$P_1 > P_2$$

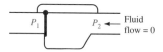

Figure 8.2 (b) No fluid flows from right to left—valve forced closed
$$P_1 < P_2$$

8.2
Historical Perspective

The diode or *rectifier* has been an important electronic device since the early 1900s. The first diodes were one of two types:

1. The vacuum diode consisted of two electrodes sealed in a glass bulb with all the air removed. One electrode, the cathode, was heated and boiled off electrons which were collected by the other electrode, the anode. Since the anode was not heated, electrons could not pass in the opposite direction and hence a diode effect occurred. This phenomenon was first discovered by Thomas A. Edison while trying to invent the light bulb and is called the Edison effect. A practical vacuum diode based on this principle was patented by John A. Fleming in 1904.

2. The other device is interesting because it was an early version of the solid state diode, called a cat's whisker. A diode is a critical element in a radio receiver, as will be illustrated in Chapter 23. Radios during the early part of this century often had a crystal which was made from a piece of natural rock, usually galena based, containing the right "stuff." The operator of the radio took a wire needle (the cat's whisker) and probed around the surface of the crystal looking for a sensitive spot in which he heard a radio signal on the earphone attached to the wire needle. This phenomenon added both mystery and challenge to radio reception, and such radio receivers were called "crystal sets."

The sensitive spots on the crystal were found to be regions of semiconducting material that formed a diode (or rectifying junction) at the junction of the metal wire and the crystal. The junction of two dissimilar materials can form a device which displays diode characteristics. Modern solid state diodes are formed in a single piece of semiconductor material by purposely structuring adjacent regions with appropriately different characteristics.

The modern version of the cat's whisker is a metal-semiconductor junction formed by depositing a thin metal film over an appropriate semiconductor layer. The metal-semiconductor junction is called a *Schottky diode,* and one application in which this type diode is used is *Schottky TTL (Transistor-Transistor-Logic)* circuits where they function to speed up the circuit operation, as will be explained in Chapter 10.

The most common semiconductor diode is an element fabricated by creating adjoining *n*-type and *p*-type semiconductor regions to form a *p–n* diode.

8.3
p- and *n*-type Semiconductors

A semiconductor material can be made either *n-type* or *p-type*. We will now describe what is meant by these terms as well as how these critical material properties are created. This information will be important in understanding how the diode and all other solid state devices—such as transistors described in Chapter 9—operate.

In the present discussion, attention will be focused on the semiconductor material, *silicon*. Most modern devices are fabricated in silicon; however, the basic principles presented could be applied to germanium, gallium arsenide, or any other semiconductor material used for making such devices.

Recall the Periodic Table of the Elements and note the position of silicon. Silicon is

Valence:	III (+3)	IV (+4)	V (+5)
	5 B Boron 10.811	6 C Carbon 12.011	7 N Nitrogen 14.0067
	13 Al Aluminum 26.98154	14 Si Silicon 28.0855	15 P Phosphorus 30.97376
	31 Ga Gallium 69.723	32 Ge Germanium 72.59	33 As Arsenic 74.9216
	49 In Indium 114.82	50 Sn Tin 118.710	51 Sb Antimony 121.75
	81 Ti Thallium 204.383	82 Pb Lead 207.2	83 Bi Bismuth 208.9804

Figure 8.3   Elements surrounding silicon in periodic table.

in the fourth column of the table and has a valence of four. A portion of this table surrounding silicon is shown in Fig. 8.3. As illustrated in Fig. 8.4(a), a single isolated neutral silicon atom has four electrons in its outer shell, each represented by a "dash" in the figure.

When millions of atoms are joined together in a lattice to form "single crystal" silicon material, each silicon atom contributes 4 electrons to the covalent bonds which connect it with its four nearest neighboring atoms. Each of the nearest neighbors also contributes an electron to each bond, making four completely filled bonds (2 electrons each) surrounding each atom—a very stable condition. The term "single crystal" means the atoms are stacked in a regular three-dimensional pattern forming a perfectly ordered structure called a *crystal lattice*. A two-dimensional representation of this is shown in Fig. 8.4(b).

This pure silicon semiconductor lattice is neutral in total charge. There is as much positive charge contributed by protons in each silicon nucleus as there is negative charge

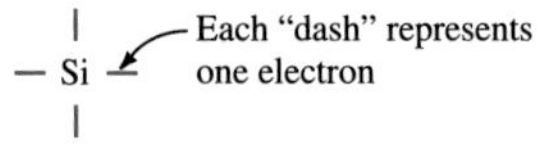

```
   |     ┌─ Each "dash" represents
 — Si ∠    one electron
   |
```

Figure 8.4 (a)   Individual silicon atom (valence = 4)

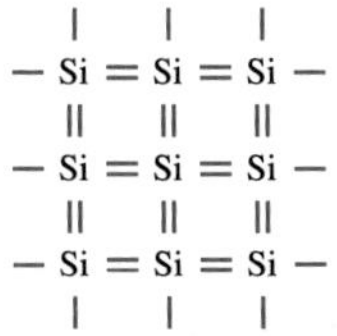

```
   |    |    |
 — Si = Si = Si —
   ‖    ‖    ‖
 — Si = Si = Si —
   ‖    ‖    ‖
 — Si = Si = Si —
   |    |    |
```

Figure 8.4 (b)   Silicon atoms in lattice

contributed by the electrons surrounding the nuclei. This is as expected since combining charge neutral atoms to make a lattice should produce a charge neutral lattice.

This silicon lattice is a relatively poor conductor of electric current. Current requires movement of charge. All the electrons in the material are tightly bound in covalent bonds, so there aren't any free electrons available that could move and create a current. Semiconductor material in this situation is termed *intrinsic*.

To create semiconductor material which can more easily conduct, it must be "doped"— yes, that's what it's called. *Doping* is the process of adding very precise amounts of a very small concentration of certain so-called impurities termed *dopants*. Actually dopants are elements from columns in the periodic table adjacent to silicon. There are two types of dopants, *donor* and *acceptor impurities;* the first is used to create *n*-type silicon, and the latter, *p*-type.

Donor impurities are so named because they *donate* extra electrons to the silicon lattice. One of the most common donors is phosphorus, an atom with a valence of 5. If we add a small portion of phosphorus to the silicon lattice, it will bind into one of the *lattice sites* in place of a silicon atom. It has 5 outer-shell electrons, four of which are used for covalent bonding with the four neighboring silicon atoms. However, the fifth electron is not used in bonding and is easily freed from the original atom by thermal energy even at room temperature. This creates a free electron in the lattice which is easily moved when a voltage is applied across the material, and the material can therefore conduct a current. The more the material is doped with phosphorus, the more free electrons are created and the more easily the material conducts, that is, its resistance is lowered. Since this doped material has free electrons, each with *negative* charge, the material is now considered *n-type*. Recall that the entire charge in the total material is still neutral, because each phosphorus atom originally has one more unit of positive charge in its nucleus than silicon and for each atom an extra electron has been contributed.

An example of a silicon lattice with extra electrons as a result of phosphorus doping is shown in Fig. 8.5.

Similarly, *p*-type material can be created by doping with atoms which have a valence of three, so-called acceptor impurities, in which free positive charge is added to the lattice. One of the most common such impurities is boron, and an example of a silicon lattice doped with boron is shown in Fig. 8.6. Notice that the trivalent boron does not have enough electrons to fill all the orbitals around it, and there remains a place (bond site) where there is an electron absent; this is called a *hole*. A hole can be shown to behave like a positively charged particle and can move through the crystal lattice by a process of "musical chairs." Hole conduction occurs by an adjacent electron jumping over and filling the original hole, but leaving a hole in the place from which it came. Then another electron near the new hole moves over and fills that hole, but leaves a new hole, and so

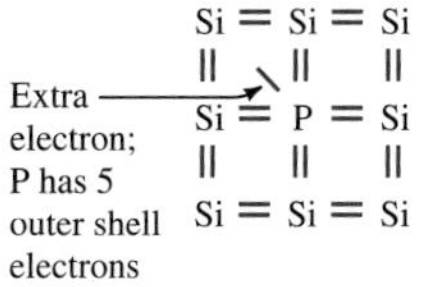

Figure 8.5   Silicon doped with phosphorus.

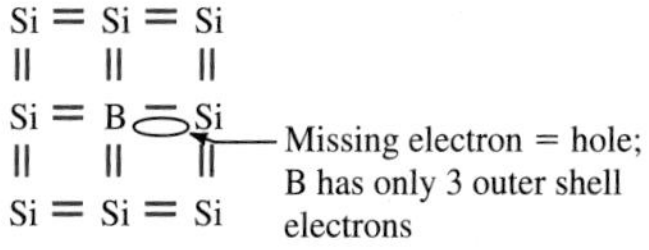

Figure 8.6   Silicon doped with boron.

on. Electrons and holes are the *particles* that conduct current in solid state electronic devices.

Since each ionized dopant atom in the lattice contributes one charge carrier at moderate doping levels, we find that in *n*-type material the free electron concentration, *n*, in electrons/cm^3 is approximately equal to the donor atom doping density, N_D, in donor atoms/cm^3, or

$$n \approx N_D \qquad (8.1a)$$

Similarly, in *p*-type material

$$p \approx N_A \qquad (8.1b)$$

where *p* is the free hole concentration in holes/cm^3 and N_A is the acceptor atom doping concentration in acceptor atoms/cm^3.

If $N_D > N_A$, the material is *n*-type and $n > p$; if $N_A > N_D$, the material is *p*-type and $p > n$. The free carrier concentrations also depend on temperature.

As stated earlier, a pure semiconductor with no dopant impurities is termed intrinsic and the free carrier concentrations are very low. In a pure silicon lattice where every atom has a valence of 4, free electrons and holes are only created in pairs; therefore, the free electron density, *n*, equals the free hole density, *p*. This special density is called the *intrinsic carrier concentration,* and designated n_i. For an intrinsic material

$$n_i = n = p$$

For silicon at 300°K, $n_i \approx 1.6 \times 10^{10}$ electrons/cm^3.

8.4

Current Conduction in Semiconductors

In a semiconductor material, there are therefore two types of mobile charge carriers that contribute to current flow. The electron is negatively charged, and the hole behaves as a positively charged particle. At temperatures above absolute zero, these free carriers are in constant random motion as a result of their thermal energy—however, their *net* motion in any particular direction is zero. An electron may start off in one direction, collide with the lattice, go off in another direction and collide again and then repeat this process over and over. The mean-free-path, the average distance traveled before a collision, for an electron in silicon is of the order of 10^{-7} cm. This random motion is illustrated in Fig. 8.7. Although the holes and electrons are in constant motion, because there is no net motion in any specific direction there is no net current flow.

There are two principle mechanisms by which charge carriers are forced to move, on average, in a particular direction and thus create an electric current: they are (1) *drift,* and (2) *diffusion.*

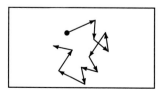

Figure 8.7   Random path of carrier in semiconductor (due to thermal energy).

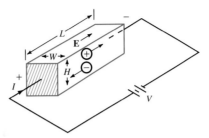

Figure 8.8   Rectangular block of semiconductor material with E-field applied.

Drift

Recall that a single electron carries a negative charge of 1.6×10^{-19} coulombs; a hole effectively carries a positive charge of the same magnitude. When an electric field is applied across a material, the resulting current is carried by both types of carriers, each moving in opposite directions. If, for example, in the block of material shown in Fig. 8.8, a voltage, V, is applied as shown from one end to the other, then an electric field vector **E** is established in the material. The magnitude of the electric field is the applied voltage divided by the distance, L, between the parallel end plates at the terminals.

$$|\mathbf{E}| = \frac{V}{L} \text{ volts/cm} \tag{8.2}$$

The direction of the electric field vector, **E**, is defined to be the direction a small positive charge would be pushed if placed in the field. The E-field in the previous figure moves positive charge (holes) away from the "+" end and toward the "−" end; simultaneously negatively charged electrons are moved in the opposite direction.

   The carriers begin drifting toward one of the end terminals and then collide with atoms in the lattice along the way and are deflected or scattered. However, the action of the field is continuous and they again start moving toward the terminal by a very indirect path yet with a net motion toward the goal. The motion of such a carrier is illustrated in Fig. 8.9.

   The effective velocity of a carrier moving by the drift action of an applied electric field is proportional to the magnitude of the field over a wide range of field magnitudes. Therefore, the magnitude of the electron drift velocity, v_n, is

$$v_n = \mu_n |\mathbf{E}| \tag{8.3a}$$

where μ_n is the electron mobility constant.

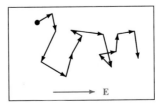

Figure 8.9   Net motion of "hole" to right due to E-field.

There is an analogous expression for holes:

$$v_p = \mu_p |\mathbf{E}| \tag{8.3b}$$

For silicon, $\mu_n = 1350$ cm^2/V-s and $\mu_p = 480$ cm^2/V-s

DRILL EXERCISE

D8.1. A cube of semiconductor material 1 cm on each side has 20 volts applied across two opposite faces. The mobility of holes in the material is 480 cm^2/V-s. Find the magnitude of the electric field in the cube, and the average drift velocity of holes.

Ans: $E = 20$ V/cm; $v_p = 9600$ cm/s.

The *conductivity* of a material is a property of the material and is a measure of the material's ability or propensity to carry electric current. The conductivity of a semiconductor is proportional to the density of electrons and holes and their respective mobilities. That is, if n is the density of electrons in electrons/cm^3 and p is the hole density in holes/cm^3, then

$$\sigma = \text{conductivity} = q(n\,\mu_n + p\,\mu_p) \tag{8.4}$$

Resistivity, ρ, generally measured in ohm-cm, is the reciprocal of conductivity.

$$\rho = \frac{1}{\sigma} \tag{8.5}$$

The resistance, R, of a material with a constant cross section can be calculated by knowing its resistivity and physical dimensions using the following relation:

$$R = \rho \frac{L}{A} \tag{8.6}$$

where A is the cross-sectional area normal to the direction of current flow and L is the length in the direction of the current.

EXAMPLE 8.1

Let us calculate the resistance from end-to-end of the rectangular bar shown in Fig. 8.8 assuming that the bar is made of a material with a resistivity of 4 ohm-cm, and its dimensions are: length, $L = 2$ cm; height, $H = 0.5$ cm; and width, $W = 0.2$ cm. Using Eq. (8.6)

$$R = \rho \frac{L}{A} = \rho \frac{L}{WH}$$

$$R = (4)\,\frac{2}{(0.5)\,(0.2)} = 80 \text{ ohms}$$

∎

DRILL EXERCISE

D8.2. The semiconductor cube in drill exercise 8.1 is determined to be *p*-type silicon with a hole concentration of 3×10^{15} holes/cm^3; the electron concentration is so low it can be ignored. Find the resistivity of the material and the resistance from one face of the cube to the other. Are these numbers always the same for any shape semiconductor?

Ans: $\rho = 4.3$ Ω-cm; $R = 4.3$ Ω; No; these answers are only the same for a cube 1 cm on a side. The resistivity is a property of the material and is independent of the shape of the bar; the resistance depends on the shape of the bar, that is, length and cross-sectional area.

The *current density,* J, in a material is defined as the current per unit cross-sectional area, and is expressed in amperes per cm ^{2}.

The current density is directly related to the applied field by a constant, the conductivity. Therefore:

$$\mathbf{J} = \sigma \mathbf{E} \tag{8.7}$$

The direction of current flow, indicated by the direction of the current density vector, is in the same direction as the electric field vector. Equation (8.7) is an alternate expression of Ohm's law.

EXAMPLE 8.2

In the rectangular bar of Fig. 8.8, assume that the material has the same properties described in Example 8.1, and that the voltage, V, applied across the bar is 12 volts. Let us determine the magnitude of the current density in the bar.

Since we have already calculated the total resistance in Example 8.1, the current, *I*, is

$$I = \frac{V}{R} = \frac{12}{80} = 0.15 \text{ amperes}$$

The current density is determined as the current per unit of cross-sectional area. Since the direction of current flow is parallel to the long axis of the bar, which is the same direction as the electric field, **E,** the cross section through which the current flows is simply the cross section of the bar. Therefore

$$J = \frac{I}{A} = \frac{0.15}{(0.5)(0.2)} = \frac{0.15}{0.1} = 1.5 \text{ amperes/cm}^2 \qquad \blacksquare$$

DRILL EXERCISE

D8.3. Verify that the same answer obtained in Example 8.2 can be obtained using Eq. (8.7) directly.

Diffusion

The second mechanism by which charge carriers move in a solid is diffusion. A *diffusion current* occurs because of the physical principle that over time particles undergoing random motion will show a net movement from a region of high concentration to a region of lower concentration. Simple examples include the diffusion of smoke particles in the

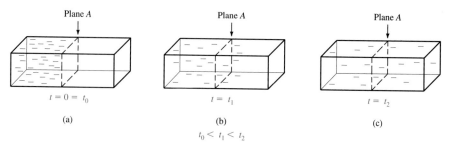

Figure 8.10 Diffusion of electrons.

air, and the diffusion of gas molecules away from regions of high concentration. Heat flow from regions of high temperature to regions of low temperature follows the same law.

In a semiconductor with no electric field applied, holes and electrons move about randomly due to thermal energy. In Fig. 8.10 the diffusion process is illustrated. Suppose at time zero, $t = 0$, a large concentration of electrons is established at the left end of the rectangular bar shown in Fig. 8.10(a). At a later time, $t = t_1$, the carriers have begun to spread out, and some have passed through "plane A" creating a net current through this plane as shown in Fig. 8.10(b). At an even later time, as illustrated in Fig. 8.10(c), the carriers will have dispersed uniformly throughout the semiconductor and the concentration difference that was the driving force behind the net movement of carriers will have been equalized; the current through plane A will have returned to zero.

The diffusion equation simply states that the diffusion current density J is directly proportional to the gradient of carrier concentration. The proportionality constant consists of two terms, the charge per carrier and a constant which describes how easily a particular carrier type diffuses in its host material, which in this case is the *Diffusion Constant for electrons*, D_n.

For illustration, consider again the rectangular bar. Current flow is parallel to the long axis of the bar, which we define as the x direction. Therefore, the one-dimensional diffusion equation for electrons states that the current density for electrons, J_n, at any cross section in the bar is given by:

$$J_n = q\, D_n\left[\frac{dn}{dx}\right] \tag{8.8}$$

where dn/dx is the electron concentration gradient at the plane under consideration.

For silicon at 300° K, $D_n = 35$ cm²/sec; there is an analogous diffusion constant for holes, $D_p = 12$ cm²/sec.

EXAMPLE 8.3

Let us determine the magnitude of the current resulting from diffusion in a region of a rectangular semiconductor bar where the electron concentration changes linearly from 3×10^{16}/cm³ to 3×10^{18}/cm³ over a distance of 0.02 cm. Assume the bar has the same physical dimensions as in Example 8.1.

The electron concentration gradient is constant over the region of interest and

$$\frac{dn}{dx} = \frac{\Delta n}{\Delta x} = \frac{(3 \times 10^{18}) - (3 \times 10^{16})}{0.02}$$

$$\frac{dn}{dx} = 1.49 \times 10^{20}/cm^4$$

Thus, the current density according to Eq. (8.8) is

$$J_n = q \, D_n \left[\frac{dn}{dx} \right] = (1.6 \times 10^{-19}) \, (35) \, (1.49 \times 10^{20}) =$$

$$8.32 \times 10^2 \text{ A/cm}^2$$

Recall that the cross-sectional area of the bar was 0.1 cm². Therefore, the current in amperes is

$$I = J \, A = (8.32 \times 10^2)(0.1) = 83.2 \text{ amperes} \qquad \blacksquare$$

Although carrier diffusion was illustrated for electrons, there is an analogous diffusion equation for holes based on the *hole diffusion constant* D_p and the hole concentration gradient. In a semiconductor we can simultaneously have holes and electrons diffusing, each according to its own diffusion constant and concentration gradient.

8.5
The *p-n* Junction Diode

A *p–n junction diode* can be created by bringing together a *p*- and *n*-type region within the same semiconductor lattice as shown in Fig. 8.11. Imagine that at the instant this junction is created, there's an *n*-type region with many free electrons joined to a *p*-type region with an abundance of free holes.

What happens? Diffusion, of course. The electrons diffuse from their region of high concentration into the *p* material where their concentration is low. Simultaneously, the high density of holes in the *p*-region drives the diffusion of holes into the *n*-region. But this initial diffusion is quickly stopped by the following fact: every time an electron diffuses into the *p*-region it leaves behind an uncovered positive charge bound in the nucleus of the dopant atom on the *n*-side; similarly, negative charge is bound in the lattice of the *p*-side as hole diffusion occurs. This process builds up charge layers as shown in Fig. 8.12

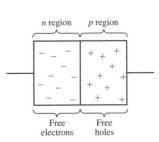

Figure 8.11   *p–n* junction at instant of formation.

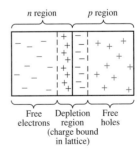

Figure 8.12   *p–n* junction with depletion region.

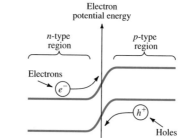

Figure 8.13 *p–n* junction potential barrier.

in the so-called *depletion region*—a region which is depleted of carriers. These charge layers prevent further diffusion.

Why is the depletion region depleted of carriers? Imagine what happens to an electron that diffuses into the depletion region; its negative charge is attracted to the new bound charge in the *n*-region and repelled from the *p*-region. It is therefore driven back to the *n*-region from which it came.

In summary, the initial diffusion process sets up two charge layers at the junction which grow to the point that they prohibit further diffusion of carriers, and create a depletion region at the junction void of any free carriers.

This charge barrier creates a state of balance with the diffusion process; any attempt for the diffusion to increase uncovers more bound charge which increases the barrier and reduces the diffusion.

This barrier can be represented as a voltage or potential barrier, and is analagous to the potential barrier that is formed between any two dissimilar conducting materials when they are joined. Thermocouples are devices used to measure temperature and consist of two dissimilar metals fused together at a point. There is a potential barrier created across the junction which can be measured and is related to temperature.

The height of the potential barrier for a *p–n* junction in equilibrium is shown in Fig. 8.13. Because electron energy is plotted as increasing upward in this figure, hole energy (opposite sign) is downward—therefore, the upper curve shows the energy barrier to electrons moving from the *n*-region to the *p*-region, and the lower curve shows the energy barrier to holes moving from the *p*-region to the *n*-region.

It is very important to note that the height of the potential barrier across a *p–n* junction can be changed by applying a voltage across the junction from an external source. The diffusion of carriers across the junction is exponentially related to the barrier height, so as the barrier height is changed by an external voltage source, V, large (exponential) changes in current, I, due to carrier diffusion are observed.

If an external voltage source is applied and:

1. the *p*-region is made more *positive* than the *n*-region, the barrier height is reduced, and carriers can easily diffuse across and the diode readily conducts a current— this is called "*forward bias*," as shown in Fig. 8.14(a).
2. the *p*-region is made more *negative* than the *n*-region, the barrier height is increased, and very few carriers can cross, and virtually no current flows—this is called "*reverse bias*," as shown in Fig. 8.14(b).

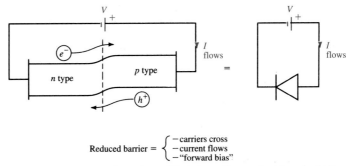

Reduced barrier = $\begin{cases} -\text{carriers cross} \\ -\text{current flows} \\ -\text{"forward bias"} \end{cases}$

Figure 8.14(a) *p–n* junction with externally applied forward bias.

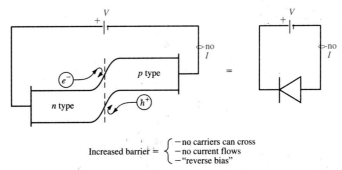

Increased barrier = $\begin{cases} -\text{no carriers can cross} \\ -\text{no current flows} \\ -\text{"reverse bias"} \end{cases}$

Figure 8.14(b) *p–n* junction with externally applied reverse bias.

A real *p–n* junction requires a small but finite foward bias voltage to be externally applied to sufficiently lower the barrier height to initiate the onset of reasonable conduction. "Reasonable" is obviously a qualitative term, but many diode manufacturers arbitrarily choose 1 microampere as the threshold of conduction and define the externally applied forward bias required to obtain 1 microampere as the diode's *"turn-on voltage,"* V_F.

For many circuit analysis calculations, we might assume the diode is *ideal*, or $V_F = 0$; such an *ideal diode* conducts with zero resistance if the diode is forward biased, and has infinite resistance if reverse biased. Several diode models are discussed in the next section along with the assumptions which make them useful.

8.6
Diode Circuit Models

There are several circuit models commonly used to describe the electronic characteristics of a real diode. The primary purpose of such models is to provide an equivalent circuit and quantitative relationships describing the key parameters. These models can then be used in analyzing circuits containing diodes in order to predict the performance of the circuits.

In this section three diode models will be described that predict the relation between the dc voltage across a diode, V_D, and the current through the diode, I_D. The three models are:

1. the ideal diode model
2. the diode equation model
3. the piecewise linear diode model

The choice of which model to use is guided primarily by two axioms: (a) what is known about the problem often favors one model over another, and (b) we use the simplest model that will provide an answer within the accuracy desired. Examples in the following sections will help make this clear.

The Ideal Diode

The *ideal diode* is simply an idealized two-terminal electronic device which performs the process of rectification perfectly; that is, an ideal diode passes current perfectly in one direction (zero resistance), and passes no current in the opposite direction (infinite resistance). Real diodes differ from this idealization slightly as we'll see later; however, this model works well for analysis of many circuits.

The symbol for the ideal diode is given in Fig. 8.15. The *p*-side of the diode is called the *anode*, and the *n*-side designated in the symbol by the straight line is called the *cathode*. For an ideal diode:

Rule 1.   If the voltage on the *p*-side is made more positive than the voltage on the *n*-side (forward bias)

then

the ideal diode conducts current as a closed switch.

Rule 2.   If the voltage on the *n*-side is made more positive than the voltage on the *p*-side (reverse bias)

then

the ideal diode will not conduct, and appears as an open switch.

A result of Rules 1 and 2 is that:

If the circuit attempts to force positive current from the *p*-side through to the *n*-side (in the direction of the arrow in the diode symbol)

then

The ideal diode conducts as a closed switch (forward bias). Otherwise, the ideal diode behaves as an open switch.

If a plot is made of the voltage applied across an ideal diode versus the resulting current, the curve lies on the negative voltage axis and the positive current axis as shown in

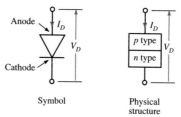

Symbol   Physical
structure

Figure 8.15   Ideal diode symbol and real diode physical structure.

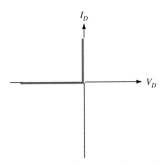

Figure 8.16   Plot of current through diode, I_D, versus voltage across diode, V_D, for an "ideal" diode.

Fig. 8.16. Any applied positive voltage results in infinite currents since the device has zero resistance under this condition; any applied negative voltage results in zero current since the device has infinite resistance in this condition. Notice that the direction of positive current conduction is the same as that of the arrow in the diode symbol.

In summary, the circuit model of the ideal diode is shown in Fig. 8.17. It behaves as a closed switch under forward bias, and an open switch under reverse bias. The analysis of circuits containing an ideal diode is not always straightforward because this component is a non-linear element and contains a discontinuity in its V–I characteristics. Therefore, linear circuit analysis must be used cautiously.

In solving circuit problems, it is necessary to determine which voltage polarity the circuit is attempting to place on the diode terminals. This can be easily done by replacing the diode with an extremely small finite "test" resistance, δR. Then if the voltage across δR is in such a direction to establish forward bias, replace the resistor with a closed switch or a short; otherwise replace it with an open circuit. We will illustrate the use of the ideal diode model with some examples.

EXAMPLE 8.4

In Fig. 8.18(a), let us compute the current, I_1, through the resistor, R, assuming an ideal diode.

First, we determine the polarity of the voltage that the circuit will establish across the diode; or equivalently, the direction that the circuit wishes to drive positive current through

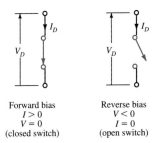

Forward bias
$I > 0$
$V = 0$
(closed switch)

Reverse bias
$V < 0$
$I = 0$
(open switch)

Figure 8.17 (a)   Ideal diode symbol

Figure 8.17 (b)   Ideal diode model.

Figure 8.17   Ideal diode symbol and its switch circuit model.

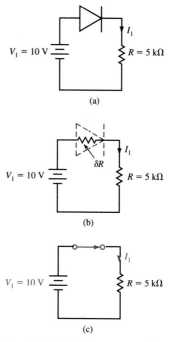

Figure 8.18 Ideal diode current in simple circuit.

the diode. To examine this, the diode is replaced with a small test resistor, δR, such that the circuit becomes that shown in Fig. 8.18(b).

It is clear in this simple circuit that the left side of the diode, p-side or anode in this case, is at a higher or more positive voltage than the other side. Therefore, the diode is forward biased and it can be replaced with a closed switch, as illustrated in Fig. 8.18(c).

We can arrive at the same conclusion by evaluating the current the circuit forces through the test resistor. From Fig. 8.18(b), the polarity of I_1 through the diode can be determined

$$I_1 = + \frac{V_1}{R + \delta R} = (+ \text{ value})$$

All terms above are positive, and the desired direction of I_1 is from left to right; or from p-side to n-side—the direction in which a diode conducts positive current.

The above analysis has indicated, in two different ways, the diode is *forward biased*, and therefore can be replaced with a closed switch.

Now with the closed switch replacing the diode, I_1 is

$$I_1 = + \frac{V_1}{R} = \frac{10}{5k} = 2 \text{ mA}$$

D8.4. In Figure D8.4(a), if V_1 is -10 V, find the current, I_1, through the resistor, R.

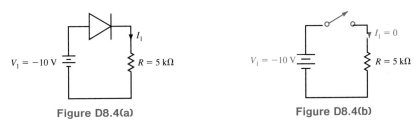

Figure D8.4(a)   Figure D8.4(b)

Ans: In this case, $I_1 = 0$; the single loop has been opened by the reverse biased diode. This is illustrated in Figs. D8.4(a) and (b).

EXAMPLE 8.5

In Fig. 8.19, if the voltage source, V_1, can be varied from 0 to 10 volts, (a) sketch a plot of the voltage at node A, V_A, as a function of V_1 for $0 < V_1 < 10$ V, and (b) calculate the current through $D2$ for two cases: (1) for $V_1 = 2$ V and (2) for $V_1 = 4$ V.

(a) The cathode of $D2$ is set at 3 volts, relative to ground, by V_2. If node A is less than 3 volts, then $D2$ is open. Thus, for any value of V_1 below 3 volts, $D1$ is conducting and $V_A = V_1$. For values of V_1 above 3 volts, $D2$ conducts as a closed switch, and node A is therefore set by V_2 at 3 V; in this range $D1$ is reverse biased, and node A stays "pinned" at 3 V. A plot of V_A as a function of V_1 is shown in Fig. 8.20.

(b) For case 1, where $V_1 = 2$ V, $D2$ is reverse biased, and therefore the current through $D2$ is zero.

For case 2, where $V_1 = 4$ V, $D1$ is not conducting, and therefore the current being supplied through the 10 kΩ resistor all flows through $D2$. Therefore

$$I_{D2} = \frac{10 - 3}{10k} = 0.7 \text{ mA}$$ ∎

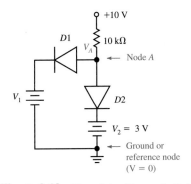

Figure 8.19 Circuit for Example 8.5.

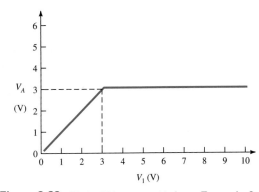

Figure 8.20 Plot of V_A versus V_1 from Example 8.5.

The Diode Equation and Model

A *real p–n junction diode* does not behave exactly like an open or closed switch; it can be shown to have a voltage–current (*V-I*) relationship over its range of normal operation which follows closely the so-called *diode equation* given as Eq. (8.9). This equation can be derived based on the assumption that carriers move by diffusion as described in section 8.4.

$$I_D = I_o(e^{\frac{qV_D}{nkT}} - 1) \tag{8.9}$$

In this equation, V_D is the externally applied voltage across the diode, I_D is the current through the diode, q is the electronic charge, T is temperature in degrees Kelvin, k is Boltzmann's constant, n is an ideality factor which for normal currents in silicon devices is approximately unity (we will assume $n = 1$ in further discussions), and I_o is called the *reverse saturation current* and equals a constant which depends on the properties of the diode. At room temperature, the constants

$$\frac{q}{kT} \cong 39 \tag{8.10}$$

The constant term I_o is made up of the product of two terms. The first term is a function of the semiconductor material properties, and is generally fixed for a particular diode fabrication process; the other term is directly proportional to junction area which is normally determined by the device layout designer, as will be explained in Chapter 13.

A plot of Eq. (8.9) using a linear current scale at room temperature and a value of $I_o = 1 \times 10^{-16}$ amperes is shown in Fig. 8.21. Notice that for applied voltages below about 0.6 volts, the current is imperceptibly close to zero.

The diode equation has some very interesting features:

For $V_D = 0$, $I_D = 0$, as expected.

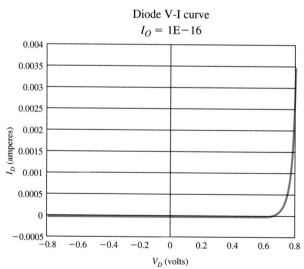

Figure 8.21 Plot of diode equation for $I_O = 1 \times 10^{16}$ A.

For V_D *positive with magnitude* $> 0.1\ V$; with slight forward bias applied, the exponential term has a positive exponent which very quickly dominates, becoming much, much larger than unity. In this forward bias condition the voltage–current relationship follows very closely the exponential function over a current range of many decades. It is often convenient to plot the forward characteristics of the diode on a plot of log I_D versus V_D, since in this region the resulting curve follows a straight line, as illustrated in Fig. 8.22.

Therefore, for *forward bias*, where

$$V_D \gg \frac{kT}{q} \quad (0.026\ V \text{ at room temperature})$$

the exponential term dominates, and

$$I_D \approx I_o e^{\frac{qV_D}{kT}} = I_o e^{39V_D} \tag{8.11}$$

In the forward bias region as the applied voltage approaches zero and also at very high bias, some deviations from this ideal exponential relationship occur.

For V_D *negative with magnitude* $> 0.1\ V$; the negative voltage makes the exponent of the exponential negative, and the reverse current flow asymptotically approaches the value of minus the constant I_o, which is a very small number. Typical silicon diodes have a value for I_o in the range of 10^{-14} A for large area junctions to 10^{-18} A for very small junctions. Schottky junctions have I_o's that depend on the materials used, but typically are in the range of 10^{-11} to 10^{-14} A.

Therefore, for *reverse bias*, where

$$-V_D \ll -kT/q$$

the exponential term is negligible compared to unity and

$$I_D \approx -I_o \tag{8.12}$$

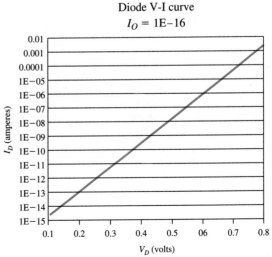

Figure 8.22   Plot of diode equation on semilog scale ($I_0 = 1 \times 10^{-16}$ A).

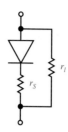

Figure 8.23 Resistors r_s and r_l are used to model diode series resistance and diode leakage resistance.

The reverse saturation current, I_o, increases rapidly with temperature. An often-used rule of thumb is that I_o approximately doubles for every 10°C rise in temperature.

Notice the primary differences between a real diode described by Eq. (8.9) and the ideal diode. In a real diode:

1. For positive applied voltage, forward bias, the bias current increases exponentially; however, some small positive voltage is required before the diode conducts reasonable current. In a silicon diode this *turn-on voltage*, V_F, is about 0.5 to 0.7 volts and is the value of V_D at which I_D reaches the microampere range. By contrast, the ideal diode has a discontinuity which occurs precisely at $V_D = 0$. This difference in modeling is very important in some circuits, where voltage changes of 0.5 to 0.7 volts make a significant difference in circuit performance.
2. For negative applied voltage, reverse bias, there is a small finite reverse current that flows and it is temperature dependent. The ideal diode had zero current for $V_D < 0$.

Series Resistance. All practical diodes also have some series resistance due to the bulk semiconductor material on either side of the junction and the resistance of the contacts made to the semiconductor material. Series resistance is only a concern in circuit analysis when the diode is forward biased and conducting significant current. Because series resistances are typically in the range of a few ohms up to a few hundred ohms, these resistances in series with the nearly infinite resistance of a reverse biased diode matter little. To model series resistance, a resistor of the proper value, r_s, is placed in series with the diode model being used, as illustrated in Fig. 8.23.

Leakage Resistance. Leakage resistance is only a factor when the diode is reverse biased and the circuit is sensitive to very small currents. Under reverse bias, a diode may conduct somewhat more current than the value $-I_o$ indicated by the diode equation. This is a result of current leakage between the diode terminals and is modeled by a very high resistance, often tens to hundreds of megohms, in parallel with the diode model. This is illustrated in Fig. 8.23 as r_l.

The resistances r_s and r_l are generally not included in a first-order calculation. They must be included in the analyses of some diode circuits when their omission would affect the predicted circuit performance. Such circuits are those operating at high forward bias currents, or those very sensitive to small reverse currents.

EXAMPLE 8.6

If a diode has an I_o of 2×10^{-15} A, let us determine (a) the voltage across this diode when it is forward biased and carrying 2 mA and (b) the change in voltage across the diode when the current is changed from 2 to 1 mA.

(a) From Eq. (8.11)

$$I_D = I_o e^{39V_D}$$

or

$$e^{39V_D} = \frac{I_D}{I_o}$$

Solving the equation for V_D yields

$$V_D = \frac{1}{39} \ln \frac{I_D}{I_o} = \frac{1}{39} \ln \frac{2 \times 10^{-3}}{2 \times 10^{-15}} \text{ V} \tag{8.13}$$

or

$$V_D = 0.026 \ln(10^{12}) = 0.718 \text{ V}$$

(b) Using the results of part (a)

$$V_D = \frac{1}{39} \ln \frac{I_D}{I_o} = 0.026 \ln \frac{1 \times 10^{-3}}{2 \times 10^{-15}} \text{ V}$$

and therefore

$$V_D = 0.026 \ln (5 \times 10^{11}) = 0.700 \text{ V}$$

Notice that reducing the current by a factor of two only changed the voltage across the diode from 0.718 V to 0.700 V, a change of 18 millivolts.

It is interesting that in such diodes the voltage across the diode always changes by 18 mV when the current is changed by a factor of two. ∎

DRILL EXERCISE

D8.5. (a) Prove for the general case that if the forward bias current of a diode is doubled, the voltage will increase by 18 mV. (b) Find the voltage change if the current is changed by a factor of 10.

Ans: (a) Hint: Use Eq. (8.13) (b) 60 mV.

EXAMPLE 8.7

A diode is rated at 100 milliwatts, which is its maximum allowed power dissipation. This limitation is generally imposed as a result of the ability of the diode's mechanical structure and packaging to remove the heat generated in the diode. If the diode is forward biased from a 5-volt dc source through a 1 kΩ resistor and the voltage across the diode is 0.65 volts, let us determine (a) the power being dissipated in the diode and (b) if the power rating would be exceeded if the same diode were put in another circuit and the forward voltage across it were measured as 0.75 V.

(a) The circuit described is shown in Fig. 8.24. The current through the resistor is

$$I_R = \frac{5 - 0.65}{1\text{k}} = 4.35 \times 10^{-3}\text{A} = 4.35 \text{ mA}$$

This same current flows through the diode, so $I_D = 4.35$ mA. The power dissipated is then

$$P = V_D I_D = (0.65)(4.35 \times 10^{-3}) = 2.83 \text{ milliwatts}$$

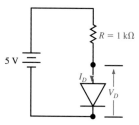

Figure 8.24   Circuit for Example 8.7.

(b) In order to calculate the power dissipated when the voltage is increased to 0.75 volts, we must first determine the new current.

From the data in part (a) we can calculate I_o using Eq. (8.11).

$$I_o = I_D e^{\frac{-qV_D}{kT}} = 4.35 \times 10^{-3} \; e^{-39(0.65)}$$

and

$$I_o = 4.26 \times 10^{-14} \text{ A}$$

We now use this value of I_o to calculate the new current, I'_D, when the voltage across this diode is 0.75 V, that is,

$$I'_D = I_o e^{\frac{qV_D}{kT}} = (4.26 \times 10^{-14})e^{39(0.75)} = 0.215 \text{ A}$$

The power dissipated under these new conditions is

$$P = V_D I_D = (0.75)(0.215) = 161 \text{ milliwatts}$$

which exceeds the maximum power dissipation allowed. If this diode were connected in a circuit under these conditions, it would burn up.   ■

Virtually every electronic component has a maximum allowed power dissipation and a maximum voltage and/or maximum current rating. They should not be exceeded.

DRILL EXERCISE

D8.6.   Given the circuit shown in Fig. 8.25 and assuming the diode characteristics are represented by the diode equation, let us find (a) the maximum (peak) instantaneous voltage appearing across the diode if the peak current through R is measured to be 1.8 mA, and (b) I_o for this diode.

Ans: (a) V_D(peak) = 0.56 V   (b) $I_o = 5.89 \times 10^{-13}$ A.

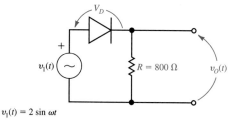

$v_1(t) = 2 \sin \omega t$

Figure 8.25   Circuit for finding peak current.

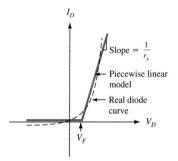

Figure 8.26 *V–I* curves for real diode and piecewise linear model.

The Piecewise Linear Diode Model

Recall the plot of I_D as a function of V_D on a linear scale as shown in Fig. 8.21 for a real diode. Again notice that the diode conducts a negligibly small current until the turn-on voltage, V_F, of the junction is reached, which is generally at about 0.7 V in silicon devices.

The real diode characteristic curve can be approximated by a model which uses two connected straight line segments. Such a model is called a piecewise linear (PL) model.

Figure 8.26 shows a real diode curve and a piecewise linear approximation to that curve. Notice that the turn-on voltage, V_F, marks the voltage at which the two line segments meet, and at which the slope is discontinuous. Modeling of the finite voltage, V_F, required to establish conduction requires including a voltage source of value V_F in the model. The portion of the characteristics that represents diode conduction has a large but finite positive slope and can be modeled by including a resistor in the model.

A circuit model that will reproduce this curve can be created using linear circuit elements and an ideal diode; such a model is shown in Fig. 8.27 and is known as the *piecewise linear circuit model*. In this piecewise linear diode equivalent circuit, the ideal diode internal to the model has infinite resistance until it becomes forward biased; this occurs when the externally applied voltage is enough to overcome V_F. Therefore, the "break" to initiate diode conduction occurs at $V_D = V_F$; we will assume $V_F = 0.7$ V unless specified otherwise.

The device has a series resistance r_s which gives the plot of the voltage-current characteristics a finite slope of $1/r_s$ in the forward biased region. The use of the PL model will be illustrated with several examples.

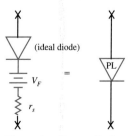

Figure 8.27 Piecewise linear diode circuit model.

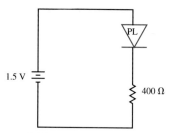

Figure 8.28 Circuit for Example 8.8.

EXAMPLE 8.8

Let us calculate the current through the 400 Ω resistor in the circuit shown in Fig. 8.28, assuming a piecewise linear model for the diode with $V_F = 0.7$ V and $r_s = 10$ Ω.

The diode is replaced with the piecewise linear circuit model, as shown in Fig. 8.29(a). Note that the ideal diode in the model is forward biased and becomes a short, or closed switch. Therefore, this circuit reduces to that shown in Fig. 8.29(b). Writing a loop equation for this circuit, we obtain

$$0 = -1.5 + 0.7 + I(10) + I(400)$$

or

$$I = (0.8)/(410) = 1.95 \text{ mA} \qquad \blacksquare$$

The piecewise linear model greatly simplifies circuit calculations in that it approximates the exponential relationship of the diode equation with linear line segments which can be modeled with linear circuit elements. It provides a significantly more accurate representation of the real diode than the ideal diode model, particularly if circuit voltages are relatively low, and the turn-on voltage, V_F, becomes a significant factor in the calculation.

EXAMPLE 8.9

If the voltage $v_x(t)$ as shown in Fig. 8.30 is applied to the circuit of Fig. 8.31(a) and the diode has a V_F of 0.7 V and an $r_s = 10$ Ω, let us plot the current through the diode in the time interval $0 \leq t \leq 2T$.

First, we replace the diode with its piecewise linear equivalent circuit as shown in Fig. 8.31(b). Note that the ideal diode in the model will remain reverse biased until $v_x(t)$ reaches

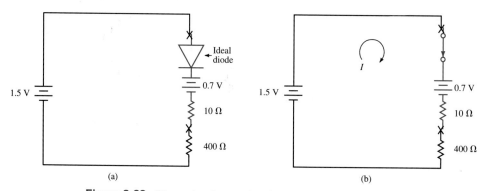

(a) (b)

Figure 8.29 Piecewise linear circuit model for Example 8.8.

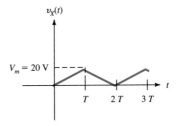

Figure 8.30 Input voltage for Example 8.9.

a value of 5.7 volts; the current I will remain zero during this time. When $v_x(t)$ exceeds 5.7 volts, positive current will begin to flow, and will increase to a maximum at time T when $v_x(t)$ is maximum. At this point, the loop equation used to determine I_{max} is

$$20 - 0.7 - 5.0 = I_{max}(100 + 10)$$

and

$$I_{max} = 130 \text{ mA}$$

After the diode initiates conduction, $v_x(t)$ increases linearly and all other elements in the circuit are linear and resistive. Hence, we expect I to increase linearly from zero to I_{max}, and therefore the plot of applied voltage $v_x(t)$ and current, I, appear as shown in Fig. 8.32. ■

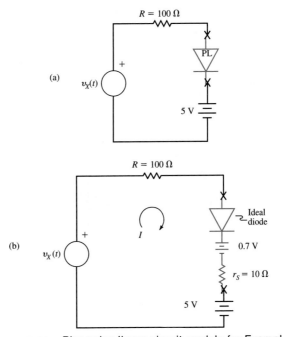

Figure 8.31 Piecewise linear circuit models for Example 8.9.

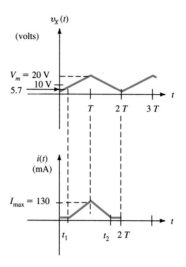

The times, t_1 and t_2 can be calculated from basic geometrical relationships:

$$\frac{t_1}{5.7} = \frac{T}{20}$$

$$t_1 = \frac{5.7\,T}{20} = 0.285\,T$$

$$t_2 = 2\,T - t_1 = 2\,T - (0.285\,T) = 1.715\,T$$

Figure 8.32   Answers for Example 8.9.

EXAMPLE 8.10

If diode $D1$ in Fig. 8.33(a) is modeled by a piecewise linear model with $V_F = 0.7$ V and $r_s = 0$, let us determine (a) the current through $R1$ and (b) the current through $D1$.

The direction of positive current is in a direction to forward bias the diode $D1$. Replacing $D1$ with the piecewise linear model assuming forward bias and $r_s = 0$ results in the circuit of Fig. 8.33(b).

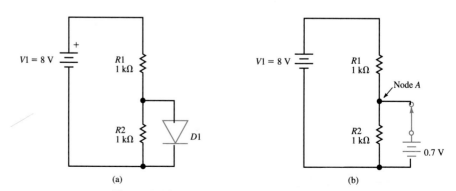

Figure 8.33   Circuit models for Example 8.10.

The current through $R1$, I_{R1} is

$$I_{R1} = \frac{V1 - 0.7}{R1} = \frac{7.3}{1\text{k}} = 7.3 \text{ mA}$$

The current through $R2$, I_{R2} is

$$I_{R2} = \frac{0.7}{R2} = \frac{0.7}{1\text{k}} = 0.7 \text{ mA}$$

The current into node A is 7.3 mA; the current out through $R2$ is 0.7 mA; therefore by Kirchhoff's current law, the current through $D1$ is

$$I_{D1} = 7.3 \text{ mA} - 0.7 \text{ mA} = 6.6 \text{ mA} \qquad \blacksquare$$

DRILL EXERCISE

D8.7. In Example 8.10, find the values of I_{R2} and I_{D1} if $V1$ is doubled.

Ans: $I_{R2} = 0.7$ mA; $I_{D1} = 14.6$ mA.

8.7
Diode Logic Gates

The diode can be used to create circuits that can perform as logic gates; such circuits provide an output that represents the logical combination of several inputs.

The AND and the OR functions have been previously described in Chapter 7. In these examples assume the diodes are ideal, and the logic levels are defined as:

$$\text{low} = 0 = \text{zero volts}$$
$$\text{high} = 1 = \text{five volts}$$

Consider the circuit shown in Fig. 8.34(a). This circuit has two inputs in the configuration shown, labeled A_1 and A_2, and output, B. Recall that for an AND gate, the only case in which the output is high is when both inputs are high. The circuit operates exactly as just described; if either input is low, the associated diode is conducting and the output voltage is pulled low. Only when both inputs are high, that is, both diodes off, is the voltage at B released and pulled high by the +5 volt source through resistor R.

The circuit shown in Fig. 8.34(b) operates in an analogous way to perform the func-

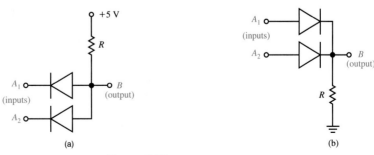

Figure 8.34   Diode logic gates.

tion of an OR gate. If either input is raised high, the associated conducting diode pulls the voltage at *B* high; only when both inputs are low, that is, both diodes off, is the output voltage, *B*, in the low state.

8.8
Power Supply Circuits

Diodes are widely used in *power supply* circuits which convert ac into dc for the purpose of operating electronic equipment. Recall that residential ac power distribution is typically done at either 110–120 volts or 220–240 V ac.

These voltages are substantially larger than the voltage required to operate most solid state electronic equipment. Digital electronic systems typically require 5 volts dc, although some new circuits require lower voltages, such as 3.3 volts. Analog electronic systems operate over a wide range of power supply voltages, but a common practice is to utilize two dc voltage supplies, one +15 volts, and a second at −15 volts, as shown in Fig. 8.35.

In designing circuits to convert household ac power into power for use in electronic circuits, the first step is generally to reduce the magnitude of the voltage from the wall receptacle, and thus a *transformer* (a device described in detail in Chapter 18) is commonly used to step down the voltage. A transformer, which consists of two separate coils of wire on a common iron core, produces an ac output voltage (across the secondary) proportional to the input voltage (across the primary); the constant of proportionality is the ratio of the number of turns of wire in the secondary coil to that in the primary coil, or the *turns ratio*. Therefore, the input part of such a circuit might look like that of Fig. 8.36, where 110 V ac is stepped down by a factor of 10 to 11 V ac. Therefore, the turns ratio for the transformer is 10 to 1. These are RMS values, so the actual amplitude of the sinusoidal wave form is a factor of $\sqrt{2}$ larger. The output of this ideal transformer can be modeled as a Thevenin's equivalent circuit with an open circuit sinusoidal voltage source of 11 V ac, and a small equivalent resistance, R_{eq}. A plot of the voltage between *A* and *B* is shown in Fig. 8.37; it is 11 V rms, 60 Hz.

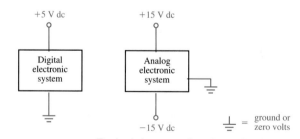

Figure 8.35   Typical dc power input requirements.

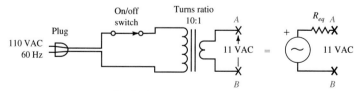

Figure 8.36   Typical input section of a power supply.

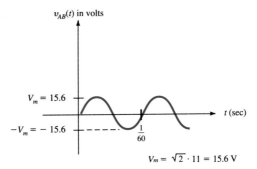

Figure 8.37   Plot of $v_{AB}(t)$.

Half-Wave Rectification

Half-wave rectification is the simplest process in which ac can be converted to dc. It is performed by a circuit in which a diode is used to clip the input ac signal excursions of one polarity to zero, while passing those of the other polarity. The following example illustrates the half-wave rectifier circuit.

EXAMPLE 8.11

If a sinusoidal voltage source, $v_1(t)$, of amplitude V_m is applied to the circuit shown in Fig. 8.38, let us draw the waveform of the voltage, $v_o(t)$. Assume an ideal diode.

Figure 8.39(a) shows the input waveform, $v_1(t)$. Figure 8.39(b) shows the calculated output waveform, $v_o(t)$.

During the positive half-cycle from time 0 to $\frac{1}{2}T$, the polarity of v_1 makes $D1$ forward biased and conducting as a closed switch, and therefore $v_o(t)$ is identical to $v_1(t)$ during this time.

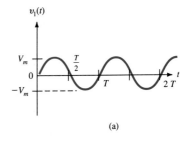

(a)

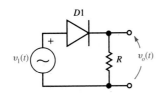

Figure 8.38   Half-wave rectifier circuit from Example 8.11.

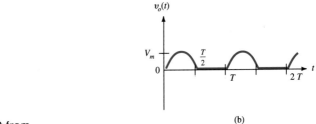

(b)

Figure 8.39   Voltage waveforms from Example 8.11.

During the negative half-cycle, from $\frac{1}{2}T$ to T, the diode $D1$ is reverse biased, and acts as an open switch. Since there is no current through the resistor R, v_o remains zero during this entire half-cycle.

It is important to notice (1) the *average* value of the input ac waveform v_1 is zero, (2) the output v_o is positive since the waveform is always greater than or equal to zero, and (3) this circuit is a simple example of a half-wave rectifier circuit, which is one type of circuit to convert ac to dc. ∎

The voltage waveform of the shape of $v_o(t)$ in Fig. 8.39(b) is termed a half-wave rectified signal. The ac voltage v_1 has been changed into an all positive dc voltage by rectification.

The dc value of a waveform is defined as its average value; therefore, for a periodic signal of period, T

$$V_{DC} = \frac{1}{T}\int_0^T v(t)dt \tag{8.14}$$

The dc value of the output voltage, $v_o(t)$, in the previous example can be calculated from Eq. (8.14)

$$V_{DC} = \frac{1}{T}\int_0^T V_m\sin\omega t\, dt$$

$$V_{DC} = \frac{1}{T}\left\{\int_0^{T/2} V_m\sin\omega t\, dt + \int_{T/2}^T 0\, dt\right\}$$

Letting $T = 2\pi$

$$V_{DC} = \frac{V_m}{2\pi}\int_0^\pi \sin x\, dx = \frac{V_m}{2\pi}[-\cos x]\big|_0^\pi \tag{8.15}$$

$$V_{DC} = \frac{V_m}{\pi}$$

EXAMPLE 8.12

Consider the half-wave rectifier circuit containing an ideal diode as shown in Fig. 8.40. We wish to find the dc value of the output voltage.

The load, R_L, represents whatever is connected to the power supply circuit. The voltage from nodes a to b is sinusoidal and has been reduced by a transformer with a turns ratio of 12:1 to 1/12 of the input value of 110 V ac.

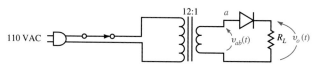

Figure 8.40 Half-wave rectifier circuit.

Thus

$$V_{ab} = 110\left(\frac{1}{12}\right) = 9.17 \text{ V rms}$$

The amplitude of this voltage, V_m, is $\sqrt{2}(9.17) = 13$ V. The voltage across R_L, designated $v_o(t)$, follows the input voltage on the positive half-cycle, and is zero when the input voltage is on a negative half-cycle. Therefore, a plot of $v_{ab}(t)$ and $v_o(t)$ appears as shown in Fig. 8.39(a) and (b) respectively.
Therefore, since $V_m = 13$ V

$$V_{DC} = \frac{13}{\pi} = 4.14 \text{ V} \qquad \blacksquare$$

DRILL EXERCISE

D8.8. A 115 V ac line is connected to a step-down transformer with a turns ratio of 18:4; if the secondary of the transformer is connected to a half-wave rectifier circuit, what is the dc voltage output of the circuit? (assume ideal diode).

Ans: 11.5 V dc.

EXAMPLE 8.13
Consider the circuit shown in Fig. 8.41 and assume the diode can be modeled with a piecewise linear model in which $V_F = 0.7$ V and $r_s = 10$ Ω. If we also assume that $v_1(t)$ is a sinusoid with an amplitude of 2 volts, as shown in Fig. 8.42(a), we wish to carefully plot $v_o(t)$.

The plot of $v_o(t)$ is shown in Fig. 8.42(b), and is derived as follows. Replacing the diode with its piecewise linear model yields the circuit shown in Fig. 8.43. The voltage source, $v_1(t)$, supplies a voltage at its peak of 2 volts. At the time this peak occurs, the appropriate loop equation is

$$2.0 - 0.7 = I_{peak}(10 + 60)$$

Therefore,

$$I_{peak} = 18.6 \text{ mA}$$

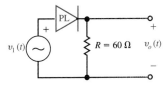

Figure 8.41 Half-wave rectifier circuit for Example 8.13 (with PL diode model).

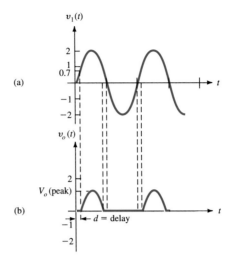

Figure 8.42 Voltage waveforms for Example 8.13.

The peak or maximum value of the voltage $v_o(t)$ is the same as the peak voltage across R. Therefore

$$V_{o(peak)} = (I_{peak})(60) = (18.6)(60) = 1.11 \text{ V}$$

This is the peak positive value of the sinusoidally varying voltage, $v_o(t)$, during the positive half-cycle. During the negative half cycle, I is zero.

Finally, in order for the diode to conduct, v_1 must be sufficiently positive to offset $V_F = 0.7$ volts; in fact, v_1 must reach 0.7 volts before the circuit forward biases the diode, and positive current can begin to flow. Therefore, there is an *offset* or *time delay* from the point where $v_1(t)$ turns positive, to the point where $v_o(t)$ becomes nonzero. Similarly, there is the same offset when $v_1(t)$ is decreasing at the end of the positive half-cycle.

Trigonometric relationships can be used to calculate the duration of this delay, d. Setting the value of v_1 equal to 0.7 volts and solving for the time d, we obtain

$$0.7 = 2 \sin (\omega d) \tag{8.16}$$

and hence

$$d = \frac{1}{\omega} \sin^{-1}(0.35)$$

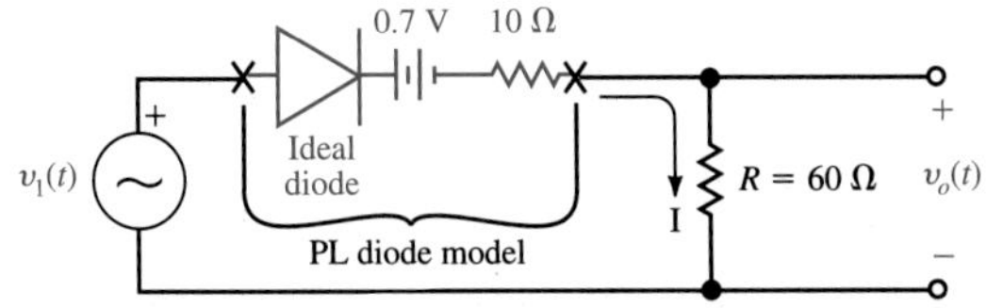

Figure 8.43 Piecewise linear circuit model for Example 8.13.

If for example the input sinusoid had a frequency of 60 Hz, or 377 radians/second, then

$$d = \frac{1}{377}\sin^{-1}(0.35) = 9.5 \times 10^{-4} \text{ sec}$$

During the portion of the half-cycle when the diode is conducting the waveform, v_o follows a sinusoidal function.   ■

Filter Circuits for Half-Wave Rectifiers

The voltage at the output of the half-wave rectifier is dc, but if the ac signal is 60 Hz, it rises and falls to zero 60 times a second. While this may be acceptable for some applications, such as some battery charger circuits, it is not very useful for powering most electronic systems. What these systems require is a dc voltage similar to the voltage supplied by a battery, that is, a constant voltage that has little or no variations with time. Real power supply systems operating from ac inputs usually do not produce completely smooth dc; however, with the addition of a *filter,* a dc signal can be closely approximated. The small variations in voltage from the filter output are called *ripple,* and good power supply circuits produce as little ripple as possible.

The output of the half-wave rectifier circuit, such as the waveform shown in Fig. 8.44(a), can be smoothed to something like that in Fig. 8.44(b) by a filter circuit, and the

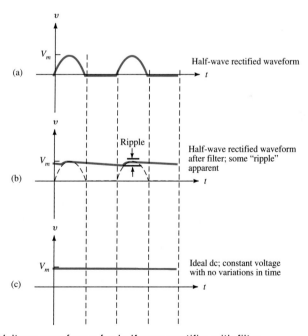

Figure 8.44   Voltage waveforms for half-wave rectifier with filter.

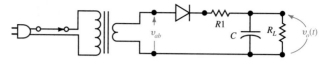

Figure 8.45   Half-wave rectifier circuit with filter.

dc value increased toward the ideal value of V_m. The ideal dc output is shown in Fig. 8.44(c). One of the most commonly used filter circuits is the RC network, low-pass filter described in Chapter 6. A power supply circuit using such a network appears in Fig. 8.45.

During positive half-cycles the diode conducts and charges the capacitor to a value of approximately V_m. R_1 is generally negligibly small and used only to prevent diode burnout due to large current transients experienced while initially charging the capacitor; it is ignored in the following analysis. When C is charged to V_m, the capacitor holds the voltage at the output high until the next charging cycle. When v_{ab} falls below the voltage across C, the diode reverse biases and disconnects from the circuit until the next cycle when v_{ab} exceeds the voltage on C. With a finite load resistance, R_L, current is supplied to the load only from C between charging cycles, and so v_o falls slightly between cycles. The fall in voltage is the exponential decay of an RC circuit of the type studied in Part I.

Therefore, the amount of *sag* in voltage between charging cycles depends on the time constant, τ, where

$$\tau = R_L C \qquad (8.17)$$

The charging voltage, $v_{ab}(t)$, and the power supply output voltage, $v_o(t)$, are shown in Fig. 8.46.

The amount of ripple can be approximated by straightforward transient circuit calculations as done in Part I, Chapter 3. Assume, for example, in this circuit that $C = 10 \ \mu F$, $R_L = 10 \ k\Omega$, and $T = 16.7$ ms which corresponds to 60 Hz ac. If v_o is charged to V_m, the expression for the decay of voltage across the capacitor is

$$v(t) = V_m e^{-\frac{t}{\tau}} \qquad (8.18)$$

where

$$\tau = R_L C = (10^4) \ (10 \times 10^{-6}) = 100 \text{ ms}$$

Evaluating the above expression at the end of a full cycle, that is, 16.7 ms, results in only

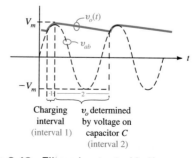

Figure 8.46   Filtered output of half-wave rectifier.

a small error if the ripple is small, because interval 2 is much longer than interval 1. Therefore

$$\frac{v_o(16.7)}{V_m} = e^{\frac{-16.7}{100}} = 0.846 = 84.6\% \tag{8.19}$$

The voltage at the end of a full cycle is 84.6% of its original value, V_m. Therefore, if V_m was 13 volts, as in Example 8.12, with the above filter it would only sag to $(13)(0.846) = 11$ volts during a full cycle. This cyclic variation from a peak of 13 volts to a minimum of 11 volts would be a 2-volt peak-to-peak ripple. Peak-to-peak means the voltage difference of the extremes. A 2-volt peak-to-peak ripple is very poor for electronic power supplies, and is presented for illustration only; good supplies can produce ripple measured in millivolts.

Often ripple is specified by the *Ripple Factor, RF*, which compares the rms value of the ripple to the magnitude of the dc voltage.

$$RF = \frac{rms\ value\ of\ ripple}{dc\ value} \tag{8.20}$$

Full-Wave Rectification

In half-wave rectifier circuits, the output is driven by the sinusodial input *only* during the positive half-cycle. A circuit using four diodes provides for charging of the output on both the positive and negative input cycles, essentially constructing the absolute value of the input wave. This circuit, called a *full-wave bridge rectifier circuit,* is shown in Fig. 8.47.

The voltage $v_{ab}(t)$ is sinusoidal with equal positive and negative voltage swings. During the positive half-cycle, diode $D1$ is forward biased and supplies current to the top of R_L; the current returns through $D2$ which also is forward biased. The other two diodes are not conducting.

During the negative half-cycle the bottom of the bridge is more positive than the top; therefore, positive current is supplied through $D3$ to the top of R_L, and returns through forward biased $D4$. During this half-cycle, $D1$ and $D2$ are reverse biased. As a result, the plots of $v_{ab}(t)$ and $v_o(t)$ for this circuit are shown in Figs. 8.48(a) and (b), respectively.

Another circuit which is sometimes used for full-wave rectification utilizes a transformer with a center tap on the output or secondary. If the center tap is taken as the ground reference, the transformer then supplies on each half-cycle a positive and a negative output, and two diodes are used to ensure that only the positive voltage swings are connected to the output. Such a circuit is shown in Fig. 8.49.

During the positive half-cycle, v_{ab1} indicates the polarity of the sinusoidal peak on

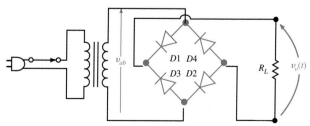

Figure 8.47   Full-wave bridge rectifier circuit.

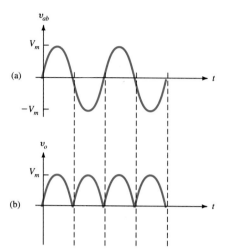

Figure 8.48   Voltage waveforms for full-wave rectifier.

each of the windings. The diode $D1$ conducts and positive current is supplied to the top of R_L. $D2$ is not conducting. During the negative half-cycle, the polarity is such that v_{ab2} represents the direction of the positive peak; during this time $D2$ conducts current to the top of R_L and $D1$ is not conducting.

It can be shown that the dc value of a full-wave rectified sinusoid, as illustrated in Fig. 8.48(b), is twice the value previously calculated for half-wave rectification. Thus

$$V_{DC} = \frac{2V_m}{\pi} \tag{8.21}$$

Capacitive filtering of the output of the full-wave rectifier circuit is much like that of the half-wave rectifier circuit. The full-wave rectifier circuit provides less ripple for the same value of filter capacitance and load resistance.

DRILL EXERCISE

D8.9.   A full-wave rectifier circuit is used to rectify an ac voltage of 20 V rms. Find the dc output voltage of the circuit. Assume no filter and ideal diodes.

Ans: 18 V.

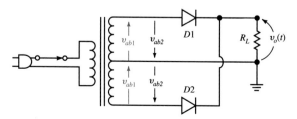

Figure 8.49   Alternate full-wave rectifier circuit.

8.9

Wave Shaping: Clippers and Clampers

There are a number of very useful circuits utilizing diodes that can be employed to alter the shape of a waveform for a particular purpose, or shift the dc level of a signal. Several of these type circuits are described in the following sections.

Clipper Circuits

Clipper circuits are used to limit the voltage excursions of a signal at some particular positive value, at a particular negative value, or both.

A simple clipper circuit is illustrated in Fig. 8.50. The voltage source v_i provides the input voltage to this circuit. The output voltage follows the input voltage exactly in this circuit, except when v_i exceeds V_A; for this latter case, $v_o = V_A$, that is

$$v_o = v_i \text{ for } v_i < V_A$$
$$v_o = V_A \text{ for } v_i \geq V_A$$

where V_A is a constant voltage.

If $v_i < V_A$, the diode is reverse biased and acts like an open switch. For this case, the input voltage, v_i, is connected directly through R to the output. Therefore, $v_o = v_i$, since there is no current through R, and therefore no voltage drop across R.

When $v_i > V_A$, the ideal diode is forward biased and the voltage source V_A is connected directly to the output, which effectively sets the output voltage at V_A.

EXAMPLE 8.14

Given the circuit of Fig. 8.50, with a sinusoidal input of amplitude 10 volts, and $V_A = 8$ V, we wish to plot the output voltage, $v_o(t)$.

The sinusoidal input voltage is shown in Fig. 8.51(a), and the circuit's output voltage, $v_o(t)$, is illustrated in Fig. 8.51(b). Notice that for all periods of time where the input exceeds 8 volts the output is clipped at 8 volts.   ∎

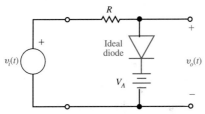

Figure 8.50 Simple clipper circuit.

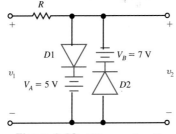

Figure 8.52 Clipper circuit.

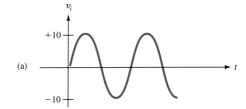

(a)

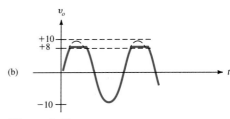

(b)

Figure 8.51 Voltage waveforms from Example 8.14 illustrating clipping.

The diode and voltage-source combination in the previous example can be placed in a similar circuit, but switched end-for-end, and used to limit the negative excursions of an input voltage. It is also possible to combine in one circuit a diode-source configuration for positive limiting with one for negative limiting; this circuit will clip the input at both positive and negative limits, each set independently by the appropriate constant voltage source. An example of this type circuit is shown schematically in Fig. 8.52. This circuit will clip positive input voltage swings at +5 volts and negative input voltage excursions at −7 volts.

Clamping Circuits

A *clamping circuit* is a circuit that produces an output which appears just as the input with one exception—the dc level or average voltage is shifted positively or negatively. Clamping circuits are also called dc restorer circuits or level shifter circuits.

For example, if the voltage waveform shown in Fig. 8.53 identified as v_i is applied to a simple clamping circuit, it might appear at the output of such a circuit as the voltage v_o shown in the same figure. In this example, the input has a dc or average value of zero; the output, v_o, has its dc level shifted lower to a new value, −A.

Circuits of this type are very useful in re-establishing a known dc level in a signal which has an unknown dc reference. This situation often occurs when signals are processed through amplifiers which are ac coupled; that is, they only amplify ac signals. For example, in a television receiver there is a clamping circuit to establish the peak value of the video signal at a known dc level. A circuit that can perform this operation is shown in Fig. 8.54(a).

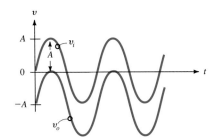

Figure 8.53 Voltage waveforms illustrating clamping.

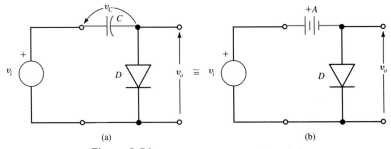

(a) (b)

Figure 8.54 Clamper circuit and model.

If the sinusoidal waveform labeled v_i in Fig. 8.53 is the input to this circuit, during the first cycle, the diode D conducts and the voltage across capacitor C charges to $+A$. The voltage across C, v_C, reaches the peak of the input voltage, and then as the voltage v_i decreases, D becomes reverse biased and C is left charged at its peak value. If this waveform is periodic, the capacitor is recharged to the peak value on every cycle, and any slow leakage of charge from the capacitor during each cycle will be ignored for now.

In this circuit, the charged capacitor acts like a dc voltage source, and the equivalent circuit of Fig. 8.54(b) can be used to analyze the circuit performance.

After a few initial "charging" cycles, when v_C has stabilized at the peak value, the diode D remains essentially open at all times.

Therefore

$$v_o = v_i - v_C = v_i - A$$

This is the waveform plotted as v_o in Fig. 8.53.

Using the same basic circuit configuration and reversing the polarity of the diode will create a dc level shift which clamps the output with the minimum of v_i at zero (instead of the maximum as in the first case). In addition, a dc voltage source placed in series with the diode can create clamping levels at any arbitrary voltage.

EXAMPLE 8.15

If the clamper circuit, shown in Fig. 8.54, has an input voltage, $v_i(t) = 12 \sin \omega t$, let us write an expression for the voltage, $v_o(t)$.

D conducts on positive swings of v_i and charges the capacitor C to a value of $v_C = 12$. The polarity is such that the positive side is on the left of the capacitor. Therefore, the output is shifted 12 volts negative relative to v_i, and $v_o(t) = [12 \sin \omega t] - 12$ V.   ∎

8.10
The Zener or Avalanche Diode

For real semiconductor diodes, *in the reverse bias region,* the reverse current remains essentially constant and extremely small until the reverse bias voltage reaches the *breakdown voltage,* called the *avalanche* or *zener breakdown voltage, V_Z.*

At this value of applied reverse voltage the diode is said to break down and it conducts a large current in the reverse direction with the voltage across the diode essentially clamped at the voltage, V_Z. As reverse bias is increased further, the increasing negative current plotted as a function of reverse voltage shows a finite slope. The inverse of this slope is a small resistance called the equivalent zener resistance, r_Z. Figure 8.55 shows a typical diode voltage–current curve including the region of zener breakdown.

In most applications a diode is biased such that it never sees a voltage large enough to cause breakdown, that is, a reverse bias larger than the zener breakdown voltage. This section discusses a very important exception, that is, *voltage regulation circuits.* The zener diode provides an effective way of clamping or limiting the voltage of a circuit at a relatively constant value to create a voltage regulation capability. Zener diodes can be purchased for this purpose, each of which has a specified breakdown voltage. Such diodes are available over a wide range of breakdown voltage values.

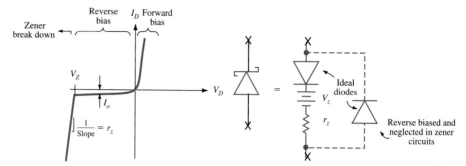

Figure 8.55   *V–I* curve for zener diode.

Figure 8.56   Zener diode symbol and circuit model.

A piecewise linear model, similar to that used for forward bias characteristics, can be employed for zener breakdown diode characteristics. In Fig. 8.56, the zener diode symbol is shown on the left and the circuit model on the right; although all diodes will break down at some voltage, the special symbol is often used for diodes specifically designed to be used in zener breakdown.

EXAMPLE 8.16

If a zener diode with a zener breakdown voltage, $V_Z = 5.0$ V and $r_Z = 100$ Ω, is connected as shown in Fig. 8.57(a), let us find the voltage from A to B and the current through the zener diode.

If we replace the zener diode with its equivalent circuit, we obtain the circuit shown in Fig. 8.57(b). The loop equation for this circuit is

$$12 - V_Z = (12 - 5) = I(2k + 100)$$

and therefore

$$I = \frac{7}{2.1k} = 3.33 \text{ mA}$$

The current, I, is positive and in the direction that would cause the diode to be in zener breakdown. The ideal diode in the zener diode model acts as a closed switch or a short.

The voltage from A to B is the sum of the voltage across r_Z and V_Z:

$$V_{AB} = I(r_Z) + V_Z$$

hence

$$V_{AB} = (3.33 \times 10^{-3})100 + 5 = 5.33 \text{ V}$$ ■

Figure 8.57   Zener diode circuit and circuit model.

The above circuit could be used, for example, in an automobile to provide the required 5 V dc operating voltage for electronic circuitry from a 12 V automobile battery.

D8.10. If a current source of 10 mA replaces the 12 V dc source in the circuit of Example 8.16, let us find the voltage from A to B and the power dissipated in the zener diode.

Ans: $V_{AB} = 6$ V. $P = 60$ milliwatts.

8.11

Graphical Solutions and Load Lines

In solving circuit problems with a nonlinear element, such as a diode, a useful technique in many cases is a graphical approach. If the characteristics of the nonlinear element are well charted, then one can superimpose a plot of the circuit response, excluding the nonlinear element, with a plot of the nonlinear characteristics of the element; where these plots intersect is the desired solution.

EXAMPLE 8.17

If a silicon diode with a forward bias voltage–current curve as shown in Fig. 8.58(a) is placed in the circuit shown in Fig. 8.59, let us find the current in this circuit.

The loop equation for this network is

$$V_D = V_1 - I_D R \tag{8.22}$$

which is the equation of a straight line on the V_D, I_D curve.

Because the constant voltage source and the resistor are linear circuit elements, the collection of possible solutions falls on this straight line. The end points of the line are easy to find. Imagine the nonlinear element acting at its two extremes. If the diode is shorted and $V_D = 0$, I_D is

$$I_D = \frac{V_1}{R} = \frac{1.5}{500} = 3 \text{ mA}$$

The point, $\{V_D = 0, I_D = 3 \text{ mA}\}$ specifies one end of a line, as shown on Fig. 8.58(b). The other extreme occurs when the nonlinear element is an open circuit and $I_D = 0$.

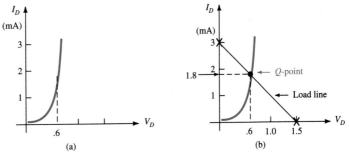

Figure 8.58   Diode *V-I* curve and load line analysis for Example 8.17.

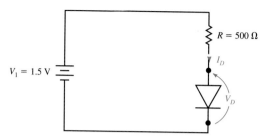

Figure 8.59 Circuit for Example 8.17.

For this case, no current flows, the voltage V_D is simply V_1, and there is no voltage drop across R. Therefore, the other end of the line is at $\{V_D = 1.5 \text{ V}, I_D = 0\}$.

These two end points specify a straight line, sometimes called a *load line*. The intersection of the nonlinear device characteristic, in this case the diode curve, and the load line is the actual *quiescent bias point*, called the *Q-point*. In this circuit the current is 1.8 mA and the voltage across the diode is 0.6 V—the bias values at the *Q*-point.

This graphical technique is often used to establish the dc bias points of transistors operating as amplifiers and will be discussed in Chapter 15. ■

8.12
Photo-Diodes and Light Emitting Diodes

Semiconductor diodes are often used to convert incident radiation to electrical current—these are called *photodiodes*. *Solar cells* are one of the most common applications. The sun's radiation creates electron-hole pairs in the depletion region of a large *p–n* diode. The electric field in this region sweeps the carriers to the terminals and a current is thus generated; the magnitude of this current is approximately proportional to the intensity of light incident on the diode.

Photodiodes are also used for light intensity meters in cameras, for sensing an interrupted light beam in burglar alarm systems, and many other applications.

Light Emitting Diodes (LEDs) are *p–n* junctions generally fabricated from special semiconductor materials; gallium-arsenide is one of the more common materials. They are useful because they allow the direct recombination of electrons and holes, thus releasing energy in the form of light. LEDs have become very useful as indicator lights on electronic display panels and for illuminating sections of digital or alphanumeric displays.

8.13
Small Signals: The Diode as an Analog Attenuator

An *attenuator* is a circuit which reduces the amplitude of a signal. It's often useful to have a means of electronically controlling the amount of ac attenuation by a separate dc control current. The diode can be utilized in just such an application.

The *p–n* junction diode has a forward bias voltage–current characteristic that approximates an exponentially increasing curve. If we calculate the slope of a line tangent to this curve at several points of increasing bias, the slope is increasingly larger at higher bias levels. This is illustrated by taking the slope at three points in Fig.8.60, labeled *A*, *B*, and *C*.

The slope at point *A* is quite low, at *B* a moderate value, and at *C* a large value. The

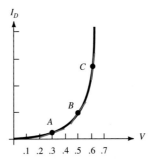

Figure 8.60   Diode *V–I* curve illustrating slope increasing with bias point.

slope at any arbitrary point can be calculated by taking the derivative of the equation for the diode curve. In the forward bias region, recall that Eq. (8.11) is

$$I_D = I_o e^{\frac{qV_D}{kT}}$$

The derivative we will define as the *small-signal diode conductance, g_d,* which represents the change in current for an incremental change in voltage.

$$g_d = \frac{dI_D}{dV_D} = \frac{d}{dV_D} I_o e^{\frac{qV_D}{kT}}$$

$$= \frac{q}{kT} I_o e^{\frac{qV_D}{kT}} = \frac{q}{kT} I_D \cong 39 I_D \qquad (8.23)$$

Thus, the slope is directly proportional to I_D, the dc current through the diode.

The inverse of conductance is resistance, and the *small-signal diode resistance* is defined as

$$r_d = \frac{dV_D}{dI_D} = \frac{1}{g_d} = \frac{1}{39 I_D} \qquad (8.24)$$

The concept of small-signal resistance and conductance is very important. The diode can be biased by a dc source to a particular operating point, for example, point *A,* or point *B* on the curve in Fig. 8.60. Then, if any small ac variations in voltage or current are allowed about that original bias point, these signal excursions follow a slope approximately tangent to the curve at that point.

The small-signal resistance, defined as the change in voltage for an incremental change in current, depends on a particular *Q*-point and is the *effective* resistance seen across the diode by the small ac signal.

Thus, by changing the dc current through the diode, we have a very predictable way to electronically establish the effective resistance seen by a small ac signal.

EXAMPLE 8.18

With reference to Fig. 8.61(a), if we assume the capacitor is sufficiently large to act as an ac short at the signal frequency, and the diode follows the diode equation, let us calculate (a) the dc voltage at v_o, and (b) the *rms* value of the ac voltage at v_o.

(a) The dc bias current through *D* and *R* is simply the current, I_{DC}, since no dc current

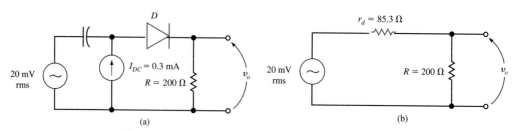

Figure 8.61   Circuit and equivalant circuit for Example 8.18

can flow through the capacitor. Therefore, the dc voltage at the output is

$$V_o = I_{DC}R = (0.3 \times 10^{-3})(200) = 0.060 \text{ V}$$

(b) Considering now only ac signals, an equivalent circuit can be constructed by considering dc current sources as open circuits, capacitors as short circuits, and the appropriate small-signal diode resistance.

If the small-signal resistance of diode D is designated r_d, then

$$r_d = \frac{1}{39I_D} = \frac{1}{(39)(0.3 \times 10^{-3})} = 85.3 \ \Omega$$

The equivalent circuit appears as shown in Fig. 8.61 (b). Using this circuit, the ac component of v_o is

$$v_o = 20\left(\frac{200}{200 + 85.3}\right) \text{ mV rms}$$

or

$$v_o = 14.0 \text{ mV rms} \qquad \blacksquare$$

8.14
Summary

- A diode is a two-terminal electronic device which conducts current if a voltage is applied with one polarity, and will not conduct current when the voltage is applied with the opposite polarity.
- The terminal of the diode which must be positive relative to the other for conduction is called the anode; the other terminal is the cathode.
- A diode biased for conduction is said to be forward biased; a diode biased for non-conduction is said to be reverse biased.
- n-type semiconductors have an excess of free electrons. In silicon, they are created by doping with impurities called donor atoms which have a valence of 5, such as phosphorus. p-type semiconductors have an excess of free holes, and are created with acceptor dopants (valence = 3), such as boron.
- Current in a semiconductor is the result of both electron (negative charge) and hole (positive charge) movement. These carriers move by drift of diffusion.
- Resistivity, ρ, is the reciprocal of conductivity, σ. $\sigma = q(n\mu_n + p\mu_p)$.
- Current density is expressed as the current per unit of cross-sectional area; $J = I/A$.
- The ideal diode is assumed to act as a closed switch when forward biased, that is, zero

resistance, and as an open switch when reverse biased, that is, infinte resistance.
- The diode equation approximates a real diode's characteristics, and is given by

$$I_D = I_o(e^{\frac{qV_D}{kT}} - 1)$$

- The piecewise linear diode model is very useful for circuit calculations and makes a two-segment straight line approximation to the real diode's *V–I* curve. This model includes the forward turn-on voltage of the diode, V_F, in series with a forward resistance.
- Diodes may be used in circuits to electronically perform the AND or OR logic functions.
- Diodes are useful in power supply circuits to perform rectification, the process of converting ac to dc.
- Filtering the rectified wave with an *RC* network reduces the ripple, the periodic voltage deviations from a constant value.
- Clipper circuits limit the voltage excursions of a signal at a particular value. Clamper circuits produce an output which is identical to the input except shifted in dc level.
- A zener diode utilizes the voltage clipping action of a *p–n* junction in breakdown to limit or regulate the voltage of power supply circuits at a precise level.
- Photodiodes are special-purpose diodes that produce a current when exposed to light. Light emmiting diodes (LEDs) are special diodes that emit light when forward biased; they are typically made with gallium arsenide (instead of silicon) and their V_F is about 2 volts.

PROBLEMS

8.1. (a) Why must the linear network analysis techniques discussed in previous chapters be applied judiciously with diode circuits? (b) In the "check valve" fluid analogy of a diode discussed in section 8.1, the case of the open valve (fluid flowing) corresponds to forward or reverse bias?

8.2. (a) Give two definitions of a "cat's whisker." (b) True or False: *n*-type dopants cause the conductivity of intrinsic silicon to increase and *p*-type dopants cause the conductivity of intrinsic silicon to decrease.

8.3. Classify the following as *n*-type dopant, *p*-type dopant, or cat.
(a) Aluminum
(b) Arsenic
(c) Boron
(d) Tin
(e) Indium
(f) Persian

8.4. Identify the samples shown in Fig. P8.4 as *n*-type or *p*-type.

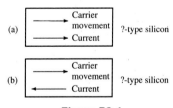

Figure P8.4

8.5. The rectangular bar in Fig. P8.5 has a resistance from end to end of 200 Ω and has a voltage applied that establishes an electric field in the bar of 5 V/cm as shown. (a) If the bar is *n*-type material, what type of carrier is carrying the majority of the current and in which direction are these carriers moving? How many carriers pass through a cross-sectional plane per second? What is the current density? What is the current? (b) same questions as in (a) if we assume the bar is p-type material.

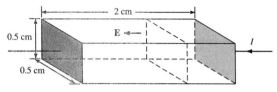

Figure P8.5

8.6. There is an electric field of 7 V/cm in a silicon material. Find the magnitude of the resulting electron and hole drift velocities.

8.7. Calculate the intrinsic conductivity of Si at room temperature (300°K). Assuming each dopant atom creates one free carrier, what is the conductivity of silicon when doped with 10^{17} atoms/cm^3 of arsenic?

8.8. A current of 1 mA flows through a rectangular solid of n-type silicon doped at a density of 10^{16} atoms/cm^3, as shown in Fig. P8.8. Calculate: (a) the magnitude of the current density, (b) the resistance of the solid from one end to the other, and (c) the magnitude and direction of the applied electric field.

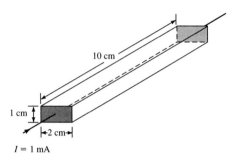

Figure P8.8

8.9. (a) Write the one-dimensional diffusion equation for holes. (b) Consider a rectangular bar of silicon with a distribution of electrons and holes as shown in Fig. P8.9(a). Find the diffusion current due to electrons, that due to holes, and the total current. (c) Repeat (b) for the distribution in Fig. P8.9(b).

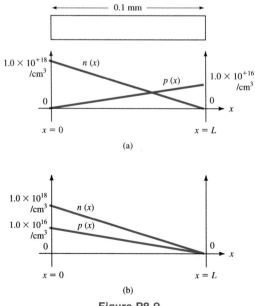

Figure P8.9

8.10. By making the appropriate choice (#1 or #2) in Fig. P8.10, save the cat. Assume an ideal diode.

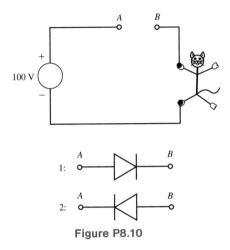

Figure P8.10

8.11. Assuming all diodes are ideal, calculate the current through R for each of the circuits shown in Fig. P8.11.

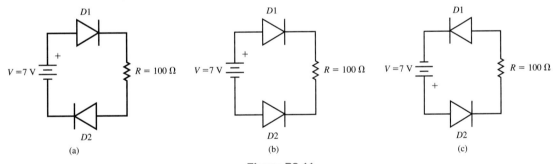

Figure P8.11

8.12. Assuming an ideal diode, in the circuit of Fig. P8.12, (a) what is the voltage across R_2? and (b) what is the current through D?

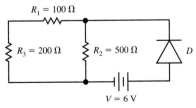

Figure P8.12

8.13. Assuming ideal diodes, in the network of Fig. P8.13, (a) what is the current through $D1$? (b) what is the current through $D2$?

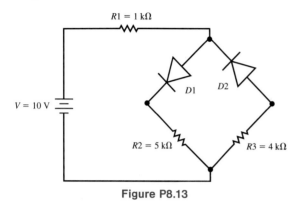

Figure P8.13

8.14. Sketch the V–I characteristics of each of the circuits in Fig. P8.14 assuming ideal diodes.

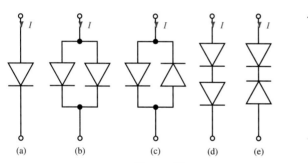

Figure P8.14

8.15. In the network of Fig. P8.15, assuming ideal diodes, calculate the current through (a) D_1, (b) D_2, (c)V_1, the source.

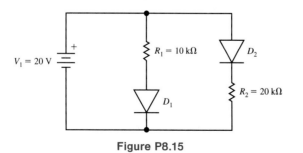

Figure P8.15

8.16. In the circuit of Fig. P8.16, assuming an ideal diode, (a) find the voltage across resistor R_1. (b) What is the current through diode D?

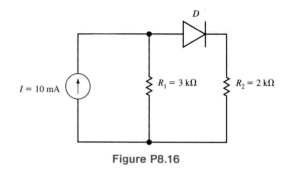

Figure P8.16

8.17. In Fig. P8.17, for $V_1 = 3$ V, (a) calculate the current through D, (b) find the voltage across resistor R_2, (c) repeat (a) and (b) for $V_1 = 10$ V.

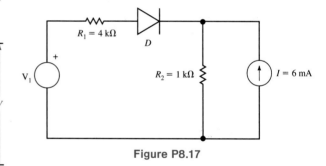

Figure P8.17

8.18. A semiconductor diode had a reverse saturation current, I_o of 3×10^{-14} A. Using the diode equation, determine and plot on a linear scale the diode current for $-3 < V_D < 0.8$ V.

8.19. A forward bias of 0.7 V causes a current of 3 mA to flow through a diode. What is the reverse saturation current, I_o, of the diode?

8.20. A diode conducts 6 mA with a forward bias of 0.6 V. What is the current through the diode at a forward bias of 0.9 V? Assume the diode equation can be used to represent the diode.

8.21. A diode which can be represented by the diode equation has a forward bias of 0.55 V and is conductiong 0.5 mA. If the voltage, V_D, is increased to 0.60 V, what is the new diode current?

8.22. Assuming ideal diodes, (a) plot the transfer characteristics of the curcuit shown in Fig. P8.22 for $0 \text{ V} \leq V_i \leq$ 6 V. (b) Plot the current in D_1 and that in D_2 over the same range of V_i.

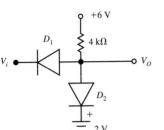

Figure P8.22

8.23. Obtain the voltage transfer characteristics (v_o versus v_1) for the circuit shown in Fig. P8.23 for $-2 \text{ V} \leq v_i \leq$ $+2$ V, using a piecewise linear diode model with $V_F = 0.6$ V and zero forward resistance.

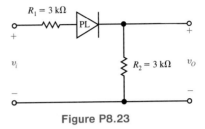

Figure P8.23

8.24. Sketch the output voltage, $v_o(t)$, for the circuit shown in Fig. P8.24 for $0 \leq t \leq 10$ ms; $v_i(t)$ is given in the same figure. Assume that the diode has: (a) ideal V–I characteristics, (b) piecewise linear V–I characteristics with $V_F = 0.6$ V, and $r_s = 50 \ \Omega$.

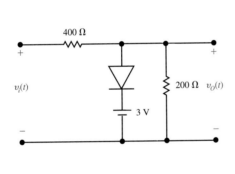
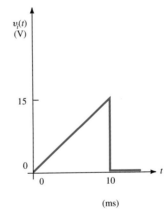

Figure P8.24

8.25. Use the piecewise linear diode model to find V_o for the following combinations of V_1 and V_2 as shown in Fig. P8.25. Assume $V_F = 0.6$ V, $r_s = 30 \ \Omega$.
(a) $V_1 = 0$, $V_2 = 0$
(b) $V_1 = 0$, $V_2 = 5$
(c) $V_1 = 5$, $V_2 = 0$
(d) $V_1 = 5$, $V_2 = 5$

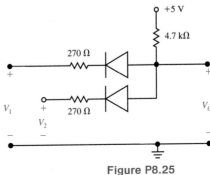

Figure P8.25

8.26. Plot the voltage across and the current through the load resistance, R_L, in the circuit shown in Fig. P8.26(a). Use the diode model shown in Fig. P8.26(b).

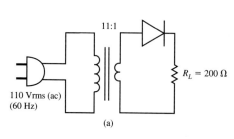

Figure P8.26

8.27. The circuit shown in Fig. P8.27 has a diode which can be modeled with a piecewise linear model in which $V_F = 0.6$ V and $r_S = 0$. (a) Plot the voltage transfer characteristics of this circuit (v_o versus v_i) for $-5 < v_i < +5$ V. (b) If a sinusoidal voltage of amplitude 5 V and frequency 60 Hz is applied as the input, plot the output voltage $v_o(t)$ and the current $i_R(t)$. (c) What is the dc voltage output in (b)?

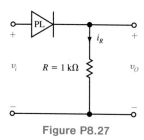

Figure P8.27

8.28. The waveform shown in Fig. P8.28 is applied as an input to the circuit in Fig. P8.27. (a) Plot the output waveform, $v_o(t)$. (b) What is the dc voltage output?

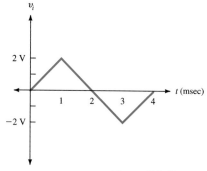

Figure P8.28

8.29. A half-wave rectifier circuit with a filter is shown in Fig. P8.29. Plot the output waveform and estimate the peak-to-peak ripple for (a) $C = 200$ μF and (b) $C = 2$ μF. Assume an ideal diode.

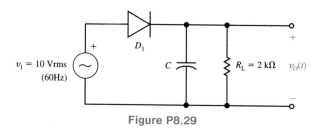

Figure P8.29

8.30. An electronic calculator operates on a dc supply of 6 V. This supply is obtained from a half-wave rectifier with an RC filter. If the calculator can be modeled as a 10 kΩ resistance and it can tolerate a maximum peak-to-peak variation of 0.1 V around the dc value, calculate the minimum value of capacitor that can be used in the filter. Use a piecewise linear diode model with $r_s = 50$ Ω, and $V_F = 0.7$ V. Assume ac input is 60 Hz.

8.31. A 4-diode full-wave rectifier circuit is shown in Fig. P8.31. Assuming the diodes are ideal, (a) plot the output voltage $v_o(t)$ and the output current $i_o(t)$. (b) Determine the maximum reverse bias on any diode in this circuit. (c) What is the dc output voltage?

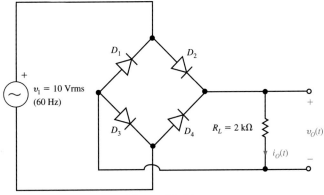

Figure P8.31

8.32. Repeat problem 8.31 using the diode model of problem 8.30.

8.33. A two-diode full-wave rectifier circuit is shown in Fig. P8.33. An *RC* network is used to filter the output waveform. Plot the output voltage $v_o(t)$ using the piecewise linear model of the diode used in problem 8.30. What is the maximum reverse bias on either diode?

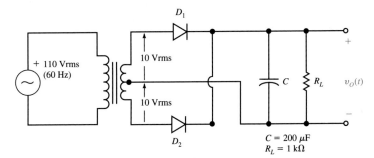

Figure P8.33

8.34. A clipping circuit is shown in Fig. P8.34. If the diode is ideal, sketch voltage transfer characteristics (v_o versus v_i) for this circuit for $-5 < v_i < +5$ V.

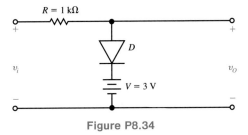

Figure P8.34

8.35. Repeat problem 8.34 assuming a piecewise linear diode model with $V_F = 0.6$ V, $r_l = \infty$ and $r_s = 40$ Ω.

8.36. The input voltage, $v_i(t) = 5 \sin(\omega t)$V, is applied to the circuit shown in Fig. P8.36. Assuming the diodes are ideal, sketch the output voltage $v_o(t)$ as a function of time.

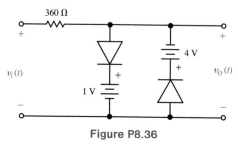

Figure P8.36

8.39. Figure P8.39(a) shows a zener diode voltage regulator circuit. Using the zener diode with the V–I curve shown in Fig. P8.39(b), (a) determine the zener diode model parameters, V_Z and r_Z, (b) obtain the voltage transfer characteristics of the circuit for $0 < v_i < 15$ V, (c) determine the power dissipated in the zener diode if $v_i = 10$ V, and (d) determine the output voltage in (c).

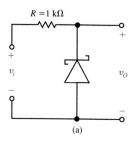

(a)

8.37. Repeat problem 8.36 using the diode model of problem 8.35.

8.38. A clamping circuit is given in Fig. P8.38; assume the diode is ideal and $v_i = 1 + 5 \sin(\omega t)$ V. (a) Sketch the output waveform, and (b) determine the dc level of the output signal.

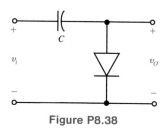

Figure P8.38

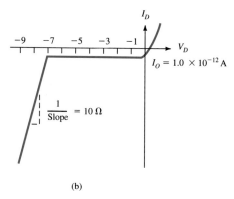

(b)

Figure P8.39

8.40. The zener diode with $V_Z = 12$ V and $r_Z = 0$ is placed in the circuit of Fig. P8.39(a). If v_i is 20 V dc, (a) what is the current through the zener diode? and (b) if a 6 kΩ load resistance is attached at the output, what is the current through the zener diode?

8.41. A diode which follows the diode equation conducts 30 mA with a forward bias of 0.8 V. (a) Determine I_o. (b) Plot the V–I characteristics of this diode. (c) This diode is placed in the circuit in Fig. P8.41; using the plot from (b), obtain the diode current and voltage graphically.

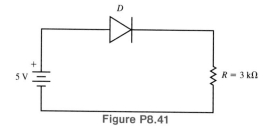

Figure P8.41

8.42. The diode in Fig. 8.42(a) has V–I characteristics shown in Fig. 8.42(b). Use a graphical method to solve the diode current and voltage in this circuit. Plot the load line on the same graph as the diode curve and clearly identify the Q-point.

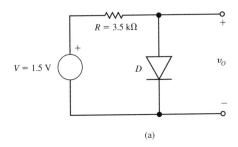

(a)

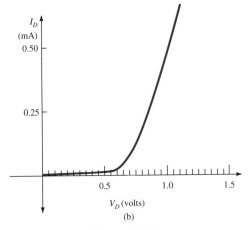

(b)

Figure P8.42

8.43. A diode which obeys the diode equation is carrying a dc current of 0.38 mA. What is the small-signal resistance of the diode?

8.44. In the circuit shown in Fig. P8.44, assume $V_F = 0.7$ V for the dc calculation. What percent of the small-signal voltage, v_s, appears across the diode?

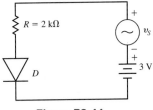

Figure P8.44

8.45. If a diode has an ac resistance of 50 Ω, what is its dc bias current?

The Transistor as a Switch: FETs and BJTs for Digital Applications

9.1

Introduction

Modern electronic systems generally utilize digital circuitry to achieve the myriad of new capabilities and functions available. The digital approach provides the possibility of digital microcomputers on a single chip—called microprocessors—to perform a large volume of data handling, calculations, and decisions, all at very high speed.

This chapter studies the basic electronic switching elements, *transistors,* used to implement digital logic gates; such gates are the building blocks of complex digital systems. A single semiconductor chip may contain millions of such transistor switching elements.

As we have seen, the diode is a very useful device capable of performing many functions. It can even be used to construct simple logic gates. However, one of the necessary functions in digital logic that the diode cannot perform is that of signal inversion, that is, changing a 0 to 1, or vice versa. The switching devices discussed in this chapter enable us to implement both simple and complex logic functions, including inversion. Although the basic principles will be first illustrated for clarity with the relay, the critical switching element now used most frequently in electronic systems is the solid state transistor.

In digital electronic systems, the *digital logic values,* that is, 1 or 0, are represented by two levels of voltage or current. The system is capable of switching various nodes from one of these levels to the other. The transistor is the electronic device that performs this switching by changing, on command, from what is essentially an open switch to a closed switch, and vice versa.

This same transistor can be used as an amplifier when incorporated in a different type of circuit, such as those described in Chapters 15 and 16. The switching behavior of the transistor, as described in the current chapter, will be described by simple circuit models.

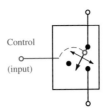

Figure 9.1   General switching device.

There are two principle types of transistors used in modern digital electronics: the *FET (Field Effect Transistor)* and the *BJT (Bipolar Junction Transistor)*. They operate on entirely different physical principles, and although they are both generally fabricated in the surface of a silicon wafer, their structure and topology are also quite different.

FETs are called *majority carrier* or *unipolar devices,* while BJTs are *minority carrier* or *bipolar devices.* They do, however, have very similar functions in digital circuits: the level of a control signal at one terminal of the device (the input) controls the flow of current between two other terminals. The value of the control voltage or current, together with some other circuit parameters, determines whether there is an open circuit or closed circuit between the other two terminals, as illustrated in Fig. 9.1.

This chapter will describe the typical physical structure of both the FET and BJT, their operation, and appropriate circuit models. These models will be useful in analyzing the performance of such devices in actual digital circuits.

First, however, the *relay* is described as an analogy of the switching transistor. The relay is currently used in some simple switching systems, such as alarms, appliance controls, and the like, as the controlling switching element. There is, however, an increasing trend to replace the relay with solid state transistor switches because of the transistor's lower cost and higher reliability.

9.2
An Analogy: The Relay as a Current Control Device

Relays, transistors, and vacuum tubes all have one thing in common: they control or regulate the flow of current in one circuit by the magnitude of the electric signal in another circuit. In digital applications this capability is used to switch current on or off.

The relay is one of the simplest mechanisms for switching an electric current. A relay consists of two primary parts: (1) an electromagnet, and (2) a set of switch contacts that are either pulled together or separated by the electromagnet when it is energized. A relay is an example of an "electromechanical" device, in that an electrical action causes mechanical motion. A sketch of a relay is shown in Fig. 9.2.

The electrical contacts that connect or disconnect the circuit are designated as either "NO" or "NC." NO stands for "normally open contacts"; these are contacts which are open or disconnected until current is directed through the electromagnet activating the magnet and pulling the contacts together. NC is the abbreviation for "normally closed contacts"; these contacts remain closed or connected until the electromagnet is energized, which pulls the contacts apart, opening the circuit.

Figure 9.3 shows a photograph of a modern relay used for control of electric power.

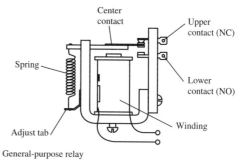

Figure 9.2 Drawing of relay.

Figure 9.3 Photograph of modern relay.

9.3

The NOT Gate: Relay Implementation

The relay can be used to implement a *NOT gate,* often called an *inverter.* Assume a relay requires +V volts to energize the coil and close the contacts; also, +V will be defined as a high logic level, and zero volts will be defined as a low level.

Consider the circuit containing the relay in Fig. 9.4. If the input (*A*) is at zero volts (logic low), the relay coil is not energized, and therefore the relay contacts are open and the output voltage (*B*) is +V (logic high). The voltage at *B* is +V because there is no current flow through *R,* and hence no voltage drop across it.

If the input (*A*) has a voltage of +V applied, the electromagnetic attraction between the energized coil and the contacts pulls the contacts closed. This action connects the out-

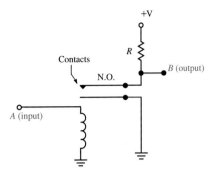

Figure 9.4   An inverter circuit using a relay.

put (*B*) to ground or zero volts. Thus, the circuit functions as an inverter: low in, high out and vice versa.

9.4
The MOSFET as a Switch

There are two primary types of field effect transistors: the *JFET (Junction FET)* and the *MOSFET (Metal-Oxide-Semiconductor FET)*. The latter is most commonly used in modern digital electronic circuits, and therefore we will concentrate here on this device. JFETs will, however, be discussed in section 9.8.

Metal-oxide-semiconductor is a description of a structure comprised of layers: metal on top, silicon dioxide in the middle, and a semiconductor, generally silicon, underneath. This is, however, of historical origin and not representative of modern structures. New technologies generally use a film of polysilicon instead of metal for the top layer, but the old designation, MOS, has been retained.

MOS transistors can be either *n-channel* or *p-channel,* each of which can be further subdivided into either *enhancement mode* or *depletion mode* as shown in Fig. 9.5. All of these transistors can operate as electronically controlled switches. For purposes of illustration, we will look first at an *n-channel, enhancement-mode device,* which is probably the most commonly used device. The definition of these terms will be made clear in the following sections. The other types of devices and their corresponding structure can be easily understood later by analogy. MOS transistors are the key switching elements in popular digital technologies such as NMOS and CMOS—described in Chapter 10.

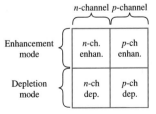

Figure 9.5   MOS transistor types.

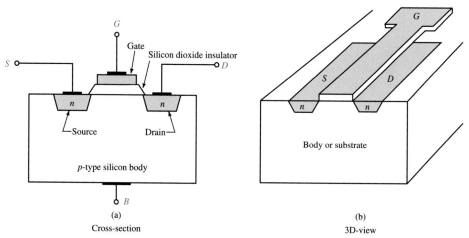

Figure 9.6   Basic structure of *n*-channel MOS transistor.

MOSFET Structure

The basic structure of a silicon *n*-channel enhancement-mode MOS transistor is shown in Fig. 9.6.

The cross-section drawing in Fig. 9.6(a) shows a *substrate* of *p*-type silicon (the *body*) into which an *n*-type *source* region, and an *n*-type *drain* region have been placed. Electrical connections are made to the source region, labeled *S,* and the drain region, labeled *D.* Covering the gap between the source and drain is a very thin layer of silicon dioxide—basically glass and a superb insulator. On top of this glass layer, called the *gate oxide* or *gate dielectric,* is a ribbon of conducting material to which the gate electrode is attached, labeled *G.* The *p*-type substrate also has an electrical contact which is labeled *B,* for body. In most digital circuits the body is electrically connected to the source and its primary purpose is to isolate one transistor from another on the same silicon chip. As shown in Fig. 9.6(a), a capacitor-like structure is formed with the gate oxide, (an insulator), and the *gate* as the top electrode and the semiconductor material (body) as the bottom electrode.

The schematic symbol used for this device is shown in Fig. 9.7(a). Sometimes the alternate symbols, shown in Figs. 9.7 (b, c), are used.

The arrow designating the body connection in the symbol is in the opposite direction for a *p*-channel device; the structure of a *p*-channel device is similar, with the *n* and *p* regions interchanged.

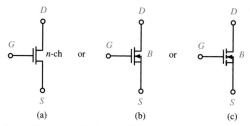

Figure 9.7   Schematic symbols for *n*-channel, enhancement-mode MOSFET.

MOSFET Device Operation

The device described in the previous section is placed in the circuit of Fig. 9.8(a) in an attempt to move electrons from the source to the drain, which is of course equivalent to moving positive current from drain to source. The voltage from gate to source, V_{GS}, is made equal to zero by connecting both the source and gate to ground as shown in Fig. 9.8(b).

In this circuit there's no apparent way current can flow between the source and drain terminals. In either direction that we might attempt to force current in the source-body-drain path, we encounter a diode oriented the wrong way for conduction. For example, V_{DS} tries to move positive current from drain to source, but the drain-body junction is reverse biased and no current flows. It is as though the MOSFET were an open switch between drain and source preventing current to flow, as shown in the circuit model of Fig. 9.8(c).

However, if a large enough positive voltage is applied to the gate of this device, current will then flow around the drain to source loop. No current flows in the gate lead because it is completely electrically insulated from the other device structures.

In order to understand this phenomenon, imagine the gate as merely the top plate of a capacitor. When we apply a positive charge to the gate, it attracts an equal and opposite charge on the other side of the dielectric at the surface. This action inverts the surface from p-type to n-type, forming an n-type conducting bridge, that is, an inversion layer or *channel* between the source and drain and—voilà, electrons can flow from the n-type source across the n-type channel to the n-type drain. In this case the device acts like a switch which is closed from source to drain. The electrons flowing from one n-region to

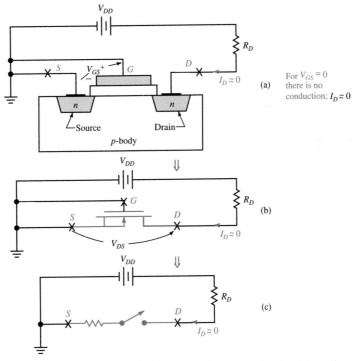

Figure 9.8   *n*-channel MOS device characteristics for $V_{GS} = 0$

another, through an induced n-type channel, are majority carriers; therefore, the MOSFET is called a majority carrier device.

Threshold Voltage, V_T

In an enhancement-mode device there is no current in the source-drain circuit when $V_{GS} = 0$. As V_{GS} is increased, some positive value of gate-to-source voltage is required before enough charge is accumulated to invert the surface below and allow conduction between source and drain. The value of V_{GS} where the drain current just begins to flow is called the *threshold voltage*, V_T. Figure 9.9 shows a typical plot of drain current as a function of V_{GS}, for a given value of V_{DS}. Note that the onset of conduction occurs at $V_{GS} = V_T$, approximately 0.4 volts in this case. Typical modern enhancement-mode devices have threshold voltages in the range of 0.3 to 0.8 volts.

A theoretical analysis based on the solid state physics of the enhancement-mode MOSFET predicts that for a constant V_{DS}, where $V_{DS} > (V_{GS} - V_T)$, the drain current of the MOSFET as shown in Fig. 9.9 follows a positively increasing curve for V_{GS} above V_T. That is

$$I_D = K (V_{GS} - V_T)^2, |V_{GS}| > |V_T| \text{ and } |V_{DS}| > |V_{GS} - V_T|$$
$$I_D = 0, |V_{GS}| \ll |V_T| \tag{9.1}$$

where V_T is the device threshold voltage. Note that I_D is independent of V_{DS} in this active region.

The constant, K, is a device parameter that is the product of two terms, one dependent on device processing and cross section, and the other dependent only on device layout or topology and in particular the channel width to length ratio (see later section on MOSFET topology).

EXAMPLE 9.1

Let us determine the value of the parameter K for the device characteristic plotted in Fig. 9.9.

The parameter, K, can be easily determined from experimentally measured device curves such as that of Fig. 9.9. For this device, the threshold voltage V_T, that is, the value of V_{GS} where current becomes nonzero, is 0.4 volts. In addition, note that at $V_{GS} = 0.6$ V, $I_D = 0.3$ mA. Therefore, evaluating K using this data point, we find that

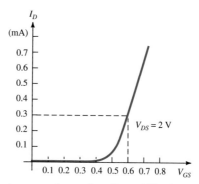

Figure 9.9   Drain current as a function of V_{GS} for n-channel device.

$$K = \frac{I_D}{(V_{GS} - V_T)^2} = \frac{0.3 \text{ mA}}{(0.6 - 0.4)^2} = 7.5 \times 10^{-3} \text{ A/V}^2$$

Note that a more accurate determination of V_T can be obtained by plotting the square root of I_D versus V_{GS}. This plot is a straight line, and its intercept on the V_{GS} axis is V_T. ∎

The Closed Switch: MOSFET in the Ohmic Region

As demonstrated, when V_{GS} is increased above the V_T of the device, the E-mode MOSFET can conduct current. This conduction is modeled in one of two ways depending on the value of V_{DS}, as illustrated in Fig. 9.10.

Case 1. Conduction through a closed switch in series with the on resistance of the device. This conduction occurs when $V_{DS} < (V_{GS} - V_T)$ and the device is said to be in the "ohmic" region. The value of I_D depends on the magnitude of the voltage source, V_{DD}, across the total resistance in the loop including the on resistance of the device.

Case 2. Conduction through a closed switch in series with a dependent current source, I_D, which is a function of only V_{GS}. This conduction occurs when $V_{DS} > (V_{GS} - V_T)$ and

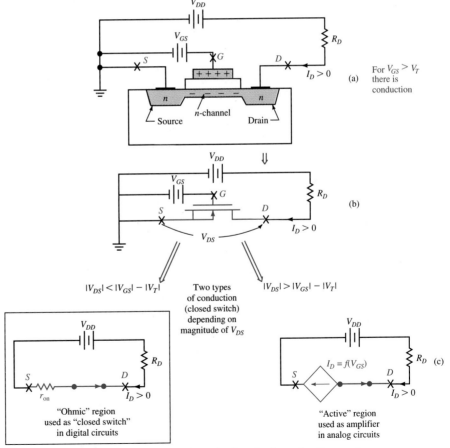

Figure 9.10 MOS transistor in the ohmic and active regions.

the device is said to be in the active region. The value of I_D for this case is set entirely by the magnitude of the current source and is independent of V_{DS}.

For switching circuits, conduction for the on transistor should be in the ohmic region, as described above in Case 1. This is generally accomplished by choosing a load resistance, R_D, sufficiently large that when the switch closes, the voltage drop across R_D, caused by I_D, is sufficient to reduce V_{DS} to a level such that the device enters the ohmic region.

Section 9.6 will clearly delineate the difference between the region of operation where I_D is independent of V_{DS}, and the ohmic region, where I_D is directly dependent on V_{DS}.

In these digital circuits if the transistor is initially off, there is no current flow, and hence $V_{DS} \approx V_{DD}$. While in this state, if a gate voltage is then applied to switch the device to the conducting state, there is a transition period during which the device moves from off through the active region to finally rest in the on condition in the ohmic region. During this transition, V_{DS} falls from its high value to a low value consistent with that of a device in the ohmic region.

The expression for I_D in the ohmic region is

$$I_D = K[2(V_{GS} - V_T)\,V_{DS} - V_{DS}^2] \tag{9.2}$$

where $|V_{DS}| < |V_{GS} - V_T|$. In the ohmic region, I_D also depends on V_{DS}. The dividing line between the ohmic region, Eq. (9.2), and the active or saturation region, Eq. (9.1), occurs where $V_{GS} - V_{DS} = V_T$.

MOSFET Circuit Model for Switching

In digital applications, the n-channel, enhancement-mode MOSFET is in the nonconducting state when $V_{GS} \ll V_T$ and is switched to the conducting state when $V_{GS} \gg V_T$ and $V_{DS} \leq V_{GS} - V_T$.

In analog circuits we will be concerned with the fine control of the current in the region between conduction and nonconduction. This will be discussed briefly in section 9.5 and in detail in Chapter 15.

In digital MOS circuits, the MOS transistor can be modeled as a device with a switch between source and drain, in series with some *on resistance*, r_{on}. The state of the switch is controlled by the magnitude of V_{GS}.

The symbol for the n-channel enhancement-mode MOS transistor is repeated in Fig. 9.11(a) and its circuit model for digital switching is given in Fig. 9.11(b), assuming it is biased in the ohmic region.

The transistor model of Fig. 9.11 is very useful in digital circuit analysis. It illustrates that when the input voltage, which is applied from gate to source, is less than the threshold voltage, the device can be modeled as an open circuit between source and drain. For the case in which the gate voltage is significantly larger than the threshold voltage, the transistor conducts current between source and drain, and has an internal resistance called the on resistance of the device, r_{on}. The value of r_{on} depends on the physical size of the transistor and its particular structure; however, typical switching devices may have on resistances from a few ohms to a few thousand ohms. The threshold voltage, V_T, and the on resistance, r_{on}, are key parameters in the analysis of digital circuits with MOS devices.

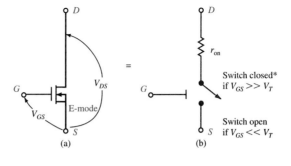

Figure 9.11 *n*-channel enhancement-mode MOS transistor.

An analytic expression for r_{on} can be obtained by first differentiating Eq. (9.2) to obtain g_{on}, the on conductance, that is

$$g_{on} = \left.\frac{\partial I_D}{\partial V_{DS}}\right|_{V_{DS}=0} = K2(V_{GS} - V_T)$$

therefore

$$r_{on} = \frac{1}{g_{on}} = \frac{1}{K2(V_{GS} - V_T)} \tag{9.3}$$

The *p*-Channel MOSFET

The *p-channel MOSFET* is the exact complement of the *n*-channel device already described. The *p*-channel device is fabricated by placing two *p*-type regions for the source and drain into an *n*-type substrate or body. As was the case with the *n*-channel device, this device has a channel structure with a gate oxide insulating layer and a gate electrode over the oxide.

As a parallel to the *n*-channel device, forcing $V_{GS} = 0$ creates a device with a non-conducting channel; however, the conducting state in a *p*-channel device requires the application of a complementary set of voltages. The drain is biased negatively relative to the source. In order to induce a conducting *p*-type channel, the gate is also made negative relative to the source to attract sufficient holes which invert the surface and create a *p*-type conducting channel as shown in Fig. 9.12.

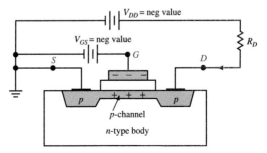

Figure 9.12 A *p*-channel MOSFET.

Therefore, the sign of V_{DS} and V_{GS} must be negative. In addition, the threshold voltage for an E-mode p-channel MOSFET is generally given as a negative value, for example, $V_T = -0.4$ V.

The circuit model for switching with a p-channel device is identical to that presented for the n-channel device, with the required sign changes to account for the negative threshold voltage and negative V_{DS}.

MOSFET Topology: The Top View

We have examined the cross section of an MOS transistor in order to understand the mechanism by which the gate-to-source voltage controls electrical conduction between the source and the drain. The threshold voltage for a given MOSFET depends on the characteristics of the cross-sectional structure—variables such as thickness of the gate oxide, the amount of doping in the substrate, etc. These parameters are generally fixed for a given fabrication technology. This implies that all devices fabricated on the same wafer will have approximately the same technology-dependent parameters such as threshold voltage, V_T.

There are, however, a number of important transistor parameters that also depend on the layout of the device, that is, the topology in the plane of the wafer, or what we call the top view. Each transistor in the circuit can therefore be customized by the size and shape of its layout. One of the most important of these layout-dependent parameters for digital applications is r_{on}.

The top view of a typical MOS device is shown in Fig. 9.13. The most important dimensions in the layout of the transistor are:

L = the *channel length*, that is, the distance from edge of source to edge of drain, usually given in microns.

W = the *channel width*, that is, the width of the source or drain region which defines the width of the active channel.

Most key transistor parameters depend on the ratio of channel width to channel length, W/L. Recall that K, the constant in the MOSFET device equation, is proportional to W/L. r_{on}, the device on resistance, is proportional to the inverse of W/L as can be seen in Eq. (9.3). Therefore, for example, given a fixed channel length, a transistor with twice the channel width would have half the value of r_{on}.

In order to achieve higher performance from a digital system, the digital logic must operate as quickly as possible. There is an intense competition between semiconductor manufacturers to make circuits which operate ever faster. When a node in the circuit is

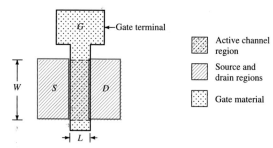

Figure 9.13   Top view of an MOS device.

switched from one state to the other, the speed-limiting barrier to these voltage changes is often an *RC* time constant. Each node has parasitic capacitance associated with it—capacitance between it and adjacent nodes, as well as capacitance to ground. Also, current through an on transistor must flow through the r_{on} of the device. Therefore, the effective node capacitance and the value of r_{on} determine this *RC* time constant, and hence the ultimate switching speed of the device.

In order to design circuits that operate faster, integrated circuit engineers take great care to reduce nodal capacitance as much as possible. They also want to reduce r_{on}, and there are two ways to change the layout to accomplish this: (1) increase *W*, which is a bad idea, because it increases source and drain capacitance and increases the area required to lay out the circuit, or (2) decrease *L*, which is a good idea.

A tremendous amount of effort is invested in developing technologies that provide the smallest effective channel length possible. The so-called *minimum feature size* generally refers to the MOS transistor channel length, because it is made as small as possible in an active switching device. Semiconductor manufacturers have been able to consistently reduce the minimum feature size over the last several decades, from about 10 microns in 1970 to about 1 micron in 1990. As of this writing, a number of manufacturers are producing commercial integrated circuits with channel lengths in the range of 0.6 to 1.0 microns. These trends in transistor manufacturing are discussed in more detail in Chapter 13.

9.5
The NOT Gate: MOSFET Implementation

The utilization of the *n*-channel enhancement-mode MOS transistor to create an inverter is shown in Fig. 9.14. This circuit configuration is used almost exclusively in digital circuits and is designated as the common source configuration, since the source terminal is grounded. The two remaining configurations will be discussed in Chapter 15. When the input signal, V_{in}, is high, the transistor is on which pulls the voltage at the output to ground. When the input is low, the transistor is off which pulls the output to a high voltage through the load resistor.

This can be clearly seen by substituting the switching equivalent circuit model in Fig. 9.11(b) for the two cases—input low and input high. Assume V_T is 1.0 volt, and consider the two cases illustrated in Fig. 9.15. In Fig. 9.15(a), when the input is low, $V_{in} \ll V_T$ and the transistor appears as an open switch. With zero current through the transistor, there is no current through R_L, and $V_{out} = +V_{DD}$.

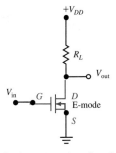

Figure 9.14   An inverter circuit using a MOSFET.

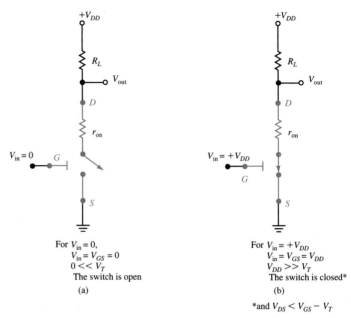

Figure 9.15   MOSFET NOT gate circuit models.

For the case in which the input is high as shown in Fig. 9.15(b), the transistor is conducting with an internal resistance from source to drain of r_{on}, and V_{out} is given by

$$V_{out} = \left[\frac{r_{on}}{r_{on} + R_L} \right] (+V_{DD}) \qquad (9.4)$$

Since the desired output is low, proper design requires that $r_{on} \ll R_L$. Furthermore, in order to ensure that this output low level is low enough to be interpreted by the next gate as a low

$$V_{out} \ll V_T$$

and therefore

$$\left[\frac{r_{on}}{r_{on} + R_L} \right] V_{DD} \ll V_T$$

EXAMPLE 9.2

In an inverter circuit with $V_{DD} = 5$ V, an n-channel enhancement-mode transistor has the following parameters: $r_{on} = 100$ Ω, and $V_T = 1$ volt. Let us determine the value of R_L required to make V_{out} in the low state 10% of V_{DD}.

From Eq. (9.4)

$$(0.1)V_{DD} = V_{out} = \left[\frac{r_{on}}{r_{on} + R_L} \right] V_{DD}$$

Solving for R_L yields

$$R_L = 900 \ \Omega$$

For $R_L > 900\ \Omega$, the value of the *"output low level"* is decreased and therefore provides better operating noise margin.   ■

9.6
MOSFET Output Curves

The *V–I* characteristics of transistors are often displayed by a family of curves on a simple chart, generally referred to as the *output curves*. The set of output curves is very useful in analyzing the transistor and associated circuitry, particularly for analog applications. For digital applications the transistor can be generally modeled using the switch model in Fig. 9.15, and output curves are not required. Nonetheless, it is helpful to understand that digital switching corresponds to moving the operating point from one extreme on the output curve to another, each end point representing one of the two possible logic states. However, there is a continuum of possible voltages and currents between these states. Output curves will now be briefly discussed; however, a detailed discussion is presented in Chapter 15.

The output curves for an *n*-channel enhancement-mode MOSFET are plots of drain current, I_D, as a function of drain-to-source voltage, V_{DS}, for several different values of V_{GS}. As you might imagine, there are no curves for $|V_{GS}| < |V_T|$, because there is no current flow for that case, and thus the bottom curve occurs for $V_{GS} = V_T$, the threshold of conduction. Figure 9.16 shows an example of the output curves for such a device with a threshold voltage of 1 volt. In this figure, V_{GS} is increased in one-volt steps to form the family of curves.

Notice that as the magnitude of V_{DS} is increased, the current, I_D, levels off, and becomes nearly constant for a given V_{GS}. This is a result of the field increasing at the drain end of the channel and pinching off the current flow.

The output curves provide a way to characterize and describe the behavior of the transistor over the full range of voltage and current swings, including the extremes representing the logic "highs" and "lows," as well as the transition region in between.

Consider again the NOT gate circuit in Fig. 9.14 and assume $V_{DD} = 5$ V and $R_L = 1\ \text{k}\Omega$. Assume also that the transistor in this figure is represented by the output curves of Fig. 9.16. This circuit contains two linear circuit elements, the voltage source, V_{DD}, and the resistor, R_L, and in addition, the transistor which is a nonlinear element.

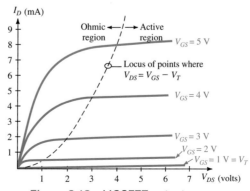

Figure 9.16   MOSFET output curves.

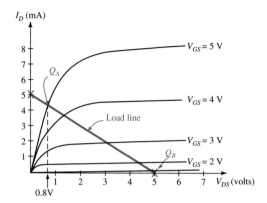

Figure 9.17   MOSFET output curves with load line.

The load line technique described in Chapter 8 can be used here to determine the circuit response. Figure 9.17 presents the same output curves with the load line added. Recall that the load line is established by first assuming the nonlinear device is a short and computing one end point of the line:

$$I_D\Big|_{V_{DS}=0} = \frac{V_{DD}}{R_L} = \frac{5}{1\text{ k}} = 5\text{ mA}$$

The other load line end point is fixed by computing the voltage V_{DS} when the device is open, that is, $I_D = 0$.

$$V_{DS}\Big|_{I_D=0} = V_{DD} = 5\text{ V}$$

The load line is drawn superimposed on the transistor's output curves by connecting the two end points.

As the input to the gate is increased to 5 volts, the Q-point (bias point) moves along the load line to the intersection of the $V_{GS} = 5$ V curve, labeled as point Q_A. At this point the output voltage, V_{DS}, reaches a lower limit of 0.8 volts. This point represents the logic "low" state, and should correspond to the "logic low" value that would be computed with the MOSFET switch model.

The "logic high" state is determined by assuming the input to the gate is "low," that is, below the threshold voltage. At this input level, the transistor ceases to conduct moving the Q-point along the load line to the point Q_B on the V_{DS} axis where the output voltage is V_{DD}.

9.7
Depletion-Mode MOSFETS

Prior discussion has focused on the enhancement-mode MOSFET; depletion-mode MOSFETS differ in structure in that a conducting channel region is deliberately created by doping the surface under the gate during device fabrication. With such a device, current can be made to flow between source and drain even with zero gate voltage ($V_{GS} = 0$). The cross-sectional structure of an n-channel depletion-mode device is shown in Fig. 9.18.

A set of typical output curves for an n-channel depletion-mode device is shown in Fig. 9.19. Notice that there is significant current flow for $V_{GS} = 0$. This device can be op-

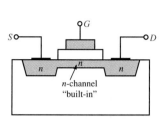

Figure 9.18 *n*-channel depletion-mode device.

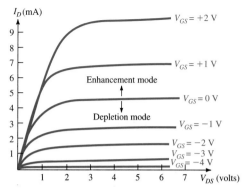

Figure 9.19 Output curves for a depletion-mode device.

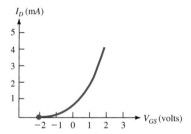

Figure 9.20 A plot of I_D-vs-V_{GS} for an *n*-channel depletion mode MOSFET.

erated in the depletion mode by making the gate region more negative, further depleting the channel of carriers and further reducing the level of current I_D. This action is illustrated in the figure for negative values of V_{GS}. The negative value of V_{GS} required to reduce I_D to zero is defined as the *pinch-off voltage*. This is analogous to a negative threshold voltage and is an indicator of a depletion-mode device. A plot of I_D versus V_{GS} for a fixed V_{DS} is shown in Fig. 9.20.

However, such a device can also be operated in the enhancement mode, where the application of a positive V_{GS} increases channel conduction, and I_D increases.

Depletion-mode structures are more complex to fabricate because they generally require additional processing steps to build in the channel region. This translates into higher production costs; therefore, depletion-mode devices are not commonly used except for special circuits.

9.8
The JFET

The *Junction Field Effect Transistor* (JFET) is a depletion-mode FET, but with a different structure than that of the MOSFET. JFETs have been used primarily in special circuit applications, such as analog circuits, where very high input resistance is required. They are generally not used as switching elements in digital electronic circuits.

The JFET has a much higher input resistance than the Bipolar Junction Transistor (BJT) and is generally more compatible with the fabrication process associated with build-

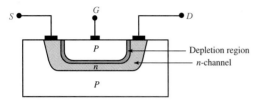

Figure 9.21   Cross section of a JFET device.

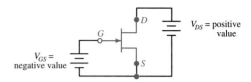

Figure 9.22   A typical JFET bias circuit.

ing BJTs than with that of MOS devices. Its current use is limited and will be discussed only briefly here.

The cross section of a typical JFET device appears as shown in Fig. 9.21. The n-channel in this structure is fabricated in a "U" shape beneath the surface of the silicon. The channel is sandwiched between two p-type layers, and the top one is connected to the gate terminal. In some instances the top and bottom p-regions are connected to the gate terminal. Recall that every p–n junction has a depletion region devoid of carriers, and the width of the depletion region is controlled by the applied voltage across the junction.

In the previous figure, if the gate is made more negative with respect to the source, the width of the depletion region grows and further blocks off a part of the conducting channel region. Therefore, current flow through the channel can be regulated by the magnitude of the applied voltage V_{GS}. For an n-channel JFET, the drain is biased positive with respect to the source, as is done with an n-channel MOSFET. The symbol for such a JFET together with the bias voltages are shown in Fig. 9.22. An example of a typical output curve for such a device is shown in Fig. 9.23. The value of V_{GS} required to just reduce the curve to $I_D = 0$ is defined as the pinch-off voltage, V_P. In this example, $V_P = -4$ volts. The gate electrode in this device is normally connected to a reverse biased p–n junction, that is, the gate junction, and therefore has a very high input resistance.

The normal range of input gate biases includes V_{GS} values from zero volts to V_p, and the device operates in this region in the depletion mode. If V_{GS} is made positive, the device can operate in the enhancement mode only for a very small range, that is, up to the onset of conduction of the p–n gate diode ($V_{GS} \approx +0.6$ V) where normal device function is lost and excessive gate current flows.

The example just described is for an n-channel JFET; there is a complementary device, the p-channel JFET, which has the same structure with p and n regions and bias voltages reversed. A p-channel JFET has a positive pinch-off voltage. For JFETs and

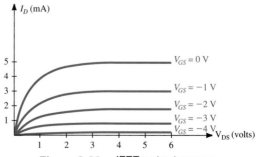

Figure 9.23   JFET output curves.

depletion-mode devices, Eq.(9.1) can be rewritten as

$$I_D = I_{DSS} \left(1 - \frac{V_{GS}}{V_P} \right)^2 \tag{9.5}$$

where I_{DSS} is the drain current when $V_{GS} = 0$.

9.9

The BJT as a Switch

The *Bipolar Junction Transistor* (BJT) is the solid state transistor that first revolutionized the field of electronics. Invented in 1947 at Bell Laboratories, its introduction into electronic systems in the late 1950s and early 1960s replaced the large, inefficient, and unreliable vacuum tube in almost every application. The MOSFET had been conceived of earlier than the BJT, but the available materials of that time lacked the purity and contamination control required to make reliable MOS devices. While the MOSFET has now emerged as the most favored device for very large integrated systems, where millions of transistors are fabricated on a single chip, the bipolar device still has many areas of application, both in digital and especially in analog systems. For an equivalent sized device, a BJT generally has the ability to provide more gain than does a MOSFET. The popularity of MOSFET circuitry for large systems is primarily the result of lower average required power per gate, a very important consideration in placing millions of such circuits on a single chip. Bipolar devices and circuits can generally switch higher currents, are less prone to disruption by outside interference and noise, and have become a standard for implementation of certain types of digital logic. TTL *(Transistor-Transistor-Logic)* and ECL *(Emitter Coupled Logic)*, both well-known classes of commercially available digital integrated circuits, utilize BJTs in their circuitry to perform the logic operations.

This section will describe the structure and operation of the BJT, and illustrate how it can be utilized as a switching device for the implementation of digital logic.

BJT Structure

The structure of a bipolar transistor consists of essentially two *p–n* junction diodes fabricated very close to each other; in fact, they are so close together that they can share a common region. Therefore, a bipolar transistor is a three-layer sandwich of alternating semiconductor material types. There are two possible variations, the *PNP transistor* and the *NPN transistor*, where these designations describe the order of the two types of materials in the device. The structure and the standard symbols for the BJT are shown in Fig. 9.24. Basic principles will be illustrated using primarily the NPN transistor; the PNP transistor will be addressed by symmetry later.

The center region of the BJT is called the *base*, labeled (*B*), and in modern bipolar transistors it is fabricated as thin as possible. One of the end regions is called the *emitter* (*E*) and has the primary purpose of emitting carriers which travel through the base and are collected at the opposite end of the transistor in the *collector* (*C*).

In modern silicon planar processing, the bipolar transistor is generally fabricated with the emitter on top of the collector with a thin base between, as illustrated in Fig. 9.25 for an NPN device. The primary carrier flow is downward from emitter to collector in the

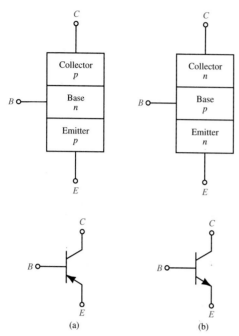

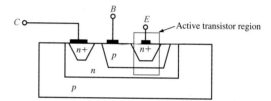

Figure 9.24 PNP and NPN transistors; structure and symbols.

Figure 9.25 Cross section of planar-processed bipolar transistor (BJT).

area labeled "active device." The lateral part of the structure serves only to make electrical connections from the surface to the layers in the vertical transistor structure.

BJT Modes of Operation: Saturation, Active, and Cutoff

Since the bipolar transistor has two junctions, and either junction can be forward or reverse biased, there are four possible bias conditions, as shown in Fig. 9.26. The junction between emitter and base is commonly called the *emitter junction;* that between collector and base, the *collector junction.*

For the case in which both junctions are reverse biased, no terminal conducts any significant current, and the transistor is said to be in *cutoff* or "off." In this condition, the

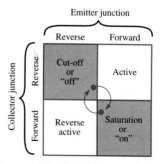

Figure 9.26 BJT bias conditions.

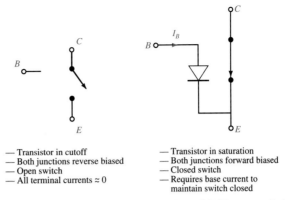

— Transistor in cutoff
— Both junctions reverse biased
— Open switch
— All terminal currents ≈ 0

— Transistor in saturation
— Both junctions forward biased
— Closed switch
— Requires base current to
  maintain switch closed

Figure 9.27   Simplified interpretation of BJT as a switch.

transistor appears as an *open switch*. When both junctions are forward biased, there is a direct *conducting path from collector to emitter*, and the transistor is said to be "on" or *saturated*. This condition corresponds to a *closed switch*. In contrast to FETs, it will be shown that a continuous supply of base current, I_B, is required to keep the switch closed. These conditions are illustrated in Fig. 9.27.

These two bias conditions, shaded in Fig. 9.26, are of primary importance for bipolar devices in digital circuits, because they represent the "on" and "off" states of the bipolar switch. However, as the figure illustrates, the bipolar device generally must pass through the other regions during switching.

If a bipolar transistor is off, and then switched to the on condition, there is a transition that occurs during which the emitter junction is forward biased, and the collector junction is reverse biased. This is called the *active region*. It is the desired bias condition when operating a bipolar transistor as an amplifier—as we will show in Chapter 15. For applications requiring switching, the transistor passes through the active region only briefly as it moves from the off condition to the fully on or saturated condition.

However, some understanding of the operation of the BJT in the active region is critical for determining the bipolar transistor's behavior as a switch. For this reason, the next section will describe briefly the operation of a BJT in the active region.

The BJT in the Active Region

Consider an NPN transistor. A circuit model of this device might first be imagined to appear as in Fig. 9.28(a), with two diodes back to back. Since the collector junction is reverse biased in the active region, that diode is dashed, indicating it's not conducting. One might incorrectly predict that the only current flow is from base to emitter; however, this is an incomplete picture.

In addition to the separate action of the two diodes, the fact that they are extremely close together and separated by only a thin base region allows an interaction between the two diodes, which is the basis of transistor action.

In this active region the emitter junction is forward biased and the potential barrier for this junction is lowered. Therefore, electrons from the *n*-type emitter diffuse into the base. The base is made very thin, and few of the injected electrons recombine with holes in the

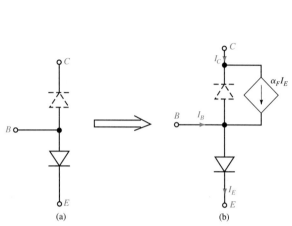

(a)   (b)

Figure 9.28   BJT model in active region.

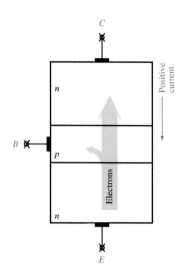

Figure 9.29   BJT carrier movement.

base. Most of the carriers survive the trip by diffusion across the base and they are swept into the collector region by the field at the collector junction.

These carriers contribute to the collector current even though the collector junction is reverse biased. As illustrated in Fig. 9.29, only a small portion of the injected electrons recombine and contribute to base current.

The electrons are minority carriers in the *p*-type base as they diffuse toward the collector, and the BJT is often referred to as a minority carrier device. In an NPN device, the dominant component of current is the electrons moving from emitter to collector, which is equivalent to positive current moving from collector to emitter.

To model this new component of collector current we add to the circuit model of Fig. 9.28(a) a dependent current source, as illustrated in Fig. 9.28(b). The magnitude of this current source is some fraction, less than unity, of the emitter current, representing the portion of the carriers that make it across the base. This source is defined as α_F times the emitter current, I_E. The *forward alpha*, α_F, is an important parameter of the transistor; it can be measured for a given device as the ratio of the collector current to emitter current for an emitter junction that is forward biased, and a collector junction that is not. Generally $V_{CB} = 0$ for this measurement. The term α_F is sometimes called the *forward common base, (CB), current transfer ratio.* It is always less than unity and has a typical value in the range of 0.97 to 0.995.

$$\alpha_F = \left.\frac{I_C}{I_E}\right|_{active\ region} \tag{9.6}$$

This phenomenon of *transistor action* is very important. Under the bias conditions just described, the current at the collector terminal of the BJT is independent of the applied collector voltage, and depends only upon the injected emitter current which is set by the current through the emitter *p–n* junction.

In order to forward bias the emitter junction, the voltage applied must be equal to the turn-on voltage of the junction diode, V_F. For the base emitter junction of a silicon transistor, we often assume that this value of $V_{BE} = V_F = 0.7$ V.

The application of Kirchhoff's current law to the three terminals of a transistor yields an important equation:

$$I_E = I_B + I_C \tag{9.7}$$

where the current directions are referenced as in Fig. 9.28(b).

EXAMPLE 9.3

Consider the NPN transistor and bias circuit shown in Fig. 9.30(a). If this transistor has $\alpha_F = 0.99$, the emitter junction forward biased at $V_{BE} \approx 0.7$ V, and $I_E = 2$ mA, let us determine (a) the collector current, (b) the collector voltage, and (c) the base current.

The emitter voltage $V_E = 0$ since this terminal is grounded.

The base voltage V_B is at the turn-on voltage of the emitter junction, which is approximately 0.7 volts in this case. The emitter junction is forward biased. A voltage source is connected directly from collector to ground, setting the collector voltage, V_C, equal to 2 volts. Since $V_C = 2$ V and $V_B \approx 0.7$ V, the collector junction is clearly reverse biased by 1.3 volts. The transistor is in the active bias condition.

Therefore, replacing the transistor symbol with the equivalent circuit for the transistor in the active region as shown in Fig. 9.30(b) yields the following.

(a) The collector current $I_C = \alpha_F I_E$ is

$$I_C = (.99)\,(2 \times 10^{-3}) = 1.98 \text{ mA}$$

(b) The collector voltage $= 2$ V

and

(c) The base current $I_B = I_E - I_C$ is

$$I_B = 2 \times 10^{-3} - 1.98 \times 10^{-3} = 0.02 \text{ mA} \qquad \blacksquare$$

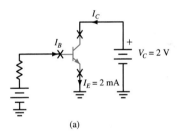

(a)

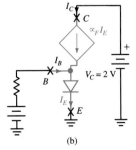

(b)

Figure 9.30   A NPN transistor circuit.

The Common Emitter (CE) Configuration

Most switching applications make use of the bipolar transistor in the common emitter configuration. This is the circuit configuration with the emitter grounded, the input or control signal applied to the device at the base, and the output taken from the collector, as illustrated in Fig. 9.31.

For this case it is convenient to develop an equivalent circuit that represents the output current, I_C, in terms of I_B rather than I_E. Recall that for the active bias region:

$$I_C = \alpha_F\, I_E \tag{9.8}$$

therefore

$$I_C = \alpha_F\, (I_B + I_C) \tag{9.9}$$

and

$$I_C\, (1 - \alpha_F) = \alpha_F\, I_B \tag{9.10}$$

hence

$$\left.\frac{I_C}{I_B}\right|_{active\ region} = \frac{\alpha_F}{1 - \alpha_F} = \beta_F \tag{9.11}$$

The parameter, β_F, the *forward common emitter current transfer ratio,* is defined as the ratio of collector to base current for active bias. This forward beta is one of the key parameters of a bipolar transistor. A common emitter equivalent circuit can now be developed analogous to that of Fig. 9.28(b), and is presented in Fig. 9.32. It can be seen from this model that for the active bias region:

$$I_C = \beta_F\, I_B \tag{9.12}$$

BJT Circuit Models for Switching

The conditions for biasing an NPN transistor in cutoff (open switch) and saturation (closed switch) will now be described, and a circuit model for each of these situations presented.

The Open Switch: BJT in Cutoff. If the emitter junction and the collector junction of a BJT are *not* forward biased, the base current is approximately zero. Under this condition, there is no collector current. The transistor's equivalent circuit is simply an open circuit between collector and emitter, analogous to an open switch as shown in Fig. 9.33(a).

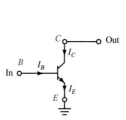

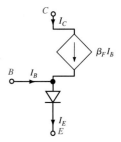

Figure 9.31 BJT common emitter configuration.

Figure 9.32 BJT common emitter equivalent circuit.

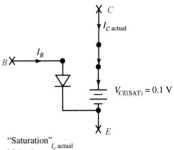

"Cut-off"

Make $I_B = 0 \Longrightarrow$ other terminal currents ≈ 0

Both junctions reverse biased

Figure 9.33(a) BJT in cutoff.

"Saturation"

Make $I_B > \dfrac{I_{c\ actual}}{\beta_F}$

then I_C determined by external circuit

Figure 9.33(b) BJT in saturation.

The Closed Switch: BJT in Saturation. The conditions required to establish a BJT in saturation are more complex than other bias conditions in that both device junctions must be forward biased. Maintaining the switch closed requires a constant base current exceeding a certain critical minimum value, called $I_{B(SAT)}$; the value of $I_{B(SAT)}$ depends on parameters of the transistor and the circuit in which the transistor is connected.

In saturation, the collector current of the BJT becomes limited by the circuit in which the device is connected in the same manner as the current through a real switch is controlled by the external circuit. If insufficient base current is provided, the device moves out of saturation and into the active region where the collector current is β_F times the supplied base current.

The critical minimum base current, $I_{B(SAT)}$, is:

$$I_{B(SAT)} = \frac{I_{C\ actual}}{\beta_F} \tag{9.13}$$

where $I_{C\ actual}$ is the collector current that would flow if the transistor were saturated (closed switch) and is determined primarily by the external circuit and bias voltages. To ensure saturation, the circuit should supply a base current larger than $I_{B(SAT)}$, that is, $I_B > I_{B(SAT)}$.

The ratio of the actual I_C to the actual I_B for a transistor in saturation is defined as the normal forced beta, β_F^*.

$$\beta_F^* = \frac{I_{C\ actual}}{I_{B\ actual}} \tag{9.14}$$

where

$$\beta_F^* < \beta_F, \quad \text{for saturated BJT}$$

A BJT in saturation appears to the circuit as a closed switch in series with a small voltage source, typically less than 0.1 volts, called $V_{CE(SAT)}$. For purposes of the present discussion, we will assume $V_{CE(SAT)} = 0.1$ V.

A circuit model of the BJT in saturation is presented in Fig. 9.33(b).

EXAMPLE 9.4

A BJT is used in the circuit of Fig. 9.34 to switch a small light bulb on and off. The bulb has a resistance of 1 kΩ and will be modeled as a 1 kΩ resistor. Assuming the transistor

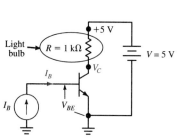

Figure 9.34   Circuit used in Example 9.4.

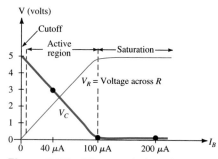

Figure 9.35   Characteristics of the BJT circuit used in Example 9.4.

has a $B_F = 50$, and the turn-on voltage of the emitter junction is 0.7 volts, let us determine the circuit's response to base currents of 0, 40, 100, and 200 μA.

To illustrate this circuit's behavior, the circuit voltages, V_C and V_R, are plotted in Fig. 9.35 for $0 \le I_B \le 200$ μA.

For: $I_B = 0$, $I_C = 0$, and there is no current through the resistor R; therefore, $V_C = 5$ V; since $I_C = 0$, the voltage across R is zero, and the transistor is in cutoff.

For: $I_B = 40$ μA

$$I_C = \beta_F(40\ \mu A) = 50(40\ \mu A) = 2\ mA$$
$$V_C = 5 - I_C R = 5 - (2\ mA)(1\ k) = 3\ V$$

At this point, the voltage across R is $5 - V_C = 5 - 3 = 2$ V.
The collector junction is still reverse biased.

$$V_{CB} = V_C - V_B = 3 - 0.7 = 2.3\ V$$

and the transistor is in the active region.

For: $I_B = 100$ μA

$$I_C = \beta_F I_B = 50(100\ \mu A) = 5\ mA$$
$$V_C = 5 - I_C R = 5 - (5\ mA)1\ k = 5 - 5 = 0$$
$$V_{CB} = V_C - V_B = 0 - 0.7 = -0.7\ V$$

The collector junction has just reached forward bias and the transistor is at the edge of saturation. The bulb, R, has the full 5 volts across it.

$I_{C(actual)} = I_{C(max)}$ and occurs when the transistor is a short; therefore,

$$I_{C\ actual} = I_{C\ max} = \frac{5}{1\ k} = 5\ mA$$

and

$$I_{B(SAT)} = \frac{I_{C\ actual}}{\beta_F} = \frac{5 \times 10^{-3}}{50} = 100\ \mu A$$

That is, 100 μA is the value of I_B required to place the device at the edge of saturation.

For $I_B = 200$ μA: This base current will drive the transistor deeper into saturation. I_C cannot exceed 5 mA, since this current is determined by the circuit when the transistor acts as a closed switch.

Therefore, $I_C = 5$ mA, $I_B = 200$ μA, and

$$\beta_F^* = \text{forced } \beta = \frac{I_{C\;actual}}{I_{B\;actual}} = \frac{5 \times 10^{-3}}{200 \times 10^{-6}} = 25$$

The *forced beta* β_F^* is always less than β_F for a BJT in saturation. The ratio

$$0 < \frac{\beta_F^*}{\beta_F} < 1$$

is an indicator of how deeply the device is driven into saturation. In this example, the ratio of forced beta to beta is 25/50 = 0.5. ■

9.10
The Ebers-Moll BJT Model

In a previous section the operation of the BJT in the active region was introduced. Recall that if the emitter junction is forward biased, an emitter current flows; however, a large portion of the electrons from the emitter cross the base and contribute to collector current. This is modeled by a dependent current source, $\alpha_F I_E$, as was shown in Fig. 9.28(b).

The analogous situation applies if the collector junction becomes forward biased. The collector junction diode would then conduct a current, and a significant portion of this current appears at the emitter terminal due to diffusion across the narrow base. We can model this current by the addition of a second dependent source, this one across the emitter diode with a value of $\alpha_R I_C$, where I_C is the current flowing through the collector diode.

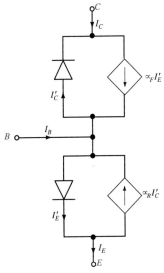

Figure 9.36 Ebers-Moll model for a BJT.

The parameter α_R is the *common base (CB) reverse current transfer ratio*. It can be measured in a real device exactly as one measures α_F, except with the emitter and collector terminals exchanged.

That is

$$\alpha_R = \frac{I_E}{I_C}\bigg|_{\text{reverse active bias}} \tag{9.15}$$

Recall that the reverse active region requires that the emitter junction be reverse biased, and the collector junction forward biased. From another viewpoint, α_R is a measure of the current transfer with the transistor operating "upside down," that is, with the collector operating as emitter, and the emitter operating as collector. Analogous to the earlier case, this measurement is usually made with $V_{BE} = 0$.

In saturation, both junctions become forward biased, and two mechanisms are modeled simultaneously: (1) the forward biased emitter is sending carriers to the collector, and (2) the forward biased collector is sending carriers to the emitter.

An equivalent circuit that includes both diodes and both dependent sources is shown in Fig. 9.36. This model was presented in 1954 in a famous paper by J. J. Ebers and J. L. Moll and is called the *Ebers-Moll (EM) model* ("Large Signal Behavior of Junction Transistors," *Proceedings IRE,* Vol. 42, Dec. 1954). It is a very powerful model in that it can describe the dc behavior of the BJT in all four regions of operation.

If we use the diode equation, that is, Eq. (8.9) with $n = 1$, to represent the current through each of these junctions, then:

$$I'_E = I_{ES}\left(e^{\frac{qV_{BE}}{kT}} - 1\right) \tag{9.16}$$

and

$$I'_C = I_{CS}\left(e^{\frac{qV_{BC}}{kT}} - 1\right) \tag{9.17}$$

and therefore

$$I_E = I_{ES}\left(e^{\frac{qV_{BE}}{kT}} - 1\right) - \alpha_R I_{CS}\left(e^{\frac{qV_{BC}}{kT}} - 1\right) \tag{9.18}$$

$$I_C = -I_{CS}\left(e^{\frac{qV_{BC}}{kT}} - 1\right) + \alpha_F I_{ES}\left(e^{\frac{qV_{BE}}{kT}} - 1\right) \tag{9.19}$$

These last two equations are known as the Ebers-Moll equations and have been widely used to describe BJT behavior. The model describing the active region presented in Fig. 9.28 is a subset of the more general Ebers-Moll model. In the active region, one diode and one source are inactive. One can construct a common emitter representation of the Ebers-Moll model, analogous to that done earlier in this chapter. The common emitter formulation uses βs instead of αs. Recalling that

$$\beta_F = \frac{\alpha_F}{1 - \alpha_F} \qquad \text{also by analogy} \qquad \beta_R = \frac{\alpha_R}{1 - \alpha_R} \tag{9.20}$$

we can derive

$$\alpha_F = \frac{\beta_F}{1 + \beta_F} \qquad \text{also by analogy} \qquad \alpha_R = \frac{\beta_R}{1 + \beta_R} \tag{9.21}$$

where β_R is the *reverse beta,* or *reverse common emitter current transfer ratio.*

Using the above relations, one can interchange between expressions involving αs and βs.

One of the very useful results of the Ebers-Moll model is that it can be used to predict a more accurate value for $V_{CE(SAT)}$. In previous analyses, we have assumed the voltage across the saturated transistor (closed switch) was 0.1 volt. The actual value can be derived from the EM model as:

$$V_{CE(SAT)} = \frac{kT}{q} \ln \left[\frac{\dfrac{1}{\alpha_R} + \dfrac{I_C}{I_B} \dfrac{1}{\beta_R}}{1 - \dfrac{I_C}{I_B} \dfrac{1}{\beta_F}} \right] \tag{9.22}$$

EXAMPLE 9.5

A BJT has a measured forward beta of 100 and a reverse beta of 2. The measured collector current of this device is 20 mA and the base current is determined to be 2 mA. Given these parameters, let us determine the saturation voltage, $V_{CE(SAT)}$.

Using Eq. (9.22), we find that

$$V_{CE(SAT)} = 0.26 \ln \left[\frac{\dfrac{3}{2} + 10\dfrac{1}{2}}{1 - 10\dfrac{1}{100}} \right] = .026 \ln \left[\frac{6.5}{0.9} \right] = 0.051 \text{ V} \qquad \blacksquare$$

9.11
BJT Output Curves

The Ebers-Moll equations can be used to predict relatively accurately the dc behavior of the BJT. Plots of these equations and the measurement of voltages and currents from actual BJTs often match well.

The output curves for a BJT in the common emitter configuration is a plot of I_C versus V_{CE} for a set of different values of I_B. An example of such a set of curves is shown in Fig. 9.37.

Notice that in the active region the collector current is almost constant for a given I_B value; the ratio of I_C/I_B is β_F, which is about 100 for this example. There is generally a small but finite slope to the output curves, which will be discussed in Chapters 15 and 16.

As the saturation region on the left side of the plot is approached, the value of I_C falls for a given I_B value, indicating a β_F^*, or forward forced beta, of less than β_F. The cutoff region is along the horizontal axis for $I_B = 0$. In this latter region, the collector current is essentially zero.

The load line approach for analysis can be used for BJTs in a similar manner to that already demonstrated for other devices.

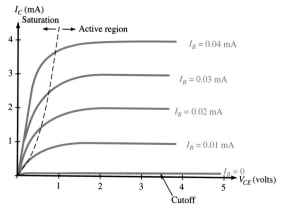

Figure 9.37   Output curves for a BJT.

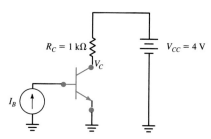

Figure 9.38   Circuit used in Example 9.6.

EXAMPLE 9.6

The circuit shown in Fig. 9.38 has a transistor with the output characteristics of Fig. 9.37. Figure 9.39 shows the output curves with the load line for this circuit superimposed. The horizontal axis intercept occurs at the voltage across the device when $I_C = 0$; this is the open circuit voltage of 4 volts. The intercept on the vertical axis is the short circuit current, the current through the device if the device acts as a perfect short; this is determined by the circuit as

$$I_{C(MAX)} = \frac{V_{CC}}{R_C} = \frac{4}{1\,k} = 4\text{ mA}$$

Now, with the load line constructed, let us determine (a) the value of I_B required to drive the transistor into saturation, and (b) the value of V_C when $I_B = 0.02$ mA.

(a) From the load line, $I_B \approx 0.03$ mA is required to move the Q-point along the load line into the saturation region on the far left.

(b) The load line intercepts the $I_B = 0.02$ mA curve at $V_C = 2$ V.   ■

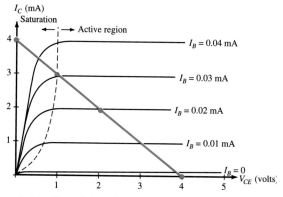

Figure 9.39   Transistor characteristics used in Example 9.6.

9.12

**The NOT Gate:
BJT Implementation**

For the circuit shown in Fig. 9.40, we assume the transistor has the following parameters: $\beta_F = 75$ and $V_{BE} = 0.7$ V.

When $V_{in} < 0.7$ V, that is, an input "low" level, the emitter junction is not forward biased; $I_B = 0$, $I_C \approx 0$, and the transistor is in cutoff. Since $I_C \approx 0$, $V_C = V_{CC} = 5$ V, the output is at a "high" level.

When $V_{in} = 5$ V, an input "high" level

$$I_B = \frac{V_{CC} - V_{BE}}{R_1} = \frac{5 - 0.7}{10 \text{ k}} = 0.43 \text{ mA}$$

and

$$I_{C(actual)} = \frac{V_{CC} - V_{CE(SAT)}}{R_2} = \frac{5 - 0.1}{500} = 9.8 \text{ mA}$$

We assume $V_{CE(SAT)} = 0.1$ V unless enough information is given to determine it more accurately.
Therefore

$$\beta_F^* = \frac{I_{C \text{ actual}}}{I_{B \text{ actual}}} = \frac{9.8 \times 10^{-3}}{0.43 \times 10^{-3}} = 23$$

$$\beta_F^* \ll \beta_F;$$

and

$$\frac{\beta_F^*}{\beta_F} = \frac{23}{75} = 0.3$$

This indicates a transistor in saturation since the base current supplied by the circuit is well in excess of that required to saturate the transistor. A good design rule of thumb is $\beta_F^*/\beta_F < 0.7$ to ensure saturation with margin. Therefore, for this circuit, $V_{out} = V_{CE(SAT)} \approx 0.1$ volt, a low level.

This circuit operates as an inverter or NOT gate with logic levels switching between 0.1 volt and 5 volts.

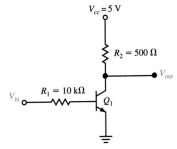

Figure 9.40   NOT gate implemented with a BJT.

9.13

Turning the Saturated Transistor Off

When the BJT is saturated, both junctions are forward biased, and the capacitances of these junctions are fully charged; the base is full of excess carriers and current is flowing. If one suddenly removes the supply current to the base, there may be a substantial delay in time before the transistor ceases to conduct from collector to emitter. This delay in turning the transistor switch off is a result of the finite time required for the excess carriers in the base to recombine in the base region and reduce the voltage which is stored on these junction capacitances. A measure of this delay is the storage time of the transistor, and in the design of switching transistors, an attempt is made to make this time as small as possible.

However, the transistor can be switched off faster using circuit techniques to remove the charge from the base more quickly. One technique is to provide a resistor from base to ground so that when the transistor is attempting to turn off, there is an external low resistance path for base charge to travel to ground, as shown in Fig. 9.41.

The time delay encountered in switching a saturated bipolar transistor off is generally the primary speed-limiting effect in bipolar logic circuits in which the transistors are allowed to saturate.

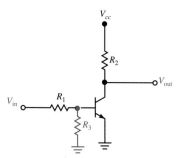

Figure 9.41 Using a resistor to speed turn-off of a BJT.

9.14

The PNP Transistor

The PNP transistor is the exact complement of the NPN transistor. It operates in a completely analogous manner, but rather with the p-type emitter providing holes, which traverse the base, and are collected by the p-type collector. To forward bias the emitter junction, the base must be made about 0.7 volts more *negative* than the emitter. Normally the collector is connected through a load resistance to the terminal of a dc voltage source more negative than the voltage at the emitter.

Considering equally sized devices, an NPN transistor generally outperforms the PNP device because the primary carriers in an NPN transistor, electrons, have a mobility higher than that of holes. Consequently, most bipolar circuits utilize NPN devices; however, some integrated circuits utilize both NPN and PNP devices in the same circuit.

9.15

Summary

- There are two types of transistors, field effect transistors (FETs) and bipolar junction transistors (BJTs).

- All transistors perform a similar function in digital circuits: the level of an input control signal at one terminal of the device controls the flow of current between two other terminals.
- The relay is an electromechanical device which can be used as an illustration of the way a transistor functions in digital electronic circuits.
- There are two primary types of FETs: MOSFETs and JFETs. MOSFETs are one of four types: either n-channel or p-channel, and each of these can be either enhancement-mode or depletion-mode.
- The enhancement-mode MOSFET acts like an open switch if $|V_{GS}| < |V_T|$, where V_T is the threshold voltage. It conducts in one of two ways if $|V_{GS}| > |V_T|$: (1) If $|V_{DS}| < |V_{GS} - V_T|$ it acts like a closed switch in series with a resistance, r_{on}; this is the ohmic region which is used in digital circuits. (2) In the other case it acts like a dependent current source controlled by V_{GS}; this is the active or saturated region which is used in amplifier circuits.
- The layout of a transistor refers to the topology of the "top view" in the plane of the silicon surface; for MOSFETs, a critical factor is W/L, the device width-to-length ratio.
- The performance of an inverter circuit using a MOSFET can be calculated using the switch model.
- MOSFET output curves are a plot of I_D versus V_{DS} for a family of V_{GS} values, as shown in Fig. 9.16.
- The load line technique is used with the MOSFET output curves to characterize the switching of an inverter.
- Depletion-mode MOSFETs have a channel fabricated in the device structure and conduct with $V_{GS} = 0$. They are used in digital MOS circuits primarily as load devices.
- The JFET is a depletion-mode FET fabricated with a p–n junction as the gate.
- The BJT can be used as a switch in digital circuits. There are two types, depending on the device structure: NPN and PNP.
- If the base current, $I_B = 0$, the BJT is in cutoff, and it acts like an open switch. If $|I_B| > 0$ the BJT conducts current in one of two possible ways depending on other bias conditions: (1) In saturation the transistor acts like a closed switch with a small voltage drop called $V_{CE(SAT)}$; this is the case most used in digital circuits. (2) In the active region it conducts like a dependent current source controlled by the base current; this is the bias condition used in amplifier circuits.
- For the BJT, $I_E = I_B + I_C$; α_F or β_F can be used to relate BJT terminal currents in the forward active region: $I_C = \beta_F I_B$, and $\beta_F = \alpha_F/(1 - \alpha_F)$.
- The BJT differs from the MOSFET in that the input terminal to the BJT connects to a p–n diode; a constant input current is required to maintain the switch closed; a MOSFET gate has a high (essentially infinite) input resistance.
- For a BJT in the active region, I_C is controlled by I_B; once saturated, I_C is controlled by the external bias circuit.
- The Ebers-Moll (EM) BJT model uses two dependent current sources and two diodes to model the large signal behavior of the transistor under all bias conditions.
- Using the EM model, an expression for $V_{CE(SAT)}$ can be developed.
- BJT output curves plot I_C versus V_{CE} for a family of values of I_B, as shown in Fig. 9.37.
- The load line technique is used with the BJT output curves to characterize the switching of an inverter.

PROBLEMS

9.1. What are the two basic types of transistors? What is the basic function of a transistor in digital circuits?

9.2. Identify the device structures shown in Fig. P9.2 as *n*-channel enhancement-mode, *n*-channel depletion-mode, *p*-channel enhancement-mode, *p*-channel depletion-mode, or none of the above.

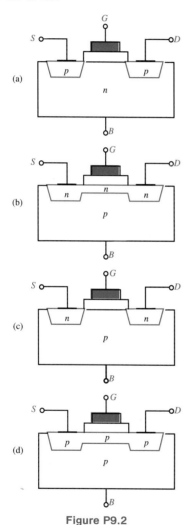

Figure P9.2

9.3. Match each characteristic, listed (a) through (n), with the appropriate structure(s), listed I through V.
(a) conducts when V_{GS} is positive
(b) hole current flows when transistor is ON

(c) electron current flows when transistor is ON
(d) conducts when $V_{GS} = 0$
(e) conducts when V_{GS} is negative
(f) turns OFF through the repulsion of electrons
(g) turns OFF through the repulsion of holes
(h) conducting channel formed during fabrication
(i) conducting channel formed by biasing the gate
(j) turns ON through the attraction of electrons
(k) turns ON through the attraction of holes
(l) the body is *p*-type
(m) the body is *n*-type
(n) the body is furry

   I. *n*-channel enhancement
  II. *n*-channel depletion
 III. *p*-channel enhancement
 IV. *p*-channel depletion
  V. cat

9.4. Sketch the cross section of a *p*-channel enhancement-mode transistor and label the source, drain, gate, and substrate contacts.

9.5. Determine the values of the parameter K and V_T for the MOSFET characteristics plotted in Fig. P9.5.

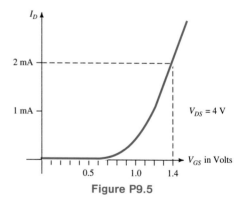

Figure P9.5

9.6. The threshold voltage of an *n*-channel enhancement mode MOSFET is 0.8 V. The MOSFET conducts a current of 2 mA with a $V_{GS} = 2$ V and $V_{DS} = 5$ V. What is the drain current when V_{GS} is doubled?

9.7. The drain currents of an *n*-channel MOSFET biased in the saturated (active) region for two values of V_{GS} are given in Table P9.7. Plot the square root of I_D versus V_{GS}

to obtain the threshold voltage and parameter K for the transistor.

$V_{GS}(V)$	$I_D(A)$
1.0	3.2×10^{-4}
2.0	1.15×10^{-2}

Table P9.7

9.8. A simple inverter using an n-channel enhancement-mode MOSFET is shown in Fig. P9.8(a). Use the transistor model shown in Fig. P9.8(b) to obtain the transfer characteristics of this inverter, a plot of V_o versus V_i.

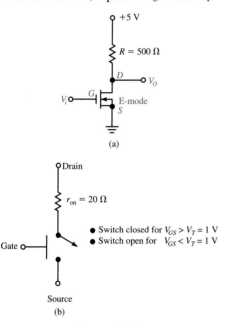

(a)

(b)

Figure P9.8

9.9. Sketch the MOSFET topology (top view) for both an enhancement-mode n-channel and p-channel MOSFET. For each case identify the types of doping (n-type or p-type) for the source, drain, and substrate (body).

9.10. A NOT gate is implemented using an n-channel enhancement-mode MOSFET as shown in Fig. P9.10 with $V_{DD} = 5$ V, $V_T = 0.9$ V, and $R_L = 1$ kΩ.
(a) If $V_{in} = 5$ V, then $r_{on} = 100$ Ω; calculate V_{out}.
(b) Find R_L if $V_{out} = 0.1$ V.

9.11. For the NOT gate and transistor described in problem 9.10(a), determine the value of K required to make $r_{on} = 100$ Ω.

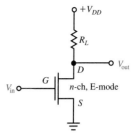

Figure P9.10

9.12. A NOT gate is modeled using an inverter similar to that shown in Fig. P9.10. For $V_{in} = V_{DD} = 5$ V and $V_{out} = 0.2$ V, determine the load resistance, R_L, and the n-channel ON source-to-drain resistance (r_{on}) if the power dissipated in the load resistance during ON state is 0.48 mW.

9.13. From the transistor output curves and the load line for an inverter plotted in Fig. P9.13, determine the values of V_{DD} and R_L.

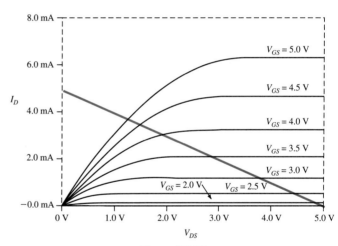

Figure P9.13

9.14. (a) On the MOSFET output curve of Fig. P9.13, sketch the curve of drain current, I_D, versus V_D for the circuit shown in Fig. P9.14. (b) What other two-terminal de-

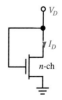

Figure P9.14

vice has a *V–I* characteristic that resembles a transistor connected this way? (c) In what region of operation is the transistor biased?

9.15. Repeat 9.14(a) for the circuit of Fig. P9.15.

Figure P9.15

9.16. The drain current of an *n*-channel depletion-mode MOSFET is plotted as a function of the gate voltage in Fig. P9.16. Obtain the value of the "pinch-off" voltage for the MOSFET.

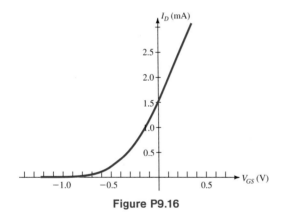

Figure P9.16

9.17. The I_D–V_{DS} characteristics of a JFET are shown in Fig. P9.17(a). Using the load line concept, compute the *Q*-point of the circuit in Fig. 9.17(b).

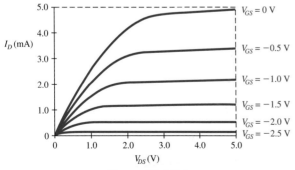

Figure P9.17(a)

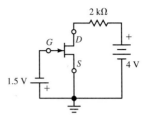

Figure P9.17(b)

9.18. For switching applications, which device terminal is usually used as the control terminal for (a) MOSFETs? (b) BJTs?

9.19. The direction of the arrow in the symbol for the PNP and NPN transistors has what significance?

9.20. Sketch the cross section of a planar-processed PNP transistor clearly showing the three terminals of the BJT.

9.21. In Fig. P9.21, determine I_C and V_C for the following values of I_B and designate the mode of transistor operation. (a) $I_B = 0$, (b) $I_B = 20$ μA, (c) $I_B = 60$ μA, (d) $I_B = 100$ μA.

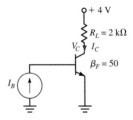

Figure P9.21

9.22. The values for α_F are measured for three transistors as $\alpha_F = 0.97$, $\alpha_F = 0.98$, and $\alpha_F = 0.99$. Calculate the corresponding values of β_F for each.

9.23. The PNP transistor in the circuit shown in Fig. P9.23 has the following characteristics: $\beta_F = 75$, and assume $V_{CE(SAT)} = -0.1$ V. (a) What value of collector current flows when the transistor is saturated? (b) If $V_{BE} = -0.6$ V, what value of V_1 is required to saturate the transistor?

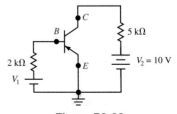

Figure P9.23

9.24. Indicate the direction of positive current (net positive charge movement) in or out of each terminal of the transistors in Fig. P9.24 assuming that both the transistors are biased in the active region of operation.

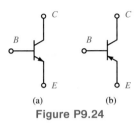

(a)   (b)

Figure P9.24

9.25. Figures P9.25 (a–f) illustrate NPN and PNP BJTs with biased junctions. For each case identify each of the junctions as forward or reverse biased. Also identify the mode of operation of the BJT, that is, cutoff, saturation, active, or reverse-active.

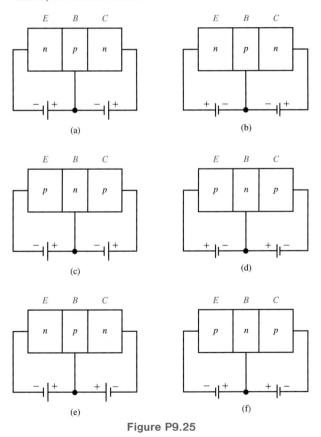

Figure P9.25

9.26. The NPN transistor in Fig. P9.26 has $\beta_F = 100$ and the emitter junction is forward biased at $V_{BE} = 0.6$ V.

(a) What is the mode of operation of the transistor? (b) Calculate the base, emitter, and collector currents.

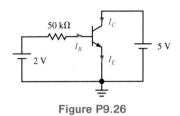

Figure P9.26

9.27. Figure P9.27 shows an NPN transistor connected in a way that it looks like a diode from the terminals; the transistor is operating in the active region. Given $V_{BE} = 0.7$ V and $\beta_F = 49$, calculate the base and collector currents.

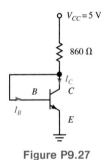

Figure P9.27

9.28. For the circuit given in Fig. P9.28, assuming $\beta_F = 100$ and $V_{BE} = 0.7$ V, (a) find V_o for $V_i = 0.8, 1.5, 2.0,$ and 2.5 V. (b) At approximately what value of V_i will the collector current be determined by the circuit bias conditions rather than the β_F relationship? What mode of operation is this? (c) For $V_i = 2.5$ V, what is the forced beta, β_F^*?

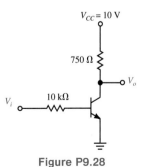

Figure P9.28

9.29. Assuming $V_{BE} = 0.7$ V, calculate the β_F for the transistor in Fig. P9.29.

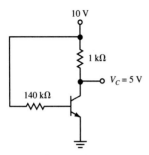

Figure P9.29

9.31. The forward and reverse βs for a BJT are measured to be 50 and 5 respectively. The collector current for the transistor is 17 mA and the base current is 1 mA. (a) Is the transistor saturated? (b) If so, determine the value of $V_{CE(SAT)}$.

9.32. A transistor with output curves shown in Fig. P9.32 is used in the circuit of Fig. P9.30 with R_2 changed to 2 kΩ. Use a graphical approach (plot the load line) to find the value of V_o if $V_i = 1.1$ V.

9.30. In the circuit shown in Fig. P9.30, assume the base emitter turn-on voltage, V_{BE}, is 0.7 V, $V_{CE(SAT)}$ is 0.1 V, and β_F is 200. Obtain the value of the output voltage for $V_i = 0.5$ V, 0.8 V, 1.0 V, and 3.5 V.

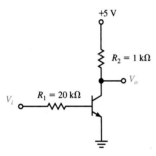

Figure P9.30

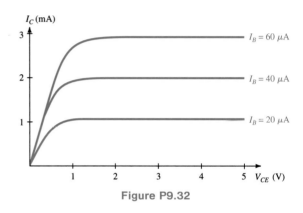

Figure P9.32

DIGITAL SYSTEMS

Digital Switching Circuits: The Logic Gate

10.1
Introduction

The "electronics revolution" has produced a stream of new products made possible in large measure by advanced electronics. Digital switching circuits are responsible for a majority of the electronic innovations which have changed, and continue to change, the way we live our lives. In the technical professions, the workplace has changed, as a result of faster personal computers, improved telecommunications, and networks.

These electronic advances are made possible by the implementation of digital electronic circuits which operate faster, are more dense, and are smaller in size than their previous counterparts. The details of how the devices are fabricated is presented in more detail in Chapter 13, which describes microelectronics technology.

The basic functions of the circuits being implemented have not changed since the earliest digital computers which were developed decades ago. There are several elementary logic functions which are accomplished by the basic logic gates. We have seen in Chapter 8 how the diode can be used to implement the AND or the OR function. We have also seen in the last chapter the use of the relay, the MOSFET, and the BJT in implementing the NOT function. This chapter demonstrates the use of these switching devices in electronic circuitry to implement additional logic functions, which are the building blocks of most complex digital circuitry. These are the NAND, NOR, EXCLUSIVE OR, and again the NOT functions. With these basic logic elements, almost any digital switching system can be constructed.

Digital electronics, including virtually all digital computers, operate with binary logic in which the voltage or currents representing the information assumes one of two possible values—either high or low—which, in turn, represent a one and a zero, respectively.

10.2

Ideal Logic Gates: Building Blocks for Digital Systems

The basic *logic gate* is a device whose output is binary (one or a zero) and predictable depending on the states of one or more inputs. Since each output or input can have only two possible values (one or a zero), a logic gate with two inputs can have four possible input combinations, four inputs would have sixteen combinations, etc. In general, the number of input combinations Z is given by the following equation, where n is the number of inputs.

$$Z = 2^n \tag{10.1}$$

We generally make a *truth table* to describe the state of the output for a given set of inputs. Recall from Chapter 7, a truth table is a chart which lists on the left all the possible input combinations and has a column on the right that shows the output that occurs for each combination of inputs. The next sections show truth tables for several of the most common gates, the NAND gate, the NOR gate, the EXCLUSIVE OR (XOR), and the NOT (inverter) gate. Truth tables for more complex logic functions are described in Chapter 11.

Also given in each of the following sections is the Boolean algebraic expression for each type of gate. In Chapter 11, Boolean algebra is explained in more detail; however, for purposes of this discussion recall that: the sum (+) represents OR, the product (•) represents AND, the bar over the top of a symbol represents the NOT operator.

The NAND Gate

The *NAND gate* is a logic gate whose output is zero only if all of its inputs are one. The standard logic symbol and the truth table for a three-input NAND gate are shown in Fig. 10.1. Note that the output is the complement or exact opposite of the truth table for the AND gate, as is apparent in the symbol which looks like that for an AND gate, followed by a small circle, which is the symbol for inversion. Also shown in the figure is the Boolean algebraic expression for the NAND gate.

The NOR Gate

The *NOR gate* has an output of zero if any one of its inputs is a one. Therefore, the only input condition in which the NOR gate has a one output occurs when all of its inputs are zero. The logic symbol, Boolean expression, and the truth table for the NOR gate are shown in Fig. 10.2. Note that this gate has an output which is the exact complement of the OR gate.

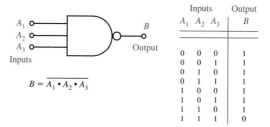

Inputs			Output
A_1	A_2	A_3	B
0	0	0	1
0	0	1	1
0	1	0	1
0	1	1	1
1	0	0	1
1	0	1	1
1	1	0	1
1	1	1	0

$$B = \overline{A_1 \cdot A_2 \cdot A_3}$$

Figure 10.1   NAND gate.

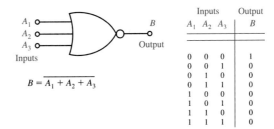

Inputs			Output
A_1	A_2	A_3	B
0	0	0	1
0	0	1	0
0	1	0	0
0	1	1	0
1	0	0	0
1	0	1	0
1	1	0	0
1	1	1	0

$$B = \overline{A_1 + A_2 + A_3}$$

Figure 10.2 NOR gate.

The Exclusive-OR Gate

The *Exclusive-OR* function (or XOR) can be constructed from more fundamental logic gates, that is, a combination of an OR gate, a NAND gate, and an AND gate. It is a circuit, however, often used with digital systems, and for convenience has been given its own logic symbol. The Exclusive-OR function is used as an "inequality comparator" because its output is high only if the inputs are not equal. An example of how a two-input XOR circuit can be implemented with basic gates, the corresponding truth table, and the special symbol adopted for the XOR gate are shown in Fig. 10.3.

The NOT Gate

The *NOT gate* was introduced in Chapter 7, and is often called the inverter. It is a special kind of gate, and is unusual in that it always has only one input and one output, and its output is the inverse or opposite of the logic state presented at its input. Therefore, the truth table for the NOT gate, or inverter, has only two lines and appears with its symbol as Fig. 7.12.

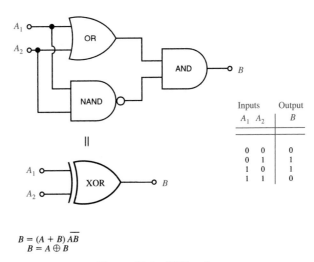

Inputs		Output
A_1	A_2	B
0	0	0
0	1	1
1	0	1
1	1	0

$$B = (A + B)\,\overline{AB}$$
$$B = A \oplus B$$

Figure 10.3 XOR gate.

10.3

Real Logic Gates: Speed, Noise Margin, and Fanout

The logic gates described in the previous section were described as ideal logic elements; however, when actual electronic hardware is used to implement real gates, there are limitations in performance that must be recognized. There are, first, the limitations associated with time; that is, it takes a finite amount of time for a real gate to respond to changes at its input. This affects the operation of the gate in several ways which will be described in this section.

Second, there are limitations associated with how the logic levels may deviate from the desired levels. Excessive loading, usually a result of too many gates connected to the output of a single gate, may adversely affect the logic levels. The concept of *noise margin* is introduced to describe the margin designed into the circuit to protect against unwanted level changes.

Finite Switching Speed

Real electronic circuits have finite capacitance, voltage, and current at each circuit node. To change the voltage on a finite capacitance with finite current requires a charging (or discharging) time related to that circuit node's *RC* time constant. Therefore, when the input to a gate changes states, there is a finite amout of time before the output actually responds. The following sections describe the manner in which these time delays are generally characterized for a typical logic gate.

Propagation Delays. *Propagation delays* are measures of how long it takes the output of a gate to respond to a transition at the input of the gate. Three parameters will be defined:

$$t_{PHL} = \text{the propagation delay high-to-low,}$$
$$t_{PLH} = \text{the propagation delay low-to-high, and}$$
$$t_P = \text{the average propagation delay.}$$

Assume for simplicity that the gate under consideration is a NOT gate, although this discussion may be applied to any gate where an input logic state change causes an output state change. When the input to a gate is given a transition from a low state to a high state, the output, after some time delay, transitions from high to low. This time difference is defined as the *propagation delay, high-to-low, t_{PHL}*. Generally, propagation delay times are measured from the 50% point, or middle, of any given transition.

Similarly, if the input transition is from a high state to a low state, the output transition from low to high occurs at a time, t_{PLH}, after the input transition. These time delays are illustrated in Fig. 10.4. In general, t_{PHL} is not equal to t_{PLH} because a circuit is usually not symmetric in its response to inputs of opposite polarity.

For convenience we often define the average propagation delay, t_p, as:

$$t_P = \frac{t_{PHL} + t_{PLH}}{2} \qquad (10.2)$$

Rise Times and Fall Times. The *rise time t_r* and the *fall time t_f* are measures of how fast the output voltage of the gate moves between the two allowed output levels (or states). This is generally measured from the 10% point to the 90% point of the full transition.

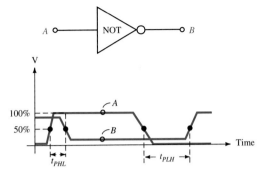

Figure 10.4   An illustration of propagation delays.

Figure 10.5 shows the output waveform for a logic gate with t_r and t_f defined. In general, t_r is not equal to t_f because most circuits are not symmetric in their ability to drive the voltage at the output node either positive or negative.

Timing and Race Conditions.   In most digital systems, a signal (for illustration consider one particular pulse) must propagate through a number of gates one after the other, before it reaches its destination. Since each gate introduces a propagation delay time, if too many such gates are strung together the total delay of the signal, that is, the sum of the average propagation delays for each of the individual gates in the string, may be excessive. Excessive delays can be disastrous to the system's performance. First, the system may not perform the desired application if it responds too slowly. However, also disastrous is the possibility of a *race condition*. This is a condition in which two pulses are intended to arrive at a destination gate in some specific order (or simultaneously), but due to each one racing through different paths in the logic with different numbers of gates, the delays stack up differently and the timing order is lost. The circuit malfunctions. Avoiding race conditions requires careful analysis of the timing delays through all critical paths, that is, the ones that can make a difference in the logic.

The *clock signal* in digital logic circuits is generally a continuous square wave, switching alternately from a high level to a low level, and is used to time and sequence the flow of digital data through a sequential digital system (see Chapters 11 and 12). In many cases, for example, every time the clock signal goes from low to high, new data are presented at the inputs of various logic circuits. Therefore, the higher the clock frequency the faster data move through the system. Computer manufacturers spend great effort to develop cir-

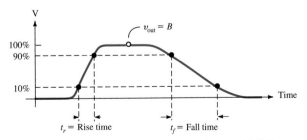

Figure 10.5   An illustration of rise time and fall time.

cuits that operate at as high a clock frequency as possible. For a given logic circuit, the upper limit, or *maximum clock frequency*, is usually determined by the accumulation of propagation delays through strings of gates in the logic. To avoid timing problems, the gate inputs should not be presented with new data before the output of the circuit has changed from the data presented in the previous clock cycle. If new data are assumed to be presented every clock cycle, then determining the total delay through the longest path in the logic will yield the maximum allowed clock frequency. The following examples illustrate these points.

EXAMPLE 10.1

Seven NOT gates, each with an average propagation delay of 7 ns, are connected as shown in Fig. 10.6. Let us determine
    (a) the total propagation delay from A to $B1$.
    (b) the total propagation delay from A to $B2$.
    In addition, if the input, A, makes a logic state change, we wish to determine (c) which output changes first, and the elapsed time before the other output change.
    (a) The logic path from A to $B1$ goes through six identical gates, each with a propagation delay of 7 ns. Therefore, the total delay is 6×7 ns or 42 ns.
    (b) Similarly, the path from A to $B2$ includes 3 gates, and therefore the total delay through that path is 21 ns.
    (c) Clearly, $B2$ changes first, and $B1$ changes $42 - 21 = 21$ ns later. ∎

EXAMPLE 10.2

A digital switching circuit operates at a clock frequency of 50 MHz. New data is presented to the circuit input on the leading edge of every clock cycle. If each gate in the network has an average propogation delay of 1.5 ns, let us determine (a) how often new data is clocked through the system, and (b) how many gates can be connected in a serial string.
    (a) The leading edge of a clock signal is defined to be the transition from a low level to a high level, and this occurs once each cycle. Therefore, the clock period, which is the inverse of frequency, tells us how often this occurs. It occurs every

$$T = \frac{1}{f} = \frac{1}{50 \text{ MHz}} = 20 \text{ ns} \tag{10.3}$$

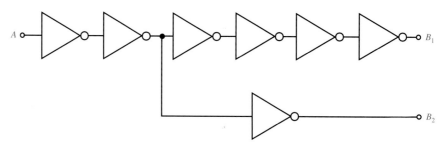

Figure 10.6   Circuit used in Example 10.1.

(b) If we require the output of a serial string of gates, each with a delay of 1.5 ns, not to exceed 20 ns, then we may only use 13 such gates, that is, 20 ns/1.5 ns, rounded down to the next integer. ■

Real Logic Levels—Noise Margin

The output of a real logic gate must produce an output signal that is unambiguous to the input of the next gate. That is, if the output is intended to be a logic 1 (high level), it must be interpreted as such by the next input. To ensure this, the minimum voltage at the output of a gate that is intended to represent a high must exceed the minimum voltage level at the input of a gate that guarantees a high will be perceived. An analogous situation applies, of course, for logic low levels as well.

The formalization of this concept is straightforward, and is illustrated in Fig. 10.7. The left side of the figure addresses the voltage levels at the input of a gate. Two specific levels are defined:

V_{IL} = the input low level = the HIGHEST voltage at the input that is always recognized by the input as a logic low.

V_{IH} = the input high level = the LOWEST voltage at the input that is always recognized by the input as a logic high.

The voltage region between these levels is the input transition region, where the input voltage is not permitted except during the brief times of transition from one level to the other. If an input voltage is applied to the gate in this transition region, the gate cannot guarantee whether it will be interpreted as a high or a low.

At the output, two voltage levels are defined:

V_{OL} = the output low level = the HIGHEST voltage level at the output that can occur when the output is intended to be low.

V_{OH} = the output high level = the LOWEST voltage level at the output that can occur when the output is intended to be high.

The above four parameters can be determined from the voltage transfer characteristics of a logic gate such as an inverter. This is illustrated in Fig. 10.8, where V_i is the input voltage to the inverter and V_o is the output voltage.

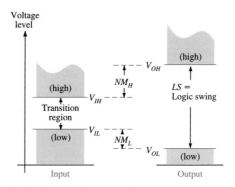

Figure 10.7   Illustration of logic levels and noise margin.

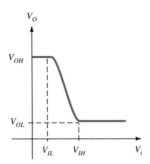

Figure 10.8 Voltage transfer characteristic.

The *transition width*, *TW*, is defined as

$$TW = V_{IH} - V_{IL} \tag{10.4}$$

An important concept is that of *noise margin*. The high and low noise margins will now be defined:

$$NM_H = \text{the high noise margin} = V_{OH} - V_{IH} \tag{10.5}$$

$$NM_L = \text{the low noise margin} = V_{IL} - V_{OL} \tag{10.6}$$

The high noise margin and the low noise margin measure the ability of the gate to properly distinguish low and high levels. In general, digital circuits with higher noise margins are more resistant to spurious interference and are more reliable in noisy environments such as automobiles and industrial applications.

The *logic swing* is the difference between the output high and low levels and is defined as:

$$LS = V_{OH} - V_{OL} \tag{10.7}$$

To be generally useful, a digital logic gate must be capable of driving more than one subsequent gate. *Fanout* is generally a specification of a digital logic gate which indicates the maximum number of similar gates that can simultaneously be connected to the output of the gate and still maintain proper operation. Fanout is an integer number, *n*, and if more than *n* gates are connected to the output of a particular gate, the circuit may not function properly.

Sometimes the term *fanout* is used merely to describe the number of gates connected to the output of a particular gate, such as, "the fanout of this gate is two." This means its output is connected directly to the input of two other gates.

Logic gates in general must accept one or more inputs from other gates. *Fanin* is a term for an integer which describes how many independent inputs a particular logic gate receives.

10.4
Basic Logic Gates Using Relays

Early computers used electromechanical relays to implement the logic functions. During the period from 1937 to 1945, two groups developed the first electronic computers with relays. Howard Aiken, a faculty member at Harvard, put together a proposal for an electromagnetic computing machine in 1937, but he couldn't get funding to build it until IBM agreed to let him try in 1939. Finally in 1944, thousands of relays and other components were installed in a frame 51 feet long and 8 feet high and was designated the Mark I. It was a large success and introduced IBM to electronic computing.

In 1937, George Stibitz of Bell Telephone Laboratories salvaged some discarded relays from the trash, took them home, and on his kitchen table connected relays together to test an idea. Using these relays and flashlight bulbs for output, he constructed a switching circuit that added two binary numbers together; he was excited to demonstrate that electronic on–off switching could be used for performing calculations. Bell Labs authorized him to develop a larger computing machine and in the summer of 1939, the last of over 400 relays was wired into place. Testing and debugging followed and the machine was made operational early in 1940. Stibitz was anxious to build a larger machine but Bell Labs management declined to fund it, believing that the $20,000 invested in the first computer was too much. With the entry of the United States into World War II the next year, government funding became available for additional work.

It soon became apparent that the relay was too large, power consuming, and unreliable to be used in very complex computing applications. Therefore, the relay was ultimately replaced with electronic switches—first the vacuum tube, and then the transistor. Modern digital computers often utilize millions of gates, and if one logic gate in the entire computer fails, the computer could crash, or give erroneous answers. Therefore, reliability of logic elements is an extremely important consideration.

Until recently, relays and other forms of electromechanical switching continued to be used in some of the logic systems that are used by the telephone company. These systems connect one telephone line to a specific second line identified by a telephone number dialed by the user. The older style rotary telephone dials send out a series of pulses as the dial unwinds, and the number of pulses is equal to the number dialed. These pulses were interpreted by the relay logic to close the proper contacts to connect the desired phones together. This function, known as a "cross-point switch" in modern telephone systems, is done by tone coding and electronic switching.

Relays are often used in simple logic applications such as some automobile systems, home burglar alarms, etc. They are described here because of their tutorial value, and the logic circuits presented in this section are directly analogous with those which use transistors. In the following examples we will define that a one is represented by a positive voltage, V, and a zero represented by ground or zero volts.

The NAND Gate: Using Relays

If we consider a two-input NAND gate, the truth table requires that both inputs be high for the output to be low; otherwise, the output is high. Figure 10.9 shows that each of the two inputs, when high, energizes a coil in a relay. The energized coil closes the contacts associated with that relay. Both relays must be energized, that is, both sets of contacts must be closed, before ground (zero volts) is connected to the output. Otherwise the output remains high since it is connected to +V through a resistor.

Note carefully that any load connected to B must not pull enough current through R to drop the voltage at the output too much when B is supposed to be high.

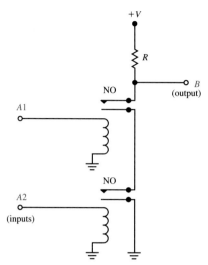

Figure 10.9   Relay implementation of a NAND gate.

The NOR Gate: Using Relays

The truth table of the two-input NOR gate requires that the output be low if either of the inputs is high. The circuit described in Fig. 10.10 accomplishes this. Since the relay contacts are connected in parallel, if either relay is energized, ground is connected to the output and the output is therefore low; the only case when the ouput is high occurs when both inputs are low—neither relay is engaged.

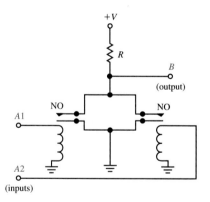

Figure 10.10   Relay implementation of a NOR gate.

10.5
Basic Logic Gates Using MOSFETS: NMOS and CMOS

The majority of today's digital electronic circuitry is implemented using MOSFET switching elements. This technology has now become more widely used than the BJT technologies which dominated during the late 1970s. Although both are still used, the MOSFET requires fewer fabrication steps to produce, and generally can operate with lower power consumption. These factors greatly influenced the rapid growth of MOS technologies.

The first commercial digital MOS-integrated circuits were fabricated in *PMOS technology*—circuits that utilized all *p-channel MOSFETs*. This rather quickly was replaced by *NMOS* or *n-channel technologies* during the later 1970s.

NMOS circuits are still utilized; however, during the early 1980s, CMOS emerged as the dominant MOS technology and remains so today. *CMOS* stands for *Complementary MOS*—circuits that require both *n*-channel and *p*-channel devices to implement each logic gate.

CMOS has one very important advantage over other technologies: when the logic is not switching, that is, quiescent, it requires no dc current to retain the information in the gate—and hence no power is used or dissipated. This is an extremely important advantage in low-power applications, such as digital watch circuitry, where the circuit must run for a long time from a small battery. It's equally important in large, complex digital circuits where the power dissipation of millions of transistors on the same chip may cause the chip temperature to rise to a level that threatens proper operation.

This section briefly presents several NMOS technologies, each with a different type of load resistance, and finally discusses CMOS technology.

NMOS with Passive Loads

The NMOS inverter, or NOT gate, with a passive or resistive load was previously shown in Fig. 9.14 and the circuit is repeated along with the voltage transfer characteristics for this gate in Fig. 10.11. This circuit utilizes an *n*-channel, enhancement-mode MOSFET as the switching element, and a resistor, R_D, as the load. The high and low output voltage levels are calculated using the MOSFET switching circuit model.

Referring to Fig. 10.11, if the input voltage to the gate is increased from zero, nothing happens to move the output from its value of V_{DD} until $V_i = V_T$. At this point the transistor begins to conduct, and the input level is V_{IL}.

As V_i is further increased, the drain current increases, and V_o begins to fall. When the transistor enters the ohmic region, the output voltage is set by the voltage division between r_{on} of the device and the load resistance R_D.

A voltage transfer curve of V_o versus V_i, similar to that shown in Fig. 10.11, can be created by mapping point by point from the output curves and loadline (as shown in section 9.6) as the Q-point moves from the OFF region to the ohmic region.

The implementation of a NAND gate using *n*-channel MOSFET switches follows exactly the same circuit topology as that used with relay switches.

The circuits for an NMOS NAND gate and NMOS NOR gate, with resistive loads, are shown in Fig. 10.12.

The NAND function requires stacking *n*-channel devices in series. For the output to be low, *both* devices must be on, and the r_{on}'s of each device are in series. These resistances in series add, and the resulting large equivalent resistance degrades the output low level. This large value of resistance must be accounted for in the gate design. Three-input NAND gates stack three transistors, and so as the number of input increases, this resistance can become a problem.

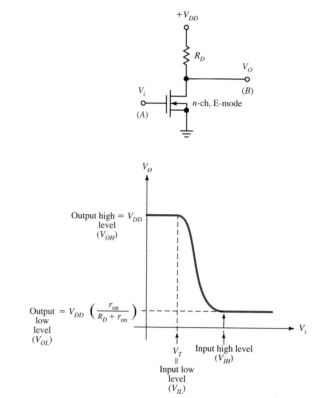

Figure 10.11   An NMOS NOT gate and its transfer characteristic.

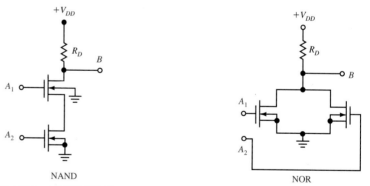

Figure 10.12(a)   NMOS NAND gate.   **Figure 10.12(b)**   NMOS NOR gate.

The NOR function has n-channel devices in parallel, and if *either* device is on, the output is low. Either device must be independently capable of pulling the output to an appropriately low level. For NOR gates with more than two inputs, additional devices are placed in parallel.

Fanout, recall, is the maximum number of gates that can be successfully driven from the output of the gate under consideration. In these MOS technologies, the input to each gate is a very high resistance, that is, the insulated gate terminal of the MOSFET. Therefore, the loading caused by additional gates is primarily capacitive, and this increased loading does not change the dc levels. Fanout limitations for MOS technologies occur primarily because of degraded switching speeds resulting from too high a load capacitance being driven with a finite transistor output current.

NMOS with Active (Enhancement-Mode) Loads

All the digital circuits described in this chapter can be constructed using the integrated circuit fabrication techniques described in Chapter 13. The fabrication of the load resistor in NMOS circuits with passive loads presents a small problem; the resistor consumes a lot of chip area. However, a MOS transistor can be biased in such a way that it electrically looks like a resistor, and as such uses less chip area than an equivalent resistor. Furthermore, it's easier to make a chip with just MOS devices than one containing both devices and resistors. Therefore, most NMOS circuitry uses active loads, or another n-channel device, properly biased, as the load resistance.

The simplest way of biasing an n-channel, enhancement-mode device as a resistive element is to connect the gate to the drain as illustrated in Fig. 10.13. The value of the resistance varies somewhat depending on the value of dc bias applied. Therefore, it's shown in the figure with an arrow through it to indicate that its value can vary. In addition, it must have at least a threshold voltage, V_T, across it before it conducts at all. These deviations from the performance of a pure resistor are not sufficient to prohibit the active load being very effectively used in NMOS logic circuits.

First, let's look in more detail at why connecting the gate to the drain creates a resistance between the terminals indicated. To illustrate this, refer to the typical output curve of the n-channel device shown in Fig. 10.14. With this connection, V_{GS} always equals V_{DS}.

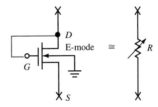

Figure 10.13 n-channel enhancement-mode device as a resistive element.

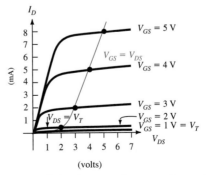

Figure 10.14   Output curves for an n-channel device.

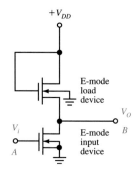

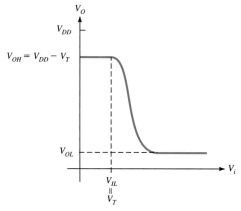

Figure 10.15   NMOS inverter with active load and its voltage transfer characteristic.

The dashed curve on the figure is the locus of all points where $V_{GS} = V_{DS}$. Notice that this curve has a positive slope for voltages above V_T, and the device corresponds to some equivalent, although variable, resistance. Using this approach, the inverter circuit and its voltage transfer characteristics appear as shown in Fig. 10.15.

Notice that the output high level, V_{OH}, is now a threshold voltage below V_{DD} because the output voltage is pulled up toward V_{DD} by the load device. When the voltage reaches a value of $V_{DD} - V_T$, the voltage across the active load is reduced to V_T and no further conduction occurs.

The output low level, V_{OL}, is again determined by the voltage division between the on-resistance of the input device, and the equivalent resistance of the load device. Recall that on-resistance can be controlled by the width to length ratio (W/L) of each MOSFET. For this type of circuit, we want the load device to have substantially higher equivalent resistance than the r_{on} of the input device. Therefore, in the layout of the inverter, the load device is typically made long and narrow (large L; small W), and the input device is made just the opposite (minimum size L; large W). The layout of such an inverter is illustrated in Fig. 10.16.

The implementation of NAND and NOR functions using active loads is analogous to the circuits with resistive loads, that is, each resistor replaced by an active enhancement-mode load as shown in Fig. 10.17. Note, an alternative MOSFET transistor symbol is used in this figure to familiarize the reader with other equivalent symbols used in the literature (refer to Fig. 9.7).

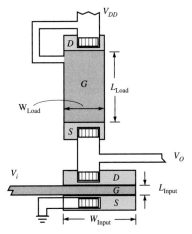

Figure 10.16   Layout of an inverter.

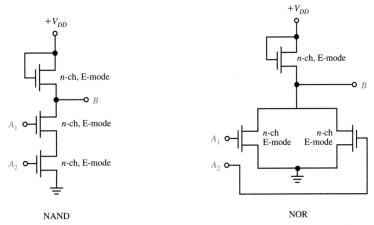

NAND

NOR

Figure 10.17   (a) NAND gate using an active enhancement-mode load; (b) NOR gate using an active enhancement-mode load. (Note: The MOSFET symbol without the body connection is often used in complex circuits for simplicity.)

EXAMPLE 10.3

The input device to an NMOS (active load) inverter has an on-resistance, $r_{on} = 100\ \Omega$, and $V_T = 1$ V. If the power supply is 5 V, let us determine (a) the output high level, V_{OH}; (b) the input low level, V_{IL}; (c) the low level noise margin, NM_L if the output low level is $V_{OL} = 0.5$ V; and (d) the effective resistance of the load device.

For an NMOS inverter circuit

(a) $V_{OH} = V_{DD} - V_T = 5 - 1 = 4$ V
(b) $V_{IL} = V_T = 1$ V
(c) $NM_L = V_{IL} - V_{OL} = 1 - 0.5 = 0.5$ V
(d) If we let r_{load} = effective resistance of load device, by voltage division

$$V_{OL} = V_{DD}\left[\frac{100\ \Omega}{100\ \Omega + r_{load}}\right]$$

Substituting into the above equation, and solving for r_{load}, we obtain

$$r_{load} = 900 \ \Omega$$

∎

NMOS with Active (Depletion-Mode) Loads

Enhancement-mode load circuitry, described in the previous section, is easier to fabricate because the load devices are made with exactly the same processing steps as the input devices. There are some advantages, however, in making the load device a depletion-mode transistor, which include improved speed and noise margin. Depletion-mode devices require additional fabrication steps to build the channel into the device, and therefore the cost to produce these parts is higher. The input device is still an n-channel enhancement-mode transistor.

Recall that a depletion-mode MOSFET with $V_{GS} = 0$ has a curve like that shown in Fig. 10.18. By connecting the gate of the depletion-mode load device directly to its source, we make $V_{GS} = 0$. Its V–I characteristics are then like those shown in the figure. It also has an equivalent resistance, the value of which depends on the dc bias or location of the bias point on this curve.

An inverter circuit utilizing this load and its corresponding voltage transfer characteristics are shown in Fig. 10.19. Notice that the output voltage can pull all the way up to V_{DD}, and that the transition is sharper than for previous circuits. These circuits operate faster because on positive output transitions, the load appears rather like a current source supplying current to the output node until right at the end when V_{DS} collapses and the load appears ohmic.

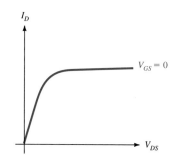

Figure 10.18 V–I curve for a depletion-mode MOSFET, $V_{GS} = 0$.

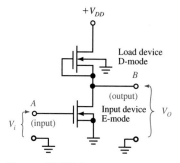

Figure 10.19(a)   Inverter circuit.

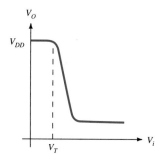

Figure 10.19(b)   Transfer characteristic.

Logic gates such as NANDs and NORs can be readily constructed using circuits similar to those used for enhancement-mode loads, if the load device is replaced with the depletion-mode load configuration.

CMOS: Complementary MOS

CMOS digital logic circuits utilize both n-channel and p-channel devices, both of which are enhancement mode devices. In CMOS logic circuits, the output voltage is either pulled high (toward V_{DD}) by one or more p-channel devices, or low (toward ground) by one or more n-channel devices. Consider a CMOS inverter as an example. The basic CMOS inverter circuit is shown in Fig. 10.20(a). The n-channel and p-channel devices are connected in series between the supply voltage V_{DD} and ground. The output is taken from the center point and hence only one of the devices is on at a time. The gates of both devices are connected together at the input. The supply voltage, V_{DD}, is often 5 V; the threshold voltages of both devices are typically about a volt or less in magnitude.

Now consider two cases. In the first case the input is high, $V_i = V_{DD}$. The switch model equivalent circuit for this case is shown in Fig. 10.20(b). The high voltage V_{DD} on the input is much larger than the threshold voltage of the n-channel device, and therefore it is on. Simultaneously, the gate and source terminals on the p-channel gate are at V_{DD} ($V_{GS} = 0$) and therefore it is off. In this situation, the output, B, is connected directly to ground through the n-channel device and V_o is zero.

Consider now the second case, when $V_i = 0$. For this case, V_{GS} for the n-channel device is zero, therefore it's off, and the p-channel device is on. The switch level equivalent circuit for this situation is shown in Fig. 10.20(c). The output is connected directly to $+V_{DD}$ through the p-channel device and thus $V_o = V_{DD}$.

In both the equivalent circuits just considered, notice that there is no dc path for current from the voltage supply source, $+V_{DD}$, to ground. Therefore, as stated earlier, current only flows in a CMOS gate during a transition from one logic state to the other.

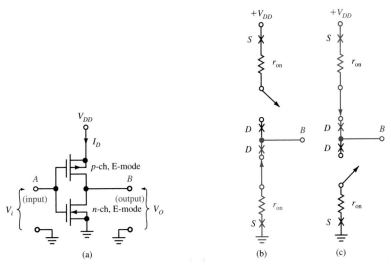

Figure 10.20 (a) CMOS inverter circuit; (b) Equivalent circuit $V_{in} = V_{DD}$; (c) Equivalent circuit $V_{in} = 0$.

Fig. 10.21 shows the voltage transfer characteristics for the CMOS inverter. It has an output high level (V_{OH}) equal to V_{DD} and an output low level (V_{OL}) at ground or zero volts. This arrangement provides maximum possible voltage swing at the output. The transition between states in a well-designed CMOS gate is steep and switches at approximately one-half of V_{DD}. These factors result in a gate with very high noise margin.

Also plotted on the previous figure is the current from the supply, I_D. It peaks near the center of the transition and, as already shown, is essentially zero at either end of the transition.

The NAND function is implemented in CMOS by stacking in series the n-channel devices and connecting in parallel the p-channel devices. Each input is connected simultaneously to one n-channel device and one p-channel device. A two-input NAND gate is illustrated in Fig. 10.22. Constructing a switch model equivalent circuit of this gate should convince the reader that it produces the NAND function at the output.

The NOR function can be produced by the opposite procedure, that is, stacking the p-channel devices in series, and placing the n-channel devices in parallel. The CMOS circuit required to implement a two-input NOR gate is shown in Fig. 10.23.

The CMOS technology also offers the possibility of implementing a special gate not easily accomplished with other technologies, the so-called *transmission gate*. The transmission gate is used as a switch that can be turned off or on to stop or pass a logic signal from one circuit to another. It is controlled by the state of a control input, labeled C.

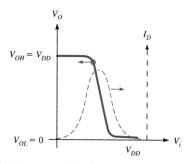

Figure 10.21   Voltage transfer characteristic for a CMOS inverter.

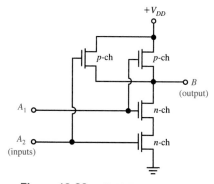

Figure 10.22   CMOS NAND gate.

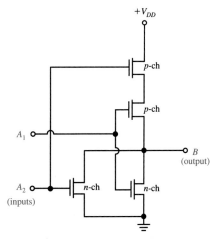

Figure 10.23 CMOS NOR gate.

The transmission gate is made by connecting an *n*-channel device in parallel with a *p*-channel device and also by constructing control logic such that both devices are simultaneously biased either off or on. Fig. 10.24 shows the circuit implementation of a transmission gate, its logic symbol, and the equivalent circuit, which is merely a switch under the control of the signal at *C*.

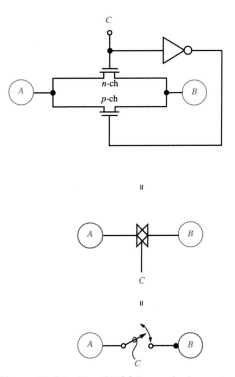

Figure 10.24 The CMOS transmission gate.

10.6

Basic Logic Gates Using BJTs: DTL, TTL, and ECL

Bipolar transistors were the first solid-state switching devices routinely used to implement digital logic circuits. The first such circuits were made in the 1950s with discrete transistors soldered together with resistors and diodes. In the early 1960s, one or two gates could be made on the same silicon chip, and the small-scale integrated circuit (see Chapter 13) became popular.

One of the bipolar logic families that emerged at this time was called DTL (*Diode-Transistor Logic*). These circuits used diodes at the input of the gate for logic operation followed by a transistor (BJT) output device for signal inversion.

The standard TTL (*Transistor-Transistor Logic*) gate was introduced about 1965 and soon replaced DTL. TTL was the principal bipolar technology for almost two decades and is still often used today. In TTL circuits, the diodes used in DTL at the gate input are replaced with a multi-emitter transistor for increased performance.

Speed and performance inprovements have been made in TTL logic over a period of years, primarily by reducing the size of the transistors and other components, which reduces the parasitic capacitances that limit switching speed. The TTL circuits used for implementing logic gates have remained virtually the same over time.

Both DTL and TTL are called *saturating* logic families, because the BJTs in the circuit are biased into the *saturated* region to achieve the effect of a closed switch. A major difficulty with saturating logic is their inherent slow switching speed. To saturate a transistor, recall both junctions must be forward biased and the base region is filled with excess carriers. In order to turn the device off, a large amount of base charge must be removed, and that takes time.

Schottky TTL was introduced about 1970 as a way of implementing TTL-like circuitry and achieving higher speed performance by preventing the devices from saturating. Schottky TTL is a nonsaturating logic family, and the transistors are never allowed to go into saturation. This is achieved by clamping the collector-base junction with a Schottky diode, so that the transistor's CB junction can never forward bias.

Another bipolar nonsaturating logic family is ECL (*Emitter Coupled Logic*). ECL is used primarily when very high speed logic circuits are required, since it is one of the fastest commercially available technologies, with propagation delay times of well under 1 ns. Unfortunately, it also uses a significant amount of power, and therefore may be impractical for extremely large logic circuits. The number of ECL gates that can be implemented on a single chip is significantly less than the number of equivalent CMOS gates. The circuits used in ECL logic use BJTs in a balanced input circuit with high gain so that small input voltage changes cause rapid output changes.

Finally, there have been a few other bipolar logic families that have been used to a lesser extent, but still may find some specialty applications. One such family is I^2L or IIL (*Integrated Injection Logic*). Invented in 1972, I^2L was a way of implementing saturating bipolar logic densely on a silicon chip using low power. Because it is a saturating logic, its speed cannot compete with modern MOS technologies; however, it does operate with a supply voltage of only one diode drop, that is, about 0.7 volts. For this reason, it has found continued use in some low-voltage applications.

We now present the implementation of the basic logic gates with the various bipolar logic families, with special emphasis on TTL because of its widespread use.

DTL: Diode-Transistor-Logic

The use of DTL logic is now rare, but understanding its operation is of great assistance in understanding TTL logic, which evolved from it. TTL has become one of the major bipolar logic families.

Consider the operation of the DTL NAND gate circuit illustrated in Fig. 10.25. Recall that the output of a NAND gate is high for every set of inputs except the case in which all inputs are high. In this latter case, the output is low. Assume the logic levels for this gate are zero volts (logic low) and five volts (logic high) and that silicon junctions can be represented by the piecewise linear model with $V_F = 0.7$. In addition, we assume the transistors have forward betas of $\beta_F = 25$. Two cases are illustrated in Fig. 10.26(a) and (b).

Case 1. Figure 10.26(a): If either or both of the inputs are grounded (logic low), then there is a current path from $+V_{CC}$ through R_B and the diodes D_1 or D_2 to ground. Node 1 is then established at 0.7 volts. In order for current to flow through $D3$ and $D4$ and into the base of $T1$, the voltage at node 1 would have to be raised to $3 \times V_F$, or $3 \times 0.7 = 2.1$ volts. Therefore, in this case, no current can flow to the base of T_1 ($I_{BT1} = 0$), and T_1 is in cutoff. With T_1, cutoff, there is no current through R_C and V_o, which is the voltage at the output B is $+5$ volts (high). The equivalent circuit for this case is shown in Fig. 10.26(a). Note that I_1 can be easily computed as

$$I_1 = \frac{V_{CC} - V_{FD2}}{R_B} = \frac{5 - 0.7}{5\,\text{k}} = 0.86\,\text{mA}$$

Case 2. Figure 10.26(b): If both inputs to this circuit are connected to $+5$ volts, then they are at a logic high. The equivalent circuit for this case is shown in Fig 10.26(b) with the transistor equivalent circuit shown for a saturated BJT. It is very important to note that BEFORE YOU ASSUME A BIPOLAR TRANSISTOR IS SATURATED, YOU SHOULD DO THE CALCULATION TO CONFIRM IT.

Diodes $D1$ and $D2$ are reverse biased, and therefore carry no current. The base current for T_1, I_{BT1}, can be calculated as:

$$I_2 = I_{BT1} = \frac{V_{CC} - V_{BE1} - V_{FD4} - V_{FD3}}{5\,\text{k}} = \frac{5 - 2.1}{5\,\text{k}} = 0.58\,\text{mA}$$

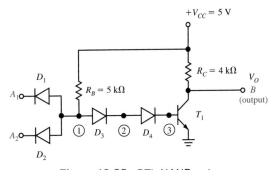

Figure 10.25   DTL NAND gate.

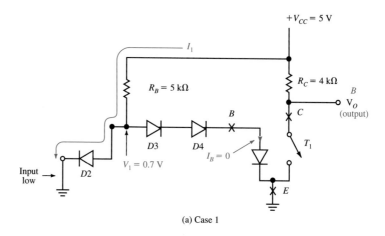

(a) Case 1

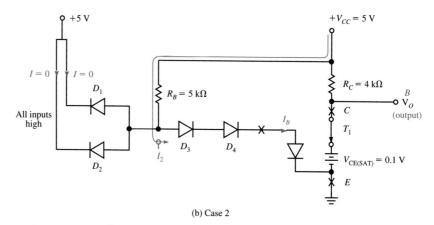

(b) Case 2

Figure 10.26   Case 1 and Case 2 for DTL NAND gate equivalent circuits.

and

$$I_{Cactual} = I_{Cmax} = I_{RC} = \frac{V_{CC} - V_{CE(SAT)}}{4\,k} = \frac{5 - 0.1}{4\,k} = 1.23 \text{ mA}$$

For saturation, I_B must be greater than $I_{B(SAT)}$ where

$$I_{B(SAT)} = \frac{I_{C\,actual}}{\beta_F} = \frac{1.23 \times 10^{-3}}{25} = 0.049 \text{ mA}$$

Notice the actual I_B forced into the base of T_1 exceeds the critical value required to saturate the transistor, that is

$$I_{BT1} > I_{B(SAT)}$$
$$0.58 \text{ mA} > 0.049 \text{ mA}$$

and therefore the transistor, T_1, is saturated, and hence the switch in the model can be legitimately closed. The output voltage at B is $V_{CE(SAT)}$, which is 0.1 volt.

EXAMPLE 10.4

Let us calculate the forced beta, β_F^* of the transistor, T_1, in the circuit of Fig. 10.26(b).

The forced beta is

$$\beta_F^* = \frac{I_{C\ actual}}{I_{B\ actual}} = \frac{I_{C\ max}}{I_2} = \frac{1.23 \times 10^{-3}}{0.58 \times 10^{-3}} = 2.12$$

$\beta_F^* \ll \beta_F$ which confirms the transistor is deeply in saturation.   ■

In calculating the maximum fanout of a gate, assume the gate under consideration is loaded with an undetermined number, n, of identical gates. Then determine the largest value for n that maintains proper operation of the gate.

EXAMPLE 10.5

Let us compute the maximum fanout of the DTL NAND gate shown in Fig. 10.25.

For the DTL gate shown, if the output is high, gates connected to this output have high inputs. This corresponds to the situation illustrated in Fig. 10.27(a). For these type gates, when the input is high, there is a reverse-biased diode in the input circuit, and the input current is zero. Therefore, adding a number of additional gates to the output in this logic state causes no problem in dc level. Each input requires negligible current and therefore the current through R_C remains approximately zero, and V_o remains high.

The other logic state, when the gate output is low, creates a more significant problem. This situation is illustrated in Fig. 10.27(b). If the output is low, each gate added to the output requires the sinking of a current I_1 to ground. The total current that must be sunk through the transistor, T_1, is

$$I_{C(TOTAL)} = I_{RC} + n\,I_1$$
$$= (1.23 \text{ mA}) + n(0.86 \text{ mA})$$

Therefore, with a fixed base drive of $I_{BT1} = 0.58$ mA, if too many gates are loaded on the output, $I_{C(TOTAL)}$ grows to the point that T_1 comes out of saturation.

At the edge of saturation

$$I_{BT1} = \frac{I_{C(TOTAL)}}{\beta_F}$$

or

$$0.58 \text{ mA} = \frac{1.23 \text{ mA} + n(0.86 \text{ mA})}{25}$$

Solving for n yields $n = 15.4$. However, in a fanout calculation, we always round down to the next whole integer, since it's impossible to have fractional numbers of gates.

Thus, n = maximum fanout = 15.   ■

TTL: Transistor Transistor Logic

The use of TTL logic has been common for nearly two decades and remains a very popular bipolar logic technology. It has become a standard for implementation of many digital systems, and it is not unusual for complex logic chips implemented in CMOS cir-

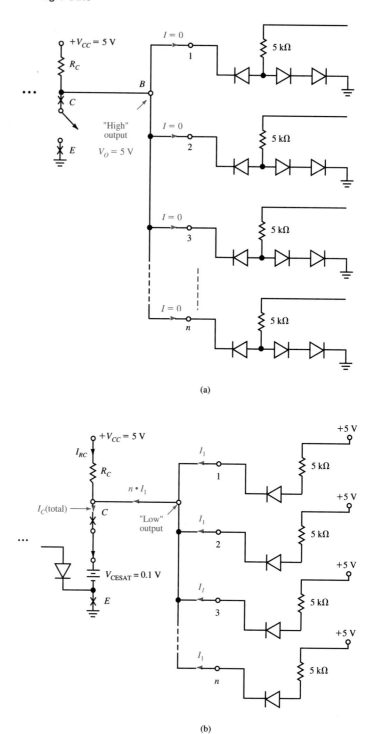

(a)

(b)

Figure 10.27   Circuits used in Example 10.5 to determine fanout.

cuitry to have special input and output circuits which mimic the logic levels of TTL at the external connections so that these chips can be directly interfaced to other TTL chips.

The basic circuitry for TTL logic was developed directly from DTL. The primary change is in the replacement of the diode input circuit that appears in DTL with a multi-emitter, NPN transistor. This transistor structure actually contains the same diode elements as in DTL, but the addition of transistor action improves the circuit's performance.

Recall that the input section of a two-input DTL NAND gate appears as in the Fig. 10.28(a). This network can be replaced with the transistor structure shown in Fig. 10.28(b) in which the p-regions of all the diodes are shared by the same p-type base region of the transistor. This multi-emitter transistor requires less silicon area to lay out than the three individual diodes, and provides faster circuit operation.

The operation of the basic TTL NAND gate, shown in Fig. 10.29, is similar to the DTL NAND gate discussed earlier. If either or both inputs are pulled low, that is, to zero volts, there is a current path through R_B and a base-emitter junction of Q_1 to the grounded input. This places node 1 at 0.7 volts. As was the case with DTL, there is a series connection of three diodes in the base circuit of Q_2: the collector junction of Q_1, diode D_1, and the base-emitter junction of Q_2. Therefore, Q_2 is off, V_o is high and the 0.7 volts at node 1 is insufficient to forward bias 3 diodes in series.

For Q_2 to turn on, the voltage at node 1 must rise to 2.1 volts ($3 \times V_F$). The voltage at node 1 tracks the *lowest* input voltage (A_1 or A_2) at a level of V_F higher than the input voltage. Therefore, node 1 reaches 2.1 volts when V_{i1} or V_{i2} rises to 1.4 volts. Thus, $V_i = 1.4$ volts is considered to be the input low level, V_{IL}, for TTL technology. For a logic gate to operate at high speed, especially if fanout is large, it must be able to rapidly charge and discharge the capacitance of the input circuits of subsequent gates attached to its output. This implies that the output circuits should readily supply or source current to raise the output voltage as required on a positive transition, and be able to sink a large current to reduce the output voltage in a negative transition. The current of Fig.10.29 does an excellent job of sinking load current when the transistor T_2 saturates. The on resistance of

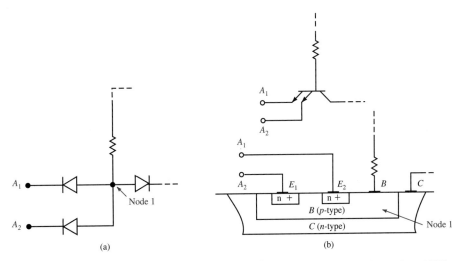

Figure 10.28 Input section of (a) two-input DTL NAND gate, and (b) two-input TTL

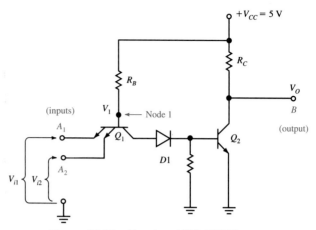

Figure 10.29   Two-input TTL NAND gate.

a saturated transistor is generally only a few ohms, and recall in the BJT switch model, we assumed it was actually zero, since there was no r_{on} placed in a series with the switch.

This circuit, however, does not perform well for positive transitions. Current to the output is supplied only by the resistor, R_C. Resistive pull-up is a relatively ineffective way of supplying positive load current. It would be better to provide a transistor from $+V_{CC}$ to the output which supplied current for positive transitions.

Changing the output circuit to a so-called "totem pole" output provides this current as shown in Fig. 10.30. This schematic of a modern two-input TTL NAND gate is the actual circuit implemented on silicon chips by many manufacturers.

To understand the operation of this circuit, consider the two cases considered previously for the DTL gate.

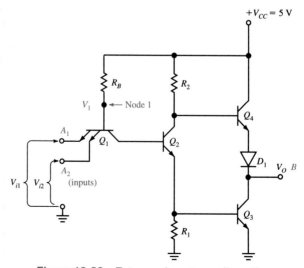

Figure 10.30   Totem pole gate configuration.

Case 1. If either or both of the inputs are grounded (logic low), then the voltage at node 1 is 0.7 volts. From node 1, there are still three silicon junctions in series: the collector base of Q_1, the emitter base of Q_2, and the base emitter junction of Q_3. Current cannot flow in this string unless node 1 reaches 2.1 volts. Therefore, $I_{BQ2} = 0$ and $I_{BQ3} = 0$ and thus, both Q_2 and Q_3 are off. Under these conditions, the output circuit reduces to that shown in Fig. 10.31. Note that current is supplied to the output load resistance and capacitance (R_L and C_L) by the transistor Q_4, and the voltage at the output, V_o, rises to the high level, V_{OH}, that is,

$$V_{OH} = V_{CC} - I_{BQ4}\, R_2 - V_{BEQ4} - V_{FD1}$$

The transistor $Q4$ remains in the active region because its collector base junction can never be forward biased. The collector is at V_{CC} and therefore, if the load current is not excessive and β_F is large

$$I_{BQ4} = \frac{I_{CQ4}}{\beta_F}$$

and the term $I_{BQ4}\, R_2$ is small.
Thus

$$V_{OH} \approx V_{CC} - V_{BEQ4} - V_{FD1}$$
$$V_{OH} \approx 5 - 0.7 - 0.7 = 3.6 \text{ V.}$$

Case 2. If both inputs to this circuit are connected high, then node 1 rises to 2.1 volts and current through R_B flows into the base of Q_2, saturating Q_2. Current also flows into the base of Q_3, saturating Q_3. Hence, the output voltage is low at a value of $V_o = V_{CE(SAT)} \approx 0.1$ V.

For this case, we would like to ensure that Q_4 is off, so that the gate is not simultaneously trying to pull the output voltage both down and up. Note that when Q_3 is on, the voltage on the base of Q_3 is $V_{FQ3} = 0.7$ V. If Q_2 is saturated and $V_{CE(SAT)} = 0.1$ V, then the voltage on the base of Q_4 is $V_{BQ4} = 0.7 + 0.1 \approx 0.8$ volts.

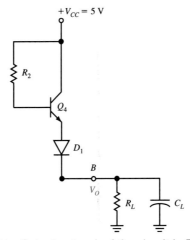

Figure 10.31   Output network of the circuit in Figure 10.30.

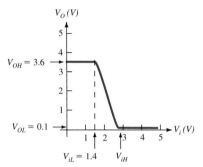

Figure 10.32 TTL gate transfer characteristic.

The output voltage is 0.1 volts. Thus, if diode $D1$ were not in the circuit, Q_4 would turn on. *But* diode $D1$ has been inserted in the circuit to add another $V_F = 0.7$ V in series with the emitter junction of Q_4. With $D1$ in the circuit, if $V_o = 0.1$ volts, the base of Q_4 must reach 1.5 volts before Q_4 is turned on, and this can't happen while Q_2 is saturated.

Thus, *either* Q_3 is on, Q_4 is off, and the output is low, *or* when Q_3 is off, Q_4 conducts current to the output. From this analysis of the TTL gate shown in Fig. 10.30, a voltage transfer plot can be developed and is shown in Fig. 10.32. Note that all the key break points are determined except V_{IH}, which would require a more thorough analysis of the circuit.

Schottky TTL—A Nonsaturating Bipolar Logic Family

Schottky TTL and so-called "low-power" Schottky TTL circuits are families of integrated circuit logic chips which have evolved from the original TTL circuits. The circuits utilized are almost identical to those discussed in the previous section on TTL. The primary difference is the addition of Schottky diodes placed in parallel with the base collector junctions of the transistors.

Recall Schottky diodes are rectifying junctions fabricated by joining a metal with a semiconductor material. The Schottky diode has a symbol shown in Fig. 10.33(a). By the proper choice of metal, these diodes can be made to turn on or conduct current at forward bias voltages in the range of 0.3 volts. This turn-on voltage is relatively small $(V_F \approx 0.3$ V), which is significantly less than the turn-on voltage of a silicon junction $(V_F = 0.7$ V). Therefore, a transistor with a Schottky diode attached, as shown in Fig. 10.33(b), can never have its base collector junction forward biased. This arrangement

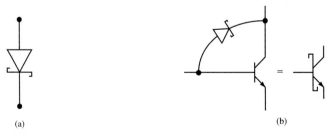

(a)   (b)

Figure 10.33 (a) Schottky diode; (b) Schottky transistor.

makes it impossible for the transistor to enter the saturation region. The transistor fabricated with this Schottky diode is called a Schottky transistor.

There are only a few additional fabrication steps required to add this diode to the chip, and therefore, such devices can be fabricated for very little additional cost. The improved performance of circuits which employ these devices is derived primarily from the increased switching speed of the transistors which are not allowed to saturate. Recall that a saturated transistor, which is modeled as a closed switch, stores a lot of charge in its base region. A relatively long time, usually measured in nanoseconds, is required to remove this charge and to switch an on transistor to the off condition. Therefore, saturating logic, in general, operates at a slower speed than comparable nonsaturating logic.

In summary, each Schottky transistor comes close to saturation, and conducts current like an on switch, but the transistor is prevented from entering saturation by the conduction of the Schottky diode. Thus, the process of switching the transistor off proceeds much more quickly.

A typical Schottky TTL NAND gate from the 54S/74S logic family is shown in Fig. 10.34.

ECL—Emitter-Coupled Logic

Another bipolar nonsaturating logic family is ECL, or Emitter-Coupled Logic. This logic is often used when extremely fast logic is required. ECL gates can have propagation delay times of less than a nanosecond and clock rates that extend up to hundreds of megahertz. The primary disadvantage of ECL logic is that these gates are power hungry and they use significantly more power per gate than most other technologies. This limits their

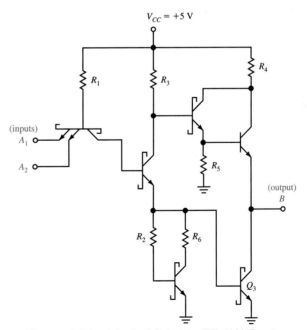

Figure 10.34   A typical Schottky TTL NAND gate.

general application in several ways. First, not as many gates can be put on a single chip as with other technologies because of the build-up of heat in the chip. Thermal managment is critical in ECL designs. Second, with reduced gate counts per chip, it takes more chips to implement a system, which means larger boards and a larger and more costly system. There are applications where the speed advantage is worth this additional cost, but generally these are special situations.

The basis of all ECL circuits is the emitter-coupled pair or as it is sometimes called, the nonsaturating current switch. This circuit configuration is basically symmetric, with two inputs and two possible outputs as illustrated in Fig. 10.35. Note that the emitters of the NPN transistors are connected together at one of the critical nodes in the circuit. Current is pulled from this node to ground by a resistor, R_E. The current comes down to this node through EITHER transistor Q_1 or Q_2, depending on which one is conducting, since only one will be conducting at a time. *The transistor with the higher base voltage will be the one conducting.* In a properly designed ECL gate, the conducting transistor remains in the active region in order to maintain the speed advantage, and the off transistor is cut off; neither device ever saturates. An equivalent circuit can be constructed by substituting the model for the bipolar transistor in the active region (from Chapter 9), resulting in the equivalent circuit of Fig. 10.36 for the case where $V_{i1} > V_{i2}$. In general, the voltage at the critical node at the top of R_E will be controlled by the higher of the V_{i1} or V_{i2}. This voltage will be one diode voltage drop, that is, $V_{BE} = 0.7$ volts, below the higher of the two input voltages. Notice that the other base emitter diode, that is, the one attached to the lower input voltage, will then have less than 0.7 volts across it; this results in zero base current and therefore zero collector current, and hence the device will be off. This transistor appears as an open switch. Therefore, there are two possible cases:

case 1:   If $V_{i1} > V_{i2}$, then Q_1 is conducting; Q_2 is off.

case 2:   If $V_{i1} < V_{i2}$, then Q_1 is off; Q_2 is conducting.

The equivalent circuit for case 1 was previously shown; the equivalent circuit for case 2 is similar, but with the transistor models exchanged.

As shown in Fig. 10.36, for case 1, the output voltage, V_{o1}, is pulled low, by the voltage drop from current through R_{C1}. This current can be calculated by computing the tran-

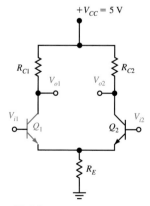

Figure 10.35   An emitter-coupled pair.

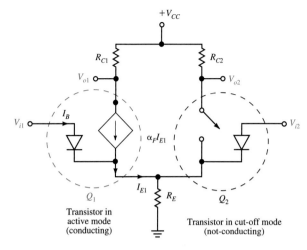

Figure 10.36 Equivalent circuit used with the network in Figure 10.35.

sistor collector current, which is $\alpha_F I_E$. The current I_E is set by the voltage at the top of R_E, that is, $V_{i1} - 0.7$, which is the voltage across R_E. On the right side, $I_{E2} = 0$. The voltage V_{o2} rises to the supply voltage, V_{CC}, because there is no current through R_{C2}, and hence no voltage drop across it.

Therefore, the current through R_E is given by:

$$I_{RE} = \frac{[V_{i1} - 0.7]}{R_E} = I_E$$

and

$$I_{C1} = \alpha_F I_E$$

Therefore

$$V_{o1} = V_{CC} - I_{C1}R_{C1} = V_{CC} - \alpha_F I_E R_{C1}$$

For case 2, just the opposite situation occurs, with V_{o1} at a voltage of V_{CC}, and V_{o2} pulled lower.

Therefore, the emitter-coupled pair acts as a current switch, steering or directing the current by providing a path for it to come down the left or the right side of the circuit depending on the relative values of the input voltages, V_{i1} and V_{i2}.

As will be seen in a later chapter, the emitter-coupled pair is also an essential circuit configuration for analog differential amplifiers.

In ECL logic applications, one input of the emitter-coupled pair is permanently connected internally to a voltage reference, that is, a constant voltage derived from a special circuit on the chip. The other input is the logic input to the circuit and this voltage is compared to the reference voltage, and the state of the current switch is determined by whether the input voltage is above or below the reference voltage. This is illustrated in the following example.

EXAMPLE 10.6

If the silicon transistors in the ECL circuit shown in Fig. 10.37 have betas of 100, let us plot the output voltages, V_{o1} and V_{o2} as a function of the input voltage, V_{i1}, as it is swept from 0 to +5 volts.

For $V_{i1} < 3$ volts, Q_1 is open and therefore there is no current flow through the resistor, R_{C1}, and V_{o1} is +5 volts. The voltage V_{o2} is determined by replacing Q_2 with the model for the transistor in the active region. The equivalent circuit for the right side of the circuit becomes that shown in Fig. 10.38.

The voltage at the critical node "A" can be determined by applying KVL to the input loop and assuming the silicon transistor has a junction turn-on voltage of 0.7 volts.

The voltage at node $A = 3 - 0.7$ V $= 2.3$ V. Therefore

$$I_E = \frac{2.3}{1\text{ k}} = 2.3 \text{ mA}$$

$$\alpha_F = \frac{\beta_F}{\beta_F + 1} = \frac{100}{101} = 0.99$$

$$I_C = \alpha_F I_E = (0.99)(2.3 \times 10^{-3}) \approx 2.3 \text{ mA}$$

and

$$V_{o2} = 5 - I_C R_{C2} = 5 - [2.3 \times 10^{-3} \times 10^3] = 2.7 \text{ V}$$

For $V_{i1} > 3$ volts, the voltages at V_{o1} and V_{o2} exchange values, as Q_1 turns on and Q_2 goes into cutoff.

Therefore, the output voltages cross at the point where the input voltage, V_{i1}, reaches 3 volts, as shown in Fig. 10.39.   ∎

The ECL gate, because of its circuit symmetry, easily provides two outputs, one of which is the complement of the other. Therefore, the logic symbol used for ECL gates often shows complementary outputs, as indicated by the symbol for an ECL inverter shown in Fig. 10.40.

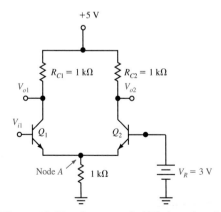

Figure 10.37   An example ECL inverter circuit.

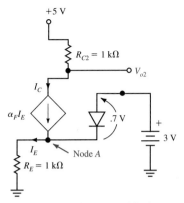

Figure 10.38   An equivalent circuit used with the network in Figure 10.37.

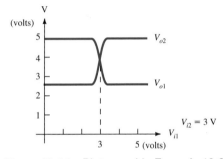

Figure 10.39   Plots used in Example 10.6.

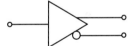

Figure 10.40   ECL inverter symbol.

Output Level Shifting

In the previous example, the output low level, V_{o1}, is calculated to be 2.7 volts. This is lower than the switching threshold of the gate (3 volts), which is set, in this circuit, by the reference voltage, and does not allow a great deal of margin. The low level output of the gate must fall below 3 volts for all conditions in order to ensure that a similar gate connected to the output of this one would always perceive the logic level to be low.

In order to improve the noise margin of ECL circuits, a level shifter stage is usually added to the basic ECL circuit previously shown. This addition is a bipolar transistor connected in a common collector configuration, that is, an emitter follower, which basically serves to shift the output voltage down by the voltage drop of one silicon diode and provide additional current gain.

EXAMPLE 10.7

We wish to calculate the logic levels and noise margins of the ECL circuit, with level shifter, shown in Fig. 10.41. Assume the β_F's of the transistors are large.

This circuit is typical of a very popular ECL family, 10K ECL. In this logic family the common convention is to establish ground, that is, zero volt reference, at the top and use a supply voltage of -5.2 V at the bottom node in the circuit.

If $V_i < V_{Ref}$, then Q_1 is off and Q_2 is conducting. The voltage at node A is a diode drop below V_{ref} or

$$V_A = -1.3 \text{ V} - 0.7 \text{ V} = -2.0 \text{ V}$$

Therefore, the current through R_E is

$$I_{RE} = \frac{[-2.0 - (-5.2)]}{780} = 4.1 \text{ mA}$$

Since $I_{E1} = 0$, then $I_{E2} = I_{RE} = 4.1$ mA.

Assuming the β_F's are large, $I_{C2} \sim I_{E2} = 4.1$ mA, and I_{B3} is negligibly small. Therefore, the current through R_{C2} is approximately I_{C2} and

$$V_{C2} = 0 - I_{C2} R_{C2} = 0 - (4.1 \times 10^{-3})(220) = -0.90 \text{ V}$$
$$V_o = V_{C2} - 0.70 = -0.90 - 0.70 = -1.60 \text{ V}$$

or

$$V_{OL} = -1.60 \text{ V}$$

When $V_i > V_{Ref}$, Q_2 is off, and since there is negligible current through R_{C2}, $V_{C2} \approx 0$, and

$$V_o = 0 - 0.70 = -0.70 \text{ V}$$

or

$$V_{OH} = -0.70 \text{ V}$$

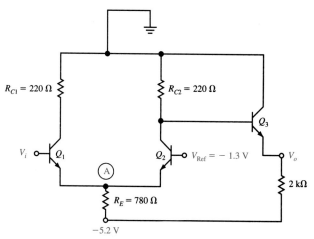

Figure 10.41 Circuit used in Example 10.7.

If input switching occurs exactly at V_{Ref}, then

$$V_{IH} = V_{IL} = -1.3 \text{ V}$$

Therefore, the low level noise margin, NM_L, and the high level noise margin, NM_H, are given by

$$NM_L = V_{IL} - V_{OL} = -1.30 - (-1.60) = 0.30 \text{ V}$$

and

$$NM_H = V_{OH} - V_{IH} = -0.70 - (-1.30) = 0.60 \text{ V}$$ ■

EXAMPLE 10.8

Let us perform a similar calculation to that shown in Example 10.7 to determine I_{RC2}; however, assume the transistors have a $\beta_F = 50$.

The current through the resistor R_{C2} is given by

$$I_{RC2} = I_{B3} + I_{C2}$$

where

$$I_{C2} = \alpha_F I_{E2} = \left[\frac{\beta_F}{1 + \beta_F} \right] I_{E2} = \frac{50}{51} \, 4.1 \text{ mA} = 4.02 \text{ mA}$$

Furthermore

$$I_{B3} = \left[\frac{1}{\beta_F + 1} \right] I_{E3} = \left[\frac{1}{\beta_F + 1} \right] \left[\frac{V_o + 5.2}{2 \text{ k}} \right]$$

and

$$V_o = V_{C2} - 0.70 = [0 - I_{RC2} R_{C2}] - 0.70 \text{ V}$$

Combining the above equations to solve for I_{RC2} yields a value of 4.06 mA. Note that this is a change of less than 1% from the value calculated in the previous example assuming the betas were infinitely large.   ■

10.7
Summary

- The building blocks of digital systems are the basic logic gates: AND, OR, and NOT (or NAND and NOR) gates.
- The number of input combinations, Z, for n inputs is:

$$Z = 2^n$$

- A truth table describes the state of the output of a logic circuit for a given set of inputs, usually all possible input combinations.
- There are a set of standard symbols used to represent each of the basic gates.
- Real logic gates require finite time for switching. Propagation delay describes the time delay a signal encounters in passing through a gate. Rise time and fall time describe the time required for the output to transition from one logic state to the other.

- A race condition is an abnormal situation in which one signal arrives at its intended destination delayed enough relative to another signal so as to cause an error; this delay is generally a result of accumulated circuit propagation delays.
- Real logic gates are designed to unambiguously recognize input voltage levels as a high or a low, specified by V_{IH} and V_{IL} respectively; they produce output levels specified by V_{OH} and V_{OL}. Noise margins are defined that indicate the magnitude of tolerance the design has for incorrectly recognizing a bit (see Fig. 10.7).
- Fanout is an integer that describes the maximum number of gates that can be connected to the output of a similar gate and ensure proper operation.
- Relays can be used to illustrate basic logic circuit functions.
- NMOS (circuits using n-channel enhancement-mode transistors) circuits are analogous to those introduced with relays, and can have resistive loads or active loads; an active load is the use of a transistor biased in a way that it can be used as a load resistor.
- Active loads can use an enhancement-mode transistor (enhancement-mode load) or a depletion-mode transistor (depletion-mode load). The voltage transfer characteristics differ depending on load type; a depletion-mode load allows the output voltage to swing all the way to the supply voltage.
- CMOS (Complementary MOS) logic circuits use a combination of n-channel and p-channel enhancement-mode MOS transistors in each gate. CMOS circuits only conduct dc current during switching; thus, they use less power on average per logic function and are used in most high-density integrated circuits.
- BJTs can be used to implement logic circuits; the NPN transistor is dominant because an NPN device performs better than an equivalent PNP. (This is primarily because electron mobility is higher than hole mobility in silicon.)
- DTL (diode-transistor-logic) is essentially a diode logic gate similar to that illustrated in Chapter 8, followed by a transistor inverter.
- TTL (transistor-transistor-logic) circuits replace the diodes in DTL with an integrated multi-emitter transistor. TTL is a very popular bipolar logic family and has been common for several decades.
- DTL and TTL circuits use BJTs which are allowed to saturate. Saturated BJTs store significant charge, and switch slowly. Schottky TTL is similar to TTL but with the addition of Schottky diodes to keep all transistors out of the saturation region and speed the operation.
- ECL (emitter-coupled-logic) uses BJTs which remain biased in the active region in both logic states. These circuits consume more power, are less dense, but are extremely fast.

PROBLEMS

10.1. (a) A logic gate has three input terminals. What is the number of possible input combinations to this logic circuit? (b) How does this change if one more input is added to the circuit?

10.2. (a) Determine the truth table for a three-input AND gate and the truth table for a three-input OR gate. (b) Determine the truth table for a logic circuit that is created by connecting an inverter to each input of the two-input AND gate as shown in Fig. P10.2(a). (c) Repeat (b) for the OR gate as shown in Fig. P10.2(b).

a NAND gate, as shown in Fig. P10.6(a). (b) Repeat (a) for a NOR gate as shown in Fig. P10.6(b).

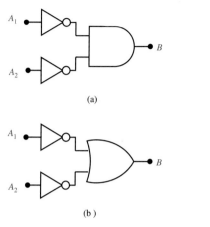

(a)

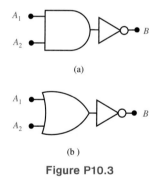

(b)

Figure P10.2

10.3. (a) Determine the truth table for the logic circuit formed by connecting the output of an AND gate to a NOT gate as shown in Fig. P10.3(a). (b) Repeat (a) for an OR gate as shown in Fig. P10.3(b). (c) Compare these results with those obtained in problem 10.2.

(a)

(b)

Figure P10.3

10.4. (a) Construct the truth table of a two-input NAND gate. (b) Compare this with the results obtained in problems 10.2 and 10.3.

10.5. (a) Construct the truth table for a two-input NOR gate. (b) Compare this with the results in problems 10.2 and 10.3.

10.6. (a) Construct the truth table for the circuit obtained by connecting together as a single input the two inputs of

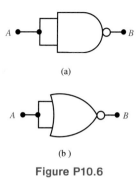

(a)

(b)

Figure P10.6

10.7. Construct the truth table for a three-input XOR-gate.

10.8. (a) Construct a two-input OR gate using a combination of two-input NAND gates. (b) Construct a two-input AND gate using a combination of two-input NOR gates.

10.9. The input and output waveforms of an inverter are shown in Fig. P10.9. Obtain the values of t_{PHL}, t_{PLH}, and t_P for the inverter.

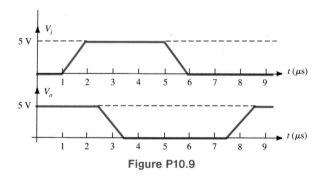

Figure P10.9

10.10. If two inverters identical to the one in problem 10.9 are cascaded, what are the values of t_{PLH}, t_{PHL}, and t_P for the combined circuit?

10.11. (a) Determine the rise and fall time of the signals shown in Fig. P10.11(a). (b) Repeat (a) for Fig. P10.11(b).

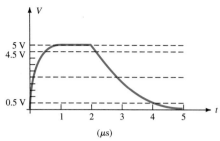

Figure P10.11(a)

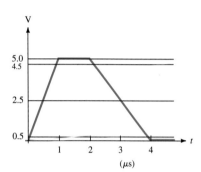

Figure P10.11(b)

10.12. A digital circuit has a clock frequency of 25 MHz. The circuit uses gates with propogation delays $t_{PHL} = t_{PLH} = t_P = 2$ ns. If data are presented at the input of a cascaded string of these gates at the leading edge of each clock cycle, what is the maximum number of gates that can be connected in series?

10.13. Obtain the noise margins NM_L and NM_H for the gate whose transfer characteristic is shown in Fig. P10.13.

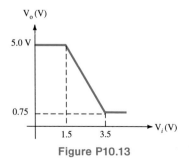

Figure P10.13

10.14. Calculate the noise margin, transition width, and logic swing of the gate whose input and output levels are shown in Fig. P10.14.

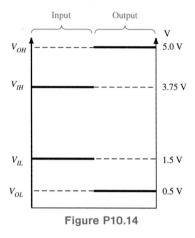

Figure P10.14

10.15. An NMOS inverter with a resistive load of $R_D = 1$ kΩ is shown in Fig. P10.15. If $V_T = 1.0$ V and $r_{on} = 100$ Ω in the ohmic region of MOSFET operation, obtain the values of V_{OH}, V_{OL}, V_{IL}, V_{IH}, NM_L, and NM_H for the inverter. Assume $V_{IL} = V_{IH}$.

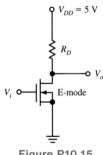

Figure P10.15

10.16. (a) Construct a circuit drawing using *n*-channel MOSFETs and resistive loads for a two-input AND gate. (b) Repeat (a) for a two-input OR gate.

10.17. (a) Design a three-input NAND gate with passive loads; show the circuit drawing including all transistors and other elements. (b) Repeat (a) for a three-input NOR gate.

10.18. An NMOS inverter with an enhancement-mode active load is shown in Fig. P10.18. (a) If $V_T = 0.8$ V for both devices, determine V_{OH} and V_{IL}. (b) If the resistance of the load device is 1.5 kΩ, what value of r_{on} for the input device is required to obtain $V_{OL} = .3$ V? (c) What value of K for the load device, K_L, is required in (b)?

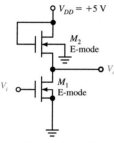

Figure P10.18

10.19. For the inverter circuit in problem 10.18, $V_T = 0.9$ V and V_{OL} is determined to be 0.2 V. If the value of r_{on} for the input transistor is 120 Ω, find the values of NM_L and the effective resistance of the load device.

10.20. For the NMOS inverter circuit given in Fig. P10.18, sketch the voltage transfer curve, V_o versus V_i. (Assume $V_T = 0.8$ V and $V_{IH} = 2.1$ V).

10.21. (a) Draw the circuit diagram of a three-input NAND gate with enhancement-mode active loads and NMOS transistors. (b) Repeat (a) for a three-input NOR gate.

10.22. The output curves for the NMOS input transistor of an enhancement-mode active load inverter are shown in Fig. P10.22; the inverter has $V_{DD} = 5$ V and the transistors have $V_T = 1.2$ V. The load device for this inverter has W/L 20% of that of the input device (and thus the value of K is 20% of that of the input device). (a) Sketch an approximate set of output curves for the load device. (b) On the plot from (a) sketch the V–I curve of the load device, that is, the points where $V_{DS} = V_{GS}$. (c) Use the results of (b) to plot a load line on the output curves of the input device. (d) Approximate the voltage transfer curve for this inverter.

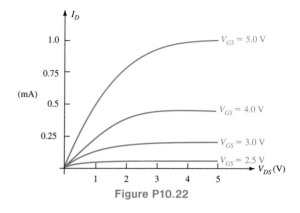

Figure P10.22

10.23. Draw a circuit diagram for each of the following gates: (a) A three-input NAND gate using a depletion-mode load (b) A two-input NOR gate using a depletion-mode load (c) A two-input AND gate using a depletion-mode load.

10.24. The load transistor of the inverter in problem 10.22 is replaced by a depletion-mode transistor. For this case, determine V_{OH} and V_{IL}.

10.25. If the depletion-mode load transistor in problem 10.24 has an output curve for $V_{GS} = 0$ shown in Fig. P10.25, use this to draw the load line and plot the approximate voltage transfer characteristics for this inverter. The input transistor's characteristics are given in Fig. P10.22.

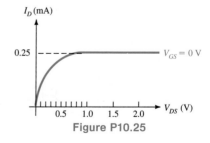

Figure P10.25

10.26. The voltage transfer characteristics of an inverter are shown in Fig. P10.26. (a) Draw the transfer characteristics of a circuit obtained by cascading two of these inverters. (b) Compare the noise margins of the cascade and a single inverter.

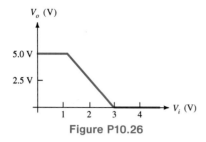

Figure P10.26

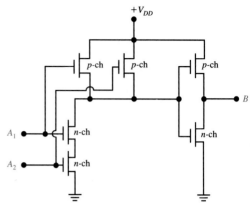

Figure P10.29

10.27. (a) Draw the circuit diagram of a three-input CMOS NAND gate. (b) Repeat (a) for a three-input CMOS NOR gate.

10.28. (a) Construct the truth table for the circuit shown in Fig. P10.28(a). (b) Construct the truth table of the logic circuit shown in Fig. P10.28(b). (c) Compare the results in (a) and (b).

10.30. Compare and contrast the TTL and ECL logic families on the basis of the following criteria:
(a) logic levels
(b) switching speed
(c) power consumption

10.31. Compare and contrast the performance of TTL bipolar logic with CMOS logic based on the following criteria:
(a) logic levels
(b) switching speeds
(c) circuit density
(d) power consumption

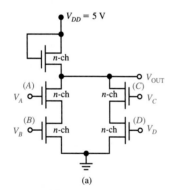

(a)

Figure P10.28a

10.32. For the NPN transistor circuit shown in Fig. P10.32, assume $V_{CE(SAT)} = 0.2$ V, $V_{BE} = 0.7$ V, $\beta_F = 50$. (a) Determine the minimum value of V_i that saturates the transistor. (b) Determine the fanout when the output is high.

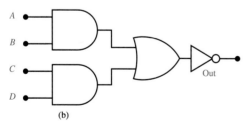

(b)

Figure P10.28b

10.29. Construct the truth table for the CMOS logic circuit given in Fig. 10.29.

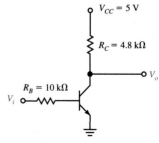

Figure P10.32

10.33. Figure P10.33 shows a two-input DTL NAND gate. For the diodes $V_F = 0.7$ V. For the BJT, assume $V_{CE(SAT)} = 0.1$ V, $V_{BE} = 0.7$ V, $\beta_F = 80$. Find (a) V_o if $V_A = V_B = 0$ V; (b) V_o if $V_A = V_B = 5$ V; (c) the forced beta, β_F^*, of the BJT in saturation

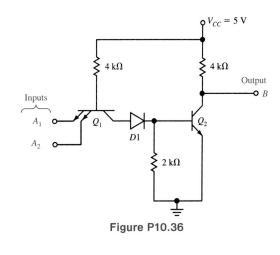

Figure P10.36

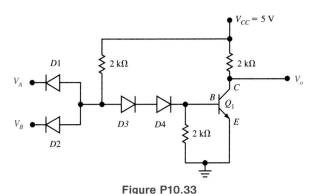

Figure P10.33

10.34. In the circuit of Fig. P10.33, Q_1 has $V_{BE} = 0.7$ V, $\beta_F = 80$ and $\beta_R = 3$. (a) Determine the actual value of V_{OL} using Eq. 9.22. (b) Determine the fanout of this gate.

10.35. (a) Construct the truth table for the circuit shown in Fig. P10.35. (b) Determine V_{IL}. (c) Determine V_{OH}.

10.37. The transistor in the circuit in Fig. P10.37 has $\beta_F = 50$, and $V_{BE} = 0.6$ V. Determine V_o, the collector, base, and diode currents for $V_i = 4$ V. Assume a Schottky turn-on voltage $V_F = 0.3$ V.

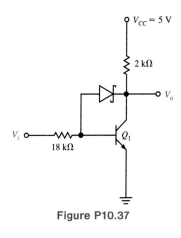

Figure P10.37

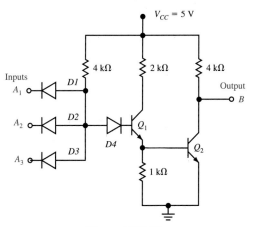

Figure P10.35

10.36. The transistors in Fig. P10.36 have $V_{BE} = 0.6$ V, $\beta_F = 120$, $\beta_R = 2$; the diode turn-on voltage is $V_F = 0.6$ V. (a) Calculate V_{IL}, V_{OL}, and V_{OH}. (b) Construct a truth table for this gate.

10.38. Figure P10.38 shows a two-input TTL NAND gate. The transistors are identical with high betas, $V_{BE} = 0.7$ V and assume $V_{CE(SAT)} = 0.25$ V; for the diode, assume $V_F = 0.6$ V. Calculate V_{OL}, V_{OH}, and V_{IL} for this circuit. Sketch the transfer characteristics of the circuit.

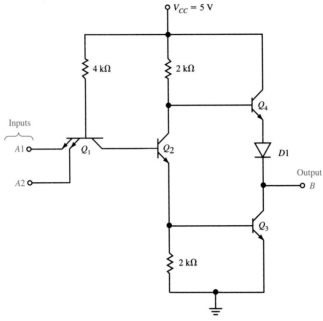

Figure P10.38

10.39. For the ECL circuit shown in Fig. P10.39, plot the output voltages V_{o1} and V_{o2} as a function of the input voltage V_i as it is swept from 0 to 5 volts. Assume $\beta_F = 125$ and $V_{BE} = 0.6$ V for the transistors.

10.40. Calculate the logic levels and noise margins of the ECL circuit with level shifter shown in Fig. P10.40. Assume that the β_F's of the transistors are large, $V_{BE} = 0.7$ V, and that $V_{IH} = V_{IL}$.

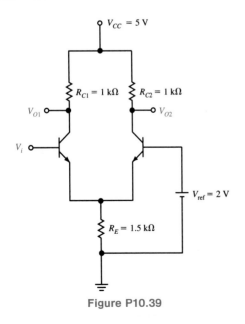

Figure P10.39

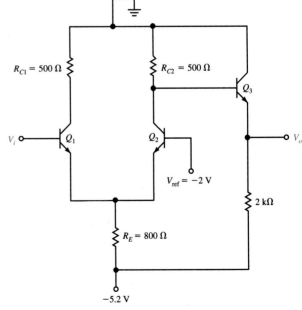

Figure P10.40

Digital Logic Circuits

11.1 Introduction

The world is going digital! As strange as that may sound, there is a great deal of truth to that statement. Today, digital technology permeates a very large segment of our lives. Essentially all long-distance communication is digital. Processing within and among computers is all done digitally. Many modern control systems employ a computer, and instruments of all types, for example, voltmeters, oscilloscopes, and the like, use digital circuits and digital readouts. In fact, if the trend toward digital watches continues, the terms *clockwise* and *counterclockwise* may not be generally understood.

Digital circuits operate in one of two discrete states: the logic 1 state or the logic 0 state. These two states are generated by the electronics. In one case the transistor circuit's output voltage is high and in the other case the output voltage is low. Furthermore, these two cases correspond to a transistor being turned off or turned on. As we have seen earlier, initially relays were used as the basic building blocks; however, at the present time, the electronic switching devices are very small, very fast, and are crammed into VLSI chips by the millions.

11.2 Number Systems

The number system that we typically use is the decimal system. This system employs ten digits and is certainly convenient for us as humans since we have ten fingers. The total number of digits allowed in a number system is called the *base* or *radix* of the system, and thus the decimal system has a base of 10.

Table 11.1 Relationship
Between Base 10 and
Base 1 Number Systems

Base 10	Base 2
0	0
1	1
2	10
3	11
4	100
5	101
6	110
7	111
8	1000
9	1001
10	1010
11	1011
12	1100
13	1101
14	1110
15	1111

Table 11.2

+	0	1		×	0	1
0	0	1		0	0	0
1	1	10		1	0	1

In digital logic circuits the output is either a 0 or a 1. Therefore, it is most conventient to use the binary number system which uses the binary digits or bits 0 and 1 and has a base of 2. A short table illustrating the relationship between decimal and binary numbers is shown in Table 11.1.

As children, we learned to add and multiply in base 10 by memorizing addition and multiplication tables. Since the decimal number system has 10 digits, we memorized a 10-by-10 array of numbers for both addition and multiplication. The binary number system however contains only two digits, and therefore binary arithmetic is much simpler. The tables for binary addition and multiplication are shown in Table 11.2.

Let us illustrate the use of these tables via the following examples.

EXAMPLE 11.1

Let us add the decimal numbers 7 and 9 in binary. Performing the addition with the aid of Tables 11.1 and 11.2, we obtain

Decimal	Binary	
	1111	Carries
7	0111	
9	1001	
16	10000	Sum

Note that in binary arithmetic $1 + 1 = 0$ with 1 to carry. ■

Binary multiplication is also a very simple operation. It is performed in the same manner as decimal multiplication. However, since the only digits are 0 and 1, it is essentially a shift and add operation, as the following example illustrates.

EXAMPLE 11.2
Let us multiply the decimal number 7 by the decimal number 5. The multiplication is

$$
\begin{array}{cll}
\text{Decimal} & \text{Binary} & \\
7 & 111 & \text{Multiplicand} \\
\times\,5 & \times\,101 & \text{Multiplier} \\
\hline
35 & 111 & \\
& 000 & \\
& 111 & \\
\hline
& 100011 & \text{Product}
\end{array}
$$

Note that for every 1 digit in the multiplier, the multiplicand is copied as a partial product. A 0 digit in the multiplier simply shifts the partial product one digit but otherwise adds nothing. ■

DRILL EXERCISE

D11.1. Given the decimal numbers 5 and 6, (a) add the two numbers in binary, and (b) multiply the two numbers in binary.

   Ans: (a) 1011; (b) 11110.

In general, we can convert a number in base $r = \alpha$ to the number in base $r = \beta$. One of the popular techniques for performing *base conversion* is the *series substitution method*. Any number in base r can be written in the format.

$$
N = \sum_{j=-m}^{n-1} a_j r^j \tag{11.1}
$$
$$
= a_{n-1}r^{n-1} + a_{n-2}r^{n-2} + \ldots a_0 r^\circ + a_{-1}r^{-1} \ldots + a_{-m}r^{-m}
$$

Base conversion using the series substitution method is performed as follows: When the terms of the series are expressed in base α, each factor of each term is converted to base β and then the series is evaluated using arithmetic in base β. The following examples illustrate the technique.

EXAMPLE 11.3
Let us convert 10011 in binary to decimal. Using the series substitution method, we obtain

$$
\begin{aligned}
10011 &= 1(2)^4 + 0(2)^3 + 0(2)^2 + 1(2)^1 + 1(2)^0 \\
&= 16 + 0 + 0 + 2 + 1 \\
&= 19
\end{aligned}
$$

Therefore, 10011 in binary is equivalent to 19 in decimal. ■

EXAMPLE 11.4

Using the series substitution method we will convert 35 in decimal to binary. The number 35 can be written as

$$35 = 3(10)^1 + 5(10)^0$$
$$= 11(1010) + 101(1)$$
$$= 100011$$

Hence, 35 in decimal is equivalent to 100011 in binary.   ■

DRILL EXERCISES

D11.2. Convert 1010101 in binary to decimal.

Ans: 85.

D11.3. Convert 28 in decimal to binary.

Ans: 11100.

11.3

Boolean Algebra

Boolean algebra provides the mathematical foundation for the analysis and design of digital systems. It is essentially the mathematics of logic circuits even though it was formulated by George Boole, an English mathematician, in 1849, many decades before digital computers were even a figment of anyone's imagination. The fundamental concepts of Boolean algebra are set forth in the following *postulates.*

P1. The algebraic system, which contains two or more elements and the operators · (AND) and + (OR), is closed.

P2. Two algebraic expressions are equal (=) if one can be replaced by the other.

P3. The unique elements 1 and 0 exist such that
  (1) $A + 0 = A$
  (2) $A \cdot 1 = A$

P4. A unique element $\overline{A}$ (the complement of A) exists such that
  (1) $A + \overline{A} = 1$
  (2) $A \cdot \overline{A} = 0$

P5. The commutative law is
  (1) $A + B = B + A$
  (2) $A \cdot B = B \cdot A$

P6. The associative law is
  (1) $A + (B + C) = (A + B) + C$
  (2) $A \cdot (B \cdot C) = (A \cdot B) \cdot C$

P7. The distributive law is
  (1) $A + (B \cdot C) = (A + B) \cdot (A + C)$
  (2) $A \cdot (B + C) = (A \cdot B) + (A \cdot C)$

The reader is cautioned to note that some of these algebraic rules are quite different than those we normally apply in real variable analysis.

The postulates which have been presented can be used to develop a number of useful theorems which further enhance our ability to manipulate logic expressions. These *theorems* are summarized in Table 11.3.

Table 11.3 Theorems of Boolean Algebra

T1	(1)	$A + A = A$	(2)	$A \cdot A = A$
T2	(1)	$1 + A = 1$	(2)	$A \cdot 0 = 0$
T3	(1)	$A + AB = A$	(2)	$A(A + B) = A$
T4	(1)	$A + \overline{A}B = A + B$	(2)	$A(\overline{A} + B) = AB$
T5	(1)	$AB + \overline{A}C + BC = AB + \overline{A}C$	(2)	$(A + B)(\overline{A} + C)(B + C) =$ $(A + B)(\overline{A} + C)$
T6	(1)	$\overline{A + B} = \overline{A}\,\overline{B}$	(2)	$\overline{AB} = \overline{A} + \overline{B}$

The postulates and theorems given above have all been stated in a (1), (2) format. These expressions are said to be *duals* of one another. The dual of an expression is found by replacing all + operators with ·, all · operators with +, all 0s with 1, and all 1s with 0. This *principle of duality* is very important in Boolean algebra.

The last theorem in the list (T6) is called *DeMorgan's Theorem* and its importance stems from the fact that it is a general technique for complementing Boolean expressions.

In the manipulation of Boolean functions, we typically have one goal: minimize the number of literals. A *literal* is any occurrence of a variable in either complemented or uncomplemented form. For example, the Boolean function

$$f(A, B, C, D) = \overline{A}BC + A(B + C + \overline{D}) + \overline{C}\,\overline{D}$$

has 9 literals.

EXAMPLE 11.5

Let us minimize the following function of four variables with ten literals.

$$f(A, B, C, D) = AB + \overline{AB}\,\overline{C} + \overline{C}\,\overline{D} + \overline{BCD}$$

$$\begin{aligned}
&= AB + \overline{C} + \overline{C}\,\overline{D} + \overline{BCD} &&\text{(T4)}\\
&= AB + \overline{C} + \overline{BCD} &&\text{(T3)}\\
&= AB + \overline{C} + B(\overline{C} + \overline{D}) &&\text{(T6)}\\
&= AB + \overline{C} + B\overline{C} + B\overline{D} &&\text{(P7)}\\
&= AB + \overline{C} + B\overline{D} &&\text{(P5,T3)}\\
&= BA + B\overline{D} + \overline{C} &&\text{(P5)}\\
&= B(A + \overline{D}) + \overline{C} &&\text{(P7)}
\end{aligned}$$

This minimum form of the function has only four literals.   ∎

As we will illustrate later, electronic circuits will be employed to realize these Boolean functions in hardware. Therefore, the fewer the number of literals in the function, the simpler the hardware realization.

D11.4. Find the minimum realization of the function $f(A, B, C) = \overline{A + B} + A\overline{B} + \overline{BC}$

Ans: $f(A, B, C) = \overline{BC}$.

11.4
Truth Tables

Another method of describing Boolean functions is the *truth table,* which displays the value of the Boolean function for all values of the variables.

EXAMPLE 11.6
Let us compute the truth table of the Boolean function

$$f(A, B, C) = AB + \overline{B}C$$

Since there are three variables, there are eight combinations which range from 000 to 111. For example, if $A = 0$, $B = 0$, $C = 0$, the function is

$$f(0, 0, 0) = 0 \cdot 0 + 1 \cdot 0$$
$$= 0$$

The truth table, which lists all combinations, is shown in Table 11.4.

Table 11.4

A B C	f(A, B, C)
0 0 0	0
0 0 1	1
0 1 0	0
0 1 1	0
1 0 0	0
1 0 1	1
1 1 0	1
1 1 1	1

■

The truth table gets its name from the fact that in the context of a truth function the 0s are replaced with Fs (False) and the 1s are replaced by Ts (True).

Truth tables play an important role in Boolean algebra for the simple reason that although there are an infinite number of ways in which to write a Boolean function, each function has one and only one truth table.

D11.5. Derive the truth table for the function $f(A, B) = A\overline{B} + \overline{A}B$

Ans:

A	B	f(A, B)
0	0	0
0	1	1
1	0	1
1	1	0

11.5
Switching Networks

The electronic devices that are used to construct the logic functions in hardware were discussed in Chapter 10. These *gates,* as they are called, realized the following functions: AND, OR, NOT, NAND and NOR. Table 11.5 provides a listing of the gates which include the operator, the symbol used to represent the gate, and the mathematical relationship between the gate's inputs and its output.

Table 11.5 Switching Devices

Operator	Symbol	Input/output relationship
AND	A_1 $f(A_1, A_2)$ A_2	$f(A_1, A_2) = A_1 A_2$
OR	A_1 $f(A_1, A_2)$ A_2	$f(A_1, A_2) = A_1 + A_2$
NOT	A $f(A)$	$f(A) = \overline{A}$
NAND	A_1 $f(A_1, A_2)$ A_2	$f(A_1, A_2) = \overline{A_1 A_2}$
NOR	A_1 $f(A_1, A_2)$ A_2	$f(A_1, A_2) = \overline{A_1 + A_2}$

At this point it is important to note that any Boolean function can be written in terms of the operators AND, OR, and NOT. Furthermore, these three operators can be completely expressed in terms of either NAND or NOR. Thus, in realizing logic functions with gates, we can use AND, OR, and NOT, or only NAND, or only NOR.

Specific logic functions within a digital system are derived through an interconnection of these gates. The gate outputs are Boolean functions, and therefore can be manipulated using Boolean algebra. In fact, it may be possible via the Boolean algebra to simplify the gate structure and thus realize the desired function with fewer gates.

EXAMPLE 11.7

Consider the gate structure in Fig. 11.1(a). Note that these gates realize the logic function

$$f(A, B, C, D) = B_1 + B_2 + B_3 + B_4 + B_5$$

$$= BC\overline{D} + \overline{B}D + \overline{A}\,\overline{C}D + BC + A\overline{C}D$$

$$= BC + \overline{B}D + \overline{A}\,\overline{C}D + A\overline{C}D \qquad \text{(T3)}$$

$$= BC + \overline{B}D + \overline{C}D \qquad \text{(P7, P4)}$$

$$= BC + (\overline{B} + \overline{C})D \qquad \text{(P7)}$$

$$= BC + \overline{BC}D \qquad \text{(T6)}$$

$$= BC + D \qquad \text{(T4)}$$

Therefore, the two gates in Fig. 11.1(b) realize the same function as do those in Fig. 11.1(a). Both gate structures have the same truth table; however, the switching network in Fig. 11.1(b) is much simpler and thus much easier to construct with electronic hardware. Note that the output of this logic circuit is independent of A. ■

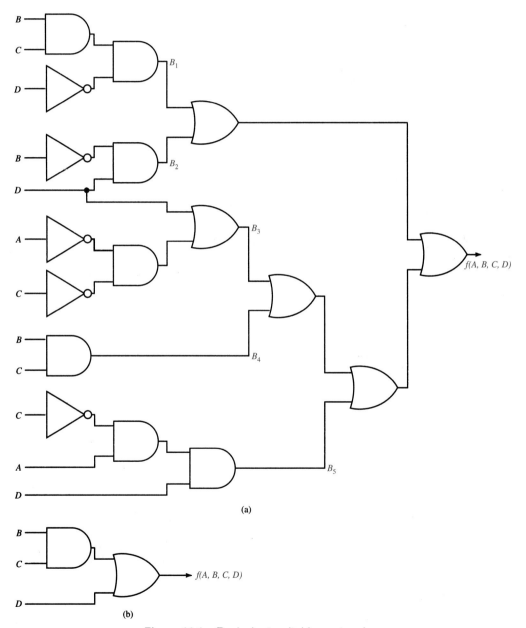

(a)

(b)

Figure 11.1   Equivalent switching networks.

11.6
Canonical Forms

Although we have listed Boolean functions in a number of ways, there are several specific forms which are of special importance. Two such forms are SOP *(Sum of Products)* and POS *(Product of Sums)*. The following examples illustrate these two forms.

EXAMPLE 11.8
The following switching functions are in SOP form:

$$f_1(A, B, C) = AB + \overline{A}B\overline{C} + AC + BC$$
$$f_2(A, B, C, D) = A\overline{B}CD + \overline{A}BC\overline{D} + AB\overline{C}\,\overline{D} + \overline{A}\,\overline{B}\,\overline{C}\,\overline{D}$$ ∎

EXAMPLE 11.9
The following switching functions are in POS form:

$$f_3(A, B, C) = (A + B)(A + \overline{B} + C)(\overline{A} + \overline{B} + \overline{C})$$
$$f_4(A, B, C, D) = (A + B + \overline{C} + D)(\overline{A} + \overline{B} + C + \overline{D})(A + \overline{B} + \overline{C} + \overline{D})$$ ∎

It is important to note that the switching functions $f_2(A, B, C, D)$ and $f_4(A, B, C, D)$ in Examples 11.8 and 11.9, respectively, have a special form, that is, every variable is present in each term in either complemented or uncomplemented form. This form is called *canonical* form. Therefore, the functions $f_2(A, B, C, D)$ and $f_4(A, B, C, D)$ are said to be in canonical SOP form and canonical POS form, respectively. Each term in canonical SOP form is called a *minterm* and each term in canonical POS form is called a *maxterm*.

The description of Boolean functions in terms of minterms and maxterms can be further simplified by coding the variables. In the minterm case, variables are coded as 1 and complements are coded as 0. Therefore, the function $f_2(A, B, C, D)$ is written as

$$f_2(A, B, C, D) = A\overline{B}CD + \overline{A}BC\overline{D} + AB\overline{C}\,\overline{D} + \overline{A}\,\overline{B}\,\overline{C}\,\overline{D}$$

The minterm code for $ABCD$ is 1011, for $\overline{A}BC\overline{D}$ is 0110, etc., and therefore the function can be expressed in the form

$$f_2(A, B, C, D) = m_{11} + m_6 + m_{12} + m_0 = \Sigma m(0, 6, 11, 12)$$

where m_i is the minterm and i is the decimal integer for the corresponding binary code.

In the maxterm case variables are coded as 0 and complements are coded as 1. Therefore, the function $f_4(A, B, C, D)$ is written as

$$f_4(A, B, C, D) = (A + B + \overline{C} + D)(\overline{A} + \overline{B} + C + \overline{D})(A + \overline{B} + \overline{C} + \overline{D})$$

The maxterm code for $(A + B + \overline{C} + D)$ is 0010, for $(\overline{A} + \overline{B} + C + \overline{D})$ is 1101, etc., and therefore the function can be expressed as

$$f_4(A, B, C, D) = (M_2)(M_{13})(M_7)$$
$$= \Pi M(2, 7, 13)$$

where M_i is the maxterm and i is the decimal integer for the corresponding binary code.

The order of the variables in the coding process is very important, that is, $f_2(A, B, C, D) = \Sigma m(0, 6, 11, 12)$ is *not* the same as $f(D, C, B, A) = \Sigma m(0, 6, 11, 12)$.

It is instructive to examine the relationship among truth tables, minterms, and maxterms. The following example serves to illustrate the connection.

EXAMPLE 11.10

Given the switching function $f(A, B, C) = \overline{A}BC + A\overline{B}C + ABC$, the minterm list form is $f(A, B, C) = \Sigma m(3, 5, 7)$ and the truth table is shown in Table 11.6.

Table 11.6

A B C	f(A, B, C)	$\overline{f}(A, B, C)$
0 0 0	0	1
0 0 1	0	1
0 1 0	0	1
0 1 1	1	0
1 0 0	0	1
1 0 1	1	0
1 1 0	0	1
1 1 1	1	0

The truth table for $\overline{f}(A, B, C)$ is also shown and can be expressed as $\overline{f}(A, B, C) = \Sigma m(0, 1, 2, 4, 6)$. Since the maxterms produce a 0 in the truth table listing, the function can also be expressed as $f(A, B, C) = \Sigma m(3, 5, 7) = \Pi M(0, 1, 2, 4, 6)$, and, of course, $\overline{f}(A, B, C) = \Sigma m(0, 1, 2, 4, 6) = \Pi M(3, 5, 7)$.   ∎

The previous example illustrates that a switching function can be immediately written in canonical form from the truth table; however, Boolean algebra can also be used to generate a canonical form.

EXAMPLE 11.11

We wish to find the canonical SOP form of the function $f(A, B, C) = A\overline{B} + \overline{B}C + A\overline{C}$. This function can be written as

$$f(A, B, C) = A\overline{B}(C + \overline{C}) + (A + \overline{A}) \overline{B}C + A(B + \overline{B}) \overline{C}$$
$$= A\overline{B}C + A\overline{B}\,\overline{C} + A\overline{B}C + \overline{A}\,\overline{B}C + AB\overline{C} + A\overline{B}\,\overline{C}$$
$$= A\overline{B}C + A\overline{B}\,\overline{C} + AB\overline{C} + \overline{A}\,\overline{B}C$$
$$= \Sigma m(3, 4, 5, 6)$$ ∎

DRILL EXERCISE ──

D11.6. Express the following switching function in both minterm and maxterm form.

$$f(A, B, C) = AB + \overline{A}\,\overline{C}$$

Ans: $f(A, B, C) = \Sigma m(0, 2, 6, 7) = \Pi M(1, 3, 4, 5)$.

Thus far our manipulation of Boolean functions has been done algebraically. It is possible, however, to graphically display the mathematical operations using *Venn diagrams*. This possibility exists because the *algebra of sets* is also a Boolean algebra in which the sets correspond to the elements and the set operations of union and intersection correspond to the Boolean operations of $+$ and $\cdot$, respectively. Sets are represented by some type of closed contour, and several basic sets together with the operations $+$ and $\cdot$ are shown in Fig. 11.2.

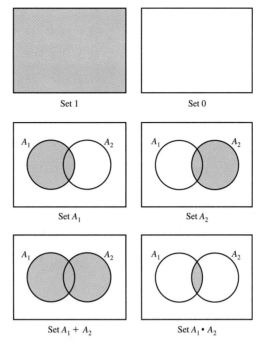

Figure 11.2 Set operations.

Venn diagrams can be used to visually demonstrate the postulates and theorems we have presented earlier. Consider, for example, the Boolean expression

$$A_1(A_2 + A_3) = A_1A_2 + A_1A_3$$

Figure 11.3 displays this expression in Venn diagrams. Note that the set $A_1(A_2 + A_3)$ is identical to the set $A_1A_2 + A_1A_3$.

Although Venn diagrams can be employed to visualize Boolean mathematical operations, much of their importance stems from the fact that they provide the basis for the minimization of switching circuits using what is called a *Karnaugh map* (K-map). We transform the Venn diagram into a K-map using Fig. 11.4 in the following manner. The Venn diagram for the three variables *A, B,* and *C* is shown in Fig. 11.4(a). Each area of the diagram within the universal set corresponds to one of the eight minterms as shown in Figures 11.4(b) and 11.4(c). Since one minterm is of no more importance than another, the diagram is reshaped to give equal area to each minterm as shown in Fig. 11.4(d). The final form of the map is shown in Fig. 11.4(e) where the *m*'s have been dropped, but the minterm number is retained in each block.

The crucial step in this transformation is the specific format of Fig. 11.4(d). If we examine m_6 in Fig. 11.4(c), we note that it is physically adjacent to m_2, m_4, and m_7. This condition has been maintained in the transformation to Figure 11.4(d). Next consider m_4 in Fig. 11.4(c). In addition to m_5 and m_6, we find that m_4 is physically adjacent to m_0. This adjacent condition can be satisfied by folding the map into a cylinder so that the left and right edges are connected. Then m_5 is adjacent to m_4, m_7, and m_1, m_0 is adjacent to m_1, m_2, and m_4, etc.

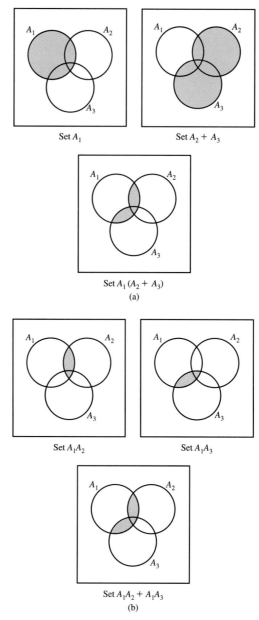

Figure 11.3   The use of Venn diagrams.

The four-variable K-map is shown in Fig. 11.5. Once again, in order to satisfy the minterm adjacent conditions the left and right edges of the K-map are considered to be the same line as are the top and bottom edges.

Finally, it is interesting to note that there exists a direct correspondence betwen the K-map and the truth table. The former has one block per minterm while the latter has one row per minterm.

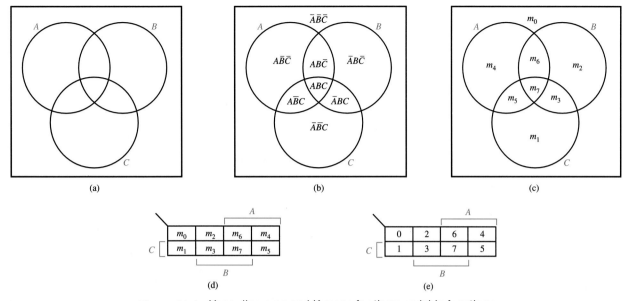

Figure 11.4   Venn diagrams and K-maps for three-variable functions.

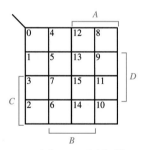

Figure 11.5   A four-variable K-map.

11.7
Function
Minimization

Terms which are logically adjacent in a Boolean function can be combined to eliminate one or more literals, thus simplifying the function. *On the K-map, terms which are logically adjacent are also physically adjacent.* Therefore, the K-map can be used as a graphical tool for Boolean function minimization as illustrated in the following example.

EXAMPLE 11.12
We wish to determine the minimum form of the switching function

$$f(A, B, C, D) = \Sigma m(3, 4, 5, 11, 15)$$

The function is shown listed on the map in Fig. 11.6(a) where in accordance with the correspondence which exists between the map and the truth table, we have placed a 1 in the

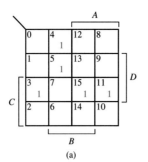
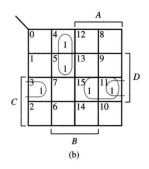

Figure 11.6   K-maps used in Example 11.12.

listed minterm blocks and ignored the 0s which would appear in the remaining blocks. The function in variable form is

$$f(A, B, C, D) = m_3 + m_4 + m_5 + m_{11} + m_{15}$$
$$= \overline{A}\,\overline{B}CD + \overline{A}B\overline{C}\,\overline{D} + \overline{A}B\overline{C}D + A\overline{B}CD + ABCD$$

where, for example, m_3 corresponds to 0011 and thus $\overline{A}\,\overline{B}CD$. Furthermore, on the map m_3 is outside of both A and B and inside both C and D. The remaining terms can be visualized in a similar manner.

Note that minterms 4 and 5 differ in only a single literal, and therefore they can be combined to yield the term $\overline{A}B\overline{C}$. Similarly, minterms 3 and 11 and minterms 11 and 15 can be combined to yield $\overline{B}CD$ and ACD, respectively. Note that minterm 11 was used twice; however, that is the rule rather than the exception, since $A = A + A + A + \dots$.

On the map, the logically adjacent terms are also physically adjacent, and are combined by drawing a loop around the minterms to be combined as shown in Fig. 11.6(b). We note, for example, that the combination of minterms 4 and 5 is outside A, inside B, and outside C; therefore, this combination yields $\overline{A}B\overline{C}$. The function representing the two remaining combinations is determined in a similar manner, and therefore the minimum function is

$$f(A, B, C, D) = \overline{A}B\overline{C} + \overline{B}CD + ACD. \qquad \blacksquare$$

In the minimization procedure, we should "cover" each minterm at least once, and then group the minterms into the largest possible blocks. When combining minterms we group them in powers of 2. A group of 2 minterms eliminates 1 literal, a group of 4 minterms eliminates 2 literals, etc.

EXAMPLE 11.13

We wish to find the minimum form for the function

$$f(A, B, C) = \Sigma m(0, 2, 3, 7)$$

The map for this function is shown in Fig. 11.7 and the minimum function is

$$f(A, B, C) = \overline{A}\,\overline{C} + BC \qquad \blacksquare$$

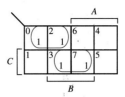

Figure 11.7 K-map for Example 11.13.

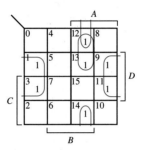

Figure 11.8 K-map for Example 11.14.

EXAMPLE 11.14

The map for the following function is shown in Fig. 11.8.

$$f(A, B, C, D) = \Sigma m(1, 3, 9, 11, 12, 13, 14)$$

Note that minterm 13 could be combined with either minterms 9 or 12. Combining minterm 13 with minterm 12 yields

$$f(A, B, C, D) = AB\overline{C} + AB\overline{D} + \overline{B}D$$ ■

DRILL EXERCISES

D11.7. Find the minimum form for the function $f(A, B, C) = \Sigma m(1, 5, 6, 7)$

Ans: $f(A, B, C) = AB + \overline{B}C.$

D11.8. Find the minimum form for the function $f(A, B, C, D) = \Sigma m(0, 1, 4, 5, 7, 9, 15)$

Ans: $f(A, B, C, D) = \overline{A}\,\overline{C} + \overline{B}\,\overline{C}D + BCD$

11.8

Combinational Logic Design

At this point we begin to apply what we have just learned to logic circuit design. The following examples will serve to illustrate the manner in which the material can be employed in a variety of applications.

EXAMPLE 11.15

We wish to design a logic circuit with the following conditions: The input is a 4-bit binary coded decimal (BCD); there is a single output line; and the circuit should produce a 1 at the output (detect) if the input is divisible by 2.

The *BCD code* is listed as follows:

0 0000	5 0101
1 0001	6 0110
2 0010	7 0111
3 0011	8 1000
4 0100	9 1001

Since the output should be 1 if the input is either 2, 4, 6, or 8, the truth table for the network is shown in Table 11.7.

Table 11.7

Input $ABCD$	Output f	Minterm Listing
0000	0	0
0001	0	1
0010	1	2
0011	0	3
0100	1	4
0101	0	5
0110	1	6
0111	0	7
1000	1	8
1001	0	9
1010	d	10
1011	d	11
1100	d	12
1101	d	13
1110	d	14
1111	d	15

An examination of this table prompts an immediate question—what are the d's? d stands for *don't care*. The BCD input can only range from 0 to 9. Therefore, the numbers 10–15 cannot occur. Since these numbers cannot occur, we "don't care" whether the output is a 1 or a 0, and therefore we use a d. When we place the function on the K-map, the d's can be 1's if they simplify the output function and 0 if they do not. The output function which results from the truth table is

$$f(A, B, C, D) = \Sigma m(2, 4, 6, 8) + d(10, 11, 12, 13, 14, 15)$$

The K-map for the function is shown in Fig. 11.9.
The simplified logic function is then

$$f(A, B, C, D) = C\overline{D} + B\overline{D} + A\overline{D}$$

This Boolean function can be written in the form

$$f(A, B, C, D) = \overline{\overline{C\overline{D} + B\overline{D} + A\overline{D}}}$$

now applying DeMorgan's Theorem the function becomes

$$f(A, B, C, D) = \overline{\overline{A\overline{D}} \cdot \overline{B\overline{D}} \cdot \overline{C\overline{D}}}$$

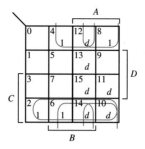

Figure 11.9   K-map for Example 11.15.

This function is directly in the form for a NAND gate realization. The logic diagram for the function using NAND gates is shown in Fig. 11.10. ∎

A code converter is a multiple input, multiple output combinational logic network which translates an input codeword into a bit pattern which represents the new codeword. The following example illustrates the design of such a device.

EXAMPLE 11.16

The *EXCESS-3 code* is a code which is obtained by adding the decimal number 3 or 0011 in binary to the corresponding BCD code. This code is listed as follows:

0	0011	5	1000
1	0100	6	1001
2	0101	7	1010
3	0110	8	1011
4	0111	9	1100

This code is *self-complementing,* for example, the complement of 9(1100) is 0(0011), the complement of 8(1011) is 1(0100), etc. We wish to obtain the logic equations for the code converter which will convert the BCD code to EXCESS-3.

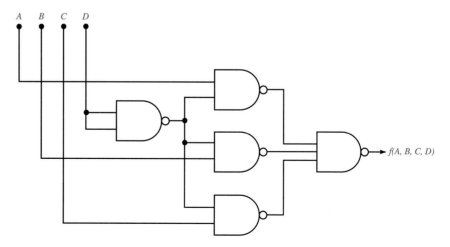

Figure 11.10   NAND gate realization of the function in Example 11.15.

The truth table for the code converter is shown in Fig. 11.11. The K-maps for the four bits in the EXCESS-3 code are shown in Fig. 11.12. The logic equations for each bit in the EXCESS-3 code are

$$B_1 = A_1 + A_2A_3 + A_2A_4$$
$$B_2 = \overline{A_2}A_3 + \overline{A_2}A_4 + A_2\overline{A_3}\,\overline{A_4}$$
$$B_3 = A_3A_4 + \overline{A_3}\,\overline{A_4}$$
$$B_4 = \overline{A_4}$$

These equations represent the 4-input, 4-output logic circuit which performs the code conversion.   ∎

Figure 11.11   Truth table for BCD to Excess-3 code converter.

Decimal	BCD				Excess-3			
	A_1	A_2	A_3	A_4	B_1	B_2	B_3	B_4
0	0	0	0	0	0	0	1	1
1	0	0	0	1	0	1	0	0
2	0	0	1	0	0	1	0	1
3	0	0	1	1	0	1	1	0
4	0	1	0	0	0	1	1	1
5	0	1	0	1	1	0	0	0
6	0	1	1	0	1	0	0	1
7	0	1	1	1	1	0	1	0
8	1	0	0	0	1	0	1	1
9	1	0	0	1	1	1	0	0
10	d	d	d	d	d	d	d	d
11	d	d	d	d	d	d	d	d
12	d	d	d	d	d	d	d	d
13	d	d	d	d	d	d	d	d
14	d	d	d	d	d	d	d	d
15	d	d	d	d	d	d	d	d

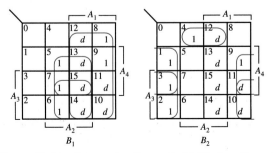

Figure 11.12   K-maps for Example 11.16. (continues)

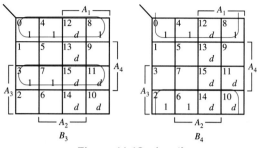

Figure 11.12   (cont)

DRILL EXERCISE

DRILL EXERCISE

D11.9. Design a logic circuit with a BCD input that detects the odd numbers 1, 3, 5, 7, and 9.

Ans: If the BCD inputs are labeled $A_1A_2A_3A_4$, then $f(A_1, A_2, A_3, A_4) = A_4$.

EXAMPLE 11.17

A factory employs two engines and each is equipped with two identical sensors. One sensor measures output speed and the other measures engine temperature. If speed and temperature are normal, the sensor outputs are low. If the speed is too fast or the temperature too high, the sensor outputs are high. We wish to design a logic circuit that will detect whenever the speed of either or both engines is too fast and the temperature of either or both engines is too high.

The logic variables are defined as follows:

$$A = \text{speed of engine 1}$$
$$B = \text{speed of engine 2}$$
$$C = \text{temperature of engine 1}$$
$$D = \text{temperature of engine 2}$$

The truth table for this example is listed in Fig. 11.13.
The minterms for the function are plotted in the K-map in Fig. 11.14. The minimized logic function which defines the logic network is

$$f(A, B, C, D) = AC + BC + AD + BD$$ ∎

DRILL EXERCISE

D11.10. Design a logic circuit with three inputs A, B, and C, such that the output is to be high only when exactly two of the inputs are high.

Ans: $f(A, B, C) = AB\overline{C} + A\overline{B}C + \overline{A}BC$.

Figure 11.13   Truth table used in Example 11.17.

A	B	C	D	f
0	0	0	0	0
0	0	0	1	0
0	0	1	0	0
0	0	1	1	0
0	1	0	0	0
0	1	0	1	1
0	1	1	0	1
0	1	1	1	1
1	0	0	0	0
1	0	0	1	1
1	0	1	0	1
1	0	1	1	1
1	1	0	0	0
1	1	0	1	1
1	1	1	0	1
1	1	1	1	1

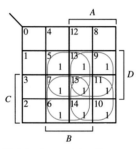

Figure 11.14   K-map for Example 11.17.

11.9
Sequential Logic Circuits

We have shown that in combinational logic circuits the output is a function of only the current input, that is, what has happened in the past has absolutely no bearing on the current output.

Sequential circuits, with their inherent *memory,* provide us with a whole new dimension in the design of logic circuits. This memory provides us with the capability to remember the past, and therefore the current output can be a function of not only the current input but prior inputs as well.

Although there are numerous examples of sequential devices, one example which we regularly encounter is the washing machine. This device has several *states,* e.g., wash, rinse, and spin dry, and they are performed in a specific order. The *output* of the machine, that is, the specific task that the machine performs, is a function of not only the machine's *present state* but its input as well. The input also transitions the machine from its present state to the *next state.*

The use of memory is clearly the key to the added capability which exists in sequential circuits. Let us now examine a general model for this new type of circuit.

The Structure of Sequential Logic Circuits

Sequential circuits can be modeled as shown in Fig. 11.15. This block diagram clearly illustrates the difference between sequential and combinational logic circuits.

As the model indicates, both the output and the next state are functions of the input and present state. Mathematically, the relationships can be expressed as follows:

$$z_i = f_1(x_1 x_2 \ldots x_n, y_1, \ldots y_p) \qquad i = 1, 2, \ldots m$$
$$Y_j = f_2(x_1 x_2 \ldots x_n, y_1, y_2 \ldots y_p) \qquad j = 1, 2, \ldots p$$
(11.2)

The equations can be written in vector notation as

$$\mathbf{z} = f_1(\mathbf{x}, \mathbf{y})$$
$$\mathbf{Y} = f_2(\mathbf{x}, \mathbf{y})$$
(11.3)

We will explore later the exact forms of the input and output signals, as well as the different types of memory devices that are typically employed in sequential circuits.

State Tables and Diagrams

State tables and *diagrams* illustrate in tabular and graphical form, respectively, the functional relationship which exists among input, present state, output, and next state. These two equivalent forms are shown in Fig. 11.16. In each case, for input *x*, the circuit will transition from present state *y* to next state *Y* with an output *z*. The following example explicitly illustrates the connection between these two equivalent forms.

EXAMPLE 11.18

The state table and equivalent state diagram for a specific sequential circuit are shown in Fig. 11.17. These equivalent descriptions indicate that if the circuit is in state *A* and the input is a 1, the circuit will transition to state *B* and produce an output of 0. Note that if the circuit is initially in state *A* and the input sequence 1001 is applied, the circuit will move from state *A* to *B* to *D* to *C* to *A* and produce an output sequence 0101. The same input applied to the circuit when it is initially in state *C* would produce the output string 1000. ■

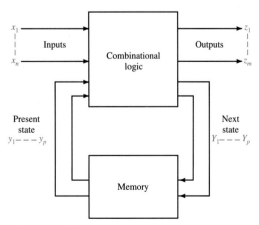

Figure 11.15   A model for a sequential circuit.

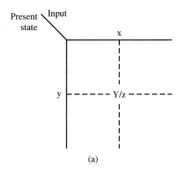

(a)

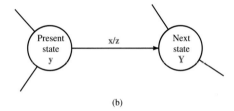

(b)

Figure 11.16 State table and equivalent state diagram.

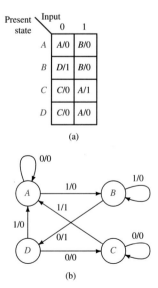

Figure 11.17 Equivalent descriptions of a sequential circuit.

D11.11. A sequential circuit is described by the state table in Fig. D11.11. If the circuit is initially in state B and the input sequence is $x = 010110$, determine the output sequence.

PS \ x	0	1
A	$A/0$	$B/1$
B	$B/0$	$A/1$

Figure D11.11

Ans: $z = 010110$.

Clocked Memory

In general, sequential circuits may operate in a variety of modes. For example, the memory elements may be clocked or unclocked. Furthermore, the circuit's input signals may be synchronous pulses, synchronous levels, asynchronous pulses, or asynchronous levels. In the material that follows, we will consider only those circuits which function under the control of a clock signal that is normally zero and contains periodic pulses defined by a $0 \rightarrow 1 \rightarrow 0$ transition. The clock signal serves to synchronize the operation of all memory elements. Therefore, when the memory elements change state, they do so in response to the *clock pulse* rather than some other input.

Devices called *flip-flops* are normally employed as memory elements in *synchronous sequential circuits*. These devices will store either a 0 or a 1. The flip-flop typically has two outputs, one of which is the normal state Q and the other is the complement $\overline{Q}$. In addition to the clock signal, clocked flip-flops have one or more control inputs. However, the effects of these control inputs are controlled by the clock signal, that is, the control inputs prepare the flip-flop for a state transition, but it is the clock pulse that actually *triggers* the change.

There are a number of *different types of flip-flops,* for example, *S–R* or set-reset, *D* or delay, *T* or trigger, and *J–K*. Each has its own operating characteristics. Since their operations are similar, we will confine our discussion here to one of them—the *J–K* flip-flop.

The block diagram for a clocked *J–K* flip-flop is shown in Fig. 11.18(a). C is the clock signal, J and K are the control signals and Q and $\overline{Q}$ are the outputs. The state table which defines the transitions under clock control is shown in Fig. 11.18(b). The operation can be summarized in the table in Fig. 11.18(c). Under the control of the clock pulse, that is, $C = 1$, $J = 0$, and $K = 0$ causes no change in the present output Q_p. $J = 0$ and $K = 1$ will *reset* the output to $Q = 0$. $J = 1$ and $K = 0$ will *set* the output to $Q = 1$. $J = 1$ and $K = 1$ *toggles* the output, that is, the output will change states from Q_p to $\overline{Q_p}$. This operation is also illustrated by the timing diagram shown in Fig. 11.19. In this diagram we assume that the transition of the flip-flop is triggered by the leading edge of the clock pulse, and we ignore all nuances associated with the timing of the signals. Note that initially the flip-flop is set, that is, $Q = 1$ (high) and at the time of the first clock pulse $J = 0$ and $K = 0$, and therefore Q remains at 1. When the second clock pulse arrives, J is still 0 but $K = 1$, and therefore the flip-flop resets to $Q = 0$ (low). At the time of the third clock pulse $J = 1$ and $K = 0$, and therefore the flip-flop is set to $Q = 1$. When the fourth clock pulse ar-

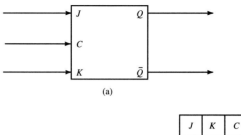

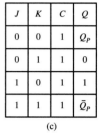

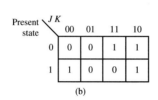

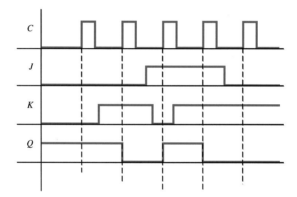

Figure 11.18   Diagrams for a clocked J–K flip-flop.

Figure 11.19   Timing diagram for a J–K flip-flop.

rives, $J = 1$ and $K = 1$, and hence the flip-flop is toggled to the opposite state $Q = 0$. Finally, during the last clock pulse $J = 0$, and $K = 1$, and therefore the flip-flop is reset to $Q = 0$. Since the flip-flop is already in this state, it simply remains there.

Analysis of Synchronous Sequential Circuits

Given a synchronous sequential circuit, we wish to determine the functional relationship which exists among the input x_k, the present state y_k, the next state y^{k+1}, and the output, z_k. The condition of the flip-flops defines the present state of the circuit. This information, when combined with the input, yields the next state and output. Recall from our earlier discussion that a state table or state diagram completely defines this relationship. In addition, since each flip-flop has two states, a circuit containing n flip-flop can be described by a state table with 2^n rows.

The following example illustrates two approaches to the *analysis procedure* and illuminates many of the facets which exist between them.

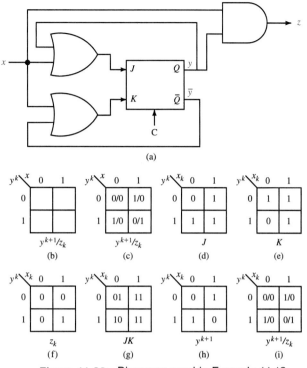

Figure 11.20   Diagrams used in Example 11.19.

EXAMPLE 11.19

Consider the network shown in Fig. 11.20(a). This synchronous sequential circuit contains one J–K flip-flop, and therefore has two states. Furthermore, since there is a single input, the state table for this network is of the form shown in Fig. 11.20(b). Assuming a combination of present states and inputs, we can complete this state table. For example, by tracing the various signals through the circuit we find that if the present state is $y^k = 0$ and the input is $x = 0$, the output is $z = 0$ and the next state, determined by $J = 0$ and $K = 1$, is $y^{k+1} = 0$. Therefore, the entry in the upper left-hand corner of the state table is $y^{k+1}/z_k = 0/0$. Also, if the present state is $y^k = 0$ and the input is $x = 1$, the output is $z = 0$ and the next state, determined by $J = 1$ and $K = 1$, is $y^{k+1} = 1$. Therefore, the entry in the upper right-hand corner of the state table is $y^{k+1}/z_k = 1/0$. The completed state table is shown in Fig. 11.20(c). Once this completed state table is known, given an initial state, the output sequence can be determined for any input string.

The analysis can also be performed from a different perspective. The logic equations which describe the network are

$$z_k = x_k y_k$$
$$J = x_k + y_k$$
$$K = x_k + \overline{y_k}$$

The K-maps for these functions are shown in Figs. 11.20(d), (e), and (f), respectively. If we transpose the data in Figs. 11.20(d) and (e), we obtain the table in Fig. 11.20(g).

This latter table defines the control inputs to the flip-flop which govern the state transition during the clock pulse. For example, if the present state is $y^k = 0$, the input is $x_k = 0$, and the control signals are $J = 0$ and $K = 0$, the next state is $y^{k+1} = 0$. Likewise, if the present state is $y^k = 0$, the input is $x_k = 1$, and the control signals are $J = 1$ and $K = 1$, the next state is $y^{k+1} = 1$. The table in Fig. 11.20(h) is completed in this manner. Finally, combining the table in Fig. 11.20(f) with that in Fig. 11.20(h) yields the state table in Fig. 11.20(i), which is identical to that in Fig. 11.20(c). ∎

DRILL EXERCISE

D11.12. Compute the state table for the synchronous sequential circuit in Fig. D11.12.

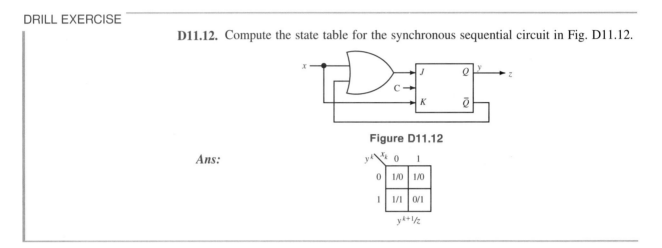

Figure D11.12

Ans:

y^k \ x_k	0	1
0	1/0	1/0
1	1/1	0/1

y^{k+1}/z

Synthesis of Synchronous Sequential Circuits

In the analysis procedure, the circuit diagram is known and we wish to determine the state table or state diagram. The *synthesis procedure,* on the other hand, typically starts with a specification of the state table or state diagram, and from it we derive the circuit.

We will find that in the synthesis procedure we essentially reverse the sequence of steps employed in analysis. As usual we will demonstrate the techniques using several examples. It is important to note that within the synthesis process there are a number of techniques that can be applied to simplify the procedure. However, we will simply note them and proceed, since an explanation of the efficient use of these techniques would be beyond our current scope.

EXAMPLE 11.20

We wish to design a synchronous sequential circuit with a single input and a single output that *recognizes* or *detects* the input sequence $x = 10$. Recognition of a particular input string can be accomplished by producing a 1 at the output whenever the specific string occurs. The importance of this example stems from the fact that this type of circuit is very useful in detecting the handshake signal sent by a printer to a computer for the purpose of acknowledging the receipt of a byte of message. As soon as the computer detects the acknowledge signal from the printer, the next byte of message can be sent. Similar applications can be found in communication systems. The circuit should be designed so that if the input string is

$$x = 001001100$$

the output will be

$$z = 000100010$$

The design process begins by converting the word description of the input/output behavior of the circuit into a state table or state diagram. We initiate the design using Fig. 11.21(a) by assuming that the circuit is in some starting state A. While in this state A, if the input is a 0, the circuit will return to state A and produce an output of 0, since this is not the first element in the string to be recognized. We are in essence marking time until the first 1 arrives at the input. When a 1 arrives and the circuit is in state A, the network moves to a new state B and produces an output of 0 as shown in Fig. 11.21(b). If a 1 arrives while the network is in state B, the circuit will remain in state B and produce an output of 0, since this is not the second element of the string to be detected. However, if the network is in state B and a 0 arrives at the input, the circuit will transition back to state A and produce an output of 1. The final state diagram is shown in Fig. 11.21(d). Note that the transition from state B to state A with an output of 1 resets the network to recognize the next input string of 10. As the final state diagram indicates, if the input string is $x = 10101010 \ldots$, the circuit will transition from state A to state B to state A to state $B \ldots$ and produce an output sequence of $01010101 \ldots$

The state table which corresponds to the state diagram in Fig. 11.21(d) is shown in Fig. 11.22(a). If we make the *state assignment* $A = 0$ and $B = 1$, the state table is changed to that shown in Fig. 11.22(b). Note that we could have made the assignment $A = 1$ and $B = 0$, which will also result in a valid circuit. Methods for making state assignments which minimize the amount of hardware are known but will not be addressed here.

The state table in Fig. 11.22(b) is split into the two tables in Fig. 11.22(c). The first table specifies the state transitions, that is, given a present state y^k and an input x, the next state y^{k+1} is known. The second table is simply a K-map for the output z. In order to realize state transitions specified in Fig. 11.22(c) using a clocked J–K flip-flop, we must now determine the proper control signals on the J and K input lines. Note that since the state table has two states, only one J–K flip-flop is required. Let us examine the upper left-hand corner of the state transition table in Fig. 11.22(c). If $y^k = 0$ and $x = 0$, the next state is $y^{k+1} = 0$. In order to affect this transition, J must be 0; however, K may be 0 or 1, and therefore we simply specify it as a don't care, d. Next consider the upper right-hand corner of the table, that is, $y^k = 0$, $x = 1$, and $y^{k+1} = 1$. In this latter case J must be 1; however, once again K may be 0 or 1, and therefore is a d. The complete tables are

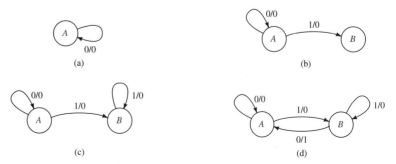

Figure 11.21   State diagram development for a 10 detector.

shown in Fig. 11.22(d). These tables are K-maps and together with the K-map for z yield the following logic equations.

$$J = x$$
$$K = \bar{x}$$
$$z = \bar{x}y^k$$

Therefore, the actual circuit derived from the equations is shown in Fig. 11.22(e).   ■

The final example will not only indicate the versatility of synchronous sequential circuits but will illustrate some additional features of the synthesis process.

EXAMPLE 11.21

We wish to design an up/down counter using clocked J–K flip-flops that counts in the range of 0, 1, 2, and 3. When $x = 0$, the circuit should count up, and when $x = 1$, the circuit should count down. The output should be designed so that it indicates the current

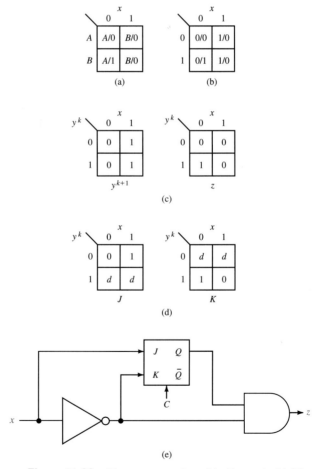

Figure 11.22   Diagrams employed in Example 11.20

count, that is, is the count 0, 1, 2, or 3. Ignoring the output for a moment we note that the state transition table will be of the form shown in Fig. 11.23(a). Note that if we make the state assignment:

$$0 \to 00$$
$$1 \to 01$$
$$2 \to 10$$
$$3 \to 11$$

the present state will indicate the current count in binary. Using this state assignment, the state transition table is shown in Fig. 11.23(b). The table in Fig. 11.23(b) is rearranged to form the K-map in Fig. 11.23(c). The K-maps for the two J–K flip-flops are shown in Figs. 11.23(d), (e), (f), and (g). The logic equations derived from the K-maps are

$$J_1 = \bar{x}y_2 + x\bar{y_2} = K_1$$

$$J_2 = 1 = K_2$$

Since the present state indicates the count, the circuit output can display the count by using two lights which are connected to the flip-flop outputs. In this manner, as the circuit counts from $0 \to 1 \to 2 \to 3$, the lights will indicate off off $\to$ off on $\to$ on off $\to$ on on. ∎

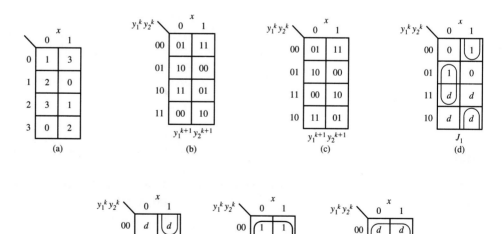

Figure 11.23   Tables used in Example 11.21.

D11.13. We wish to find the logic circuit for the state table shown in Fig. D11.13. Derive the logic equations for J, K, and z using the state assignment $A = 0$ and $B = 1$.

x	0	1
A	$B/1$	$A/0$
B	$A/0$	$B/1$

Figure D11.13

Ans: $z = \bar{x}\bar{y} + xy$, $J = k = \bar{x}$.

11.10
Summary

- The base or radix of a number system is the total number of digits allowed in the system, for example, the binary system has two digits.
- Boolean algebra provides the mathematical foundation for the analysis and design of logic circuits.
- The truth table is a method of describing Boolean functions.
- Logic functions can be realized using either AND, OR, and NOT, or NAND or NOR gates.
- Logic functions can be expressed in either a canonical sum-of-products (SOP) form or a canonical product-of-sums (POS) form. Each term in canonical SOP form is called a minterm and each term in a canonical POS form is called a maxterm.
- Venn diagrams can be used to visually demonstrate Boolean postulates and theorems.
- Karnaugh maps can be used to minimize Boolean logic functions.
- Sequential logic circuits are described by state tables or state diagrams.
- Flip-flops are a useful form of digital memory.
- Synchronous sequential circuits operate under the control of a clock.

PROBLEMS

11.1. If $A = 0111$, $B = 1010$, $C = 0101$, and $D = 1011$, find
(a) $A + B$
(b) $C + D$

11.2. Given the data in problem 11.1, find
(a) $A + C$
(b) $B + D$

11.3. Using the data in problem 11.1, find
(a) $B + C$
(b) $A + D$

11.4. If $A = 111$, $B = 011$, $C = 110$, and $D = 010$, find
(a) $A \times D$
(b) $C \times D$

11.5. Given the data in problem 11.4, find
(a) $B \times D$
(b) $B \times C$

11.6. Using the data in problem 11.4, determine
(a) $A \times C$
(b) $A \times B$

11.7. Given the data in problem 11.1, find
(a) $A \times B$
(b) $C \times D$

11.8. Using the data in problem 11.1, compute
(a) $A \times C$
(b) $B \times D$

11.9. Using the data in problem 11.1, find
(a) $B \times C$
(b) $A \times D$

11.10. Convert the following numbers in base 2 to base 10:
(a) 1100
(b) 10101

11.11. Convert the following numbers in binary to decimal:
(a) 1111
(b) 10010

11.12. Convert the following numbers from base 2 to base 10:
(a) 101010
(b) 110111

11.13. Convert the following numbers in binary to decimal:
(a) 1100011
(b) 110110111

11.14. Convert the following numbers from decimal to binary:
(a) 7
(b) 13

11.15. Convert the following numbers from base 10 to base 2:
(a) 19
(b) 25

11.16. Convert the following numbers from decimal to binary:
(a) 48
(b) 57

11.17. Convert the following numbers from base 10 to base 2:
(a) 101
(b) 172

11.18. Use the Boolean postulates and theorems to simplify the following expressions:
(a) $f(A, B, C) = A\,\overline{B} + \overline{B}C + A\overline{B}$
(b) $f(A, B, C) = \overline{A}B\overline{C} + ABD + \overline{A}BC$

11.19. Compute the truth table for the following functions:
(a) $f(A, B, C) = B\overline{C} + \overline{B}C$
(b) $f(A, B, C) = \overline{B} + C$

11.20. Compute the truth table for the following functions:
(a) $f(A, B, C) = \overline{AB} + C$
(b) $f(A, B, C) = \overline{AB} + \overline{B}C + A\overline{B}$

11.21. Use a Venn diagram to show that $f(A, B, C) = \Sigma m(1, 2, 5, 6) = B\overline{C} + \overline{B}C$

11.22. Using a Venn diagram, show that $f(A, B, C) = \Sigma m(0, 1, 3, 4, 5, 7) = C + \overline{B}$

11.23. Use a Venn diagram to demonstrate that $f(A, B, C) = BC + \overline{A}\,\overline{B} + A\overline{B}C = C + \overline{A}\overline{B}$

11.24. Find the minimum form for the Boolean functions
(a) $f(A, B, C) = \Sigma m(0, 1, 3, 4, 5, 7)$
(b) $f(A, B, C) = \Sigma m(1, 2, 5, 6)$

11.25. Find the minimum form for the Boolean functions
(a) $f(A, B, C) = \Sigma m(1, 2, 5, 7)$
(b) $f(A, B, C) = \Sigma m(0, 1, 3, 6, 7)$

11.26. Expand the following function into a set of minterms and then use the K-map to find a minimum form.
$$f(A, B, C) = \overline{A}\overline{B} + \overline{B}C + A\overline{B}$$

11.27. Expand the following function into a set of minterms and then use the K-map to find the minimum realization.
$$f(A, B, C) = BC + \overline{A}\overline{B} + A\overline{B}C$$

11.28. Expand the following function into a set of minterms and then use the K-map to find the minimum realization.
$$f(A, B, C, D) = \overline{A}B\overline{C} + ABD + \overline{A}BC$$

11.29. Expand the following function into a set of minterms and then use the K-map to find the minimum realization.
$$f(A, B, C, D) = \overline{A}\overline{B}D + \overline{A}BCD + AB\overline{C} + A\overline{B}D$$

11.30. Expand the following function into a set of minterms and then use the K-map to find the minimum realization.

$$f(A, B, C, D) = \overline{B}\,\overline{C}D + \overline{A}B\overline{C}D + A\overline{C}D + \overline{B}CD + BCD$$

11.31. Find a minimum form for the switching functions
(a) $f(A, B, C, D) = \Sigma m(1, 4, 7, 9, 12, 15)$
(b) $f(A, B, C, D) = \Sigma m(4, 6, 7, 12, 14, 15)$

11.32. Find the minimum form for the switching functions
(a) $f(A, B, C, D) = \Sigma m(2, 3, 11, 13, 14, 15)$
(b) $f(A, B, C, D) = \Sigma m(0, 2, 3, 7, 9, 13)$

11.33. Find the minimum form for the switching functions
(a) $f(A, B, C, D) = \Sigma m(0, 2, 4, 10, 12, 14)$
(b) $f(A, B, C, D) = \Sigma m(3, 4, 6, 7, 11, 12, 13, 15)$

11.34. Find the minimum form for the switching functions
(a) $f(A, B, C, D) = \Sigma m(1, 4, 5, 7, 9, 15)$
(b) $f(A, B, C, D) = \Sigma m(0, 2, 5, 7, 8, 10, 13, 15)$

11.35. Design a logic circuit with a BCD input that detects all numbers that are greater than or equal to 7.

11.36. Design a logic circuit with a BCD input that detects the decimal numbers 3, 4, and 5.

11.37. In the network in Fig. P11.37, circuit A inputs certain bit patterns to circuit B. A bit pattern containing exactly two 1s is an error and should trigger the alarm circuit. Design an alarm circuit which produces a 1 if exactly two 1s appear in the input bit pattern to circuit B.

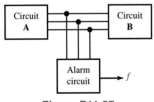

Figure P11.37

11.38. A half adder circuit has two input bits a_i and b_i. The output is a sum bit s_i and a carry bit c_{i+1}. Derive the logic equations for the sum and carry bits that are necessary for designing a half adder circuit.

11.39. The 4221 code is a weighted code in which the four bit positions are assigned the weights (4221). The code is listed below.

0 – 0000		5 – 0111	
1 – 0001		6 – 1100	
2 – 0010		7 – 1101	
3 – 0011		8 – 1110	
4 – 1000		9 – 1111	

Design a BCD to 4221 code converter.

11.40. Design a logic circuit with three inputs A, B, and C, such that the output will be high when at least two of the inputs are high.

11.41. Design a logic circuit with four inputs A, B, C, and D such that the output will be high only when an even number of the inputs, greater than zero, are high.

11.42. A logic circuit has four input lines, A, B, C, and D, and A represents the MSB and D represents the LSB. Design a logic circuit in such a way that the output is high if the input signal is outside the mid range of the input numbers, that is, in the range $(0)_{10}$ to $(3)_{10}$ or $(12)_{10}$ to $(15)_{10}$.

11.43. Three sensors are strategically placed along a production line. The sensor outputs are low if all measurements are in the normal range. Measurements outside the normal range produce a high output. The sensors are used to trigger an alarm if any two of the measurements are simultaneously high. Design a logic circuit which will produce a high signal for the alarm.

11.44. A tank of fluid employed in a chemical process is being monitored with three sensors. The sensors measure temperature, pressure, and fluid level. If all the sensor measurements are in the normal range, the sensor outputs are low. If the measurements are outside the normal range, the sensor outputs are high. Design a logic circuit that will produce a high signal for an alarm under the following conditions:

- pressure and temperature are too high
- fluid level is too high and either pressure or temperature or both are too high

11.45. An industrial plant has two reservoirs that are used to temporarily store fluid A and fluid B. Each reservoir is equipped with two sensors—one detects the incoming flow rate and the other detects the fluid level. Assume the flow rate sensor outputs are low when the flow rate is satisfactory and high when the flow rate is too high. In addition, assume the fluid level sensor outputs are low when the fluid level is satisfactory and high when the level is too high. Design a detector circuit that will produce a high output whenever the fluid level in either or both reservoirs is too

high and the flow rate for either of both reservoirs is too high.

11.46. Compute the state table for the synchronous sequential circuit shown in Fig. P11.46.

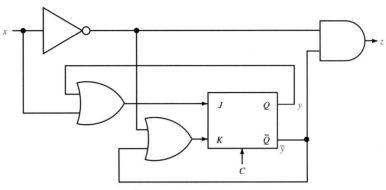

Figure P11.46

11.47. Compute the state table for the synchronous sequential circuit shown in Fig. P11.47.

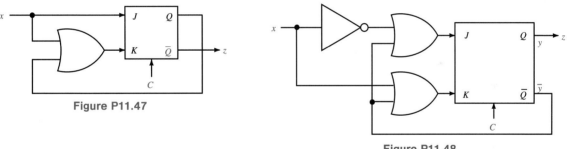

Figure P11.47

11.48. Determine the state table for the synchronous sequential circuit shown in Fig. P11.48.

Figure P11.48

11.49. Determine the state table for the synchronous sequential circuit shown in Fig. P11.49.

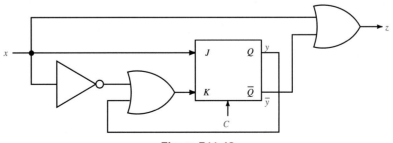

Figure P11.49

11.50. Compute the state table for the synchronous sequential circuit shown in Fig. P11.50.

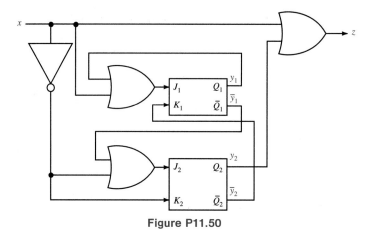

Figure P11.50

11.51. Compute the state table for the synchronous sequential circuit shown in Fig. P11.51.

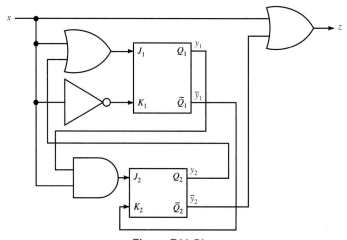

Figure P11.51

11.52. Find the logic equations for a *J–K* flip-flop realization of the state table in Fig. P11.52 using the state assignment $A = 00$, $B = 01$, $C = 11$, and $D = 10$.

	x	
	0	1
A	A/0	B/0
B	C/0	B/0
C	D/0	B/0
D	A/0	B/1

Figure P11.52

11.53. Find the logic equations for a *J–K* flip-flop realization of the state table in Fig. P11.53 using the state assignment $A = 00$, $B = 01$, $C = 11$, and $D = 10$.

	x	
	0	1
A	A/0	B/0
B	C/0	D/0
C	A/1	B/0
D	A/0	D/0

Figure P11.53

11.54. Determine the state diagram for a synchronous sequential circuit which has a single input and single output that will recognize the input string 0000 with overlap, that is, if

$$x = 0000010000$$
$$z = 0001100001$$

11.55. Determine the state diagram for a synchronous sequential circuit which has a single input and single output that will recognize the input string 0110 with overlap, that is, if

$$x = 101101100$$
$$z = 000010010$$

11.56. Determine the state diagram for the synchronous sequential circuit which has a single input and a single output that will recognize the input string 1001 with overlap, that is, if

$$x = 10100100100$$
$$z = 00000100100$$

11.57. Determine the state diagram for the synchronous sequential circuit which has a single input and single output that will recognize the input string of exactly one 1 followed by two 0s.

12

Microprocessor Systems

12.1
Introduction

It is believed that the microprocessor had its genesis with a programmable logic unit equipped with a small attendant memory. The evolution of this technology continued until the early 1970s, when the first microprocessors appeared. Their appearance signaled a new era in computing technology. Their notoriety stemmed from the fact that they are extremely powerful and very versatile. They are very small, consume little power, and yet are capable of processing in the range of 300 MIPS (million instructions per second).

In the material which follows, we will describe the basic components of any computer system, the two most important classes of microcomputers, the elements that should be considered in microprocessor system integration, and finally, a discussion of an IBM PC or compatible as an example of microprocessor system integration for engineering applications.

12.2
The Generic System

The basic computer architecture is shown in Fig. 12.1. Note that the system is composed of three fundamental units: *memory, central processing unit* (CPU), and *input/output* (I/O).

The memory stores the program (instructions) to be executed, the addresses which identify the location of actual data, and the data itself. These items are typically stored in semiconductor *random access memory* (RAM). *Read only memory* (ROM) is a specially fabricated memory which cannot be altered and normally contains such things as tables and programs that will not change during processing. The memory cells, each of which can store a single bit, are normally grouped into word lengths of 8, 16, and 32 bits. Each 8-bit grouping is called a *byte.* Each cell contains a semiconductor transistor circuit which

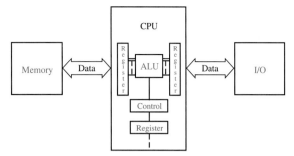

Figure 12.1   The basic computer elements.

can alternately be switched on and off. As indicated in the previous chapters, these two states of the transistor represent the binary numbers 0 and 1. In addition to the memory cells, the memory also contains the attendant circuitry necessary for both the read and write operations.

A *static* memory cell stores one bit of information in two "cross-coupled" gates as in a flip-flop described in Chapter 11. A group of flip-flops designed to store a word is called a *register*. *Dynamic* semiconductor memory stores each bit by the amount of charge placed on a small capacitor. This type memory is more dense (more cells per unit area of silicon) than static memory and is often used for large memory arrays such as RAM.

The computer's engine is the central processing unit (CPU) or simply the *processor*. The processor consists of a *control unit*, an *arithmetic-logic unit*, and *registers*. The typical registers within this unit are the program counter, instruction register, scratch pad register where intermediate results are stored, and the accumulator where results are normally accumulated. While the memory stores such things as programs, data, and final results, it is the processor that executes the program, and it does so in a sequential fashion. For example, it begins by reading the first instruction in the program. If this instruction requires data, the data are fetched from memory, manipulated in some arithmetic operation within the ALU, and the results are returned to memory. The ALU is the critical logic circuitry in the processor which actually performs the fundamental operations, such as addition, subtraction, and other Boolean operations. The actual list of operations, for example, adding two numbers, which the processor is designed to execute is called the *instruction set*.

The input/output (I/O) unit is the interface to the real world. As such, it must deal with all human instructions or instrumentation data. If the input data are not in digital form, the I/O unit will convert it into a bit stream. Some typical signals in analog form are video and speech as well as signals such as temperature, pressure, and humidity. The input device may be a device such as a keyboard or a sensor, and the output may be typically a cathode ray tube (CRT) or a printer.

The units communicate among themselves over what is called a *bus*. The bus is nothing more than an interconnecting pathway for data and instructions. Each bus has an associated bus width, the number of bits that are transferred in parallel over the bus.

It is interesting to note that although the processor and memory are very complicated circuits, they are typically less expensive components because they are made with VLSI circuits and handle only digital information. I/O units, on the other hand, are normally much simpler devices; however, they typically contain electromechanical or electromagnetic components and many of them are large and consume a considerable amount of power in comparison to the microprocessor and memory.

The hardware is typically very small and the various components are physically located close to one another. The switching speeds are becoming so fast that the travel time for the pulses, representing the bit stream, may soon become a limiting factor. For example, since the electronic components are switching in the nanosecond range, let us consider the distance traveled by a pulse in one nanosecond. Assuming that the pulse travels at the speed of light, the distance traveled is

$$3 \times 10^8 \text{ m/s} \times 10^{-9} \text{ s/ns}$$
$$= 0.3 \text{ m/ns or approximately 1 foot/ns}$$

Since the components within a system are continuously passing data back and forth, it is mandatory that the components be close to one another so that signals arrive within the proper time frame.

Software is a program or group of programs which lists the instructions that the computer is to perform. Specialized engineering tasks may require custom software, written for a specific application. The hardware is generally relatively inexpensive in comparison to the development of custom software which is often very labor intensive. Such software must be designed, tested, certified, and maintained, all of which requires a considerable amount of time and money. General purpose (generic) software applications, on the other hand, are often very cost-effective.

The internal software for a system is used to control and manage the basic units. *Memory management* allocates and recaptures blocks of memory, *processor management* controls the execution of all tasks, and *I/O management* controls the input/output devices.

In general, *programming* (writing software) is the process by which a sequence of instructions is specified in order to perform a particular task. Application programs are typically written in some specific language such as FORTRAN, Pascal, C, C ++ , BASIC, and the like. However, regardless of the language used by the programmer to formulate the problem for a computer solution, the computer will employ a *compiler* to convert this language into *machine language* which is a sequence of fundamental instructions (operational codes or op-codes) that determine the states of registers within the CPU, and controls the flow of instructions and data which determine the future states of registers.

12.3
Microcomputers

Having described the basic architecture, we now examine what is perhaps the two most important classes of microcomputer systems—the *personal computer* and the *workstation*.

Personal Computers

Personal computers have become an integral part of a number of practical applications such as database management and process control. There are several important platforms that are used today. Two of the most widely utilized are the Apple machine, such as the Macintosh, a powerful and popular microcomputer system, and the IBM PC and its clones and compatibles (computers that run the same software and accept the same plug-in printed circuit boards). Both of these systems have become major computational platforms used in industry, government, and the academic community. For purposes of illustration here we will select the IBM PC-based platform as the vehicle to discuss personal computer hardware, software, and applications. This system has widespread acceptance, is applicable to a wide spectrum of engineering applications, and is economically viable from a hardware and software standpoint.

The heart of a personal computer is the *microprocessor.* The IBM series of PCs employ the Intel 86 family of microprocessors as CPUs. The software in this family is downward compatible, that is, the code developed for older processors can still be used in the newer ones. The first chip of the 86 family was the 8088 introduced in 1978. This chip became the CPU for the IBM PC and XT and their compatibles. The 8088 was essentially a low-cost version of the 8086 chip obtained by shrinking the 16-bit wide data bus of the 8086 to an 8-bit bus. The internal computation of the 8088 chip is still performed using 16 bits. The next generation of the IBM PC, the AT, used the more advanced 80286 chip, which also employed a 16-bit CPU and used the ISA (Industrial Standard Architecture) bus. However, when IBM introduced the PS/2 (Personal System 2), it used the 80386, which is a 32-bit chip with a faster bus—Micro Channel Architecture (MCA)—which is not compatible with the ISA bus. Therefore, the PC clone manufacturers cooperated to develop the ESIA (Extended Industrial Standard Architecture) bus, which is compatible with both the ISA and the MCA.

It is important to note that the Intel family of microprocessors belongs to a category known as CISC (*Complex Instruction Set Computer*). In the CISC, the instructions are decoded into microcode, which is executed to generate hardware operations such as sending addresses to the address bus and reading in data from the data bus. The advantage of the CISC is that one instruction can perform a number of operations, and therefore the size of the memory required to store the instructions is reduced. The disadvantage of this architecture results from the fact that the execution of the instructions is through microcode, and hence the speed of the CISC is relatively slow in comparison with the hardwired microprocessor, which is called a RISC (*Reduced Instruction Set Computer*). However, this disadvantage is minimized by processors such as the 32-Bit Intel 486™ DX2 and Pentium™ processors, which are rated at about 32 and 63 SPECmarks, respectively (a standard measurement of computer speed). These microprocessors are designed to combine the advantages of both the RISC and CISC and thus reduce the speed gap which exists between personal computers and the workstations described below.

Workstations

The first generation of RISC was launched in the mid-1970s when researchers at IBM discovered an interesting fact—a typical program requires only about 20% of the instructions to be resident in the microprocessor during 80% of the computations. As a result of this discovery, a Model 801 computer was built using a set of simple instructions and a hard-wired control unit in place of the microcode in a CISC machine. During the same time frame, researchers at the University of California at Berkeley were developing the *Scalable Processor Architecture* (SPARC), while those at Stanford University were developing the *Microprocessor without Interlocking Pipeline Stages* (MIPS).

The RISC design is fundamental to most workstations. RISC is used in order to increase the speed of computation. RISC uses a very simple instruction set implemented in a hard-wired control unit, a large set of registers, and a hierarchical memory composed of slower dynamic RAM and a very fast cache, to reduce the memory access time. Cache is a small amount of very fast semiconductor memory placed between the main memory and the processor. It can greatly enhance the speed of the processor's operation by storing information most likely to be needed in a quickly accessible location. Thus, the RISC microprocessor can execute one instruction per clock cycle. Furthermore, the RISC often uses a multistage pipeline design. This pipeline architecture is similar to an assembly line in which each stage in the assembly line is analogous to a stage in the pipeline. In this mode, multiple instructions can be simultaneously executed. This parallel operation coupled with the simple instruction design makes the pipeline simple and efficient.

The second generation of workstations resulted from the first generation of RISC research. The 32-bit SPARC stations 1 and 2 from Sun Microsystems, and the 32-bit R2000 and R3000 from MIPS, are examples of second generation machines. The speed of the second generation of RISC processors is about 10 to 20 SPECmarks. It appears that the success of these workstations provided the pressure for major CISC minicomputer maker DEC and mainframe CISC computer maker IBM to develop RISC-based workstations.

The third generation of RISC microprocessors employs two advanced techniques: the *superpipeline* and the *superscalar*. The superpipeline design uses more stages so that the workload is further divided to increase the speed. The superscalar design uses multiple pipelines to perform jobs in parallel and thus increase throughput.

There are a variety of third generation workstations manufactured by, for example, DEC, HP, IBM, MIPS, Silicon Graphics, and SUN that run in the 70 to 300 SPECmark range. The speed is proportional to clock rate, and in the near future will be in the range of 500 SPECmarks or 1 Giga Flop (1 billion floating point operations per second).

EXAMPLE 12.1

Given that a superscalar RISC computer is capable of performing several tasks simultaneously in order to reduce computation time, we wish to determine if it can increase the speed of a job which has inherent parallelism only.

The superscalar RISC computer has a high speed rating; however, it is not useful for all computations. Users have to actually run the job and measure the speed of computation in order to determine the true speed. ■

DRILL EXERCISE

D12.1. Since the superpipelined RISC processor divides a job into multiple stages, it does not require parallelism for fast computation as does the superscalar RISC processor; hence, is it suitable for any computation?

Ans: No. As an analogy, if you have only one car to assemble, you don't need an assembly line. The same is true for jobs running on the superpipelined RISC.

System Software

The system software for a microprocessor system consists of a number of very distinct and important components.

Operating System. An operating system is generally a machine language program which provides the direct interface between the CPU and all other devices; it controls all I/O devices attached to the microprocessor, and provides the necessary environment for the machine to run other programs. For example, programs such as compilers, assemblers, graphical user interfaces (GUI), and word processors run under an operating system. The two most popular *operating systems* are DOS for IBM PCs and their compatibles and UNIX for workstations.

Compiler. *Compilers* are used to translate programs written in high-level languages into machine codes, which are binary codes that the computer can execute. There are a variety of high-level languages; for example, C, FORTRAN, PASCAL, BASIC, and COBOL, and each one has its own particular application regime. FORTRAN is used in scientific applications, COBOL is used in business applications, and C is very versatile and very effective in almost all applications, particularly the computer industry. Most language compilers employed in PCs or workstations adhere to an American National Standards Institute (ANSI) standard.

Assembler. The *assembler* translates the program from assembly language to machine code. The assembly language is a mnemonic representation of the binary machine code, and writing programs in an assembly language is a direct way of controlling the operations of a microprocessor. Therefore, assembly language programs are an effective vehicle for talking directly to hardware and manipulating bit streams. For example, in modern automobiles, sensors are attached to such things as the engine, fuel tank, and doors. These sensors supply information to a microprocessor which controls the various warning signals. The status of these automobile components can be sensed bit by bit, for example, one bit represents the status of a door, and the data are easily manipulated by an assembly language program.

Graphical User Interface. A *graphical user interface* or GUI greatly simplifies the use of computers. Using a mouse, one need only point and click on the icons. The GUI simply accepts the user's commands and translates them into operating system operations. Most workstations and high-end personal computers employ a GUI. For example, Microsoft Windows is a popular GUI used in IBM PCs and their compatibles, and Sun SPARC stations use Open Windows.

12.4
Microprocessor System Integration Considerations

Because of their versatile hardware and software, personal computers and workstations can be used in a wide variety of applications. These applications span a spectrum that ranges from general computing to control of a very specific process in a plant. Since there is an enormous amount of hardware and software support available for these existing machines, there is generally no reason to design a microprocessor-based system from scratch. Therefore, rather than invest in the development cost of an application-oriented system, it is much easier to employ an existing system which has been modified for the specific application. This approach utilizes, to the fullest extent, the enormous investment which the computer industry has made in their existing systems, and reduces the development time required to reach the target application. A brief summary of the items to be considered in microprocessor system integration is listed below.

Speed Requirements

The current vintage of personal computers and workstations has architectures that are so complicated it is difficult to predict the speed at which a particular software package will be processed. However, an extrapolation of the experience gained in other applications is often a viable basis for selecting a computer model. In addition, other specifications may also be helpful in arriving at a correct decision; for example, the sampling rate of the input signals, the bandwidth of a closed loop control system, and the like. In this context it is important to note that while a high-end workstation has a better capability for floating point (non-integer) number processing, it has roughly the same capability for integer calculations as a high-end personal computer.

Some good advice on computer selection, to satisfy speed requirements, can also be gleaned from benchmark reports for similar applications in computer trade magazines. In general, it is difficult to select a machine that will satisfy precisely the stated requirements, and therefore, it is always prudent to assume a worst-case scenario in order to leave some design margin.

Interface Design

Because of the tremendous proliferation of computers, most scientific devices and instruments available today have been designed so that they can easily be interfaced to a computer. Perhaps the two most popular standard interface specifications are the *serial interface* (*RS-232C*) for low-speed devices and instruments such as those used for temperature measurement, and the *IEEE-488 interface* which is suitable for medium-speed instruments such as digitizers and waveform generators.

It is important to note that there are a significant number of standard plug-in printed circuit boards for personal computers and workstations that are designed for various applications. As an example, an analog-to-digital (A/D) converter card with a variety of sampling rates and on-board software support can be purchased and plugged into a computer in order to interface with an analog signal. Therefore, by carefully selecting off-the-shelf interface hardware, interface development time can be kept to a minimum.

Software Tools

The major cost in designing an application-oriented computer system is the software development. Interface hardware is inexpensive by comparison. Therefore, the two most critical items in computer selection are software support and computer speed. Most interface hardware comes with some software support, and therefore, priority consideration should be given to the interface hardware with the best software support.

In order to tailor existing software to a specific target application, a compiler or an assembler is required for the software development. In addition, a graphical user interface may be needed to report the results of the development effort. Therefore, a knowledge base of the existing software tools is important in expediting the development.

12.5

System Integration Using an IBM PC or Compatible

Because the IBM PC and its compatibles are relatively inexpensive systems, have excellent hardware interface support, a wide variety of applications software, and are widely used, we select this platform as an example of microprocessor system integration for engineering applications. A similar approach is applicable for other machines.

In our analysis we will consider the three main issues presented above: speed considerations, the hardware interfaces, and the available software tools.

Speed Considerations

The speed of a computer depends upon the speed of its various components. Therefore, we will briefly discuss speed and its relationship to the processor, memory design, bus, coprocessor, and hard disk.

Processor. Table 12.1 lists some of the important characteristics of the processors currently employed in IBM PCs and their compatibles.

The speed of the microprocessor is roughly proportional to the speed of the clock. The SX chips have only a 16-bit external data bus which slows down the memory and I/O access time, because two clock cycles are required to produce a 32-bit word. The clock circuitry in the Intel 486 DX chip is standard (i.e., it does not have a clock doubling circuit); therefore, it has the same internal and external clocks. The 486 DX2 chip uses a clock doubling technique to increase the internal clock speed. This technique increases microprocessor speed without increasing the cost of the memory and I/O for matching the speed

Table 12.1 Processor Characteristics

Intel processor name	No. of internal data bits	No. of external data bits	No. of address bits	Internal clock speed (MHz)	External clock speed (MHz)
386 SX	32	16	32	20	20
				25	25
386 DX	32	32	32	25	25
				33	33
				40	40
486 SX	32	16	32	20	20
				25	25
486 DX2	32	32	32	50	25
				66	33
486 DX4	32	32	32	100	25
486 DX	32	32	32	33	33
				50	50
Pentinum	32	64	32	60	60
				66	66
				90	90
				100	100

of the processor. The Intel 486 (including SX, DX, and DX2) contains 8k of on-chip cache memory to serve as a buffer for memory access.

EXAMPLE 12.2

If the Intel 486 has 32 address lines, how much memory can it accommodate?

Since the chip has 32 address lines and the smallest addressable memory location is a byte, it can accommodate 2^{32}. 4 gigabytes of memory. ■

DRILL EXERCISE

> **D12.2.** A chip has 20 address lines. How much memory can this chip accommodate?
>
> *Ans:* $2^{20} = 1$ megabyte.

Memory Design. A hierarchical memory, including a small amount of very fast *cache,* that is, static RAM (SRAM), and a large amount of *main memory* consisting of dynamic RAM (DRAM), is a design widely used to achieve a better performance/cost trade-off for high-speed microprocessors. The cache controller selectively moves the information most likely to be used by the processor from the DRAM to the cache in order to reduce the memory access time. If all the data needed by the CPU is in cache (a cache "hit"), the data can be accessed in about 10 nanoseconds (SRAM access time). If the cache controller is unable to have the required information resident in cache when needed by the CPU, the access time is about 80 nanoseconds (DRAM access time). The cache controller maintains coherency between the cache and main memory, and has built-in logic to minimize the probability of a CPU request for data that is not resident in cache, that is, a cache "miss."

Cache memory typically comes in the following sizes: 32k, 64k, 128k, and 256k. The more cache available, the less probability of a cache miss, and the faster the speed of computation. The amount of cache required will, of course, depend upon the particular application. The main memory may typically be 2, 4, 8, or 16 Mbytes. In order to run a large code, such as Microsoft Windows, at least 4-Mbyte of memory is necessary to reduce the waiting time required to move codes between the hard disk and main memory. Upgrading the main memory for large codes will normally lead to a speed improvement.

Intel has included an 8k cache in the 486 microprocessor. Having cache memory physically present on chip reduces the distance for communication and further decreases the memory access time. In addition, the on-chip cache has a 16-byte wide data path to the CPU, which further enhances the communication bandwidth. Therefore, the Intel 486 microprocessor has a three-level hierarchical memory consisting of the on-chip cache, external cache, and main memory. The Intel 386 microprocessor without the on-chip cache has only a two-level memory.

EXAMPLE 12.3

A hierarchical memory system consists of high-speed cache memory (static RAM) and low-speed dynamic RAM. Is the purpose of this hierarchial structure (1) to store in the cache the information needed for performing the instruction in the next clock cycle, or (2) to temporarily transfer from the dynamic RAM to the cache the information with the highest probability of being requested in the next clock cycle?

The answer to this question is (2). Since there is no mechanism for knowing exactly what information is needed during the next clock cycle, a hierarchial memory is designed from a statistical point of view to place in the cache the information most likely to be requested. When the information needed is not in the cache—what is called a *cache miss*—the processor will take extra time to move the needed information from the dynamic RAM to the cache. ∎

DRILL EXERCISE

D12.3. Is it true that the memory access speed is crucial to computer speed because a high-speed processor has nothing to work on if the memory cannot provide the data for the CPU in time?

Ans: Yes.

Bus. Computers, including PCs and workstations, have a *mother board* where the CPU resides. This mother board has a number of expansion slots in which users can plug in such things as a memory card, parallel and serial I/O card, graphics card, and the like. The expansion slots on the mother board are connected together by a bus. This bus consists of data, address, and control buses. For example, a 32-bit wide data bus means 32 lines for data transfer between CPU, memory, and I/O. The address bus is used to send an address, and the control bus is used to handle the protocols of data flow, interrupts from I/O devices, direct memory accesses, etc.

The three bus specifications employed in IBM PCs and compatibles are the ISA, MCA, and EISA. An overview of the major characteristics of these three buses is given in Table 12.2. The PC, XT, and AT buses have only 20 and 24 address bits, respectively, permitting only 1 and 16 Mbytes of memory, respectively. The ISA is used for an 8-bit wide data bus for the 88 processor and a 16-bit bus for the 286 processor. However, the MCA and EISA have a 32-bit wide data bus to match the 32-bit Intel 386 and 486 microprocessors. It is interesting to note that although the ISA is a slow bus, it is still a widely used bus for these processors because of its low cost and rich support in terms of plug-in cards. When making a purchasing decision, one must consider not only the speed requirement, but the cost as well.

Coprocessor. (Floating Point Unit) For applications such as sophisticated graphics and solving differential equations, the computer must process numerous floating point number computations. A very large or very small number must be represented as a floating point number in scientific notation in order to be stored and processed by a microprocesso

Table 12.2 Bus Characteristics

Bus name	No. of address bits	No. of data bits	Peak data transfer rate (Mbytes/sec)
ISA of PC XT	20	8	0.5
ISA of PC AT	24	16	0.8
MCA	32	32	20
EISA	32	32	33

with a finite number of bits. For example, the number 4.35×10^{27} cannot be represented by a 32-bit number because the maximum number represented by a 32-bit number is $2^{32} - 1$; however, it can be represented by a floating point number in which 4.35 is the mantissa and 27 is the exponent. Floating point number computation is slow, even when using the best library in a microprocessor, without a coprocessor, because a microprocessor has only an integer arithmetic unit. Hence, a *coprocessor* for floating point number processing significantly increases the computing speed for programs requiring this kind of computation.

A floating point standard, IEEE 754, was established by a number of coprocessor chip manufacturers. This standard specifies the formats of *single-precision* (32-bit) and *double-precision* (64-bit) floating point numbers. The operations include addition, substraction, multiplication, division, remainder, and square root. In addition, most of the chip makers implement transcendental functions such as sine, cosine, tangent, arctangent, log, and exponential functions in their chips. The 386 uses the 387 coprocessor (or its clones) or the Weitek ABACUS (3167) for floating point number processing. The Intel 486 SX uses the 487SX math coprocessor; however, the Intel 486 DX or DX2 and Pentinum processors use an on-chip floating point processor. The Intel 486 and Pentinum processor designs significantly improve the speed of computation because they reduce the communication time between the CPU and the coprocessor.

Hard Disk.   A hard disk stores data magnetically on spinning platters in an electromechanical system (the hard disk drive). Applications which require frequent access to large amounts of data or frequent transfers of data to and from hard disks, need to employ a high speed interface with a hard disk. Current hard disks have an access time on the order of milliseconds due to the limitations of the mechanical system; this is very slow compared to semiconductor memory. Perhaps the best way to improve the disk data transfer rate is to select a computer with a disk cache. The *disk cache* is in essence a hierarchical memory design in which the cache controller moves into the cache the information most likely to be requested by the processor. This approach prevents the processor from having to run at the slow speed defined by the mechanical access time of the disk.

EXAMPLE 12.4

When we choose a computer for a high-speed application, we consider (1) the processor speed only, that is, buy the computer with the highest clock rate, or (2) the processor speed, the amount of cache memory, bus speed, the hard disk access time, and disk cache to achieve the optimal speed for the particular application.

The prudent approach is that described by (2) and the considerations are typically application dependent, for example:

(a) high-speed requirement for input/output data to an external instrument: bus speed is important, and Direct Memory Access (DMA) should be used to maximize the speed.
(b) high-speed requirement for running a reasonably sized (100k range) code: cache size needs to be chosen appropriately. Although there is no way to predict the performance without actually running the code, it is always important to consider the worst case and leave a safe margin for the computer requirements.
(c) high-speed requirement for storing or reading data to/from a hard disk: a high-speed hard disk and disk cache should be considered to meet the requirement. For example, a

system used to monitor and record current changes in a motor and a system for recording the temperature of a motor have different requirements for processor and hard disk speeds, since the current changes rapidly but the temperature does not.   ■

D12.4. If a microprocessor has a 16-bit wide data bus, how many memory accesses are required for a 32-bit word?

Ans: Two.

Interface Design

Interface Cards. There are a number of interface cards that can be used in a variety of applications. Some of the more popular cards are listed in Table 12.3. In addition to the list provided in the table, there are cards for digitizers, digital oscilloscopes, waveform generators, stepping motor controllers, local area network controllers, and the like.

Speed Considerations. Speed is also a primary consideration in interface design. For example, a real-time control system, such as a plant process controller, must be capable of responding to external conditions within some required time frame. The interface card must acquire the external data, send it to the CPU, receive the CPU commands, and forward these commands to the device under control within some specified time limit. There

Table 12.3 The Characteristics of Interface Cards

Name of the interface card	Description	Application
A to D (A/D)	To convert analog signals to digital signals	To input external analog signals for digital processing
D to A (D/A)	To convert digital signals to analog signals	To convert a digital signal to the analog world, for example, driving an actuator
Serial interface (RS-232C)	To convert a parallel digital signal to a serial digital signal	(1) To connect to serial printers and MODEMs (2) To connect to low-speed instruments
Parallel interface	To communicate with external devices in parallel digital signals	(1) To connect to parallel printers (2) To connect to instruments and devices
IEEE-488 bus (or GPIB)	To communicate with a set of instruments with the same interface	To set up an automatic measurement or control system by a network of instruments
Frame grabber	To digitize images from cameras	To input images for digital image processing

are two primary methods for handling the I/O: *Interrupt,* and *Direct Memory Access* (DMA).

The interrupt, which is a hardware signal, is used by an interface card to request the CPU to perform a subroutine, that is, an interrupt service routine stored in main memory, in order to input or output data. As an example, suppose that a personal computer is used to process a number of jobs, one of which is to monitor the temperature in a room. The computer turns on the compressor to cool the air when the temperature in the room rises above one threshold, and turns the compressor off when the temperature falls below another threshold. This operation is performed in the following manner. The CPU receives the interrupt request from an interface card which senses the temperature periodically. When the interrupt is received, the CPU completes the current operation for the machine code under execution, saves the information necessary to resume the interrupted job, and then acknowledges the request from the interface card with a hardware signal. When the interface card receives the acknowledge signal, an abbreviated representation of the name of the subroutine for performing the input operations is sent to the CPU, which begins executing the subroutine. This subroutine first saves information which would otherwise be destroyed, and then inputs the temperature from the interface card and checks to see if it is necessary to turn on or turn off the compressor. If required, a signal is sent to the compressor. The subroutine then restores the saved information and lets the CPU resume processing of the original interrupted job.

Interrupt I/O is an efficient way to handle the input and output while processing other jobs, because the CPU will stop computing only when the interrupt request is received. The overhead associated with the interrupt involves such things as the time required to save the information necessary to resume the interrupted job, and the time required to save and restore the information that the interrupt service routine destroys. Therefore, interrupt I/O is suitable for medium-speed I/O because the sampling rate of external signals is typically less than ten thousand samples per second.

Some interface cards have a built-in RAM or a *FIFO* (first-in-first-out) *buffer.* This memory, which allows data to be saved in the interface cards, permits the CPU to read all the data in the RAM or FIFO by performing the interrupt service routine only once, and thus reduces the number of interrupts issued by the interface card. Therefore, the built-in memory for interface cards results in less interrupt overhead and a corresponding improvement in overall performance.

When the overhead of an interrupt service is so high that it is impossible to receive all of the data and process it simultaneously, an interface card with memory can be used to perform block data transfer (instead of one word at a time) so that the frequency of interrupts can be reduced in order to reduce the overhead.

If a high-speed I/O transfer rate is required to process a high volume of information in real time, then DMA should be used to reduce the I/O overhead. The DMA controller is a special processor for handling the direct data transfer between memory and I/O without using the CPU. For example, suppose that a speech signal is inputted at 20 kilosamples/sec by DMA. Whenever a block of speech data is sampled, converted to digital data, and stored in the buffer of an interface card with a *DMA controller,* a DMA request is issued by the interface card. As a result of this request, the CPU suspends its current operation, acknowledges the DMA request, and disconnects itself from the bus as shown

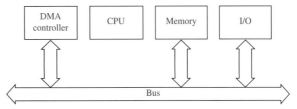

Figure 12.2   CPU suspends current operation and DMA takes control of the bus.

in Fig. 12.2. The DMA controller then takes control of the bus and issues the control signals necessary to move the digital data from the buffer to memory. As soon as a block of data in the interface card has been moved to memory, the DMA controller releases control of the bus, and the CPU reconnects to the bus and resumes processing. It is not necessary for the DMA controller to execute the machine codes for performing a data transfer; however, the CPU has to fetch the machine codes and execute them to perform the same task. Therefore, the DMA is about one order of magnitude faster than the CPU for I/O data transfer.

EXAMPLE 12.5

The Direct Memory Access (DMA) technique is used when (1) a significant amount of data must be transferred to memory from an instrument with a high speed requirement, or (2) a significant amount of data needs to be manipulated by a CPU with a high speed requirement.

The DMA process is used to input or output data between memory and I/O with a high speed requirement. During the DMA process, the CPU is in an idle state and the DMA controller takes control of the bus to handle the data transfer. Therefore, the CPU cannot use the DMA to manipulate data.   ■

DRILL EXERCISE

D12.5. In a manner similar to an interrupt process, does the CPU save the current job information in order to resume the job after the DMA is finished?

Ans: No. The CPU suspends its operation and lets the DMA controller have control of the bus; thus, the registers inside the CPU are not altered by the DMA and hence, there is no need to save this information. This is one of the reasons that the DMA process is faster than the interrupt process.

EXAMPLE 12.6

What factor need not be considered in selecting an A/D or D/A converter card: (1) if the card has a DMA interface for high-speed data transfer, (2) if the card has the data conversion rate for the application, (3) if the card has on-board memory to minimize the data transfer time, (4) if the card has software support to save time for software development, or (5) how the A/D or D/A chips are enabled through an address decoder.

The DMA interface is crucial for high-speed A/D or D/A operations.

The data conversion rate has to meet the speed requirement.

The on-board memory reduces the strict requirement for the CPU to perform an output/input operation within a fixed period because the on-board memory serves as a buffer.

The appropriate software support can conserve a significant amount of development time.

The minor detail of decoding circuitry is not a consideration for choosing an A/D or D/A card. ∎

Software Tools

Software Tools for I/O. Table 12.4 provides a summary of software packages used for interfacing and controlling external devices. These packages can be used not only to input, output, and control, but to record, display, and analyze as well. The appropriate use of these packages significantly enhances project development. When the packages cannot satisfy the needs of a project, then programs must be written to perform the necessary tasks. Special languages for data acquisition and analysis can be used to shorten the de-

Table 12.4 Interface/Control Packages

Software Package	Description
Laboratory data acquisition and control	Real-time data acquisition (A/D or D/A), process control (open or closed loop), analysis and curve fitting routine for experimental data.
Data acquisition and analysis programming language	Programming language with utilities for data acquisition (A/D or D/A), analysis, graphics, and statistics.
Menu-driven data acquisition and analysis	Menu driven (without programming) for data acquisition, analysis graphics, and statistics.
Menu driven for control of IEEE-488	Menu-driven data acquisition, analysis, and graphics for IEEE-488 instruments.
Real-time data acquisition and display system	High-speed data acquisition and graphics display to replace a chart recorder.
High-speed data to hard disk transfer	High-speed data transfer to/from hard disk or virtual (RAM) disk.
Industrial control and monitoring	Process control and alarming, data acquisition and real-time display.
Storage scope emulator	Storage oscilloscope emulation and signal analysis.
C, BASIC, and PASCAL utility package	Utilities for C, BASIC, and PASCAL to generate real-time graphics and to perform data acquisition and PID control.

velopment time; however, if the task is very complicated, the use of a general language is perhaps a better tool for programming. There exist some utility packages which provide extremely useful libraries capable of reducing the effort involved in handling data acquisition and graphics; hence, programmers need only focus on the particular application at hand.

Language Considerations. C is one of the most popular languages for developing applications in industry because it is a high-level language with the capability of a low-level assembly language. Programmers can use the C language to write and debug code much faster than with an assembly language; and in addition, they have the capability to interact with the hardware directly, just as they would with an assembly language. However, compilers are still incapable of generating machine codes that run as fast as the human-written assembly language. Most C compilers for an IBM PC can accept a mixed code composed of both C and assembly language statements so that the time-critical portion of the code can be written in assembly language and the remaining part can be written in C. This dual approach yields the most efficient code, since the running speed and the development time are both optimized.

EXAMPLE 12.7

In which case would we have to write a program in assembly language even though this is time consuming? (a) the program is short, (b) the program requires a high speed, (c) the program needs to talk with hardware directly, (d) the program is easy to debug, or (e) the program requires bit operations.

High-level languages (such as C and FORTRAN) are easy to write and debug; C language can talk with hardware directly and can perform bit operations similar to the assembly language. However, when speed is an absolute requirement for the program, human beings can generally produce better codes using assembly language than present C or FORTRAN compilers can produce.   ■

DRILL EXERCISE

D12.6. Does a program written in C language and compiled to machine code require less memory for storage than an assembly code written manually for the same purpose?

Ans: No. In general, the assembly code directly programs the microprocessor and takes less memory for code storage than the machine code generated by a C compiler for the same purpose.

EXAMPLE 12.8

Which of the following features is not characteristic of C language? (a) bit operations, (b) I/O operations, (c) looping operations, (d) talking with hardware directly, (e) less readability than assembly code.

The answer is (e), the last feature. High-level languages are closer to human language; hence, they are easier to read.   ■

D12.7. A statement in C language is equivalent to several instructions in an assembly language. Therefore, does the machine code generated from C language take less space than that written directly in assembly language?

Ans: No. The code in an assembly language written by human beings usually is in a compact logic and takes less memory.

C++ is an *object-oriented language* built upon the C language, that is, C is a subset of C++, and is suitable for the development of large-scale codes. The computer industry has shifted from C to C++ because of the significant improvement in productivity. The object-oriented approach makes the code reusable for all future projects and reduces the amount of software maintenance. The significant coding enhancement provided by C++ is the use of classes, which consist of data and functions. A class can be used as many times as possible by employing different data, so that a class for a given set of data is treated independently and is called an object. Therefore, a program can be viewed as the message flow between a number of objects. The use of C++ will result in a significant reduction in development costs, and therefore, it is always a viable software tool.

EXAMPLE 12.9

A large project team is undertaking the development of software for a major interdisciplinary project. Which language is the most suitable and why? (a) C language, (b) C++ language, (c) assembly language, (d) FORTRAN language, or (e) COBOL language.

C++ language, an object-oriented programming language, is better than a procedural language (such as C, Assembly, FORTRAN, or COBOL) in handling the communication between objects (data and functions) so that programmers can treat each object as a black box and handle only the parameters for communication to each object. C++ is even closer to human language than C language. ■

EXAMPLE 12.10

Since the cost of software development is generally the major part of a computer system's costs for a specific application, such as the process control of a new chemical plant, the strategy is (1) to buy codes suitable for this particular application and modify them to fit the needs using the appropriate language, or (2) to choose an appropriate language and develop the software from scratch to have an efficient system.

For all the reasons discussed above, the first strategy is best. ■

12.6
Summary

- The basic computer architecture consists of a memory, central processing unit (CPU), and input/output (I/O).
- Memory may consist of both Random Access Memory (RAM) and Read Only Memory (ROM).
- The CPU normally consists of a control unit, an arithmetic logic unit (ALU), and registers.

- The I/O units may be a cathode ray tube (CRT), keyboard, printer, and the like.
- Two important classes of microcomputers are personal computers (PCs) and workstations.
- Two types of microprocessors are CISC (Complex Instruction Set Computer) and RISC (Reduced Instruction Set Computer).
- The operating system, compiler, assembler, and graphical user interface are types of system software.
- Cache memory is very high-speed memory located close to the processor to reduce memory access time.
- A bus is a communication highway within the computer.
- Two methods used for I/O are interrupt and direct memory access (DMA).

PROBLEMS

12.1. If a cache memory is present, the memory speed is determined by the static RAM access time. (Yes/No)

12.2. Based upon the information provided in Table 12.1, the Intel 386 SX may be used to design a low-speed and low-cost system. (Yes/No)

12.3. The reason that the DMA controller can move data faster than the CPU is that the DMA only knows to move data between specified locations, whereas the CPU must fetch and execute machine codes to decide where to move data. (Yes/No)

12.4. What is the range of an unsigned integer in a 64-bit representation?

12.5. The maximum speed of the serial interface is 19200 baud (bits per second). How many bytes can this serial interface transfer in a second?

12.6. For a fixed rate interrupt, for example, a clock in a personal computer, what is the best way to generate a fixed rate interrupt to the CPU? (1) a software code running in the CPU, (2) a programmable timer chip, or (3) the development of a special electronic circuit.

12.7. The assembly language is a direct way to program a hardware operation inside a microprocessor since each instruction can be translated into a binary machine code (one to several bytes), and each machine code performs a specific operation among the microprocessor, memory, and I/O. (Yes/No)

12.8. Is it possible to write part of a code, requiring speed, using an assembly language, and part of the same code, requiring no speed, using C language to save development time for programming and debugging? (Yes/No)

12.9. The superpipelined RISC processor uses hardware to divide the computation into a few stages; hence, it saves the computation time required for adding two integers together. (Yes/No)

12.10. How long does it take to fill an 8k cache memory if the system memory clock is 25 MHz and the cache stores one byte on each clock cycle?

12.11. A RAM (Random Access Memory) has 22 address lines, and an access time of 12 ns. a) How many bits of information can be stored? b) How many bytes of information can be stored? c) If this memory is organized so that one byte of information is available at the output terminals simultaneously, what is the minimum time required to access 8 bits?

12.12. A DRAM is operating in a microprocessor system with a memory clock of 12 MHz. This memory is organized so only one bit is available at a time (a "by-one" organization).
(a) If this memory is a 4Meg memory (meaning it is capable of storing just over 4 million bits), how many address lines are required?
(b) how much time is required to retrieve a 32 bit word from this memory?

12.13. The following seven steps are given in random order. Arrange them in the proper sequence to describe the steps a CPU might take to service an interrupt signal generated by pressing a key on a keyboard, and storing the key code in main memory.

(a) Acknowledge interrupt signal from the keyboard controller.

(b) Jump to a special sequence of machine language codes called the "interrupt service routine" for the keyboard input.

(c) Store the key code in main memory.

(d) Resume execution of the main program sequence.

(e) Load the CPU program counter and other registers from reserved memory locations.

(f) Fetch the key code from the keyboard controller and temporarily store the code in an internal register.

(g) Store the program counter and other internal CPU register information in reserved memory locations.

12.14. Assume the single precision standard for a 32-bit word is as follows: The highest order bit (MSB) is the sign bit ("0" is +), followed by 8 bits for the exponent in excess 127 format (that is, subtract 127 from stored exponent to obtain actual exponent), followed by 23 bits for the mantissa. Write the 32-bit binary word used to represent each of the following:

(a) 2,381

(b) −2,381

(c) 1,284,348

(d) 6×10^{23}

(e) 4.38426×10^{18}

(f) 4.38426×10^{-18}

12.15. A modern super video display has 1280×1024 pixels of full color. The binary information for this display array is usually stored in local memory (VRAM) on the video interface card. Calculate and compare the VRAM memory requirements for this display if it supports

(a) 65,536 colors

(b) 16,777,216 colors

12.16. A certain processor system performs DMA transfers from main memory to video memory. The memory bus is 32 bits wide (32-bit words) and has a bus clock rate of 25 MHz. What is the fastest time a 1024 word memory transfer could be performed?

12.17. A processor system performs DMA transfers from main memory to a cache in 64 word sectors. The memory bus is 32 bits wide (32-bit words) and has a bus clock rate of 16 MHz. Assume that the CPU with a CPU clock rate of 66 MHz calls for 3 words of data from the cache. Assuming no cache "misses" and starting from an empty cache

(a) How long will it take to fill one sector in the cache

(b) How long will it take for the CPU to obtain the 3 words of data from the cache

(c) How much time is required for the entire process?

12.18. Calculate the answer for problem 12.17 assuming the ratio of cache "hits" to "misses" is $2:1$.

Microelectronics Design and Manufacturing

13.1
Introduction

Recent innovations in microelectronic technologies have drastically changed the way in which electronic systems are implemented. *Microelectronics,* as the name implies, is the design, development, and manufacture of electronic circuits and systems that are extremely small and very complex. It is possible, for example, with microelectronic technology today to build a complete memory circuit no bigger than your fingernail which contains computer memory logic capable of storing tens of millions of bits of information. A circuit made using microelectronic technologies is termed an *integrated circuit* (IC) or a *chip,* and the entire circuit is constructed on one substrate.

While it may be unlikely that you will become a microelectronics specialist, it is highly possible that at some point in your career you may become a customer of an IC design or semiconductor manufacturing organization. You may be a member of a project team that has a need for some form of advanced electronics, and the development or purchase of ICs will be an important consideration. This chapter will provide you with a general understanding of IC design and manufacturing, and enable you to intelligently and effectively communicate with the IC design and process engineers in order to accomplish your goals.

There are a wide variety of microelectronic technologies available, as shown in Fig. 13.1. *Film-integrated circuits* are not as dense, that is, there are fewer circuits per unit area, as semiconductor ICs, and therefore they are not described here. Semiconductor ICs are almost always made using a silicon substrate or base layer, although some other materials, such as gallium arsenide, are sometimes used for special applications. Semicon-

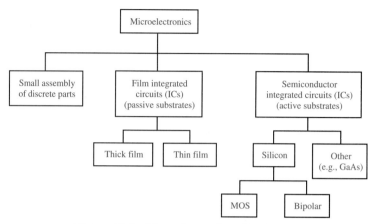

Figure 13.1 Microelectronic technologies.

ductor ICs are sometimes called *monolithic* ICs, since they are fabricated on single crystal, or monolithic, material. *Monolithic* comes from the Greek meaning "a single stone."

The invention of the transistor in 1947 by Bell Laboratories and the invention of the integrated circuit in 1958 separate the history of electronics into three "eras." Prior to 1947, all electronic circuits were made using *vacuum tubes*. These large devices, usually about the size of a small light bulb, produced quite a bit of heat, were inefficient, unreliable, and required several hundred volts to operate. A typical vacuum tube placed beside a dime to show the scale is presented in Fig. 13.2. During this era radios and other electronic equipment were large and expensive. The first large digital computers were implemented using vacuum tubes and these systems filled entire rooms and required an air conditioning system as large as the room housing the computer. The computing power of such a system was less than that now available in a typical personal computer.

In the 1950s and 1960s the *transistor* replaced the vacuum tube in most applications, and circuits were implemented using discrete transistors—individual transistors mounted in a package with wire connections. *Printed circuit boards* (PCBs) were used to connect the transistors together. In PCB technology, an insulating board has a layer of conducting material, usually copper, laminated to its surface. The copper layer is then etched in a pattern so that the remaining copper pattern provides the electrical interconnections between individual components which are soldered into holes drilled through the board, as was shown in Fig. 8.1(b). PCBs are still used to physically support and interconnect ICs, as illustrated in the modern computer board shown in Fig. 13.3.

The notion that transistors could be "integrated" with other components to achieve an entire circuit function on a single piece of silicon material was a benchmark in the development of electronic circuitry. In 1958, Jack Kilby of Texas Instruments combined, for the first time on one substrate, active solid state devices, or transistors, with the surrounding circuitry. Shortly following this development, Robert Noyce of Fairchild Semiconductor

Figure 13.2 Standard vacuum tube.

Corporation completed another version of a complete circuit on a single piece of semiconductor material. Kilby and Noyce, although they were working for different companies, are considered to be the co-inventors of the integrated circuit.

From these embryonic developments, the semiconductor industry has constantly focused on increasing the complexity of the circuits implemented on a single chip and concurrently reducing the dimensions of the components on each chip, both of which have had the effect of increasing circuit performance.

The fabrication process for ICs simultaneously produces a matrix of identical chips on a circular slab of semiconductor (usually silicon) material called a *wafer*. A wafer or "slice," as it's sometimes called, is fabricated by cutting a "boule" of single-crystal silicon into thin slices much like one slices a loaf of bread. Boules are cylinders of pure silicon material solidified from molten silicon. Generally the boule is ground flat along one side to identify the orientation of the crystal planes within the single crystal material. Thus, each wafer, when sliced from the boule, has a "flat" section on its circumference as shown in Fig. 13.4.

Each of these slices or wafers will serve as a substrate material on which an array of identical integrated circuits will be fabricated as illustrated in Fig. 13.5. Over the past two decades, the diameter of silicon wafers used in manufacturing has increased as the ability to manufacture and process pure silicon material of increasing diameters has improved. In the early 1970s, one-inch diameter silicon wafers were common; in the 1990s companies are utilizing wafer fabrication equipment compatible with 6-inch, 8-inch, and even larger diameter wafers. Since the area of a silicon wafer increases as the square of its di-

Figure 13.3   Modern computer board—VME960MX-SBC. (Courtesy of Tronix Product Development)

Figure 13.4   Wafer sliced from a silicon boule.

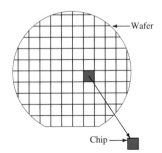

Figure 13.5   Array of IC chips on a wafer.

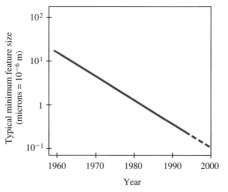

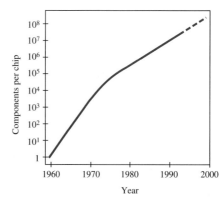

Figure 13.6 IC minimum feature size as a function of time.

Figure 13.7 Growth in the number of components per chip.

ameter increases, larger wafer diameters significantly increase the number of circuits that can be made from processing a single wafer.

Simultaneously with the increase in wafer diameter over the last several decades, improvements in photolithography have allowed the *minimum feature size,* that is, the physical size of the smallest design feature, to be continuously reduced. This dimension is commonly the gate length of an MOS transistor, or the width of the smallest metal contact. It has decreased from a typical value of 10 microns in the 1970s, to 3 microns in the 1980s, to less than a single micron in the 1990s, as illustrated in Fig. 13.6.

The smaller feature size, combined with improved processing of larger silicon areas, has led to a continuous increase in the maximum number of components that can be placed on a chip. Figure 13.7 illustrates the exponential growth in the number of components per chip over the last 30 years, a relation which has come to be known as *Moore's law.*

A series of acronyms have been used to describe the approximate level of circuit complexity in an IC chip. Most often these are defined by the number of logic gates contained on the chip as illustrated in Table 13.1.

At the end of *wafer processing*—a series of many individual processing steps—a typical silicon wafer contains an array of completed identical circuits. Each circuit is electronically tested on the wafer for functionality before being cut from the wafer and packaged as an individual chip. The entire manufacturing process is illustrated in Fig. 13.8.

Later sections in this chapter will describe some details of wafer processing.

Table 13-1 IC Circuit Complexity

Acronym and Meaning	Approximate Number of Gates
SSI—Small-Scale Integration	1–10
MSI—Medium-Scale Integration	10–200
LSI—Large-Scale Integration	200–2,000
VLSI—Very Large Scale Integration	2,000–10,000
ULSI—Ultra Large Scale Integration	above 10,000

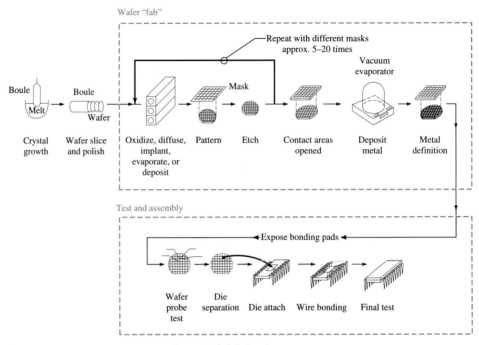

Figure 13.8 Typical IC fabrication process sequence.

13.2

Integrated Circuit Technologies and Structures

As illustrated in Fig. 13.9, integrated circuits can generally be classified in two large groups, depending upon their circuit function: *analog* or *digital*. They can also be grouped by the technology used to implement the transistors in the circuits: *bipolar* or *MOS*. There are, however, a few *mixed mode* circuits that contain both analog and digital functions.

Analog circuits have been fabricated principally using circuits with bipolar transistors, at least until recently. The famous 741 op-amp chip, the popular 555 timer chip, and the

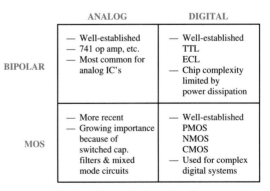

	ANALOG	DIGITAL
BIPOLAR	— Well-established — 741 op amp, etc. — Most common for analog IC's	— Well-established TTL ECL — Chip complexity limited by power dissipation
MOS	— More recent — Growing importance because of switched cap. filters & mixed mode circuits	— Well-established PMOS NMOS CMOS — Used for complex digital systems

Figure 13.9 IC classification.

565 phase-locked-loop chip are all analog (linear) chips implemented using bipolar IC technology. MOS analog circuits are comparatively new in their development. The primary force advancing the development of MOS analog circuits is the growing requirement to integrate analog functions together with digital circuits on a single chip. This has been made easier with the development of *switched-capacitor circuits,* that is, a new type of analog circuit that achieves precision by matching circuit capacitors. This new technology uses the same type MOS transistors used in digital circuits.

Popular digital integrated circuits have been fabricated using bipolar circuits for over 20 years. For example, small- and medium-scale (SSI and MSI) TTL circuits, including the famous 7400 series of chips, and ECL circuits, utilize bipolar transistors.

Because bipolar circuits require more power per transistor than MOS devices, they are not as well suited for large, complex digital integrated circuits, that is, VLSI and beyond. The growing importance of very dense microprocessor chips and large memory chips, each containing millions of transistors, has driven most high-density digital logic to utilize MOS transistors.

Recall that bipolar transistors come in one of two types, either NPN or PNP. The structure of an NPN bipolar transistor is shown in Fig. 13.10(a). MOS transistors are either *n*-channel or *p*-channel; the structure of an *n*-channel MOS transistor is shown in Fig. 13.10(b).

These structures are created in an integrated-circuit manufacturing line often involving hundreds of processing steps. Fortunately, there are only a few *types* of process steps; they are just repeated (with slight changes) in groups over and over to fabricate the integrated circuit. A description of the types of steps or IC processing technologies is detailed in the next section.

One of the most critical processes is called *photolithography.* This is a photographic-like process in which the pattern or image on a master photographic plate, called the *mask,* is transferred to a photographically sensitive layer and applied to the wafer. In this manner, the lateral dimensions of integrated devices are determined.

The vertical dimensions, or depth of the structures from the silicon surface, are determined by other process steps.

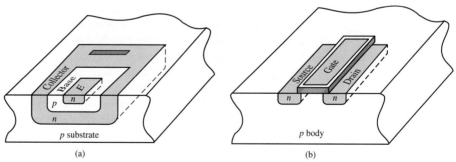

Figure 13.10   Physical structure of (a) NPN bipolar transistor and (b) *n*-channel MOS transistor.

13.3

**Fabrication
Technologies**

Modern IC manufacturing facilities are impressive operations. The *wafer fabrication* (wafer fab) process is carried out in a clean-room environment to minimize the detrimental effects of dust, pollen, or other airborne particulates. A single dust particle looks like a boulder when building structures with dimensions of less than a micron. Manufacturing personnel are wrapped head to foot in white smocks and hoods, and they can be seen adjusting large equipment through glass panels in the walls of the manufacturing areas. There are fewer people than you might expect, because robotics has replaced most human operations. Wafers glide gently along on tracks from one operation to another, and stack up in teflon carriers at certain points as shown in Fig. 13.11.

The wafer fab operations begin with clean, polished wafers of silicon, and proceed through a sequence of process steps that create an array of identical integrated circuit chips on each wafer. The final steps will include *wafer probe,* where metal probes much like needles, are made to touch each of the electrical contacts on the silicon surface to provide power to each circuit and test its operation. The circuits that don't work are marked as "bad" by a drop of ink and then the wafer is sawed along the lanes between chips (called the scribe lines) and the good chips are mounted in packages. Each wafer also contains a few "drop-ins" which are special circuits with test devices and circuits that are useful in characterizing the process.

As stated earlier, there are hundreds of process steps that must be completed one after the other in the fabrication of a modern integrated circuit; however, fortunately most of these steps are variations of only three basic types of steps:

1. Grown layers and deposited layers
2. Photolithography
3. Doping

Grown Layers and Deposited Layers

An important class of process steps in the fabrication of an integrated circuit is the group of steps in which layers of different materials are created on the surface of a wafer. There are two primary ways in which this is done: (1) *grown layers,* and (2) *deposited layers.*

Silicon has become *the* semiconductor material most utilized in IC fabrication for many reasons; however, one of the most important is its compatibility with planar processing, which is the widely used method of silicon processing described in this chapter. Silicon when oxidized becomes silicon dioxide, which is essentially glass—a very good electrical insulator, and a barrier to chemical impurities. This glasslike barrier can be used to protect the silicon material underneath. Therefore, *grown oxides* refer to process steps whereby a bare silicon surface is exposed to an oxygen-rich environment at high temperature, and oxygen atoms combine with silicon in the surface to create a surface layer of pure silicon dioxide. This process of growing an oxide layer is used, for example, to produce the gate oxide or insulating layer under the gate of MOS field effect transistors. Grown oxides can be utilized for other steps in the fabrication process as well.

The process of *oxidation* is generally performed in an oxidation furnace similar to that

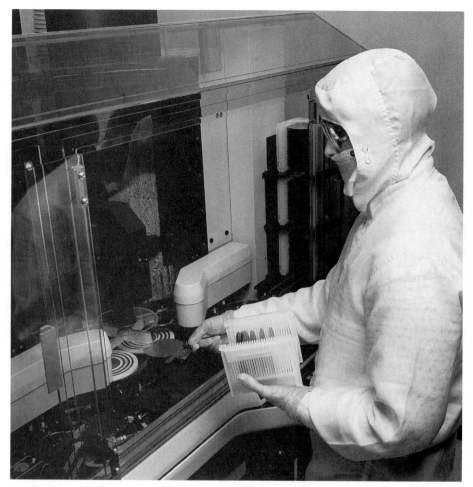

Figure 13.11 Wafer transport in an IC fabrication facility. (Courtesy of Honeywell)

shown in Fig. 13.12. Typically these furnaces have large cylindrical tubes of quartz surrounded by heating elements that will provide an internal temperature inside the quartz tube in the range of 1,000 degrees centigrade. Oxygen, in the form of oxygen gas or steam, is caused to flow through the tube from one end to the other. Silicon wafers are lined up on edge in what is called a quartz boat, much like a dish rack with slots which support the wafers. This configuration keeps them separated from each other so oxygen can flow over the surface of each wafer. The boat, loaded with as many as 50 to 100 wafers, is slowly slid into the core of the quartz tube and the oxidation process begins. The temperature of the wafers and the length of time in the furnace determine the thickness of the grown oxide as well as the oxide quality.

Deposited layers are produced in two ways: (1) they result from a wafer being placed in a chamber in which a chemical reaction directly deposits on the surface of the wafer a layer of the desired material, or (2) a material is evaporated and then condensed on the

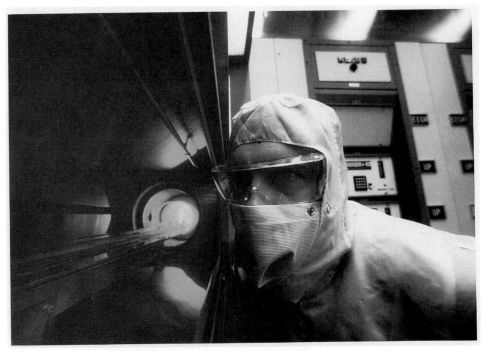

Figure 13.12   Oxidation furnace. (Courtesy of Honeywell)

surface of the wafer. As an example, deposited oxides are quite common primarily because a layer of SiO_2 can be deposited at a much higher rate than a grown oxide, and therefore, production time is reduced. Deposited oxides, however, generally do not have the same quality or uniformity as grown oxides.

Other materials are commonly deposited as well. Polysilicon (silicon which is not single crystal, but has a random orientation of the silicon atoms) is a deposited layer commonly used for gate electrodes in MOS transistors. One of the final steps in integrated circuit fabrication is the deposition of metal layers which are then patterned to produce the conductive interconnections that connect the transistors and other components into a functioning circuit. Aluminum metal is one of the most common metal layers used for interconnections and is usually evaporated and then condensed on the wafer surface to form a thin film of metal. This metal film is then etched to form a pattern of interconnections which replace wires in a conventional circuit.

Typically, deposited layers are put down on a number of wafers simultaneously in a vacuum chamber, an airtight chamber from which the air and other impurities are removed.

Deposited oxide layers and polysilicon layers are generally the result of chemical reactions which are carried out in reaction chambers. The product of this reaction is the desired material which coats the surface of wafers in the chamber.

Metal layers are generally deposited by evaporating the metal inside a vacuum chamber. The metal atoms spread from the evaporation source in all directions and coat the surface of the wafers, as well as everything else placed in the vacuum chamber, as it condenses. Metal is evaporated by elevating the temperature of the metal to its boiling point in one of two ways. For aluminum metal, which has a relatively low vaporization temperature, this is sometimes done by placing aluminum on a tungsten filament that has been heated "white-hot" by an electric current—a process called *thermal evaporation.* Another method, *E-beam evaporation,* is performed by heating a crucible of metal to the vaporization point by focusing an intense beam of electrons on the center of a pool of the desired metal in a ceramic crucible.

Sputtering is another commonly used method for depositing thin films of materials. Sputtering uses ion bombardment in a gas plasma to free atoms from a disk of "target" material and deposit them on the desired substrate.

Photolithography

It is truly an engineering marvel that millions of transistors can be placed side by side on a silicon chip no bigger than a dime, and what's more they are all interconnected in a complex circuit which operates as a memory chip, a microcomputer, or serves some other complex function. The fundamental process by which these complex structures are created depends on what is essentially a photographic process called *photolithography.*

The IC structure is built up in layers, each successive step adding material or modifying particular selected areas on the surface of the chip. The complex pattern or structure defined for each layer is first created on what is called a *mask.* A mask is analogous to the photographic negative and contains the image of the pattern to be created. If you are familiar with black and white photography, the photolithographic process is very much like printing black and white photographs. The mask contains a high contrast clear and opaque pattern generally on a plate of glass. The creation of this pattern will be discussed in a later section on the design and layout of integrated circuits. Figure 13.13 shows a typical mask used to produce the pattern for one layer in the fabrication of a single simple integrated circuit; a mask contains this pattern repeated in an array.

In the photolithographic process, a photosensitive (light-sensitive) layer is first applied to the wafer. The photosensitive layer is placed on the wafer by applying a glob of viscous liquid called *photoresist* at the center of the wafer, and then spinning the wafer at a high rate, spreading the photosensitive emulsion out into a uniform thin layer across the surface of the wafer. This layer is then dried and baked prior to exposure.

A high-intensity light is then directed through the mask and onto the photosensitive layer on the wafer. The mask serves to block or to transmit the parallel beam of light and thus expose or not expose portions of the photosensitive layer with the same pattern as that on the mask, as shown in Fig. 13.14.

Photoresist is generally an organic polymer, and there are two basic types, positive and negative. With negative photoresist, the areas exposed to the light become polymerized and very tough. Therefore after developing, these areas remain on the surface, and the nonexposed areas wash or "clear." This process produces a pattern in photoresist which is the inverse of the pattern on the mask. Positive photoresist works in the opposite way.

Exposure of the photosensitive layer can be done by a "contact print," in which the mask is put in direct contact with the photoemulsion on the wafer surface. A proximity print is similar to a contact print but there is a very small gap between the surface of the photoresist and the mask. This latter approach prevents scratching the mask and prolongs

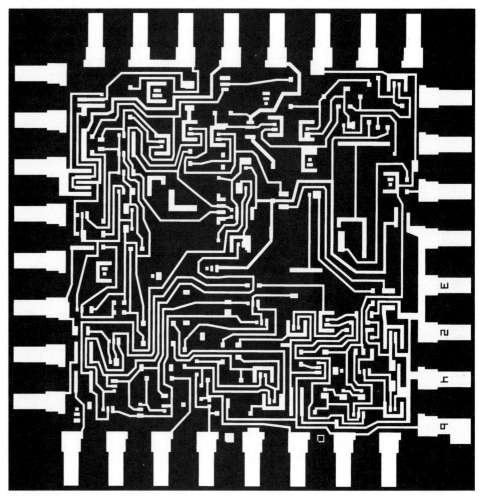

Figure 13.13 Typical mask pattern for single chip (mask contains an array of chip patterns).

its useful life. A more common approach in modern IC fabrication is that of projection printing, in which the image from the mask is projected through very precise optics onto the surface of the wafer by machines called "steppers." This optical approach provides the additional benefit of being able to further reduce the size of the image from the mask via the projection optics.

There is an increasing difficulty with the optical lithographic approach just described. Engineers are designing ever smaller device structures, and currently the dimensions of the features being produced are approaching the wavelength of visible light—a fundamental limitation. Therefore, new photolithographic systems using ultraviolet light and even shorter wavelengths, such as X-rays, are being investigated.

Therefore, in summary, the photolithographic process consists of the following steps:

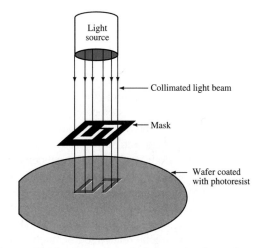

Figure 13.14   Exposure of photoresist through mask.

1. A light-sensitive emulsion known as photoresist is spun onto the surface of a wafer. The wafer is then dried and baked.
2. A mask is placed between a collimated light source and the surface of the wafer, and the wafer is illuminated for a controlled period of time to properly expose the photosensitive layer on the surface of the wafer with the desired pattern.
3. The wafer and exposed emulsion is then developed, washed, and dried, which results in the exposed areas (assuming positive photoresist) being removed, and non-exposed areas remaining firmly attached to the surface of the wafer.
4. Generally, the next step involves dipping the wafer with the patterned photo-emulsion in an appropriate acid which will etch away the parts of the surface layer which are exposed and leave the particular areas covered with photoresist untouched.

This process therefore transfers the pattern defined on the mask to the surface of the wafer.

Doping

Doping is the process of adding small and controlled amounts of impurities to semiconductor material to change the electrical characteristics of the semiconductor. Doping is used to create *n*-type or *p*-type semiconductor regions as required to make the device structures described in the previous chapters. Doping silicon with acceptor impurities creates *p*-type semiconductor material, and donor impurities create *n*-type semiconductor material. Recall the most commonly used donor impurity is phosphorus, and one of the most common acceptor impurities is boron.

These impurities can be placed in the semiconductor surface by one of two principle techniques: diffusion or ion implantation. *Diffusion* is a process in which the wafers are put in boats very similar to those used for oxidation, and slid into a similar quartz tube furnace. A gas containing the desired dopant impurity is passed through the quartz tube, and at very high temperatures these impurity atoms diffuse into the surface of the silicon wafer. Diffusion is typically done at temperatures in the range of 900–1100 degrees centigrade. The depth to which the impurities diffuse into the silicon surface is determined by the diffusion temperature and the length of time the wafers are left in the diffusion furnace. The higher the temperature the faster the rate of impurity diffusion, and of course the longer the wafer is left in the furnace the deeper the impurity atoms diffuse.

One of the limitations of diffusion is that the highest concentration of impurity atoms is always at the surface of the wafer. *Ion implantation* offers a means of injecting impurity atoms into a silicon wafer with some control over the depth of maximum concentration. An ion-implanter is basically a particle accelerator which is used to accelerate impurity atoms to a very high velocity. In an ion-implanter a beam of impurity ions (such as boron or phosphorus) is created and directed onto the surface of a silicon wafer. The ions shoot into the surface of the wafer, much like bullets being shot into a target, and dope the silicon with the desired impurity. Ion implantation offers generally more control and precision over the amount of doping deposited. Ion-beam current can be measured and is directly related to the number of dopant ions per second that are being injected into the silicon. The depth of penetration of the ion beam is related to the energy of the beam; a high-energy beam will penetrate deeper into the silicon wafer than a low-energy beam. Therefore, the depth of maximum doping can be adjusted by the energy, or accelerating voltage, of the ion-implanter's ion beam.

In general, ion implantation cannot produce doped layers as deep as the diffusion process. Therefore, ion implantation is used for doping most modern transistor structures which are themselves relatively shallow devices. When deep junctions are required, diffusion processes are still utilized.

A Simple Example of IC Processing

Most modern integrated circuits are produced by a particular combination of the basic process steps just described together with the appropriate set of masks and choice of material layers. In order to illustrate the use of these processes to create an electronic device structure, a fabrication sequence capable of producing two resistors in a silicon substrate will now be presented:

1. First, a layer of silicon dioxide is grown uniformly across the surface of an *n*-type wafer. This layer will become a "masking layer" for doping impurities to be introduced later. The wafer cross section is shown in Fig. 13.15(a).
2. The IC designer creates a "layout" for the two resistors which is transformed into the mask shown in Fig. 13.15(b). This mask defines the shape of the regions on the wafer surface that will ultimately become the resistors.
3. Next a layer of photoresist is spun onto the wafer, creating a photosensitive emulsion on top of the silicon dioxide layer, as shown by the cross section drawing in Fig. 13.15(c).
4. The mask is placed over the wafer, and an intense light source is used to expose the photoresist, which is then developed, washed, and baked creating an organic polymeric film on the surface of the wafer with the same pattern as that originally contained on the mask. Assuming negative photoresist was used, the opaque areas on the mask have been transferred to open "windows" or holes in the emulsion layer, as shown in Fig. 13.15(d).

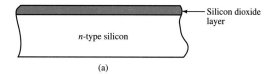

(a)

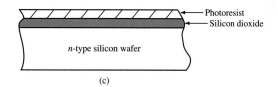

Mask pattern for   (b)   Mask with repeated
single two-resistor   pattern
circuit

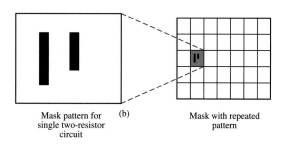

(c)

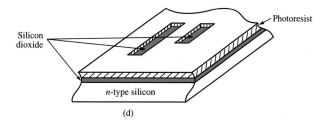

(d)

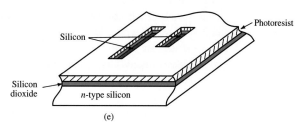

(e)

Figure 13.15   IC processing steps for fabricating semiconductor resistors.

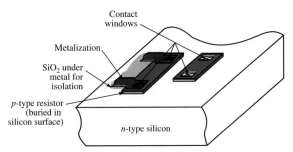

Figure 13.16 Metal connections to completed IC resistors.

5. The wafer is now dipped into hydrofluoric acid, an acid that etches silicon dioxide, but doesn't significantly attack either the photoresist or pure silicon. The acid etches away the silicon dioxide layer only in the areas where there was a window in the photoresist layer, since the silicon dioxide under the photoresist is protected from the acid. The same pattern has now been transferred to the silicon dioxide layer as shown in Fig. 13.15(e).

6. The residual photoresist is now stripped away, and the wafer is placed in a diffusion furnace containing a flow of gas rich in boron (a *p*-type dopant). The boron diffuses into the *n*-type wafer only in the areas where a "hole" has been etched through the silicon dioxide surface layer. In only these regions the *n*-type substrate is converted to *p*-type by the higher concentration of *p*-type dopant, and these areas will become the resistors. The time the wafer spends in the diffusion furnace is critical; the longer the diffusion time, the deeper the *p*-type material diffuses, and the thicker the resulting resistor.

7. The two elongated regions made in the previous step will function as resistors if we can make electrical contact to the ends of each one. This is done by again growing a silicon dioxide layer across the surface, and then using photolithography and another mask to etch small holes at each end called "contact windows." Then a metal layer is deposited across the entire surface. Photoresist is spun on top of the metal, and a third mask is used to pattern the metal. By dipping the wafer in a metal etching solution, a metal pattern is created.

This completes the fabrication of this simple, two-resistor, integrated circuit. The final product is shown in Fig. 13.16, where metal is shown contacting the end of only one of the resistors. Of course, an array of these circuits would actually be created simultaneously across a wafer, although the drawings in this example showed just one circuit for convenience. In addition, real IC processes generally contain hundreds of individual process steps, many more than illustrated in this simple example. Furthermore, a modern IC process may involve 10 to 20 mask layers.

13.4
Resistivity and Sheet Resistance

As illustrated in the example of the previous section, a sequence of processing steps will produce a resistive layer. There exist simple mechanisms for describing the electrical behavior of such layers.

First, recall that the *resistivity* of a homogeneous material is a volumetric indicator of the resistive properties of the material itself and does not depend on the physical geometry of the material in any way. In general, the more dopant atoms added to a semiconductor, the more free carriers available, and therefore the material conducts more easily, or has a lower resistivity. There are charts and formulas available in the literature for calculating the amount of doping required to achieve a certain resistivity.

Recall from Chapter 8 that a rectangular block of semiconductor of length L, and cross-sectional area A, has a net resistance in ohms of R, where

$$R = \rho \left(\frac{L}{A} \right) \tag{13.1}$$

The *sheet resistance* is a measure of the resistive characteristics of a very thin sheet of resistive material as is often created in the fabrication of an IC. Consider a thin resistive layer which is SQUARE ($L = W$) and has a thickness t, as shown in Fig. 13.17. The resistance of this special case is

$$R_{sh} = \rho \left(\frac{L}{A} \right) = \rho \left(\frac{L}{Lt} \right) = \frac{\rho}{t} \tag{13.2}$$

Notice that the resistance of this square sheet depends only on two parameters, both of which are properties of the material layer: its bulk resistivity and the layer thickness. For any fabricated layer these two parameters are constant over the surface of the wafer. Therefore, this particular resistance is a useful quantity to determine and is given the special name *sheet resistance*, R_{sh}; it is the resistance from one edge to the opposite edge of a square piece of the material layer. The units of sheet resistance are ohms per square.

Using this approach, the total resistance of any rectangular region of length L and width W can be readily obtained by counting the number of *square* blocks that can be placed in the rectangular region. If N blocks can be placed in the region, then the resistance of the region is:

$$R = R_{sh} N \tag{13.3}$$

Note that for a simple rectangular region, N is just L/W; however, if the width changes, counting squares is critical, as illustrated in Example 13.2.

EXAMPLE 13.1

Figure 13.18 shows the top view of a diffused resistor which is 6.5 squares long. Let us determine the value of this resistor if it is fabricated from a diffused layer with a sheet resistance of 300 Ω/square.

The total resistance is the product of the sheet resistance and the number of squares:

$$R = R_{sh} \text{ (number of squares)} = R_{sh} \left(\frac{L}{W} \right)$$

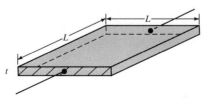

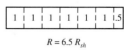

$R = 6.5 R_{sh}$

Figure 13.17 A square sheet of resistive material.

Figure 13.18 Diffused resistor configuration used in Example 13.1.

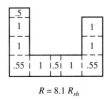

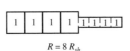

$$R = 8\,R_{sh}$$

$$R = 8.1\,R_{sh}$$

Figure 13.19 Diffused resistor configuration used in Example 13.2.

Figure 13.20 Diffused resistor configuration used in Example 13.3.

Therefore

$$R = R_{sh}\,(6.5) = (300)\,(6.5) = 1950\ \Omega \qquad \blacksquare$$

EXAMPLE 13.2

Figure 13.19 shows the top view of a resistor made with the same diffused layer as Example 13.1. Let us determine its resistance.

The total number of squares is 8; therefore

$$R = R_{sh}\,(8) = (300)\,(8) = 2400\ \Omega \qquad \blacksquare$$

If rectangular resistors have a right angle in the pattern, the corner square has an equivalent resistance of 0.55 squares. The resistance of serpentine diffused resistors can be calculated using this fact.

EXAMPLE 13.3

Figure 13.20 shows a drawing of a resistor with two 90-degree turns. Let us calculate the total number of squares contained in this resistor.

First the corner squares are marked, and contribute 0.55 squares each. Next, the three linear portions are marked off by making squares as long as the width. The figure indicates that the total number of squares is 8.1. $\qquad \blacksquare$

13.5

IC Design and Layout

The design and layout of an integrated circuit is the first phase in the development of a new chip. This is the sequence of events that begins with formulation of the basic concepts about what the chip should be able to do, and ends with a design layout which can be converted to masks for actual chip fabrication. In executing this process, the IC designer will make use of a number of computer tools (or simulators) that will simulate the operation of the chip; this enables the designer to have a high level of confidence that the new chip will actually do what it's supposed to do, before the first silicon is fabricated.

The flow chart in Fig. 13.21 illustrates the various sequences of steps in this process; it will help to refer to this chart throughout the remaining discussion. The process begins with the formalization of the functional specifications. This involves writing down all the functions that are to be required of the new chip. Quite often this is an iterative process between the IC designer and the customer or user of the chip to determine exactly what is needed.

Based on these functional requirements an architecture for the integrated system or subsystem is developed. This is a block diagram in which each block represents a func-

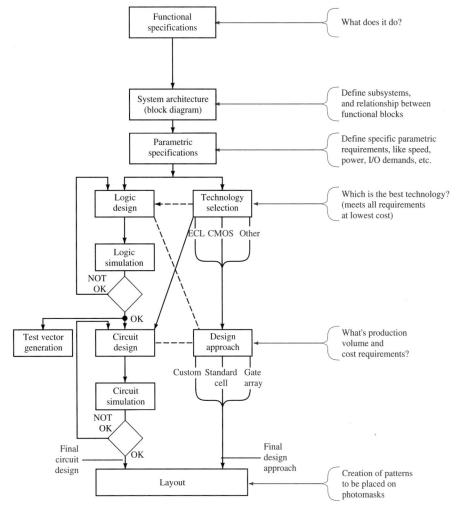

Figure 13.21 Flow chart for IC design and layout.

tional unit (like memory, arithmetic logic unit, etc.), and the blocks are connected by paths that show the proper flow of information. As an example, Fig. 13.22 shows the block diagram of a typical digital system on a chip.

As the architecture is developed, a more rigorous set of specifications, called the parametric specifications, must be developed. These deal with the required operational details of the proposed new chip, such as, What will be the maximum power dissipation allowed? How fast must the chip execute the functions described by the architecture? What are the specific Input/Output (I/O) requirements such as minimum output drive current?, etc. Determining accurate answers to these types of questions will significantly affect the results of the next step, which is technology selection.

Technology selection is an attempt to best match the needs of the proposed chip with the capabilities of each of the technologies. CMOS is by far the most commonly used dig-

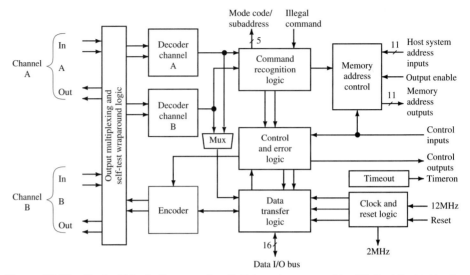

Figure 13.22 Typical block diagram of a digital system on a chip. (United Technologies Microelectronics Center, Inc. UT1553B; Courtesy of United Technologies Microelectronics Center, Inc.)

ital technology for relatively complex integrated systems. It is relatively fast, very power-efficient, and offers the lowest cost per circuit function. ECL is faster, but uses more power and costs more; therefore, it is used only when speed requirements demand it. There are other less used technologies which find some application in special situations.

The logic design is a critical step in the design process. It is derived from the system architecture and the parametric specifications. The logic design is the definition of all the logic gates (like ANDs, NORs, etc.) and all the connections between the gates required to implement the functions of the new chip. A typical example of a chip's logic design is shown in Fig. 13.23. There is some interaction between the logic design and the technology selection, as certain technologies more easily offer certain types of gates. For example, ECL gates often provide at little added cost both the output variable and its complement. This may simplify some logic circuits.

A logic diagram is generated for the entire chip which can be used as input for a *logic simulator computer program.* Such a program will enable the designer to model the behavior of the logic circuit by putting in various test signals, called test vectors, and observing the predicted output. Another function with which the computer generally assists is the development of an appropriate set of test signals that will ensure that most of the gates in the chip have been exercised. This step, called *test vector generation,* is often performed by software on the same system as the logic simulator. Care must be taken to design a logic circuit which can easily be tested by the application of signals at the external pins.

Logic design is usually an iterative process in which the logic simulator produces results, some of which are not exactly as expected, and then the logic diagram is modified to improve the design. One critical function of a logic simulator is to determine possible race conditions or potential signal timing problems. If one signal must arrive at a certain

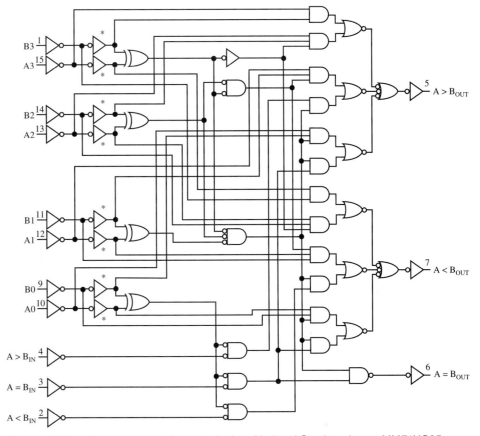

Figure 13.23 Example of chip's logic design, National Semiconductor MM74HC85. (Courtesy of National Semiconductor)

gate ahead of another, this must be known and the circuit design and layout done to ensure that it occurs.

The *circuit design* is the construction of the schematic drawing of the chip at the transistor level and shows the electrical connections between every transistor, and any important parasitic elements, such as capacitance between metal interconnections. Figure 13.24 shows a portion of the circuit design of a CMOS integrated circuit. Because of the complexity of modern chips, a circuit drawing for the entire chip becomes very large and unwieldy; therefore, circuit documentation is often done only for certain blocks of circuitry at a time. The performance of these blocks is simulated using *computer circuit simulators* and the results obtained are used to determine the proper parametrics in the logic simulation.

The complete circuit design is developed from the logic design and the particular technology chosen for implementation. Circuit simulations of the final circuits provide an excellent way to confirm the design integrity before committing to chip fabrication. The accuracy of the circuit simulation depends entirely on how well the circuit and the technology

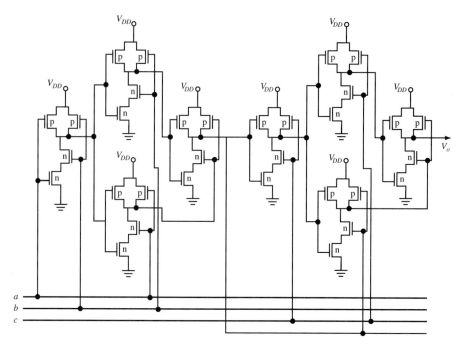

Figure 13.24 CMOS IC circuit for full adder.

are modeled. Most modern IC companies invest heavily in developing accurate computer tools for the simulation of new designs since it is far less expensive to perform multiple simulations than to produce a batch of ICs which don't work. The circuit simulation process also tends to be iterative, and improvements in the circuit design may be done as a result of the simulated performance.

There is another major design decision in the development process, that of *design approach.* The design approach described here is the choice of (1) *full-custom*, or (2) *semi-custom,* of which there are two primary types, *standard cells* or *gate arrays.*

The choice of the design approach is driven as much by economic factors as by technical ones. The total anticipated production volume is a critical factor in this determination. It is generally desirable to minimize the total cost of the project. Full-custom designs require more initial cost, but lower production cost at high volume. Therefore, this approach is generally preferable for high production volume circuits, like memory chips. Gate arrays are at the other end of the spectrum in that they have lower design cost but use silicon less efficiently and therefore have higher cost per unit. They are best for low-volume production. The standard cell design approach is somewhere in between. The various design approaches will be discussed in more detail in the next section.

Finally, *layout* is the process by which the complete design is translated to geometric shapes which will be used to generate the light and dark areas on the masks. Layout of ICs is done on a computer system which makes use of the hierarchical nature of these designs. Once the layout for a certain type of gate is done, it can be repeated everywhere that same type of gate is utilized. Such computer layout systems often have *routers* which keep track of the connections required between circuit elements and determine a possible path for the interconnecting metal. Layout also has an iterative component, because it is impossible to accurately know all the parasitic parameters, such as capacitance on a particular metal line, until the layout is done and it is determined exactly how long that line will be. The conclusion of the layout process is a computer file, and usually graphic plots, that show the pattern to be placed on each of the masks to be used for the actual chip fabrication.

13.6

Design Approach: Custom, Standard Cells, and Gate Arrays

The primary factors affecting the selection of the best design approach are economic as well as technical. First the three approaches will be described and then the economic considerations discussed.

A *custom design,* often called full-custom design, is one in which each mask level is basically hand-crafted to achieve the optimum layout in terms of density and performance. A computer is often used to assist the designer in such layouts. With this approach there are no initial restrictions on the assignment of the silicon area. The designer can squeeze or mold functional blocks to fit any "floor plan" or partitioning of chip surface by function that seems best. An example of the floor plan of a full-custom chip, the Intel Pentium™ Processor, is shown in Fig. 13.25. In the layout of this type chip, the various circuit functions are blocked into areas that provide just enough area for the required function and minimize the length of metal connection paths between blocks. The "sculpting" of a custom design is very intensive and requires a large investment of engineering hours. In completing a custom design, great effort is invested in trying to minimize the total chip area. It is expected that such designs will be produced in high quantities, and therefore even a small reduction in area, which translates directly into cost reduction, will add up to large savings. Therefore, custom chips have the largest initial design cost, make more efficient use of the silicon, are smaller and thus have the lowest unit cost in production.

The other design approaches are often called *semi-custom,* and include *standard cells* and *gate arrays.* A standard cell design is a more structured design approach in which the chip area is initially partitioned into rows dedicated to logic separated by spaces or lanes in which interconnections are routed. Logic blocks are predefined and the layouts for the blocks or *cells* stored in a computer database or library. The layouts all have the same physical height, and all inputs to these cells are at the top and bottom. Therefore, cells can be chosen from the library to implement the logic desired, and the cells are then lined up in "cell rows." The common height for each cell makes a uniform row, no matter which cells are selected.

The primary advantage to this design approach is that by structuring the layout in this way, the design process is much more easily automated. Most standard cell computer layout systems contain a library of available logic blocks or cells. Once the logic design is entered into the computer, the computer software will select the order of cells in each row in order to minimize the interconnection lengths, and will use the interconnection information from the logic design to place and connect all lines which interconnect the various logic cells. Typically, metal is used for interconnections in one direction, and polysilicon is used for interconnection lines running normal to the metal lines.

Typical cells in a standard cell library consist of basic logic primitives, such as NAND

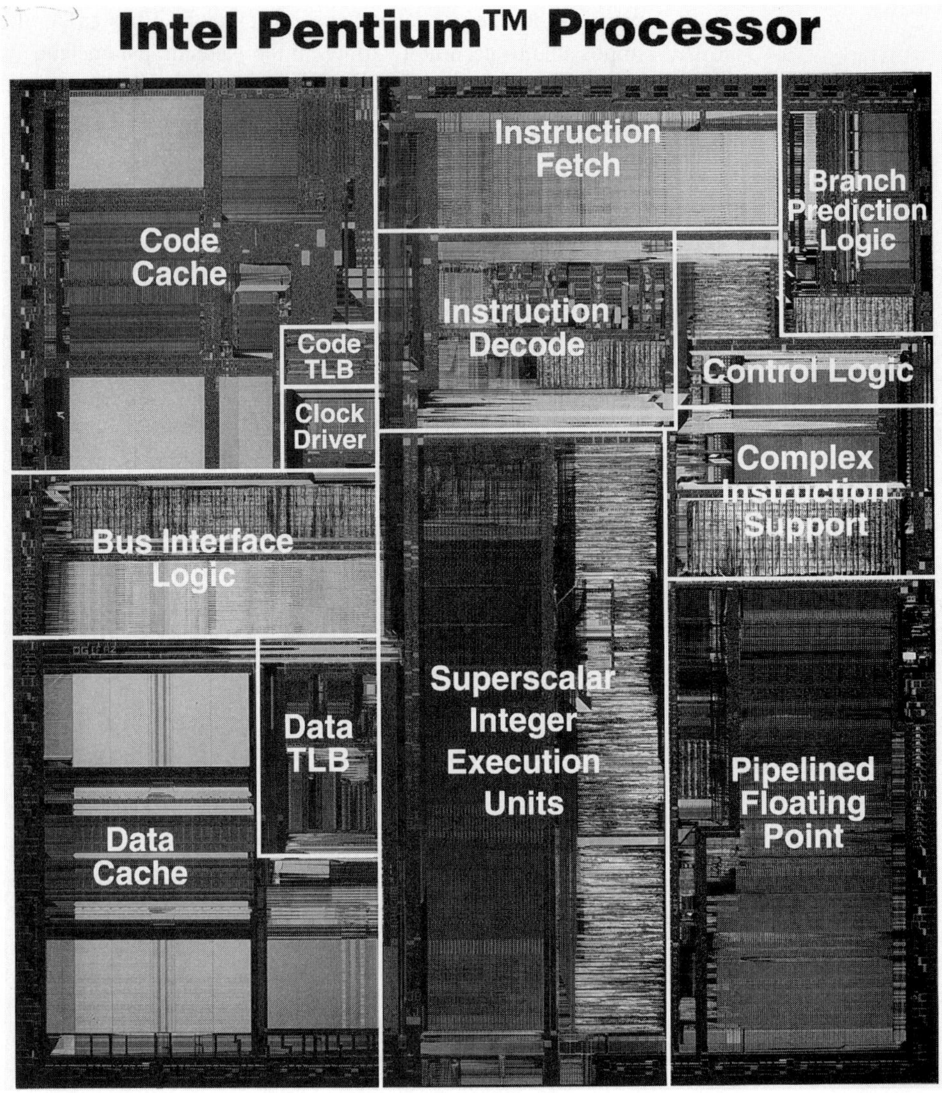

Intel Pentium™ Processor

Figure 13.25   Floor plan for the Intel Pentium™ Processor. (Courtesy of Intel; Pentium is a trademark of Intel Corporation)

gates and NOR gates, as well as *macros,* which are larger, more complex functions, such as memory registers. Figure 13.26 shows the layout of a standard cell chip. A completed standard cell design contains a unique pattern for each mask level used in producing the integrated circuit, just like a full-custom design.

The gate array approach is different in that the customization of the layout to achieve a unique circuit is done on only a few levels; the majority of the mask levels are generic and common to all designs. The concept is that wafers are produced containing chips each with a large array of transistors, all unconnected. Wafers are then stockpiled just prior to the metalization step, which will connect the individual transistors together to form the

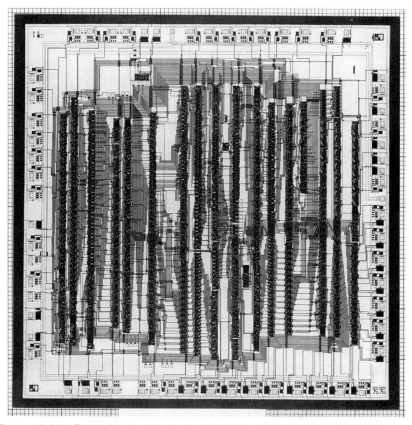

Figure 13.26   Example of a standard cell chip layout. (Courtesy of GEC Plessey)

desired circuit. A gate array design for a specific application consists of only a few custom mask levels: typically one or two levels of metalization and corresponding contact windows. Figure 13.27 shows the wafer processing flow for gate arrays.

The gate array design process is generally highly automated, much like the standard cell approach. If one inputs the desired logic diagram into the gate array design software, it will select patterns of metal necessary to create the various logic blocks from a library of choices as well as route connecting lines between all the logic blocks. The automation

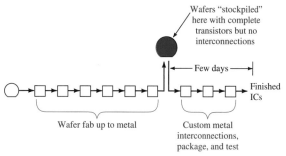

Figure 13.27   Wafer processing flow for a gate array.

reduces the design cost, and since only 3 or 4 custom mask levels are required, the cost of producing a small lot of these devices is much less than the other approaches.

Thus, a gate array chip typically has a few custom metal levels on top of a standard base substrate consisting of a matrix of thousands of transistors. It is impossible to utilize all of the transistors on a gate array, and therefore there is significant layout inefficiency. Typically a gate array chip has areas of transistor arrays for creating logic, and wiring channels between the arrays for routing the interconnecting metal. A gate array chip will be much larger than a chip in which the same function is implemented using a custom or standard cell approach.

Another advantage of the gate array approach is the short turnaround time required from the completion of design to delivery of silicon parts. In other approaches, the wafer fabrication process cannot even begin until the design is complete. However, with the gate array approach, the wafers are processed through most of the steps with the generic masks that create arrays of transistors, and are stockpiled just prior to the few custom mask levels at the end of the process. This means a gate array circuit can be completed in a much shorter time, as short as a few days, instead of weeks or months.

The optimum design approach for a particular application depends on many factors, but the most important is the anticipated production volume. In computing the total cost per chip, the design costs are amortized over the number of chips produced. Therefore, if a large number of chips will be produced, then a more expensive design process, such as full custom, may be viable.

EXAMPLE 13.4

Assume the cost of fabricating 4-inch diameter CMOS wafers is $400 each, and wafer yields are at 60%. Wafer yield is the number of *good* chips divided by the total number of chips on a wafer. A gate array design for a particular logic circuit requires 0.75 cm^2 of area per chip, and the initial design cost is $35,000. A full-custom design of the same circuit would require a design expenditure of $130,000, but would reduce the chip area by 20%. Let us determine the preferred design approach if the total required production is (a) 10,000 parts and (b) 1 million parts. (Assume here that the 60% yield applies for both approaches.)

(a) Let us first determine the number of good gate array chips that can be produced on a wafer. The area of the wafer is:

$$A = \pi r^2 = 3.14 \, (2)^2 = 12.56 \text{ square inches}$$

Now, converting square inches to square centimeters, we obtain

$$A = 12.56 \, (2.54)^2 = 81.03 \; cm^2$$

The total number of possible chips on the wafer is then

$$N = \frac{A}{A_{chip}} = \frac{81.03}{0.75} = 108 \; chips$$

Notice that in the above calculation, we have rounded down to the next whole number; fractional chips don't count. Actually the number is slightly less than that calculated above because partial chips at the edges are not functional.

The number of good chips, N_{good}, is 60% of the total number:

$$N_{good} = 0.60 \, (108) = 64 \; good \; chips \; per \; wafer$$

The production cost, P, per chip is therefore

$$P = \frac{400}{64} = \$6.25 \; per \; chip$$

At a production volume of 10,000 units, the amortized design cost per chip, D, is

$$D = \frac{35,000}{10,000} = \$3.50$$

Therefore the total cost, C, to produce each gate array chip is

$$C = P + D = 6.25 + 3.50 = \$9.75 \; each$$

A similar calculation for a production quantity of 1 million gives $C = \$6.29$.

   Note that we have neglected packaging and final test yield loss by assuming that 100% of the chips from the wafer are put in packages and test good. This number can be high but it's never 100%. The costs associated with packaging and testing are discussed in the next section.

(b) Using the same process for the custom design approach produces the results summarized in the following table:

	gate array	custom
10,000 parts	$9.75	$17.94
1 million parts	$6.29	$ 5.07

Notice that the gate array is much cheaper than a custom design in smaller quantities; however, the more expensive design costs of the custom chip pay off at large volumes.   ■

   The relationships which have been demonstrated in the example can be summarized in the graph of Fig. 13.28, which shows the total chip production cost as a function of total production volume. In this figure, the crossover volume where a full-custom design is less expensive than a gate array occurs at about 50,000 parts, which as of this writing is about the right order of magnitude for most CMOS circuits. A standard cell approach falls between these two curves at low volume but falls below the gate array curve at lower quantities, and therefore it is a good compromise for intermediate volumes. New technologies in field-programmable gate arrays and logic arrays offer other cost-effective alternatives for low-volume production.

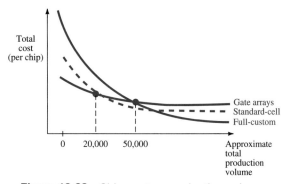

Figure 13.28   Chip cost-vs-production volume.

13.7

IC Manufacturing: Yield, Wafer Cost, Packaging, and Testing

Yield is one of the most important parameters in the manufacture of integrated circuits in that it can make the difference between profits and bankruptcy for an IC company. In the example given in the previous section, the wafer yield was assumed to be constant at 60% even when the chip size was reduced. However, this assumption is not accurate, in that for a given process, yield is directly related to chip size—the smaller the chip area the better the yield. For example, most yield models assume there is a distribution of defects across the wafer surface. Larger chip areas have a greater chance of including a defect and thus become nonfunctional. Figure 13.29 illustrates a certain distribution of defects on the wafer. For the smallest chip size, there are many defect-free chips; however, as the chip size is increased, a larger percentage of the total chips contain a defect, and thus yield decreases.

The probability, *Pr,* that a chip of area *A* has no defects can be estimated by a Poisson distribution:

$$Pr = e^{-AD} \tag{13.4}$$

where the *defect density, D,* is the number of defects on average per unit area, which is typically a number less than 2 per cm^2.

EXAMPLE 13.5
Let us calculate the defect density for the chip described in Example 13.4. If we assume a Poisson distribution, then

$$Pr = e^{-AD}$$

Using the data for both yield and chip area, we obtain

$$0.60 = e^{-0.75D}$$

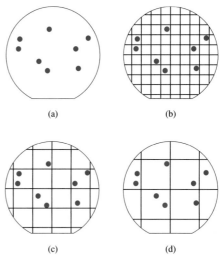

(a)   (b)

(c)   (d)

Figure 13.29 Typical defect distribution on a wafer.

Solving for *D,* we find that

$$D = 0.68 \text{ defects/cm}^2 \qquad \blacksquare$$

EXAMPLE 13.6

In Example 13.4, we assumed that when the chip size was reduced by 20%, the yield remained unchanged at 60%. Let us use the results of Example 13.5 to calculate a more accurate estimate of anticipated yield.

The new data indicates that

$$Pr = e^{-(0.6).68} = 67\%$$

Therefore, the yield increases as the chip size is reduced. $\qquad \blacksquare$

Wafer production cost was also assumed constant in the simplified case described in Example 13.4. Generally, most IC manufacturing operations can provide some discount on wafer fabrication pricing for larger quantities. There are the normal economies of scale that make it indeed less expensive to produce a larger number of the same product.

Finally, packaging and testing must be included in cost estimates. *Packaging* consists of cutting the wafer into individual die, mounting each one in a separate package, and connecting the pads on the IC to the package pins. The electrical connections are usually done by a process called *wire bonding* in which small-diameter gold or aluminum wires (typically 0.7 mils in diameter) are stitched from the chip to the package.

Integrated circuit packages are generally made of either plastic or ceramic. Plastic is less expensive in large volume, but often requires expensive initial tooling. Ceramic packages are more often used for low-volume production and higher reliability configurations.

Testing of integrated circuits is a nonnegligible part of the total cost. The first electrical testing occurs on the finished wafer but before the individual chips are separated. The results of this test, called the *wafer probe* test, determine the wafer yield. The parts undergo *final test* once they are mounted in the package. There is some loss during packaging, although packaging yields should be reasonably high (above 90%).

The facilities required to perform these tests and the software to produce the test vectors are expensive and often complex. Integrated circuit test engineering should start as a part of the design, with *testability* designed into the circuit. This means that the designer and the customer must plan ahead to determine how the circuit will be tested to ensure that all the logic elements are functional in the final part.

13.8
Summary

- Modern microelectronic devices are primarily fabricated using film deposition techniques (thick or thin films) on passive substrates, or semiconductor technology with active substrates. Semiconductor circuits are primarily fabricated on silicon substrates called wafers, although other semiconductor materials, for example, gallium arsenide, are used for special purposes.
- The control or switching of a current is an essential process for electronic circuits. This was accomplished by the vacuum tube until the 1950s and early 1960s when transistor technology replaced them in most applications. The integrated circuit (IC) which was invented in 1958 marked an important milestone in electronics; it enabled a complete circuit to be fabricated in a single semiconductor substrate.

- A wafer contains an array of chips (individual circuits) which are separated after fabrication.
- Wafer size has increased from a diameter of one inch in the 1960s to 8 inches in the 1990s. Minimum feature size has decreased from approximately 10 microns in the 1970s to less than a micron in the 1990s. Components per chip have increased from about 100 in the 1960s to 10^8 in the 1990s. The exponential growth in chip complexity with time is known as Moore's law.
- Acronyms are sometimes used to describe the scale of integration or chip complexity, for example, VLSI (very large scale integration) typically refers to a chip with 2,000 to 10,000 gates.
- IC chips can contain analog or digital circuitry using either MOS or BJT transistors. "Mixed-mode" circuits contain analog and digital functions.
- ICs are fabricated in a clean-room environment. There are three basic types of fabrication processes: (1) grown or deposited layers, (2) photolithography, and (3) doping.
- Sheet resistance is a convenient way of characterizing layers of resistive material. It is used to calculate the resistance of integrated resistors from the resistor's layout dimensions.
- IC design and layout is a complex iterative process with computer simulations used to verify logic function, circuit function, and layout. The selection of the best technology and design approach depends on technical and economic factors.
- The design approach for integrated digital systems includes: custom, standard cells, and gate arrays. Anticipated production volume is a critical factor in selecting the optimum approach.
- Manufacturing issues such as yield, wafer fabrication cost, packaging, and testing are critical for the success of an IC development.

PROBLEMS

13.1. (a) Describe the way in which electronic switching was accomplished in each of the three "eras" of electronics. (b) What year was the integrated circuit invented and who are considered its co-inventors?

13.2. (a) What is the purpose of grinding a flat on one side of a silicon boule? (b) What is the name given to a slice taken from a silicon boule?

13.3. (a) By approximately what factor did IC complexity, measured by components per chip, increase from 1970 to 1990? (b) Explain Moore's law and write an analytic expression for it that applies between 1980 and 1990.

13.4. Give a primary reason why silicon has become the most popular semiconductor material for IC substrates.

13.5. What is a "grown oxide"? Describe how it is produced.

13.6. For each of the following types of layers, designate whether it is typically a grown or deposited layer: (a) gate oxide, (b) polysilicon, (c) metal film.

13.7. Briefly describe the purpose of each of the following in the photolithography process: (a) mask, (b) photoresist.

13.8. Briefly describe the two primary ways in which doping is accomplished in semiconductor processing.

13.9. Construct a flow chart which outlines the entire sequence of steps used to fabricate the "Simple Example of IC Processing" described in this chapter.

13.10. A doped polysilicon layer is deposited on the oxidized surface of a silicon wafer. The resistivity of the deposited layer is 2 Ω-cm and it is 10 microns thick. (a) Determine the sheet resistance of this layer. (b) What is the resistance of a resistor etched in this layer which has a length of 80 microns and a width of 4 microns.

13.11. The sheet resistance of the layer used to make the resistors in Fig. P13.11 is 100 $\Omega/\square$. Find the resistance value for each of the resistor layouts.

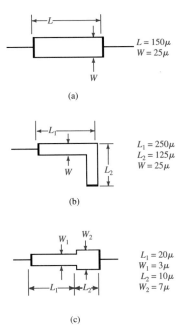

(a)

(b)

(c)

Figure P13.11

13.12. Compare and contrast CMOS and ECL technologies for implementing integrated digital systems.

13.13. (a) List the three basic IC design approaches. (b) Which is the most cost effective at very high production volume? Which is best at very low production volume?

13.14. A 6-inch (150-mm) diameter wafer is used to fabricate IC chips with dimensions of 0.8 cm $\times$ 0.9 cm. Assuming a wafer yield of 73%, (a) how many good die will be produced by this wafer? (b) what's the defect density?

13.15. (a) For the IC in problem 13.14, if the packaging and final test yield is 85%, how many good finished parts are produced from the starting wafer? (b) If the wafer fabrication cost is $450 per wafer, and packaging and final test cost is $1.00 per chip, what is the total cost of producing each part (neglecting design costs)?

13.16. Using a Poisson yield model and a defect density of 0.5 defects/cm^2, what is the probability of a chip having no defects assuming it is the same size as the chip in problem 13.14?

13.17. Using a Poisson model, if the wafer fabrication yield is improved by 10% on a chip which has an area of 1.2 cm^2, then how much was the defect density improved?

ANALOG SYSTEMS

An Example of an Advanced Integrated Circuit, The Intel Pentium^{TM} Processor,
Contains 3.1 Million Transistors *(for floor plan see Figure 13.25)*
(Photograph Courtesy of Intel; Pentium is a Trademark of Intel Corporation)

Modern Computer
Board—VME960MX-
SBC

*(Photograph Courtesy of
Tronix Product
Development)*

Array of Complex IC
Chips on a Wafer

*(Photograph Courtesy of
Leti)*

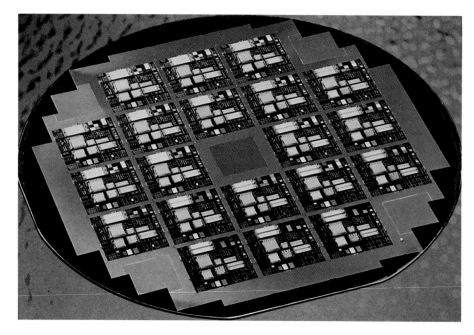

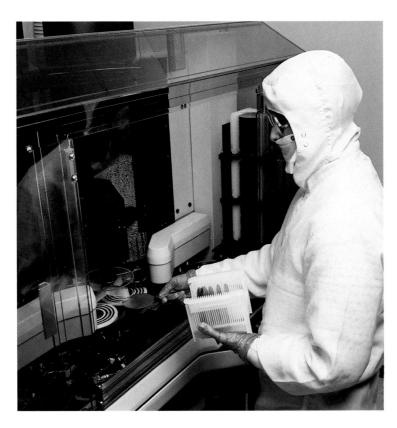

Wafer Transport in an IC Fabrication Facility

(Photograph Courtesy of Honeywell)

Oxidation Furnace

(Photograph Courtesy of Honeywell)

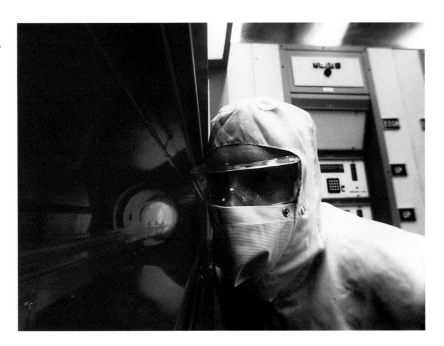

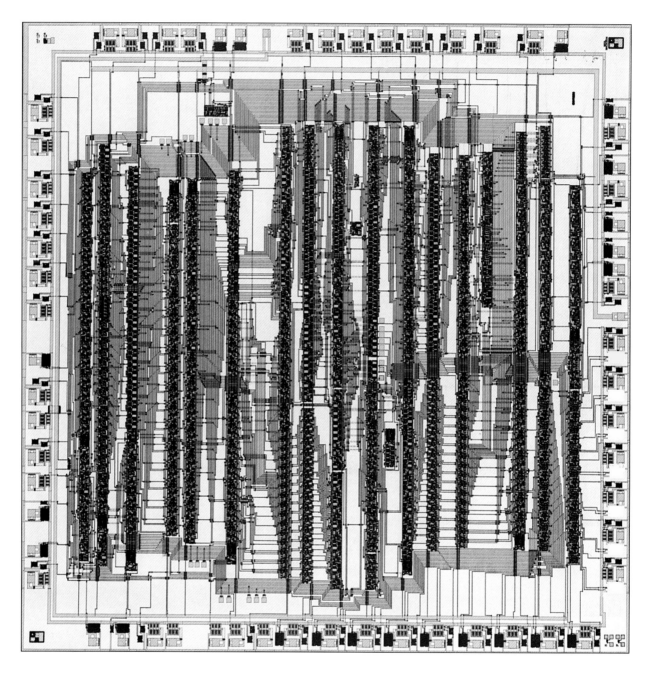

Example of a Standard Cell Chip Layout

(Photograph Courtesy of GEC Plessey)

Op Amps: Introduction to Amplifiers and Analog Systems

14.1

Introduction

Analog systems process information by continuous variations of a signal's amplitude, frequency, or phase. As described in Chapter 7, the quantities most readily observed in nature are analog, or continuously variable, such as temperature, pressure, velocity, etc. These quantities can, and typically do, vary in value continuously with time. As a result, sensors which measure these quantities generally produce an electronic signal which is analog and varies in proportion to the magnitude of the parameter being measured. An analog amplifier or signal conditioner is almost always required following the sensor to boost the strength of the signal. The designer of an electronic system which utilizes such signals may choose to convert this analog signal to its digital representation by an analog-to-digital (A/D) converter, or the designer may utilize an analog system which processes the information directly in analog form. In general, more complex systems convert the information to digital format so the data can be easily processed by a digital computer, such as a microprocessor. In simpler systems, and certain special applications, analog systems are used. A wide variety of circuit applications and functions can be achieved using strictly analog circuitry; in fact, until the invention of the microprocessor in the mid-1970s, almost all practical electronic systems were analog, with the exception of the large mainframe digital computer.

The *operational amplifier* (op amp) described in this chapter is one of the basic building blocks for constructing analog electronic systems. It can be used in combination with various external components for a wide variety of purposes.

The amplitude of a signal is a very critical parameter, because it represents the intensity or strength of the signal; if the amplitude goes to zero the signal no longer exists.

477

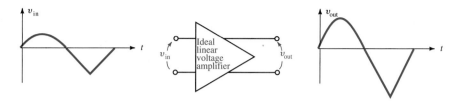

Figure 14.1   A linear amplifier and its characteristics.

As introduced in Chapter 7, an *amplifier* is an electronic device which produces an output signal that is an exact copy or replica of the input signal increased in amplitude as illustrated in Fig. 14.1.

A linear amplifier produces an output which is a constant *A*, the gain of the amplifier times the input signal. The gain, *A*, is assumed to be greater than unity; otherwise the output would be less than the input, and this result is not called amplification but attenuation, which is a decrease in signal amplitude. An ideal linear *voltage amplifier* has an output voltage, v_{out}, given by Eq. (14.1), where v_{in} is the input voltage and A_v is the voltage gain.

$$v_{out} = A_v\, v_{in} \tag{14.1}$$

If A_v is absolutely constant, the output signal is a perfect reproduction of the input, that is, only the magnitude has changed. In many real systems the voltage gain changes slightly as a function of either input signal level or some other variable, and these changes produce output distortion. *Distortion* is a measure of how much the output signal differs from the shape of the input signal. Therefore, the better the amplifier, the lower the distortion.

Voltage amplifiers are the most common amplifier configuration used in electronic analog systems; however, current amplifiers can be constructed which perform in an analogous way. The output current of an ideal linear *current amplifier* is directly proportional to the input current as illustrated in Fig. 14.2, where A_i is the current gain.

Since amplifiers always produce output signals larger than their input, an external supply of energy is required. Amplifiers generally utilize a dc power source applied to its "power supply" terminals, and the symbol for a voltage amplifier with power supply connections appears in Fig. 14.3. In general, the power supply must provide both a positive dc voltage and a negative dc voltage to the amplifier, as shown in Fig. 14.3. These voltages are referenced to ground, and the reference node is defined as zero volts. Notice also in this figure that the signal output is referenced to ground. The amplifier output signal shows only one wire since the other connection is made to ground.

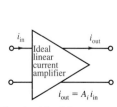

Figure 14.2   An ideal linear current amplifier.

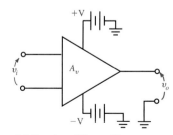

Figure 14.3   Amplifier with power supply connections.

14.2
The Ideal Op Amp

The *op amp*, or operational amplifier, was first used in analog computers which simulated complex systems defined by mathematical expressions involving differential and integral equations. Analog computers are rarely used today. However, in modern analog systems, op amps remain a very powerful and versatile component which are used to implement many different analog circuit functions. The widespread use of op amps in electronic circuitry has been greatly facilitated by the integration of complete op amp circuits on single silicon chips, making them extremely cost-effective. A high-performance op amp can be purchased for about a dollar, and can be used to implement circuit functions for many applications, such as amplification, the addition of signals, absolute value, integration, differentiation, etc.

An op amp is a voltage amplifier with a differential input and a very large voltage gain. A typical symbol for an op amp is shown in Fig. 14.4(a), and a circuit representation of this op amp is shown in Fig. 14.4(b). The term *differential input* means the output voltage responds only to the *difference* in voltage between the two inputs, which are designated as v_+ and v_-, the *noninverting* and *inverting* inputs, respectively. Therefore, if both inputs change together, there is no response. The circuit representation in Fig. 14.4(b) indicates that the output voltage v_o is directly proportional to the difference in voltage between the inputs v_+ and v_-. The differential input resistance, r_i, is the resistance between the two inputs and is typically extremely large.

We would like for the output voltage v_o to be an exact replica of the voltage produced by the dependent voltage source in the equivalent circuit, and therefore the output resistance r_o should be quite small.

Therefore, an *ideal op amp* is one which has the following characteristics:

1. The voltage gain A_v is infinite
2. The input resistance r_i is assumed to be infinite
3. Since r_i is infinite, the current into either input is always zero
4. The output resistance r_o is zero

The equivalent circuit of the ideal op amp is illustrated in Fig. 14.5.

In a practical circuit the voltage gain, A_v, of the op amp is made as high as possible, and typically ranges from several thousand to several hundred thousand. We often assume in the analysis of op amp circuits that A_v is large enough to be assumed infinite, and the external performance of the circuit is then controlled by the values of other external components.

One of the most common uses of the op amp is that of an *inverting amplifier* in which two resistors are connected to an op amp to create an amplifier circuit with a gain which

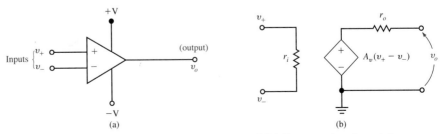

Figure 14.4   (a) Op amp symbol, and (b) Op amp circuit model

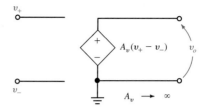

Figure 14.5   Op amp equivalent circuit.

depends only on the *ratio* of the two resistors. This circuit, described in detail in the next section, is practical and widely used.

14.3
The Inverting Amplifier

The addition of two external resistors, R_1 and R_2, to the op amp creates a new circuit, called the inverting op amp circuit, which is shown in Fig. 14.6(a). Note that resistor R_2 is in the so-called "feedback" position which connects a part of the output voltage back to the inverting input. The voltage gain of this new circuit will be defined as A_F, and

$$A_F = \frac{v_o}{v_i} \tag{14.2}$$

Substituting the circuit model for the ideal op amp shown in Fig. 14.5 into the inverting op amp circuit configuration yields the circuit in Fig. 14.6(b). We will first analyze the circuit assuming a finite value of A_v and then investigate the result of taking the limit as $A_v \to \infty$, that is, an ideal op amp. In section 14.5 we will develop a set of simplified assumptions that can be used in solving most problems with ideal op amps.

From the circuit in Fig. 14.6(b), we note that the current through the feedback resistor, i_f, can be expressed as

$$i_f = \frac{(v_o - v_i)}{R_2 + R_1} \tag{14.3}$$

Likewise, the voltage at the inverting input, v_-, can be written as

$$v_- = v_i + R_1 i_f \tag{14.4}$$

Combining Eqs. (14.3) and (14.4) yields the expression

$$v_- = v_i + R_1 \left[\frac{v_o - v_i}{R_2 + R_1} \right] \tag{14.5}$$

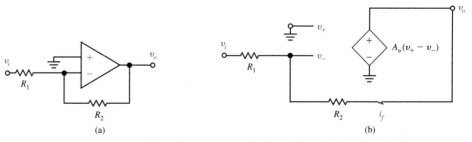

Figure 14.6   Inverting op amp circuit and equivalent circuit.

Since v_+ is connected directly to ground, that is, $v_+ = 0$ V, the output voltage, v_o, is

$$v_o = -A_v \, v_-$$

$$v_o = -A_v \left[v_i + R_1 \left[\frac{v_o - v_i}{R_2 + R_1} \right] \right] \tag{14.6}$$

The voltage gain of this circuit, A_F, is equal to the ratio, v_o/v_i, and hence

$$A_F = \frac{v_o}{v_i} = - \frac{1 - \left[\dfrac{R_1}{R_1 + R_2} \right]}{\dfrac{1}{A_v} + \left[\dfrac{R_1}{R_1 + R_2} \right]} \tag{14.7}$$

We can now apply a very useful and simplifying assumption. If we let A_v become infinitely large, the first term in the denominator approaches zero and the gain of the circuit is reduced to the expression

$$A_F = \frac{v_o}{v_i} = - \frac{R_2}{R_1} \tag{14.8}$$

Note that the voltage gain of the complete circuit, A_F, is controlled entirely by the ratio of the two resistors and is independent of A_v, provided that A_v is very large.

The control of the circuit voltage gain, A_F, by the *ratio* of two resistors is extremely important. There are a number of factors that can induce variations in resistor values, one of the most common being variations in ambient temperature. Any factor that causes a variation in *both* resistors simultaneously, such as an increase in temperature, will cancel out, since the circuit gain depends only on the ratio of the values. In this argument we have tacitly assumed that both resistors were fabricated in a similar way so that their percentage variation with temperature for a given temperature change is the same. This technique for compensating the circuit for temperature variations is widely used and enables circuits to be designed with stable performance over a wide range of environmental conditions.

It is interesting to use the previous results to calculate the voltage at the inverting input, v_-. If we solve Eq. (14.8)—a result that was derived assuming A_v is infinitely large—for v_o, and substitute this expression into Eq. (14.5), we reach the interesting result that under these conditions, $v_- = 0$. That is, the voltage at the inverting input v_- is driven to exactly the same voltage as that at the noninverting input v_+, which in this case is zero volts.

14.4
The Principle of Negative Feedback

The *principle of negative feedback* is clearly illustrated in the operation of the inverting amplifier described in the previous section. Re-examining the circuit in Fig. 14.6(b), we note that the noninverting v_+ input is connected to ground and a part of the output signal is connected to the inverting input by a feedback path through the resistor network R_2 and R_1. *Therefore, the feedback to the inverting input (v_-) drives the output of the amplifier to the voltage required to establish the inverting input (v_-) at the same voltage as that of the noninverting input (v_+).* In the situation described above, the inverting input voltage is driven to zero volts, that is, ground potential, the same voltage to which the noninverting input is connected. This, of course, assumes the gain of the op amp itself, previously defined as A_v or the open loop gain, is extremely large.

Let us now qualitatively explain the operation of negative feedback and then use the result to simplify the analysis of op amp circuits.

The operation of negative feedback will be illustrated using the inverting amplifier of Fig. 14.6. Recall that an op amp is a differential amplifier and hence the output is proportional to the difference between v_+ and v_-. Because the gain is extremely high, or for purposes of discussion here, infinite, if v_+ is slightly greater than v_- the output of the amplifier will attempt to approach a very large positive voltage. Conversely, if v_+ is slightly less than v_- the output voltage will attempt to approach a large negative voltage. It is the resistors in the circuit which provide negative feedback and control the output response in a predictable and manageable way, as illustrated in the next paragraph.

Suppose a positive voltage is applied to the input, v_i. This places a positive voltage on the v_- input. Since the v_+ input is connected to zero volts, the v_- input becomes more positive than the v_+ input. This condition drives the output voltage toward a large negative value. However, this output voltage is coupled back to the v_- input through the network of R_2 and R_1 and lowers the voltage at that node until it just reaches zero — or exactly the same voltage as is connected to the v_+ input. If the output voltage attempted to go more negative, the voltages at the input would reverse in polarity, and halt or reverse the negative voltage swing of the output. *Therefore, the output voltage stabilizes at a value which exactly feeds back the proper voltage level to place the v_- input at the same voltage as the v_+ input.*

Using this concept, and assuming an ideal op amp, the gain A_F of the circuit in Fig. 14.6 can be calculated in an alternative and much simpler way.

From Fig. 14.6(b), we can write the following loop equation:

$$v_o - v_i = i_f(R_2 + R_1) \tag{14.9}$$

and

$$i_f R_1 = v_- - v_i \tag{14.10}$$

From the above discussion, we know the feedback will cause v_- to be driven to zero since $v_+ = 0$.

Therefore

$$i_f = -\frac{v_i}{R_1} \tag{14.11}$$

Substituting Eq. (14.11) into (14.9) yields

$$v_o - v_i = -\frac{v_i}{R_1}(R_2 + R_1)$$

from which we obtain

$$A_F = \frac{v_o}{v_i} = -\frac{R_2}{R_1} \tag{14.12}$$

Notice that this is the same result obtained earlier, but derived by a much simpler procedure.

It is interesting to note that the v_- node becomes a "virtual ground," in that, since the v_+ node is grounded, v_- is always driven to ground potential. Therefore, calculation of the input resistance of the circuit, R_i, is simple, since it is just the resistance seen looking into the circuit at v_i with v_- grounded, that is

$$R_i = R_1 \qquad (14.13)$$

EXAMPLE 14.1

Let us assume an ideal op amp is used in the circuit of Fig. 14.6(a), with $R_1 = 2$ kΩ and $R_2 = 80$ kΩ. For this network let us determine the voltage gain A_F, which is sometimes called the *closed loop gain,* (a) if the op amp gain is infinite and (b) if the gain of the op amp is 100.

(a) For this case

$$A_F = -\frac{R_2}{R_1} = -\frac{80 \text{ k}}{2 \text{ k}} = -40$$

(b) For the case when A_v cannot be considered infinitely large, the simpler expression Eq. (14.8) is not valid, and Eq. (14.7) must be used. Substituting the circuit values into this equation yields

$$A_F = -\frac{1 - \dfrac{2 \text{ k}}{82 \text{ k}}}{\dfrac{1}{100} + \dfrac{2 \text{ k}}{82 \text{ k}}} = -28.4$$

Notice that in this latter case, A_F was reduced substantially—from 40 to 28.4 when A_v was reduced from infinity to 100. ∎

14.5

The Noninverting Amplifier

When an ideal operational amplifier is combined with two resistors in the manner shown in Fig. 14.7, the result is a *noninverting amplifier* circuit. In order to calculate the voltage gain of this circuit configuration, the assumption of an ideal op amp described in the previous section will be utilized. This assumption includes two primary considerations, which may be used in solving any op amp problem when an ideal op amp is assumed. These considerations are:

1. The input resistance of the op amp is infinite, therefore, no current flows into either the v_+ or v_- inputs.
2. The gain of the op amp is infinite, therefore the feedback drives the voltage, v_-, to exactly the same value as v_+.

From Fig. 14.7, assuming no current flows into the inverting input, that is, it appears like an open circuit, we can write the following expression for v_- in terms of the output voltage v_o. Because of the negative feedback

$$v_+ = v_- = v_i = \frac{R_1}{R_1 + R_2} v_o$$

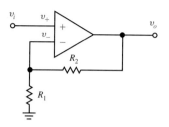

Figure 14.7   Noninverting op amp circuit

and therefore

$$A_F = \frac{v_o}{v_i} = 1 + \frac{R_2}{R_1} \qquad (14.14)$$

Note that the voltage gain for this circuit configuration has a positive sign and therefore does not invert the input signal. Again the magnitude of the voltage gain, A_F, is set by the ratio of the two external resistors. Once again, this configuration provides cancellation of the effect of variations in resistor values which are common to both resistors. Finally, the minimum value of voltage gain for the circuit configuration is unity, which is achieved when $R_2 = 0$. Recall that the inverting amplifier can have a voltage gain less than unity, although this is rarely done because it is not often useful to have a circuit producing an output smaller than its input.

EXAMPLE 14.2

Given the circuit of Fig. 14.7, if R_1 is 10 kΩ, what value of R_2 is required to produce a circuit with a voltage gain of 25?

Substituting the known values in Eq. (14.14) yields

$$25 = 1 + \frac{R_2}{10 \text{ k}}$$

Solving for R_2 we obtain $R_2 = 240$ kΩ.   ■

14.6
The Unity-Gain Buffer

There are situations in the design of analog systems that require a unity-gain buffer. The *unity-gain buffer* is a circuit which has a voltage gain of unity; that is, the output voltage is the same as the input voltage. There is, however, an important difference between the output and input of this circuit. The output resistance of this op amp circuit is very low. Think of this as a small resistance in series with the dependent voltage source controlling the output voltage. This configuration enables the circuit to force the voltage at the output to the correct value, almost independent of the load connected to the circuit. The input to the circuit is an extremely high resistance and therefore will present negligible load to the source of the input voltage.

The noninverting amplifier circuit configuration described in the previous section will produce unity gain when R_2 is set to zero. This is the equivalent of a short between the output and the inverting input, and therefore the resistor R_1 becomes inconsequential and can be dropped from the circuit. We are left with the unity-gain buffer circuit shown in Fig. 14.8.

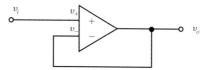

Figure 14.8 Unity-gain buffer op amp circuit.

This circuit has a voltage gain A_F of unity. Therefore

$$v_o = v_i \qquad (14.15)$$

In summary, the unity-gain buffer or amplifier is a special case of the noninverting amplifier in which the entire output voltage v_o is fed back to the inverting input v_-. Since the voltage gain of this circuit is unity, its primary function is an impedance converter or transformer that can receive a voltage from a high-impedance source, and provide an output of the same voltage at a very low output impedance level; this circuit provides substantial current gain. Circuits such as this are often used for "line drivers" which send the signal down a long length of wire or cable.

DRILL EXERCISES

D14.1. Find the value of I_o in the network of Fig. D14.1; assume an ideal op amp.

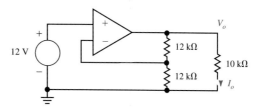

Figure D14.1

Ans: $I_o = 8.4$ mA.

D14.2. Fig. D14.2 is a differential voltage amplifier, often called an instrumentation amplifier (also see Problem 14.13). Such circuits are often used to amplify the signal from a sensor, which may be differential, such as that from a bridge circuit, and pro-

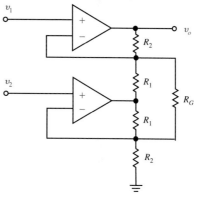

Figure D14.2

vide a single-ended voltage (referenced to ground) at its output for application to an analog-to-digital converter, for example. Assuming ideal op amps are used, what is an expression for the output voltage in terms of the two inputs?

Ans:

$$v_o = (v_1 - v_2)\left[1 + \frac{R_2}{R_1} + \frac{2R_2}{R_G}\right]$$

14.7
The Summing Circuit

The circuit shown in Fig. 14.9 is similar to the circuit for the inverting amplifier shown in Fig. 14.6. The primary difference is that this circuit has two input voltages, v_{iA} and v_{iB}, and two corresponding input resistors, R_{1A} and R_{1B}.

This circuit acts as a *summing circuit* and will provide an output voltage which is related to the sum, or the addition, of the values of the two input voltages. This circuit has the same basic topology as the inverting amplifier circuit and retains a -1 multiplier (inverting) factor in the expression for voltage gain. Furthermore, as we shall see, each input voltage is weighted in the sum by a resistor ratio.

In analyzing this circuit we will again assume the op amp is ideal, and therefore v_- is forced to zero ($v_- = v_+ = 0$). With this assumption, we can easily write expressions for i_A, i_B, and i_f as follows.

$$i_A = \frac{v_{iA}}{R_{1A}}; \ i_B = \frac{v_{iB}}{R_{1B}}; \ i_f = \frac{v_o}{R_2} \qquad (14.16)$$

Applying Kirchhoff's law at the inverting input node results in the following expression:

$$i_A + i_B = -i_f \qquad (14.17)$$

Combining Eqs. (14.16) and (14.17) leads to the following result:

$$\frac{v_{iA}}{R_{1A}} + \frac{v_{iB}}{R_{1B}} = -\frac{v_o}{R_2}$$

Therefore

$$v_o = -\left[\frac{R_2}{R_{1A}} v_{iA} + \frac{R_2}{R_{1B}} v_{iB}\right] \qquad (14.18)$$

This equation defines the output of a summing circuit with two input voltages and two corresponding input resistors. Summing circuits with three or more input voltages and re-

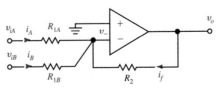

Figure 14.9 Summing op amp circuit.

sistors can be designed by a straightforward extension of these results. Once again, notice that the relative gain of each of the inputs to the sum is controlled by the ratio of two resistor values.

14.8
The Integrator

In the circuit of Fig. 14.10, a capacitor is placed in the feedback position of an inverting op amp circuit. This circuit will provide an output voltage proportional to the integral of the input voltage. To illustrate this fact, we again assume an ideal op amp, and v_- is made zero by the feedback. Then

$$i_f = C\frac{dv_o}{dt}; \quad i_1 = \frac{v_i}{R_1}; \quad i_f = -i_1$$

Combining these equations leads to the expression

$$\frac{v_i}{R_1} = -C\frac{dv_o}{dt} \tag{14.19}$$

Solving this equation for v_o yields

$$v_o = -\frac{1}{R_1C}\int_0^t v_i\,dt + v_o(0) \tag{14.20}$$

The term $v_o(0)$ is the output voltage at time zero resulting from any initial charge on the capacitor.

EXAMPLE 14.3

An *integrator circuit* of the type shown in Fig. 14.10 has the following component values: $C = 100\ \mu F$, and $R_1 = 30\ k\Omega$. If the voltage, v_i, at the input of this circuit is given by the plot shown in Fig. 14.11(a), derive an expression for, and construct a plot of, the output voltage, v_o. Assume $v_o(0) = 0$.

Using Eq. (14.20)

$$v_0(t) = -\frac{1}{3}\,2t \text{ for } 0 < t < 2 \text{ seconds}$$

At $t = 2$ seconds, $v_o(2) = -1.33$ volts.
For the next interval

$$v_o(t) = -\frac{1}{3}\,[-4(t-2)] - 1.33 \text{ for } 2 < t < 3 \text{ seconds}$$

A plot of the output voltage is given in Fig. 14.11(b).　∎

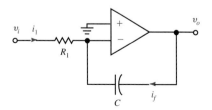

Figure 14.10 Op amp integrator circuit.

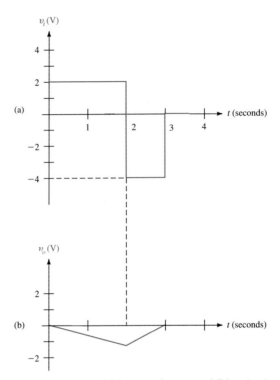

Figure 14.11   Integrator (a) input voltage, and (b) output voltage.

14.9
The Active Filter

As indicated in Chapter 6, a filter is a network which is frequency-selective in that it responds to certain signal frequencies differently than others. While a passive filter uses only passive components, such as capacitors, resistors, and inductors, an active filter utilizes amplifiers, usually op amps, and by doing so can often eliminate the need for costly and bulky inductors and simultaneously achieve higher performance.

When considering only resistive elements in the feedback and input signal paths, and an ideal op amp, the closed loop voltage gain, A_F, is real. With capacitors and/or inductors in op amp circuits we can, in an analogous way, consider the impedance of circuit elements, and obtain a complex voltage gain, $\mathbf{A}_F$.

As a simple example, consider the low-pass active filter circuit of Fig. 14.12. The cir-

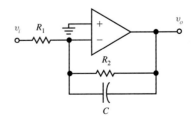

Figure 14.12   Low-pass filter circuit.

cuit is an inverting op amp configuration, however, the RC network in the feedback path is frequency-selective.

The complex voltage gain of this circuit, $\mathbf{A}_F$, can be computed by taking the ratio of two impedances, the impedance of the feedback branch to that of the input branch. The complex voltage gain is the ratio of the phasor transforms of the output to the input voltage. That is

$$\mathbf{A}_F = \frac{\mathbf{v}_o}{\mathbf{v}_i} = -\frac{\mathbf{Z}_2}{\mathbf{Z}_1} \tag{14.21}$$

where

$$\mathbf{Z}_2 = \frac{1}{j\omega C + \left(\dfrac{1}{R_2}\right)} \tag{14.22}$$

and

$$\mathbf{Z}_1 = R_1 \tag{14.23}$$

Therefore

$$\mathbf{A}_F = -\frac{\left[\dfrac{R_2}{R_1}\right]}{1 + j\omega R_2 C} \tag{14.24}$$

The frequency response of this circuit is plotted in Fig. 14.13. At low frequencies the gain of the circuit is constant at R_2/R_1. The high cutoff frequency is the value at which the magnitude of the voltage gain is reduced to 0.707 of its maximum. This value corresponds to a reduction in power by one-half, that is, the half-power point. For this circuit ω_{HI} is

$$\omega_{HI} = \frac{1}{R_2 C} \tag{14.25}$$

This simple circuit does not produce a very sharp rolloff in gain with frequency, since it has a single-pole response. However, more complex circuits with multiple op amps can produce filter responses with very accurate and sharp response characteristics. By placing frequency selective circuits in the input path, and/or feedback path, low-pass, high-pass, and band-pass filters can be created.

It has been assumed here that the frequency response of the op amp itself extends beyond the frequency operation of the filter; otherwise, the frequency dependent characteristics of the op amp must also be included in determining the circuit's response.

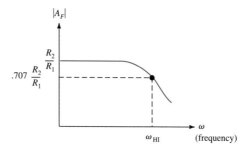

Figure 14.13 Low-pass filter frequency response.

14.10
The Current-to-Voltage Converter

There are some applications for which a circuit is needed that will transform an input current to an output voltage. To accomplish this, a simple op amp circuit can be used with a resistor in the feedback path as shown in Fig. 14.14. The input current i_i cannot flow into the op amp since its input resistance is infinite; therefore, this current flows entirely through R_F. Hence

$$i_i = -i_f \tag{14.26}$$

The voltage at v_- is zero, and therefore

$$v_o = -i_i R_F \tag{14.27}$$

It is impossible to refer to the "voltage gain" or "current gain" of this circuit, because the input is a current and the output is a voltage. The parameter that describes how much the output voltage changes for a given change in input current is the *trans-resistance*. *Trans* means the voltage and current are not measured at the same place.

One such use of this type circuit is in an electronic illumination (light) intensity meter. Photodiodes are devices that generate an output current approximately proportional to the incident radiation, such as light, on them. If a photodiode is connected to the input of a *current-to-voltage converter* circuit, the output voltage will be proportional to incident light intensity.

EXAMPLE 14.4
A photodiode designed for visible light has a sensitivity of 25 μA per milliwatt of incident radiation. If this diode is used in the light meter circuit of Fig. 14.15, calculate the value of R_F needed to make the magnitude of the output voltage equal 4 volts when the incident radiation deposits 50 milliwatts at the detector.

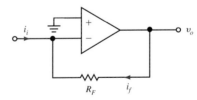

Figure 14.14   Current-to-voltage converter circuit.

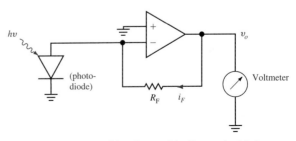

Figure 14.15   Circuit used in Example 14.4.

The detector photodiode produces 25 μA per milliwatt of radiation; therefore, at 50 milliwatts, it produces:

$$i_{PH} = (50)\,(25 \times 10^{-6}) = 1.25 \text{ mA}$$

In order to produce a magnitude of 4 volts at the output of the circuit

$$R_F = \frac{4}{1.25 \times 10^{-3}} = 3.2\text{k}\Omega$$

The photodiode produces positive current flow from the anode, and therefore in this circuit, positive current flows inward and opposite to the assumed direction of i_f. Therefore, in operation, the output voltage goes more negative as light intensity is increased, reaching -4 volts with 50 milliwatts of incident radiation. The photodiode could be reversed end-for-end, and the output voltage would go positive with increasing light intensity. ∎

14.11
Summary

- A linear amplifier is an electronic circuit which produces an output signal that is an exact copy of the input signal but multiplied by a constant greater than unity, that is, increased in amplitude.
- For an ideal voltage amplifier

$$v_{out} = A_v\, v_{in}$$

- A differential amplifier amplifies only the *difference* in voltage between two inputs.
- An ideal op amp is a differential amplifier with: (1) infinite voltage gain, (2) infinite input resistance (and hence no input current), and (3) zero output resistance.
- The op amp can be combined with external components to construct a wide variety of useful and practical circuits. The availability of inexpensive IC op amps makes these type circuits relatively cost-effective and simple to implement.
- The inverting amplifier configuration (see Fig. 14.6) requires two external resistors. If an ideal op amp is used, the closed loop voltage gain of the circuit, A_F, is given by

$$A_F = -\frac{R_2}{R_1}$$

If the voltage gain of the op amp, A_v, is finite, then A_F will be less than the maximum value given by the resistor ratio.

- Negative feedback in an ideal op amp circuit drives the output voltage of the amplifier to the voltage required to place the "−" input at the same value as that at the "+" input.
- The noninverting amplifier circuit has a closed loop gain set by the ratio of two resistors and a minimum value of +1.
- Other op amp circuit configurations are capable of implementing a voltage summing circuit, the integrator, the active filter, and current-to-voltage converter.

PROBLEMS

14.1. The input and output signals of an amplifier are shown in Fig. P14.1. (a) Find the voltage gain of the amplifier. (b) Is it a linear amplifier? Explain your answer.

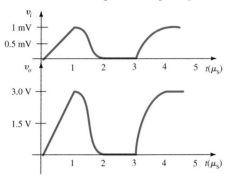

Figure P14.1

14.2. (a) Determine the voltage gain of the inverting amplifier circuit shown in Fig. P14.2 using the op amp model of Fig. 14.5. Assume $A_v = 50$, $R_2 = 10$ kΩ, $R_1 = 800$ Ω. (b) Repeat (a) with $A_v = 500$. (c) Repeat (a) taking the limit as $A_v \rightarrow \infty$. (Hint: consider the applicability of Eq. 14.7.)

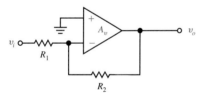

Figure P14.2

14.3. For the circuit shown in Fig. P14.2, find the voltage gain by using the op amp model in Fig. 14.5 and letting $A_v = 800$, $R_1 = 100$ Ω, and $R_2 = 3.3$ kΩ.

14.4. Calculate the voltage gain, A_F, of the circuit shown in Fig. P14.4; assume an ideal op amp.

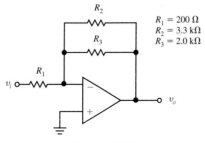

$R_1 = 200$ Ω
$R_2 = 3.3$ kΩ
$R_3 = 2.0$ kΩ

Figure P14.4

14.5. Determine the voltage gain, A_F, for the circuit of Fig. P14.5, assuming an ideal op amp.

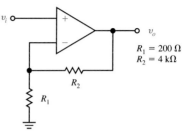

$R_1 = 200$ Ω
$R_2 = 4$ kΩ

Figure P14.5

14.6. Determine the voltage gain, A_F, for the circuit of Fig. P14.6, assuming an ideal op amp.

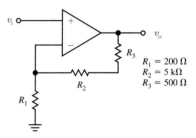

$R_1 = 200$ Ω
$R_2 = 5$ kΩ
$R_3 = 500$ Ω

Figure P14.6

14.7. Determine the voltage gain for the circuit of Fig. P14.6 if a resistor $R_4 = 2$ kΩ is added from the junction of R_2 and R_3 to ground; assume an ideal op amp.

14.8. Determine the value of R_2 required to establish the voltage gain of the circuit in Fig. P14.8 at $A_F = 30$; assume an ideal op amp.

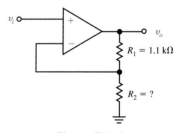

$R_1 = 1.1$ kΩ

$R_2 = ?$

Figure P14.8

14.9. Find the closed loop voltage gain, A_F, for the circuit in Fig. P14.9, assuming an ideal op amp.

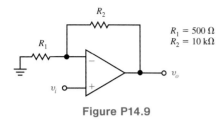

$R_1 = 500 \ \Omega$
$R_2 = 10 \ k\Omega$

Figure P14.9

14.10. Negative feedback in an ideal op amp circuit establishes the "−" input at what voltage?

14.11. Calculate the voltage gain of the circuit shown in Fig. P14.11, assuming an ideal op amp. Are there values of R that should be avoided?

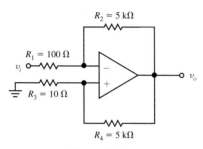

Figure P14.11

14.12. The signal from an electronic pressure transducer (sensor) has a maximum value of 5 mV, corresponding to maximum pressure. We would like to amplify this signal to a maximum of 2 volts. If an op amp circuit similar to that of Fig. P14.9 is used and $R_1 = 0.5 \ k\Omega$, what is the proper value for R_2? (Assume an ideal op amp.)

14.13. For the circuit shown in Fig. P14.13, show that if $R_1/R_2 = R_3/R_4$, then $v_o = (R_2/R_1) (v_1 - v_2)$, assuming an ideal op amp.

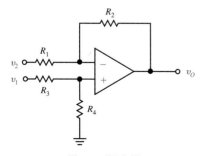

Figure P14.13

14.14. For the circuit in Fig. P14.14, determine an expression for the output voltage, v_o, in terms of the input voltages, v_A, v_B, and v_C; assume an ideal op amp.

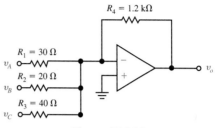

Figure P14.14

14.15. Using an ideal op amp, draw a circuit to implement the following function: $v_o = -[2v_A + 7 \ v_B]$.

14.16. A circuit with an ideal op amp is shown in Fig. P14.16(a); let $C = 220 \ \mu F$ and $R = 10 \ k\Omega$. (a) Plot the input waveform, v_i versus time, for the output, v_o shown in Fig. P14.16(b). (b) If the input is connected to a constant voltage, V_1, show that the output is the solution of the equation:

$$C \frac{d \ v_o(t)}{dt} + \frac{V_1}{R} = 0$$

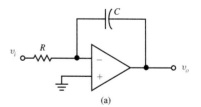

(a)

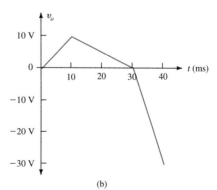

(b)

Figure P14.16

14.17. Design a circuit to obtain the solution of the following differential equation:

$$10\,\frac{d\,v(t)}{dt} + \frac{5}{7} = 0$$

14.18. The circuit shown in Fig. P14.18(a) is obtained by interchanging the capacitor and resistor in an integrator circuit. (a) Show that the circuit acts as a differentiator, that is

$$v_o(t) = -RC\,\frac{d\,v_i(t)}{dt}$$

(b) Sketch the output voltage for the circuit, $v_o(t)$, as a function of time if the input voltage, $v_i(t)$, is given in Fig. P14.18(b); assume $R = 10$ kΩ, $C = 0.22$ μF, and an ideal op amp.

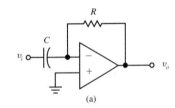

(a)

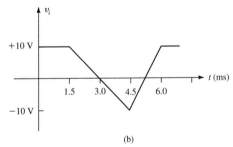

(b)

Figure P14.18

14.19. Plot the magnitude of the closed loop gain, A_F, of the circuit presented in Fig. P14.19 as a function of frequency; let $R_1 = 1$ kΩ, $R_2 = 5$ kΩ, $C = 20$ μF, and assume the op amp is ideal.

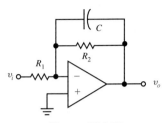

Figure P14.19

14.20. Plot the magnitude of the closed loop gain, A_F of the circuit presented in Fig. P14.20 as a function of frequency; let $R_1 = 2$ kΩ, $C_1 = 30$ μF, $R_2 = 3$ kΩ, $C_2 = 2$ μF, and assume the op amp is ideal.

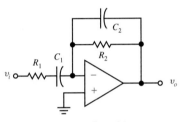

Figure P14.20

14.21. The circuit shown in Fig. P14.21 is an amplifier circuit that has a closed loop gain which can be controlled by the position of a switch. Obtain the gain (a) when the switch, S_1, is closed and (b) when the switch, S_1, is open.

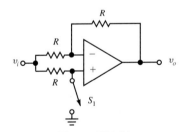

Figure P14.21

14.22. The waveforms shown in Fig. P14.22(a) are the input and output voltages of the op amp circuit of Fig. P14.22(b). Specify the components and their values that should be placed in the boxes labeled A and B in order to obtain the given performance. (Assume the desired circuit input resistance is 1 kΩ.)

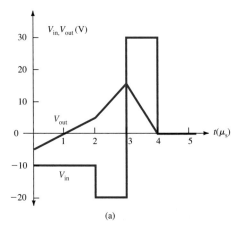

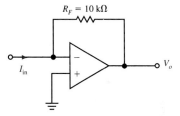

Figure P14.22

14.23. Determine the maximum gain and bandwidth of the filter shown in Fig. P14.23; assume an ideal op amp.

14.24. The circuit given in Fig. P14.24 is a current-to-voltage converter. Determine the transresistance of this circuit; assume an ideal op amp.

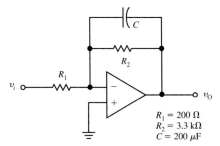

$R_1 = 200\ \Omega$
$R_2 = 3.3\ \text{k}\Omega$
$C = 200\ \mu\text{F}$

Figure P14.23

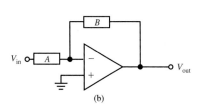

Figure P14.24

14.25. The circuit of Fig. P14.25 is an electronic radiation detector. The photodiode has a sensitivity of 10 μA/mW of incident radiation. If the voltmeter reads 10 V, determine the power being deposited by the radiation.

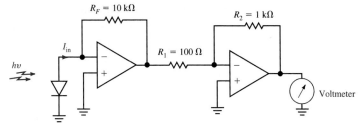

Figure P14.25

The Large-Signal Analysis of MOSFETs and BJTs for Analog Applications

15.1
Introduction

Transistors are the devices that perform the amplification of electronic signals in most modern systems, and as such are critical elements in analog circuits. We have already seen in Chapter 9 how these same devices can perform as electronic switches in digital circuits. In this chapter we will explore how transistors are biased and how they can be used to make amplifiers and other circuits for processing analog signals.

There are a number of terms commonly used to describe various types of amplifiers and analog systems. A *small-signal amplifier* is one in which the signal voltage variations are reasonably small, and over this small range the nonlinearities of the transistor can be approximated by linear circuit models. From a practical standpoint, this generally requires that the voltage swings not exceed the range of a few hundreds of millivolts, unless feedback is used to linearize the response as is done in many op amp circuits.

A *large-signal amplifier* is one in which the voltage swings are large, that is, several volts or more, and the device nonlinearities cannot be ignored. Typically these circuits are used to provide high levels of signal output, such as a power amplifier driving a stereo speaker. The design and analysis of these circuits is often done graphically, using plots of the voltage-current characteristics of the transistor over their entire operating voltage range. The transistor output curves provide such graphical information for large-signal analysis.

MOSFET transistor output curves were introduced in Chapter 9. Figure 9.16 illustrated an example of an output curve for an *n*-channel, enhancement-mode MOSFET with a threshold voltage of 1 volt. Output curves provide the voltage-current relationships for a transistor over the entire range of practical voltages and currents for that device. These curves are useful in understanding how large-signal amplifiers operate.

There are also several sections in Chapter 9 that deal with amplifiers. In section 9.9 the BJT in the active region is discussed and the bias region appropriate for analog amplifiers identified. The common emitter circuit configuration is discussed, and in section 9.10 the Ebers-Moll model is presented as a large-signal model for the BJT, which is often used in the analysis and modeling of bipolar transistor amplifiers. Finally, BJT output curves are introduced in section 9.11 and they form the basis for the large-signal graphical analysis performed in this chapter.

A *single-stage amplifier* is one in which there is only one amplifying element. By combining several single-stage amplifier circuits, we produce a *multi-stage amplifier.* *Audio amplifiers* are designed to amplify signals that include frequencies perceptible to the human ear, that is, approximately 30 Hz to 15,000 Hz. A *video amplifier* is designed to amplify the signal frequencies required for television imaging (see Chapter 23).

15.2
dc Biasing

The first step in designing or analyzing any amplifier is to consider the biasing. The biasing network is comprised of the power supply and the passive circuit elements surrounding the transistor that provide the correct dc levels at the terminals. This is termed setting the Q-point or quiescent operating bias point, that is, the terminal voltages and currents with NO signal applied. Much of this chapter addresses methods for establishing proper dc bias.

A good bias circuit must not only establish the correct dc levels, but must maintain them even when confronted with sources of variation, such as

1. Changing temperature. Solid state devices tend to be temperature sensitive, and therefore, the bias circuit should maintain the proper dc levels even if transistor parameters shift with temperature changes.
2. Variations in transistor characteristics resulting from the random spread in parameters in manufacturing. Normal manufacturing lines do not produce identical products and hence there is a distribution of transistor characteristics for devices produced on the same line at different times. A well-designed bias circuit can produce consistent circuit performance even if individual transistor characteristics vary.

15.3
Large-Signal Amplifiers: Using MOSFETs

Consider a transistor with the output characteristics measured and plotted as shown in Fig. 15.1. This device is an enhancement-mode MOSFET with a threshold voltage of 1.5 volts. The transistor's response is represented by a family of output curves. There are actually an infinite number of curves since V_{GS} can vary continuously—one for each possible value of V_{GS}. For convenience, we only plot a curve for incremental values of V_{GS}; however, one can interpolate curves between those plotted.

The Common Source Amplifier

Let us now place the transistor with characteristics shown in Fig. 15.1 in a *common source bias circuit* as illustrated in Fig. 15.2. Note that the total resistance in series with the transistor from source to drain is $R_D = 2$ kΩ. A dc load line is therefore established which

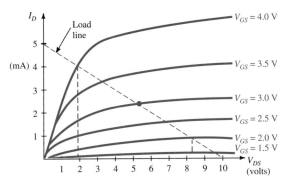

Figure 15.1 Enhancement-mode MOSFET transistor output characteristics.

connects the points: $V_{DS} = 10$ V, which is the value of V_{DD} on the horizontal axis, and $I_D = 10$ V/2 k$\Omega = 5$ mA on the vertical axis. The load line is sketched on the output curves of Fig. 15.1 as a dashed line.

This circuit is called a *fixed bias circuit* because the gate-to-source voltage is fixed by the resistor network consisting of R_1 and R_2. Since no dc current flows into the gate of a MOSFET

$$V_{GS} = V_{DD} \left[\frac{R_2}{R_1 + R_2} \right] \tag{15.1}$$

A loop equation, including R_D and the transistor from drain to source, is

$$V_{DD} = I_D R_D + V_{DS} \tag{15.2}$$

The current I_D in real devices is dependent primarily on V_{DS}, V_{GS}, and it is nonlinear. These relationships are provided graphically by the large-signal transistor characteristics expressed on the output curves. The use of the load line, which is a plot of R_D's characteristics, superimposed on the transistor's curves, can provide a graphical solution to Eq. (15.2). The solution satisfies the following conditions:

- In order to satisfy the load resistor's V–I characteristics, the bias point must fall somewhere on the load line.
- In order to satisfy the transistor requirements, the bias point must be located on an output curve corresponding to the appropriate value of V_{GS}.
- The intersection of the two curves provides the solution.

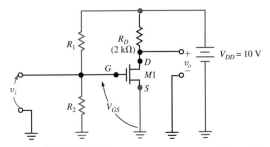

Figure 15.2 MOSFET in a common source bias circuit.

EXAMPLE 15.1

Given the transistor network in Fig. 15.2 and the transistor characteristics shown in Fig. 15.1, let us determine (a) the value of V_{DS} if $V_{GS} = 2.0$ V, (b) the value of V_{DS} if $V_{GS} = 4.0$ V, (c) the best place to locate the Q-point, and (d) the values of R_1 and R_2 required for the Q-point selected.

(a) From Fig. 15.1 a V_{GS} of 2.0 V yields a V_{DS} of 8.2 V.

(b) Figure 15.1 illustrates that a V_{GS} of 4.0 V yields a V_{DS} of 1.9 V. The results for (a) and (b) are obtained by locating the intersection of the appropriate output curve and the load line.

(c) Since this circuit is an amplifier, we would like for the output signal to swing equal voltage excursions on either side of the Q-point. We have tacitly assumed that our input signal will have equal positive and negative voltage swings. Therefore, placing the Q-point near the center of the range where the load line intersects the operating curves is generally best. For this example, let's place the Q-point at the point indicated, where $V_{GS} = 3.0$ V.

(d) Now using Eq. (15.1), we can compute values for the resistor network, R_1 and R_2.

$$3.0 = \left[\frac{R_2}{R_1 + R_2} \right] 10$$

If we choose one of the values, for example, $R_2 = 10$ kΩ, then solving for R_1 yields $R_1 = 23$ kΩ. ∎

The operation of this circuit as a large-signal amplifier can be easily seen from Fig. 15.3. The same output curves are reproduced with the 2 kΩ load line and the Q-point set at

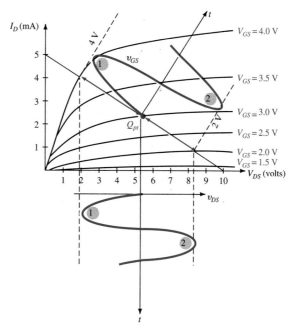

Figure 15.3 Large-signal common source amplifier operation.

$V_{GS} = 3$ V. Observe what happens if the input voltage to the circuit, v_i, is changed: the voltage v_i, which is applied directly gate to source, that is, $v_i = V_{GS}$, is increased sinusoidally from 3 volts up to a peak of 4 volts and back to 3 volts again. The actual bias point moves up and down the load line following the movement of V_{GS}. In a similar manner, the output voltage of this circuit, v_o, which is the same as V_{DS}, moves from just over 5 volts down to 1.9 volts, and back up again. On the figure, we refer to the gate-to-source and drain-to-source voltages during these *swings* as v_{GS} and v_{DS}, respectively, indicating that the TOTAL instantaneous value is the dc bias plus the signal.

During the second half-cycle, the input change continues sinusoidally, symmetrically down to 2 volts and back up to 3 volts again, while the output voltage swings positively to a peak of 8.2 volts, and then returns to its quiescent value of just over 5 volts.

Notice that this circuit provides signal gain. The input voltage changed 2 volts peak-to-peak, while the output of the circuit changed from a low of 1.9 volts to a high of 8.2 volts, or a difference of 6.3 volts peak-to-peak. Therefore, the *large-signal voltage gain* of the circuit is

$$A_v = \frac{\Delta v_o}{\Delta v_i} = \frac{-6.3}{2} = -3.15$$

The *negative sign* indicates that the amplifier inverts the signal so that an increase in v_i produces a decrease in v_o.

The Self-Bias Circuit

Another bias circuit configuration often used is called the *self-bias* circuit. This circuit is useful for devices that require a negative gate-to-source bias voltage, such as a depletion-mode *n*-channel device. The circuit achieves a negative gate-to-source voltage by raising the source voltage higher than that at the gate.

As an example, consider the MOSFET transistor with the output curves shown in Fig. 15.4. The circuit of Fig. 15.5 will self-bias this transistor.

First the Q-point will be placed on the $V_{GS} = -1$ V curve at its intersection with a 2 kΩ load line. The Q-point location and the dc load line are also shown in Fig. 15.4

The quiescent dc voltage on the gate of the device is zero, since there is no dc cur-

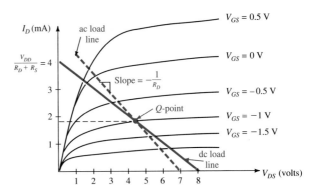

Figure 15.4 Output curves for depletion-mode MOSFET transistor.

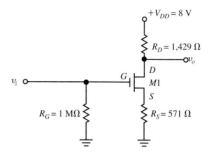

Figure 15.5 Depletion-mode MOSFET transistor amplifier using self-bias circuit.

rent flow through R_G. The voltage across R_S is given by the expression

$$V_S = I_D R_S$$

and thus

$$V_{GS} = -I_D R_S \tag{15.3}$$

Since the resistor R_S has been added in the drain-source current loop, the dc load line now has a slope of $-1/(R_D + R_S)$ and the loop equation becomes

$$V_{DD} = I_D(R_D + R_S) + V_{DS} \tag{15.4}$$

The desired Q-point is located at a quiescent drain current of 1.75 mA. Therefore, Eq. (15.3) yields

$$V_{GS} = -1 = -R_S(1.75 \times 10^{-3})$$

and hence

$$R_S = 571 \ \Omega$$

To obtain the load line shown in the figure, we note that

$$R_D + R_S = 2 \ \text{k}\Omega$$

and therefore

$$R_D = 2 \ \text{k} - 571 = 1429 \ \Omega$$

Note that there is a potential problem with the circuit of Fig. 15.5. The addition of R_S tends to lower the voltage gain of the circuit. The output voltage is the sum of V_{DS} and the voltage across R_S. As V_{DS} decreases and thus I_D increases, the voltage drop across R_S increases, thus canceling some of the output change.

This problem can be avoided by adding a *bypass capacitor* across R_S, as shown in Fig. 15.6. Bypass capacitors are assumed to be large enough in value to have a very low reactance at the minimum signal frequency, and therefore are considered to be a short circuit to ac. The capacitor C_S effectively shorts R_S and therefore we can construct an *ac load line* which has a slope of $-1/R_D$, and passes through the Q-point. This ac load line is labeled as such in Fig. 15.4. Furthermore, this new circuit has *coupling capacitors, C_1* and C_2, which couple the ac signal in and out of the circuit but block the dc levels. Coupling capacitors also are chosen to act as ac shorts at the minimum signal frequency. The input signal, v_i, moves the bias point on the ac load line. The computed voltage gain is -1.8.

Therefore, the procedure for large-signal analysis of MOSFET circuits can be summarized as follows.

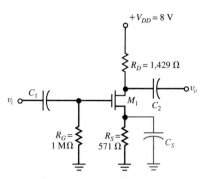

Figure 15.6 Transistor amplifier with a source bypass capacitor.

1. Draw the dc load line on the output curves of the transistor. This load line intersects the voltage axis at V_{DD}, and intersects the current axis at V_{DD} divided by the total resistance in series with the S and D terminals of the transistor.
2. Locate the Q-point on this load line. This is done by calculating the gate-to-source voltage, V_{GS}, applied by the bias circuit. The Q-point is the intersection of the dc load line with the appropriate V_{GS} curve.
3. If there is a bypass capacitor in the circuit shunting an element in the S-D loop, then construct an ac load line. The ac load line passes through the Q-point and has a slope given by

$$slope = -\frac{1}{total\ resistance\ within\ S\text{-}D\ loop\ not\ shorted\ by\ C}$$

In most cases, this will be simply $-1/R_D$.

If there are no bypass capacitors, then the ac load line is the same as the dc load line.

4. Calculate the large-signal voltage gain. This is done by first assuming reasonable variations in the input signal, that is, use variations in V_{GS} which fit the output curves. Now graphically determine the extent of the resulting output signal excursion, that is, maximum and minimum voltages, or deviations from the initial Q-point, as the bias point moves up and down the ac load line. The large-signal voltage gain is then

$$A_v = \frac{\Delta v_o}{\Delta v_i} \tag{15.5}$$

Note that in this section we used n-channel transistors. A p-channel transistor can be used in an analogous manner and the same analysis techniques applied. The only difference is the polarity of the required bias voltages.

Large-signal amplifiers with MOSFETs should be used cautiously because FETs have an inherent nonlinearity. These devices do, however, work well in small-signal applications, or in more complex circuits which employ feedback to linearize their response. From the output curves of Fig. 15.1, we note that as V_{GS} is increased by equal increments, the drain current does not increase in equal increments. This nonlinearity can introduce distortion in the output of the amplifier's response.

15.4

Other FET Amplifier Configurations

There are three basic circuit configurations for FET amplifiers: common source (CS), common gate (CG) and common drain (CD).

The MOSFET circuits discussed in the previous section were configured so that the input signal was applied between the gate and source, and the output connected between drain and source. The source was the common terminal. These are all common source amplifiers. There are, however, two other possibilities: common gate and common drain, and all three circuit configurations are illustrated in Fig. 15.7.

The large-signal analysis of these two remaining circuit configurations follows the same procedure as that outlined for the common source configuration in the previous section.

The Common Gate Amplifier

The large-signal analysis of this configuration is illustrated by the following example.

EXAMPLE 15.2

Consider the *common gate amplifier circuit* shown in Fig. 15.8. Assume that the transistor is an *n*-channel depletion-mode device that can also be operated in the enhancement mode and has CS output curves as shown in Fig. 15.9. For this network we wish to establish the dc bias conditions and then estimate the large-signal voltage gain.

The bias circuit is essentially the same as the self-bias circuit previously described. The dc current I_D flows through R_D, the transistor from D to S, and R_S. Recall that as positive current flows from drain to source, electrons are actually flowing from source to drain.

The voltage at the gate is always zero in this circuit because the gate is grounded. In addition, the dc quiescent voltage at the source, V_S, is $I_D R_S$. Therefore

$$V_{GS} = 0 - I_D R_S$$

If we select the Q-point as shown in Fig. 15.9, then $I_D = 2$ mA and $V_{DS} = 6$ volts. At this Q-point, $V_{GS} = -1$ volt. Since $V_G = 0$, and $V_{GS} = -1$ volt, $V_S = +1$ volt. Therefore, the required value of R_S is

$$R_S = \frac{V_S}{I_D} = \frac{1}{2 \times 10^{-3}} = 500 \ \Omega$$

In addition, at the Q-point

$$V_D = V_S + V_{DS} = 1 + 6 = 7 \text{ volts}$$

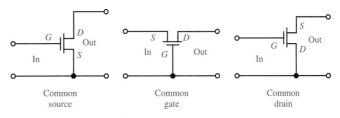

Figure 15.7   Transistor amplifier configurations.

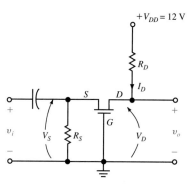

Figure 15.8 A common gate amplifier circuit.

Since the dc voltage across R_D and the dc current through it are now known

$$R_D = \frac{V_{DD} - V_D}{I_D} = \frac{12 - 7}{2 \times 10^{-3}} = 2.5 \text{ k}\Omega$$

The Q-point is plotted on Fig. 15.9, and this completes the dc analysis of the bias circuit.

In order to estimate how this circuit performs as a large-signal amplifier, let us consider the circuit's response to changes in input voltage. If we let v_i vary + and − 1 volt, then V_S moves from its quiescent value of +1 volt up to a maximum of +2 volts and down to a minimum of 0 volts. Recall that in this circuit, $V_{GS} = -V_S$, and therefore during this same time, V_{GS} changes from a quiescent value of −1 volt, to an extreme of −2 volts and then to 0 volts during the second half-cycle. The source-to-gate voltage swings cause I_D to vary from about 1.4 mA to 2.9 mA. Since we are changing V_{GS} and thus V_S externally, the actual bias point trajectory in Fig. 15.9 can be found by subtracting V_S (given) from V_D (calculated below) to determine values for V_{DS}.

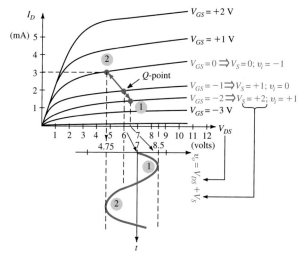

Figure 15.9 Output curves and large-signal response for the amplifier circuit in Fig. 15.8.

The output voltage, V_D, is given by

$$V_D = V_{DD} - I_D R_D$$

therefore, the output voltage swings positively to a maximum value of $12 - (1.4 \text{ mA})$ $(2.5 \text{ k}\Omega) = 8.5$ volts, and down to a minimum of $12 - (2.9 \text{ mA})(2.5 \text{ k}\Omega) = 4.75$ volts.

The peak-to-peak change in output voltage is $8.5 - 4.75 = 3.75$ volts. The input voltage change which causes this output swing is 2 volts peak-to-peak. Therefore, the large-signal voltage gain is $3.75/2 = 1.89$. ∎

Recall that in *common source* (CS) circuits, the signal is inverted; that is, a positive change in the input voltage produces a negative change in the output voltage. Therefore, in CS circuits the voltage gain should have a "$-$" sign to indicate signal inversion.

The *common gate* (CG) circuit just analyzed in Example 15.2 demonstrates a positive change in output for a positive change in the input voltage, and therefore its gain expression should carry a "$+$" sign. The common drain circuit also has a "$+$" gain. Therefore, the only FET circuit configuration that inverts the signal is the common source.

The Common Drain Amplifier

The *common drain* (CD) amplifier is sometimes called a *source follower,* a typical example of which is shown in Fig. D15.1. A calculation of the voltage gain of this circuit will reveal that it is slightly less than unity, and the approximation is often made that it is simply "one." This circuit is therefore not useful for providing increased signal voltage levels, but is primarily used for impedance conversion. The input resistance of this circuit can be made quite high, and its output resistance can be made quite low. Since the current gain of the circuit can be quite high, it is useful for driving low-resistance loads. This circuit configuration will be examined in more detail in the next chapter.

DRILL EXERCISE

D15.1. Consider the common drain amplifier circuit shown in Fig. D15.1. Note that the drain is connected directly to the supply, that is, an ac ground. The calculation of dc bias conditions for this circuit is similar to that of the self-bias circuit previously described. Assuming that the transistor is described by the output curves of Fig. 15.9, find the value of R_S that would place the quiescent dc source current at 1.0 mA. (Hint: Source current and drain current are identical.)

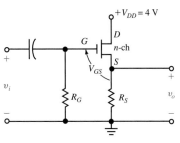

Figure D15.1

Ans: $R_S = 2 \text{ k}\Omega$.

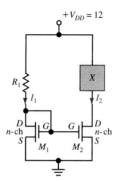

Figure 15.10   MOSFET current mirror circuit.

Current Mirrors

Some modern integrated analog systems use a bias circuit called a *current mirror.* This circuit uses two transistors in a common source configuration and takes advantage of the fact that transistors made together on the same chip have very similar or matched characteristics. The basic circuit for a current mirror is shown in Fig. 15.10. We can analyze this circuit by examining the two halves. The left half *sets up a current* which is mirrored to the right half in order to bias the device or circuit in the box labeled "*X.*"

Assume that *n*-channel enhancement-mode transistors are used which have output curves as shown in Fig. 15.11. The MOSFET on the left, *M*1, has a connection directly from its gate to its drain, which forces the following condition to be always true:

$$V_{GS} = V_{DS} \tag{15.6}$$

The solid curve in Fig. 15.11 is the locus of all points that satisfy the above relationship, and is the *V–I* curve for the MOSFET connected this way. Note that this is a rapidly increasing curve reminiscent of the diode curve.

Suppose we wish to set the current down the left side, I_1, at 2.5 mA. Then the curve indicates that the voltage across the device *M*1 is 3 volts, and R_1 must be

$$R_1 = \frac{V_{DD} - V_{DS}}{I_1} = \frac{12 - 3}{2.5 \times 10^{-3}} = 3.6 \text{ k}\Omega$$

This current I_1 through M_1 establishes $V_{DS} = V_{GS} = 3$ volts, and this voltage is applied directly across the gate to source of a matched transistor, *M*2. Naturally, it produces

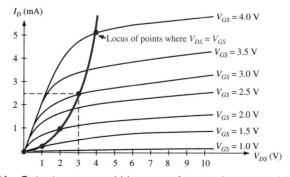

Figure 15.11   Output curves and bias curve for transistors used in Fig. 15.10.

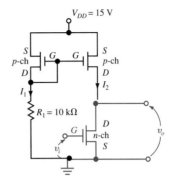

Figure 15.12   A common source amplifier circuit with current mirror biasing.

a bias on *M2* that yields the same drain current, and therefore $I_2 = 2.5$ mA. This is the bias current that is pulled from the circuit in the box, and hence this bias configuration is sometimes called a current *sink*.

The analogous circuit with *p*-channel transistors can be used to create a current *source*, as illustrated in the next example. Note carefully that the voltage gain of a CS amplifier increases as the ac load resistance increases. An ideal current source has an infinite resistance, and if such a device existed, and could be used as the load resistance, the CS circuit would produce infinite gain. Of course, in reality, there is no *ideal* current source. However, the current mirror circuit has a relatively high resistance and can be used with circuits to produce very high voltage gains.

EXAMPLE 15.3

A current source (current mirror) circuit is used both to provide dc bias and to serve as the load resistance for an *n*-channel common source amplifier as shown in Fig. 15.12.

If the voltage drop across the *p*-channel device on the left side is 1.5 volts, what dc current is the current mirror delivering to the *n*-channel device, that is, what is I_2?

Ans:

$$I_1 = \frac{15 - 1.5}{10 \text{ k}} = 1.35 \text{ mA}$$

$I_2 = I_1 = 1.35$ mA.

The calculation of the voltage gain of this type circuit can be performed using small-signal models as presented in the next chapter.   ■

15.5
Power and Voltage Limits of MOSFETs

Large-signal amplifiers generally have large voltage and/or current swings. There are physical limits on the ability of actual MOSFETs to withstand high voltages and currents, and therefore real devices typically have the following parameters specified:

1. Maximum drain current
2. Maximum *D–S* voltage

3. Maximum power dissipation
4. Maximum *G–S* voltage

Maximum drain current is specified so that the designer will maintain the operating current in a range that will not destroy the device. The physical size of the transistor and the cross section of the current-carrying connections place an upper limit on device current. The drain current safe operating area is designated on a transistor's output curve as everything below I_{Dmax}, as illustrated in Fig. 15.13.

Similarly, there is a maximum safe operating voltage, V_{DSmax}. This requirement exists because higher voltages generate fields which may cause breakdown in insulating layers or reverse biased junctions. The upper limit of the safe region on the V_{DS} axis is also illustrated in Fig. 15.13.

Finally, there is an upper limit on power dissipation due to the finite ability of the transistor, its substrate, and package to remove heat from the device. Power dissipated is the product of voltage and current so that

$$P = I_D V_{DS} \text{ watts} \tag{15.7}$$

This is a hyperbolic function. As an example, the contour for a maximum power of 80 milliwatts is also plotted on the curve in Fig. 15.13.

Therefore, for the device described by the curves in Fig. 15.13

$$I_{Dmax} = 80 \text{ mA}; \ V_{DSmax} = 8 \text{ volts}; \ P_{max} = 80 \text{ mW}$$

The safe operating area is the clear area under the maximum power contour. The *Q*-point and the load lines should all be designed to fall in the safe operating area. It is worth noting that on a typical load line, the *Q*-point is often closest to maximum power dissipation near the center of the line. A family of power dissipation contours can be generated on one plot so that the power dissipated at any bias point can be easily estimated.

In addition, the maximum specified V_{GS} should not be exceeded. If the gate voltage is too large, it causes electrical breakdown on the insulating gate oxide.

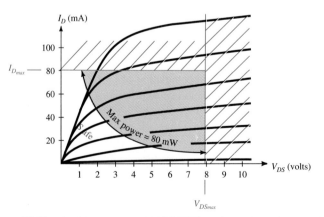

Figure 15.13   An illustration of MOSFET's safe region of operation.

15.6
Large-Signal Amplifiers: Using BJTs

As was described in Chapter 9, section 9.11, the output curves of the common emitter (CE) bipolar junction transistor are very useful in predicting the large-signal behavior of such devices in circuits. The analysis of BJT circuits follows a similar procedure to that illustrated earlier in this chapter for MOSFETs. The analysis techniques will be illustrated with examples.

In order to construct dc load lines for BJT circuits, we proceed in a manner which is analogous to that previously described for MOSFETs. Reference to figures 9.37 to 9.39 may be useful.

1. Imagine the transistor is an open circuit such that $I_C = 0$. Then determine the voltage across the device. Generally, this voltage will be the supply voltage. This voltage is the intercept on the V_{CE} axis.
2. Imagine the transistor is a short from C to E such that $V_{CE} = 0$. Then determine the current through the short. Generally, this current will be the supply voltage divided by the total resistance in series with the collector emitter loop. This is the current value which is the intercept on the I_C axis.
3. Connect the two intercepts with a straight line to form the dc load line. This approach is viable for any circuit configuration.

The Common Emitter Amplifier

Consider the following example for a *common emitter amplifier.*

EXAMPLE 15.4

An NPN transistor has the output characteristics shown in Fig. 15.14. This device is placed in the common emitter circuit of Fig. 15.15.

The output curves do not provide information about the input characteristics of a device; however, for a BJT in the active region we can generally make a simplifying assumption. The input is applied across the base emitter junction, and in the active region this junction is forward biased. For silicon devices, the dc voltage across the base emitter junction is approximated by the condition

$$V_{BE} = \text{turn-on voltage} = 0.7 \text{ volts}$$

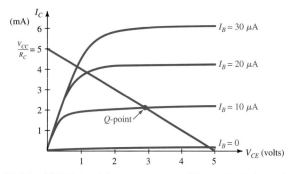

Figure 15.14   NPN transistor common emitter output characteristics.

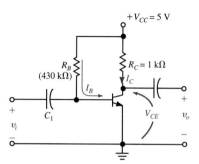

Figure 15.15   A common emitter NPN BJT amplifier circuit.

The dc load line has been drawn on the BJT output curves, with an intercept on the V_{CE} axis at 5 volts, and an intercept on the current axis at $V_{CC}/R_C = 5$ mA.

We can determine the location of the Q-point by a simple calculation of the base current. The dc base current for this device, I_B, flows from V_{CC} through R_B and into the base of the device. Since $V_B = V_{BE} = 0.7$ V

$$I_B = \frac{V_{CC} - V_{BE}}{R_B} = \frac{5 - 0.7}{430\,\text{k}} = 10\ \mu\text{A}$$

Therefore, the Q-point lies on the intersection of the dc load line and the $I_B = 10\ \mu\text{A}$ output curve. From the curve, all the dc bias conditions are now known, that is $V_{CE} = 3$ volts and $I_C = 2$ mA.   ∎

In amplifiers, the BJTs are always biased in the active region making the collector junction reverse biased. For this case, β_R is not a factor, and it is common convention to refer to just β for β_F. Therefore, $\beta \equiv \beta_F$.

EXAMPLE 15.5

Let us determine the beta (β) of the transistor in Example 15.4. The beta at the Q-point is defined as the ratio of collector current to emitter current, that is

$$\beta = \frac{I_C}{I_B} \quad \text{at the Q-point; } V_{CE} = constant$$

$$\beta = \frac{2 \times 10^{-3}}{10 \times 10^{-6}} = 200 \qquad \blacksquare$$

The spacing between the output curves for a given family of I_B's indicates the beta. The beta in real transistors has some bias current dependence; however, for convenience in calculations, beta is often assumed to be a constant for a given transistor. It is often measured at $V_{CB} = 0$.

We can make the following additional observations about the above example:

1. As a large-signal amplifier, the output voltage can swing positively all the way up to 5 volts when $I_B = 0$, and down to about 1 volt with a base drive current of 30 μA.
2. Note that both the output voltage and the current I_C change approximately linearly with changes in the input CURRENT, I_B. However, the input is applied across the

emitter base *p–n* junction and recall that I_B increases exponentially as V_B increases. In addition, the voltage V_B changes directly with v_i since coupling capacitor C_1 acts as an ac short. Therefore, as a large-signal VOLTAGE amplifier, this circuit would be highly nonlinear—in fact, the output voltage would vary approximately exponentially with input voltage. If the input signal is very small, a linear approximation is valid, and this circuit can be used as a small-signal amplifier. Alternatively, an input series resistance will linearize the response with the penalty of decreased voltage gain.

3. Bipolar transistors manufactured on the same line can have relatively large variations in beta from one batch to another, sometimes as much as 50% to 100%. Imagine the problem of trying to manufacture the circuit of Example 15.4 if the transistors varied in beta from 100 to 400. The circuit provides a specific base current, 10 μA; however

$$\text{if } \beta = 100, I_C = 100 \ (10 \ \mu A) = 1 \text{ mA, and}$$
$$\text{if } \beta = 400, I_C = 400 \ (10 \ \mu A) = 4 \text{ mA}$$

With this circuit the Q-point is very sensitive to the value of β and in this example, β variations can move the Q-point from near cutoff to near saturation. This is unacceptable for a circuit intended to be manufactured in quantity.

4. *Thermal runaway* could affect this circuit. Thermal runaway is a catastrophic cycle which can ultimately destroy a transistor. The power dissipated in a device during operation increases the device's temperature. This temperature increase can cause an increase in base leakage current, which gets multiplied by β and appears as a larger increase in collector current. The situation is aggravated by the fact that beta normally increases with increasing temperature, which also increases the collector current. The increased current, in turn, increases the power dissipated in the device and increases the heating, and the cycle continues, increasing the temperature further. If left unchecked, the device can self-destruct.

The BJT Self-Bias Circuit

The use of a bias circuit, which is designed to stabilize the collector current instead of the base current, will solve the problems outlined above. The *BJT self-bias circuit* of Fig. 15.16 greatly reduces the effects of β variations and temperature on the quiescent oper-

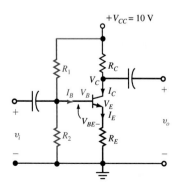

Figure 15.16   A BJT self-bias circuit.

ating point. In this circuit, the collector current is determined by the voltage across a resistor, R_E, which is placed in series with the emitter.

To analyze this circuit, first consider the voltage divider, R_1 and R_2. Its purpose is to establish a known dc voltage, V_B, at the base of the transistor. For simplicity, we will now assume that the value of $(R_1 + R_2)$ is small enough that the current down through the divider is much greater than I_B. That is

$$\frac{V_{CC}}{R_1 + R_2} \gg I_B \qquad (15.8)$$

This condition ensures that changes in I_B will not significantly change the value of V_B, and therefore

$$V_B = V_{CC} \frac{R_2}{R_1 + R_2} \qquad (15.9)$$

In addition, note that

$$V_B = V_{BE} + I_E R_E \qquad (15.10)$$

Equations (15.8–15.10) provide a simple solution to many problems using the self-bias circuit.

EXAMPLE 15.6

Consider the circuit of Fig. 15.16. We wish to find the value of I_E assuming the following component values:

$$\begin{array}{ll} R_1 = 3 \text{ k}\Omega & R_C = 4 \text{ k}\Omega \\ R_2 = 1 \text{ k}\Omega & R_E = 1.5 \text{ k}\Omega \end{array}$$

First, let us consider the voltage divider circuit comprised of R_1 and R_2. Using Eq. (15.9)

$$V_B = V_{CC} \frac{R_2}{R_1 + R_2} = 10 \frac{1 \text{ k}}{1 \text{ k} + 3 \text{ k}} = 2.5 \text{ volts}$$

Then from Eq. (15.10) we find that

$$V_B = 2.5 = V_{BE} + I_E (1.5 \text{ k})$$

The value of V_{BE} for a BJT in the active region is assumed to be the junction turn-on voltage, $V_{BE} = 0.7$ volts. Therefore

$$I_E = \frac{2.5 - 0.7}{1.5 \text{ k}} = 1.2 \text{ mA}$$

Note that the value of β never entered this calculation; this is because the assumption expressed in Eq. (15.8) indicates that I_B does not affect the value of V_B. Let's now arbitrarily assume a β value of 100 and check to see if this assumption was justified.

The current through the voltage divider is approximately

$$I_{R1} = I_{R2} = \frac{V_{CC}}{R_1 + R_2} = \frac{10}{1 \text{ k} + 3 \text{ k}} = 2.5 \text{ mA}$$

And the base current is approximately

$$I_B = \frac{I_E}{\beta + 1} = \frac{1.2 \times 10^{-3}}{101} = 0.012 \text{ mA}$$

Therefore, the original assumption, expressed in Eq. (15.8), is a justifiable one. ■

BJT Self-Bias Circuit—Complete Analysis

If the assumption made in Eq. (15.8) is not valid, then the calculations become more cumbersome, but not difficult. First consider the voltage divider consisting of R_1 and R_2, as shown in Fig. 15.17(a). A Thevenin equivalent for this circuit is shown in Fig. 15.17(b). The expressions for the Thevenin equivalent voltage and resistance are

$$V_{BT} = V_{CC} \frac{R_2}{R_1 + R_2} \tag{15.11}$$

$$R_{BT} = R_1 \| R_2 = \frac{R_1 R_2}{R_1 + R_2} \tag{15.12}$$

Using the Thevenin equivalent circuit in Fig. 15.17(b), the bias circuit of Fig. 15.16 can be redrawn as shown in Fig. 15.18. The loop equation around the base emitter loop is

$$V_{BT} - I_B R_{BT} - V_{BE} - I_E R_E = 0 \tag{15.13}$$

The equation for I_B is

$$I_B = \frac{I_E}{\beta + 1} \tag{15.14}$$

and combining Eqs. (15.13) and (15.14) produces one equation with one unknown. This approach can be used to determine the bias conditions for any BJT self-bias circuit.

EXAMPLE 15.7

Consider the self-bias circuit of Fig. 15.16. The transistor has a $\beta = 25$ and the resistors have the following values:

$$\begin{array}{ll} R_1 = 45 \text{ k}\Omega & R_C = 3 \text{ k}\Omega \\ R_2 = 30 \text{ k}\Omega & R_E = 2 \text{ k}\Omega \end{array}$$

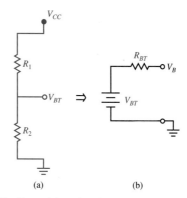

(a)   (b)

Figure 15.17   Base bias circuit and its Thevenin equivalent.

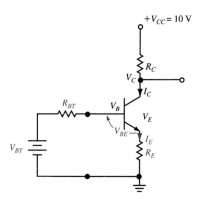

Figure 15.18   An equivalent circuit for the network in Fig. 15.16.

Let us determine both I_C and V_C.

Since the condition of Eq. (15.8) may not be met, we will perform the analysis by developing a Thevenin equivalent of the input circuit. This approach is always valid.

From Eqs. (15.11) and (15.12), we find that

$$V_{BT} = V_{CC} \frac{R_2}{R_1 + R_2} = 10 \left[\frac{30 \text{ k}}{30 \text{ k} + 45 \text{ k}} \right] = 4 \text{ V}$$

and

$$R_{BT} = \frac{(30 \text{ k})(45 \text{ k})}{30 \text{ k} + 45 \text{ k}} = 18 \text{ k}\Omega$$

The KVL equation around the base emitter loop is

$$V_{BT} - \frac{I_E}{\beta + 1} R_{BT} - V_{BE} - R_E I_E = 0$$

Using a value of $V_{BE} = 0.7$ V and solving for I_E yields the value

$$I_E = 1.23 \text{ mA}$$

Knowing I_E, we can now compute I_C as

$$I_C = \frac{\beta}{\beta + 1} I_E = \left[\frac{25}{26} \right] 1.23 \times 10^{-3} = 1.18 \text{ mA}$$

and then

$$V_C = V_{CC} - I_C R_C = 10 - (1.18 \times 10^{-3}) (3 \text{ k}) = 6.46 \text{ V} \qquad \blacksquare$$

EXAMPLE 15.8

Consider the circuit shown in Fig. 15.19. The transistor in this network is a PNP device described by the output curves shown in Fig. 15.20.

Let us identify the Q-point, construct both the dc and the ac load lines on the output curves, and determine the current gain.

The dc load line is constructed as shown. The reader can verify that the condition of Eq. (15.8) is satisfied and therefore V_B can be determined by Eq. (15.9) as $V_B = -4$ volts;

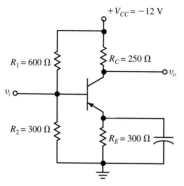

Figure 15.19 A PNP common emitter transistor amplifier circuit.

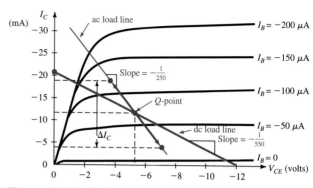

Figure 15.20 Output curves and load lines for the transistor in the network in Fig. 15.19.

therefore, $V_E = -4 - (-0.7) = -3.3$ volts and

$$I_E = \frac{-3.3}{300} = -11 \text{ mA} \approx I_C$$

Interpolating, one can judge that the value of I_B at the Q-point is approximately -75 μA; having located the Q-point, the ac load line with the slope shown in the figure is constructed.

The large-signal current gain of this circuit can be determined by

$$A_i = \frac{\Delta I_{out}}{\Delta I_{in}} = \frac{\Delta I_C}{\Delta I_B}$$

If we assume the input signal changes the base current ± 50 μA, the bias point moves up and down the ac load line to the extremes shown in Fig. 15.20. The corresponding total change in output current is determined from the plot as

$$\Delta I_C = (18 - 4) \times 10^{-3} = 14 \text{ mA}$$

Therefore

$$A_i = \frac{14 \times 10^{-3}}{100 \times 10^{-6}} = 140$$

■

The voltage gain calculation for BJTs requires the input resistance of the device in order to determine input current changes from input voltage changes. For this reason, a detailed discussion of ac voltage gain for BJT circuits will be deferred until the next chapter.

15.7
Other BJT Amplifier Configurations

The BJT circuits examined in the previous section have all had the signal input applied from base to emitter, and the output taken from collector to emitter. These are all common emitter circuits. As was the case with MOSFETs, there are three possible bias circuit configurations for BJTs: common emitter (CE), common base (CB), and common collector (CC). These three circuit configurations are illustrated in Fig. 15.21. The meth-

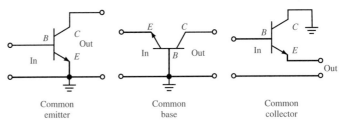

Figure 15.21   BJT circuit configurations.

ods for calculating dc bias for the other configurations follow the same principles as those outlined for the common emitter.

The Common Base Amplifier

Consider the following example for a *common base amplifier.*

EXAMPLE 15.9
Given the common base BJT bias circuit shown in Fig. 15.22, let us calculate the location of the Q-point, assuming β is large, if

$$R_1 = 6\,k\Omega \qquad R_C = 4.5\,k\Omega$$
$$R_2 = 1\,k\Omega \qquad R_E = 1.5\,k\Omega$$

First, note carefully that this is exactly the same dc circuit as that previously discussed in the common emitter circuit configuration. The network is drawn differently to emphasize its different ac configuration: signal *in* on the emitter and *out* on the collector, with a common, ac grounded base. V_B is computed in the same manner in which it was determined in the CE configuration by using Eq. (15.9):

$$V_B = \left[\frac{1\,k}{7\,k}\right] 20 = 2.86 \text{ volts}$$

Therefore

$$I_E = \frac{2.86 - 0.7}{1.5\,k} = 1.44 \text{ mA}$$

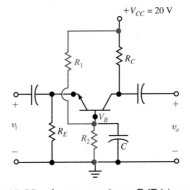

Figure 15.22   A common base BJT bias circuit.

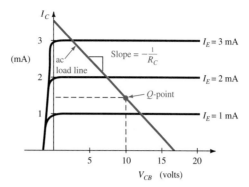

Figure 15.23 CB output curves and ac load line for circuit in Example 15.9.

And hence

$$V_C = 20 - (1.44 \times 10^{-3}) \, (4.5 \text{ k}) = 13.5 \text{ volts}$$

I_C is approximately equal to I_E if the β is reasonably high. The Q-point is now described by the set of dc terminal voltages and currents above. ∎

The large-signal analysis of CB BJT circuits often makes use of a set of output curves called *"common base output curves"* which plot I_C as a function of V_{CB} for a family of I_E values, as shown in Fig. 15.23. Note that in the active region $I_C = \alpha_F I_E$ and $\alpha_F \approx 1$ and hence the curve for each value of I_E is approximately at the same value of I_C.

Assume that the transistor in Example 15.9 is described by the CB output curves of Fig. 15.23. The Q-point can be located, as shown in the figure, based on the dc bias calculations in the example, that is

$$V_{CBQ} = V_C - V_B = 13.5 - 2.86 = 10.6 \text{ V}$$
$$I_{CQ} = 1.44 \text{ mA}$$

We will now construct the approximate ac load line through this Q-point. The base is at ac ground because of the bypass capacitor; the only resistance in the circuit from the collector to the supply is R_C; and therefore, the slope is $-1/R_C$ or $-1/4.5$ k.

As the input ac signal changes the value of I_E, the bias point moves up and down the ac load line. The corresponding values of V_{CB} can thus be determined graphically.

The Common Collector Amplifier

Consider the following example of a *common collector amplifier.*

EXAMPLE 15.10
The network in Fig. 15.24 is an NPN common collector amplifier, sometimes called an emitter follower. The ac gain of an emitter follower circuit is slightly less than unity, and like the common drain circuit does not provide increased signal voltage. It is used primarily for impedance transformations.

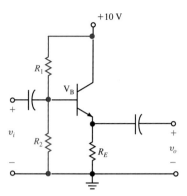

Figure 15.24 An NPN common collector amplifier.

Let us now calculate the dc value of I_C for this common collector amplifier, assuming the following values:

$$R_1 = 4 \text{ k}\Omega$$
$$R_2 = 2 \text{ k}\Omega \qquad R_E = 1.0 \text{ k}\Omega$$

Since Eq. (15.8) is valid

$$V_B = V_{CC}\left[\frac{R_2}{R_1 + R_2}\right] = 10\left[\frac{2 \text{ k}}{4 \text{ k} + 2 \text{ k}}\right] = 3.33 \text{ V}$$

$$V_E = V_B - V_{BE} = 3.33 - 0.7 = 2.63 \text{ V}$$

$$I_E = \frac{V_E}{R_E} = \frac{2.63}{1 \text{ k}} = 2.63 \text{ mA}$$

and

$$I_C = I_E = 2.63 \text{ mA} \qquad \blacksquare$$

Current Mirrors Using BJTs

The *current mirror bias circuit* illustrated earlier with MOSFETs can also be implemented using bipolar transistors. The following example illustrates the technique.

EXAMPLE 15.11

Let us calculate the current pulled from the block labeled X in the circuit of Fig. 15.25. Assume the NPN transistors are matched and have large β's.

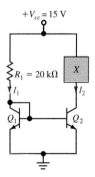

Figure 15.25 A BJT current mirror circuit.

The left side of the circuit establishes the current. Transistor Q_1 has a connection from base to collector—a configuration known as a *diode-connected transistor*. This configuration acts like a diode, and conducts with a voltage drop of $V_{BE} = 0.7$ volts. The current down the left side of the circuit is therefore

$$I_1 = \frac{V_{CC} - 0.7}{R_1} = \frac{15 - 0.7}{20 \text{ k}} = 0.72 \text{ mA}$$

Because the transistors are matched, the V_{BE} of Q_1, when connected across the base emitter of Q_2, generates the same current down the right side of the circuit. Therefore

$$I_2 \approx I_1 = 0.72 \text{ mA}$$
■

15.8
Power and Voltage Limits of BJTs

The voltage, current, and power limitations for bipolar transistors are similar to those described for MOSFETs in section 15.5. Typically, a bipolar transistor will have specified: (1) maximum collector current, (2) maximum *C-E* voltage, (3) maximum power dissipation, and often (4) maximum base current.

These real physical limitations of a device place restrictions on the safe operating area of the output curves. For example, there is a *do not exceed* limit on V_{CE}, designated as V_{CEmax}, imposed by the breakdown voltage limit of the device. The maximum current, I_{Cmax}, is a limit imposed by the size of the device and the connections to it. The power dissipated in a BJT is

$$P = I_C V_{CE} \tag{15.15}$$

The maximum power allowed, P_{max}, is determined by the ability of the device to effectively dissipate heat. The safe region is restricted to a similar shaped area as that illustrated in Fig. 15.13 for MOSFETs, but on axes labeled I_C and V_{CE}.

15.9
Thyristors and Silicon-Controlled Rectifiers (SCRs)

As we have seen, diodes and transistors can be used to switch currents *on* and *off*. A diode is a simple switch that changes conductivity in response to the polarity of the applied voltage. A BJT can also be used to switch currents through control of its base current. However, if large currents must be switched, then a large base current is continuously required to hold the transistor *on*.

Both the thyristor and the SCR are *four-layer* structures consisting of alternating semiconductor types (*pnpn*). These devices can control large amounts of power with only a small amount of control energy. Once these devices are switched *on* they can hold themselves in the *on* state until the applied voltage being switched is reversed or removed. For this reason they are often used in applications which require switching power from a few watts to many kilowatts, such as sophisticated power supply circuits and motor speed control circuits.

The cross section of a *pnpn* device is shown in Fig. 15.26(a). The operation of this device is best understood by considering it to be two transistors (an NPN and a PNP) merged together and connected as shown in Fig. 15.26(b). A schematic circuit drawing of this configuration is shown in Fig. 15.26(c).

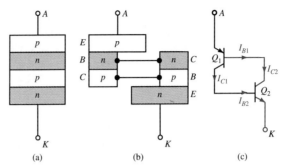

Figure 15.26   Model of a *pnpn* device as two transistors.

From the circuit drawing, we note that if Q_1 is *off*, that is, not conducting, then there is no base current for Q_2 and it is also *off*. Likewise, if Q_2 is *off* there is no base current for Q_1, and it is *off*. If both transistors are *off* there is no conduction from anode (A) to cathode (K).

However, the introduction of any significant base current in either device creates a snowball effect and positive feedback proceeds to lock the device in a full conducting mode. For example, if I_{B2} is the base current of Q_2, then $\beta_2 I_{B2}$ flows in the collector of Q_2, which is the same as the base current of Q_1. The collector current of Q_1, which is the same as I_{B2}, is $\beta_1 I_{C2}$. Therefore, note that when the product of the two betas exceeds unity, that is $\beta_1\beta_2 > 1$, and there is any available base current, a regenerative process occurs which turns *on* both transistors and current can readily flow from A to K.

Thyristors

The symbol for a *thyristor* is shown in Fig. 15.27(a) and a typical *V–I* curve for the device is shown in Fig. 15.27(b). Notice that when V_{AK} is negative, the device behaves much like a reverse biased diode since two of the junctions in the series stack are reverse biased. It conducts minimal current until V_B, the breakdown voltage, is exceeded. However, for positive values of V_{AK}, there is minimal conduction until V_{BO}, the *breakover voltage*, is reached. At this point, there is an avalanche breakdown of the *center* junction which supplies base current to both *transistors* in the model; both transistors are switched to the *on* state and the device conducts readily from A to K. Conduction will continue until either the current is reduced below the holding current, I_H, or the polarity of V_{AK} is reversed.

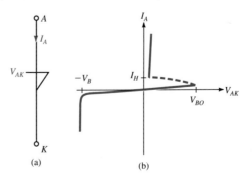

Figure 15.27   A thyristor and its *V–I* curve.

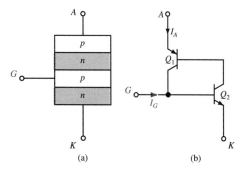

Figure 15.28   An SCR and its equivalent circuit.

Silicon-Controlled Rectifiers

Silicon-controlled rectifiers, or SCRs, are similar to thyristors, with the exception that a control connection, called the gate, G (NOT to be confused with the Gate of an FET), is added as shown in Fig. 15.28(a). The terminal G provides an external path through which base current can be added to initiate switching the device to the *on* state. The equivalent circuit for an SCR is shown in Fig. 15.28(b).

The symbol used for the SCR is given in Fig. 15.29(a) and typical V–I curves for the device are shown in Fig. 15.29(b). Note that if $I_G = 0$, the SCR and the thyristor have the same characteristics. Increasing the value of I_G simply lowers the value of V_{AK} at which the device switches *on*.

Motor Speed Control

One of the most common applications of the SCR is in *motor speed control.* In this simple example a sinusoidal supply voltage is applied to a dc motor through an SCR as shown in the circuit of Fig. 15.30(a). The SCR provides rectification, that is, conversion to dc, by only allowing current to flow in one direction. However, more importantly, the average current through the motor can be regulated by the fraction of time that the SCR is switched *on* during the positive cycle of the current waveform. This time fraction is usually expressed as a phase angle, called the *conduction angle, θ. The firing angle, ϕ,* is $180° - \theta$.

The SCR can be switched *on* via a firing pulse of gate current, $i_G(t)$. The pulse length

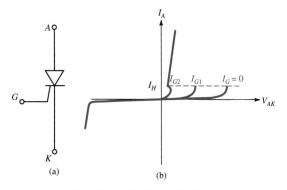

Figure 15.29   An SCR and its V–I curves.

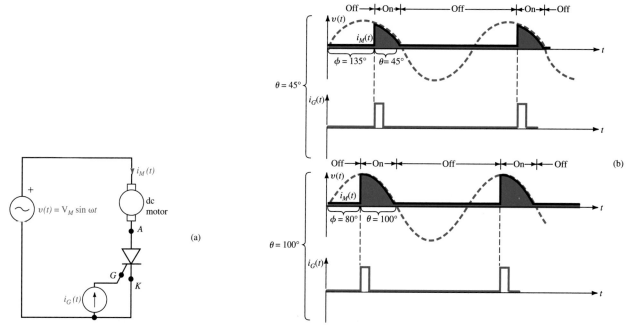

Figure 15.30   (a) A motor speed control circuit; (b) A plot of gate current for two different firing angles.

is not important because the SCR will remain in the conducting state until the polarity of the applied voltage reverses.

Figure 15.30(b) plots the applied voltage, $v(t)$, the motor current, $i_M(t)$, and the gate current, $i_G(t)$ for two different *firing angles*. The average current delivered to the motor is proportional to the area under the $i_M(t)$ curve. Notice that the average current delivered to the motor is much larger for a conduction angle of 100 degrees than for one of 45 degrees. By varying the phase, or the time of the gate firing pulse relative to the sinusoidal supply voltage, the average current and hence the speed of the motor can be controlled.

15.10
Summary

- A small-signal amplifier is one in which the signal voltage variations are relatively small, and the nonlinearities of the transistor's response can be approximated by linear circuit models.
- A large-signal amplifier has large-signal voltage swings and the nonlinearities of the transistor must be considered. Typically, this analysis is done by graphical techniques such as plotting the load line on the output curves.
- A single-stage amplifier is one in which there is only one amplifying element, typically a transistor; a multi-stage amplifier is obtained by cascading more than one single-stage amplifier together in order to obtain performance not possible from a single stage.
- The first step in the analysis of any amplifier circuit is to determine the dc bias conditions at the Q-point—the transistor's terminal voltages and currents with no signal applied.
- For a common source MOSFET amplifier, the Q-point is located on the output curves at the intersection of the dc load line and the appropriate V_{GS} curve. The output volt-

age swing can then be determined by plotting the movement of the bias point up and down the load line (deviations from the Q-point) as the input signal changes V_{GS}.

- If a bypass capacitor is used, then an ac load line must be plotted, and the signal excursions determined by movement of the bias point up and down the ac load line.
- The other MOSFET circuit configurations are considered. The common gate circuit is analyzed in a similar manner. The common drain circuit is discussed and will be analyzed further in the next chapter.
- The current mirror circuit provides a means of establishing a known dc bias in a MOSFET by utilizing the property of matched transistors on a single integrated circuit chip.
- Real MOSFETs have finite limits for applied voltages, currents, and power; circuits should be designed to operate well within these limits.
- The large-signal analysis of BJT circuits follows a similar procedure to that for MOSFETs. For a common emitter amplifier the Q-point is located on the output curves at the intersection of the dc load line and the appropriate base current curve. The output voltage swing is determined by plotting the movement of the bias point up and down the load line, that is, plotting the deviations from the Q-point, as the input signal changes I_B.
- If a bypass capacitor is used, an ac load line must be plotted and the bias point moved on the ac load line.
- The other BJT circuit configurations, common base, and common collector are described and dc bias calculations illustrated. Similar large-signal techniques can be applied. The small-signal gain of these circuits is described in the next chapter.
- Real BJTs, like MOSFETs, have finite limits for applied voltages, currents, and power.
- Thyristors and silicon-controlled rectifiers (SCRs) are four-layer structures (*pnpn*) which can control large currents with small input (control) power. These devices are commonly used in power control circuits, such as setting the speed of a motor.

PROBLEMS

15.1. Describe the difference(s) between an amplifier classified as a small-signal amplifier and a large-signal amplifier.

15.2. The circuit of an *n*-channel enhancement-mode MOSFET amplifier is shown in Fig. P15.2. (a) If we wish to establish $V_{GS} = 2$ V, find R_2 if $R_1 = 150$ kΩ. (b) For this circuit would we expect V_T to be more or less than 2 V?

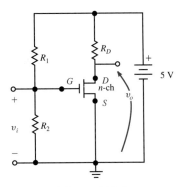

Figure P15.2

15.3. The output curves of an *n*-channel enhancement-mode MOSFET are shown in Fig. P15.3. This transistor is placed in the common source amplifier of Fig. P15.2 with $R_1 = 100$ kΩ, $R_2 = 30$ kΩ, and $R_D = 1.25$ kΩ. (a) Construct the load line for this circuit. (b) Obtain the quiescent operating point, that is, *Q*-point. (c) If v_i is a sinusoidal signal of amplitude 1 V, sketch the variation of the output voltage v_o about the operating point. (d) Find the large-signal voltage gain. (e) Describe any possible distortion of the output signal. If we reduce the amplitude of the input signal, would there be less distortion?

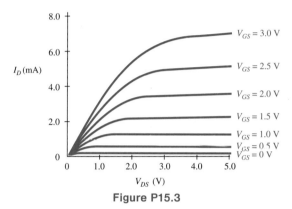

Figure P15.3

15.4. Describe the function of the following in an amplifier circuit: (a) coupling capacitor, (b) bypass capacitor.

15.5. What is the difference between an ac load line and a dc load line? Under what conditions are the two identical?

15.6. The output curves of an *n*-channel MOSFET are shown in Fig. P15.6(a). This transistor is placed in the circuit of Fig. P15.6(b). (a) What mode device is this MOSFET? (b) If we wish to put the *Q*-point on the $V_{GS} = 0.5$ V curve, what value of R_S should be selected? Hint: First compute V_G; then write an equation for V_{GS} and use it to solve for R_S. (c) Construct the dc load line on the output curves. (d) Obtain the *Q*-point of the circuit. (e) Construct the ac load line. (f) Sketch the output voltage for a triangular input voltage of 1 V peak-to-peak. (g) Find the large-signal voltage gain.

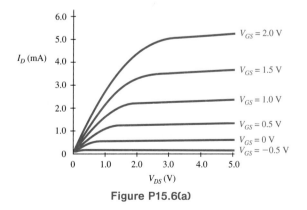

Figure P15.6(a)

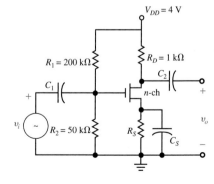

Figure P15.6(b)

15.7. Figure P15.7 shows a common source amplifier circuit using an *n*-channel depletion-mode MOSFET in a self-biasing configuration. (a) Why is this configuration self-biasing? If I_D is measured to be 1.2 mA, determine V_{GS}. (b) Calculate the slope of the dc load line. (c) Calculate the slope of the ac load line.

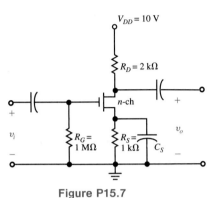

Figure P15.7

15.8. Figure P15.8 shows a common gate amplifier configuration using an *n*-channel depletion-mode transistor. The *Q*-point for this circuit is chosen to be $I_D = 5$ mA, $V_{DS} = 5$ V, and $V_{GS} = -2$ V. (a) Calculate the values of resistances R_S and R_D. (b) What is the slope of the dc load line? (c) What is the slope of the ac load line?

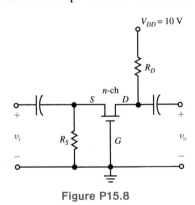

Figure P15.8

15.9. (a) Draw the circuit for a common drain amplifier similar to that shown in Fig. D15.1 with $R_S = 750$ Ω, and $R_G = 2$ MΩ. (b) If the *Q*-point is chosen to be $I_D = 2$ mA, find V_{GS}. (c) What is the approximate voltage gain of this circuit?

15.10. In Fig. P15.10, the voltage drop across M_1 is 2 V. (a) Explain the operation of the circuit and determine I_1. (b) Find I_2.

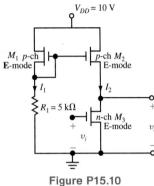

Figure P15.10

15.11. List the factors that impose limits on a MOSFET's region of safe operation. How would you expect each of these to be related to the physical size or structure of the MOSFET?

15.12. Plot the constant power contour for 1 mW and 4 mW on the MOSFET output curves of Fig. P15.6(a).

15.13. Figure P15.13 (a) shows a BJT common emitter amplifier circuit; Figure P15.13. (b) shows the output curves for the transistor. Assuming $V_{BE} = 0.7$ V, (a) draw the dc load line. (b) Determine the *Q*-point. (c) At the *Q*-point, what is the value of V_{CE} and I_C? (d) Determine the transistor's β.

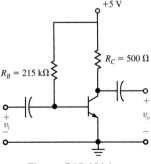

Figure P15.13(a)

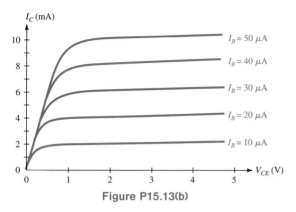

Figure P15.13(b)

15.14. Figure P15.14 shows a BJT common emitter circuit and curves. The Q-point is chosen to be $I_C = 30$ mA and $V_{CE} = 5$ V; assume $V_{BE} = 0.7$ V. (a) Draw the dc load line. (b) Calculate the value of R_C. (c) Calculate the value of R_B. (d) Calculate the value of transistor's β. (e) If $v_s(t)$ is sinusoidal with an amplitude of 2.5 V, show on the load line the range of movement of the bias point, and sketch a plot of $v_{CE}(t)$. (f) Calculate the value of the large-signal voltage gain.

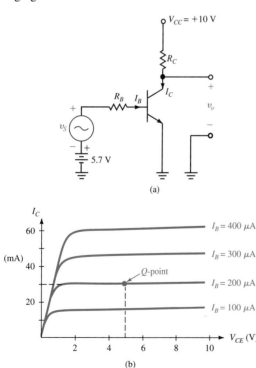

Figure P15.14

15.15. Figure P15.15 shows a BJT common emitter self-bias circuit. Assuming $I_B \ll I_1$, $\beta = 250$, and $V_{BE} = 0.7$ V, (a) calculate the values of I_E, I_C, and I_B. (b) Calculate the value of V_{CE}. (c) If v_i is increased by 0.1 V, what's the effect on v_o?

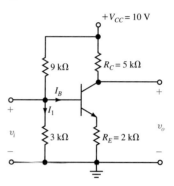

Figure P15.15

15.16. Figure P15.16 shows a BJT common emitter self-bias amplifier circuit with a PNP transistor. Assuming $V_{BE} = -0.7$ V and $\beta = 120$, (a) find the value for R_1 that makes $I_E = -1.2$ mA. (b) Find the value for R_C that will make $V_C = -6$ V. (c) Determine the slope of the dc load line. (d) Where would we place a bypass capacitor in this circuit and why?

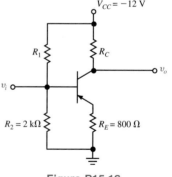

Figure P15.16

15.17. Assume that the circuit of Fig. P15.15 utilizes the transistor characterized by the output curves in Fig. P15.13(b), $V_{CC} = 5$ V, $R_E = 180$ Ω, $R_C = 360$ Ω and all other values are the same: $V_{BE} = 0.7$ V. (a) Draw the dc load line. (b) Draw a bypass capacitor in this circuit and draw the ac load line.

15.18. The common base amplifier shown in Fig. 15.22 is biased with $V_{CC} = 10$ V, $R_1 = R_2 = 3$ kΩ, and $R_E = 4$ kΩ. The transistor has a $\beta = 100$ and $V_{BE} = 0.7$ V. (a) Determine I_E and I_C. (b) Determine the value of R_C that will place the collector quiescent voltage at 7 V.

15.19. The common collector BJT bias circuit of Fig. 15.24 has $R_1 = 6$ kΩ, $R_2 = 4$ kΩ, and $R_E = 2.7$ kΩ. Determine the dc emitter current assuming $V_{BE} = 0.7$ V and β is large.

15.20. In the circuit shown in Fig. P15.20, transistors Q_1 and Q_2 provide dc bias to Q_3. What is the dc collector current of Q_2?

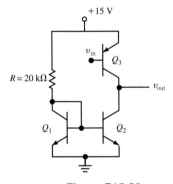

Figure P15.20

15.21. (a) On the transistor output curves shown in Fig. P15.13(b), plot three constant power contours: 4 mW, 10 mW, and 30 mW. (b) If we wish to limit the power dissipated in this device to 10 mW, draw an acceptable load line that would give large swings in V_{CE}. (c) If we wished to obtain large output current swings, would you design a different load line?

15.22. Draw a constant power curve for $P = 200$ mW on the output curves for the transistor shown in Fig. P15.14. If this is the maximum specified value for power dissipation, is the Q-point selected in problem 15.14 acceptable?

15.23. A thyristor with $V_{BO} = 5$ V and $I_H = 10$ mA is placed in the circuit of Fig. P15.23(a). The voltage source $v_1(t)$ ramps up to 10 volts over 10 seconds, and back down again as shown in Fig. 15.23(b). Plot the current $i(t)$ over the same 20 seconds.

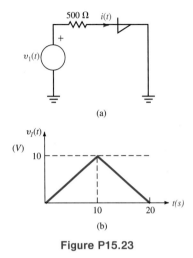

Figure P15.23

15.24. The circuit shown in Fig. P15.24 can be used as a timer circuit. Switch S_1 is closed at $t = 0$ and the capacitor is initially uncharged; at time t_1 the thyristor will fire and turn on the small indicator light which has a resistance of 100 Ω. The thyristor has $V_{BO} = 20$ V and $I_H = 30$ mA. Find time t_1.

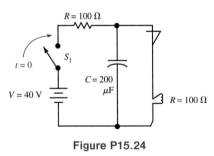

Figure P15.24

15.25. For the SCR motor speed controller shown in Fig. 15.30, sketch the waveforms for the gate current and the motor current as a function of time for firing angles of 30° and 160°.

15.26. Figure P15.26 is a circuit in which switch S_1 is used in the normal mode of operation to control Q_1 which switches a large current in load R_L. If there is an accidental short across the load (simulated by closing S_2), the increased current in Q_1 and R_{sense} causes the SCR to fire, and this lowers the base voltage of Q_1, turning it off. Assume the SCR fires at $V_G = 0.7$ V, $V_{BE} = 0.7$ V, and $\beta = 500$. (a) Determine the normal current through R_L (and Q_1) when S_1 closes. (b) If there is a short across R_L, at what load current will the circuit "shut-down" (turn off) Q_1? (c) What must be done to reactivate the circuit?

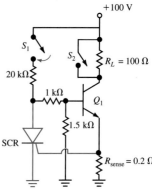

Figure P15.26

16

The Small-Signal Analysis of MOSFETs and BJTs

16.1

Introduction

In this chapter we will develop small-signal models for MOSFETs and BJTs and apply them in the solution of transistor circuit problems. The purpose of a small-signal model is to simplify the calculations involved in determining the ac response of a circuit.

A *small-signal model* is a *linearized model,* which presumes that the signal variations about the Q-point are sufficiently small so that all nonlinearities can be ignored. This assumption enables us to use a model with normal linear circuit elements, like resistors, dependent sources, etc.

We will also assume in this chapter that these circuits operate in the *mid-band frequency range.* This range of signal frequencies is high enough that all coupling capacitors and bypass capacitors have sufficiently low impedance to be considered as ac shorts. In addition, the impedance of all device junction capacitances and parasitic capacitances are sufficiently high so that they can be considered as open circuits and ignored. It is in this *mid-band range* that most amplifiers are designed to operate. The ac gain of the circuit is generally constant over this range of frequencies, and is at its maximum value. The gain will generally decrease or "roll off" at both ends of the mid-band frequency region.

The concept that a nonlinear device can be modeled over a limited range by a linear element was first introduced in section 8.13.

The signal input to a BJT is normally applied to the base-emitter *pn* junction. Therefore we must develop a large and small signal model of the diode which we will then apply to modeling the transistor.

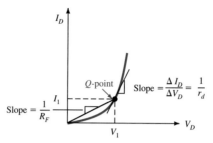

Figure 16.1   Illustration of dc and ac resistance on the V–I characteristic of a diode.

Diode dc Resistance

Recall that we considered a *pn* junction diode's large-signal V–I characteristics as described by a nonlinear relationship, expressed by the diode equation presented in Chapter 8 as Eq. (8.9); assuming $n = 1$

$$I_D = I_o [e^{\frac{qV_D}{kT}} - 1] \tag{16.1}$$

A plot of the V–I characteristic predicted by this equation is shown in Fig. 16.1. Now assume there is a dc bias established so the diode conducts a current, I_1. The diode curve then enables us to determine the diode dc voltage, V_1. The pair of coordinates V_1 and I_1 identifies the Q-point, and the *static forward resistance, R_F*, is a dc resistance which is obtained from the expression

$$R_F = \frac{V_1}{I_1} \tag{16.2}$$

Diode Small-Signal (Dynamic or ac) Resistance

We can now consider the effect of a small signal applied to the diode. If we allow the voltage across the diode to vary up and down a small amount about the Q-point, the resulting current will also vary a small amount approximately *following a line tangent to the curve at the Q-point*. The relationship between the voltage variations, ΔV_D, and the current variations, ΔI_D, can be expressed as a ratio, which we call the incremental or *small-signal resistance, r_d*.

The slope of a line tangent to the V–I curve at the Q-point is proportional to the inverse of r_d, that is

$$r_d \approx \frac{\Delta V_D}{\Delta I_D} \bigg|_{at\ the\ Q\text{-}point} \tag{16.3}$$

and the slope at the Q-point is

$$slope = \frac{1}{r_d} = \frac{i_d}{v_d} \tag{16.4}$$

where v_d and i_d are the small-signal or incremental changes in voltage and current, which are by definition changes from the Q-point.

An expression for r_d was developed in Chapter 8 by differentiating the diode equation. This important result is

$$r_d = \frac{1}{\dfrac{\partial I_D}{\partial V_D}} = \frac{kT}{qI_D} \qquad (16.5)$$

which at room temperature becomes

$$r_d = \frac{0.026}{I_D} \; \Omega \qquad (16.6)$$

EXAMPLE 16.1

Let us determine the ac resistance of a silicon diode carrying 3.5 mA at room temperature.

From Eq. (16.6) the resistance is

$$r_d = \frac{0.026}{I_D} = \frac{0.026}{3.5 \times 10^{-3}} = 7.43 \; \Omega \qquad \blacksquare$$

The dc bias circuits, such as those described in the previous chapter, establish the desired dc quiescent operating condition, or Q-point, of transistor amplifier circuits. If the applied signal causes small variations from this bias point, then small-signal linear models can be used for analysis, as will be illustrated in this chapter.

Nomenclature Review

The following table briefly reviews the meaning of uppercase and lowercase letters in describing the voltages and currents in small-signal analysis.

	Diode	MOSFET	BJT
Instantaneous total value	v_D, i_D	v_{GS}, i_D	v_{BE}, i_C
Instantaneous signal component	v_d, i_d	v_{gs}, i_d	v_{be}, i_c
Quiescent or dc value	V_D, I_D	V_{GS}, I_D	V_{BE}, I_C
Supply voltage (dc magnitude)	V, I	V_{DD}, V_{GG}	V_{CC}, V_{BB}

16.2
General Linear Two-Port Models

Transistors, both MOSFETs and BJTs, can be modeled as special cases of general two-port linear networks. A *port* is defined as a pair of connections, and thus a two-port network has a pair of wires (connections) for the input and another pair for the output as shown in Fig. 16.2(a).

The mathematical description of a two-port network can be accomplished in several fully equivalent forms. These are commonly referred to as models using: *y parameters* (admittance parameters), *z parameters* (impedance parameters), or *h parameters* (hybrid parameters).

In each case, two equations can be used to relate the input voltage and current to the output voltage and current using four parameters.

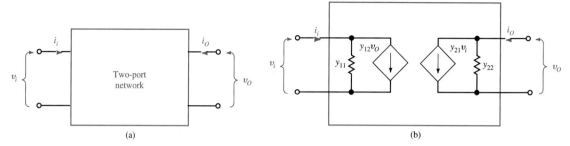

Figure 16.2 (a) A two-port network, and (b) two-port network using y parameters.

Using y parameters:

$$i_i = v_i y_{11} + v_o y_{12}$$
$$i_o = v_i y_{21} + v_o y_{22}$$

(16.7)

Using z parameters:

$$v_i = i_i z_{11} + i_o z_{12}$$
$$v_o = i_i z_{21} + i_o z_{22}$$

(16.8)

Using h parameters:

$$v_i = i_i h_{11} + v_o h_{12}$$
$$i_o = i_i h_{21} + v_o h_{22}$$

(16.9)

A dimensional analysis will confirm that all the y parameters have units of admittance and all the z parameters have units of impedance. The units for the hybrid parameters are as follows: h_{11}—impedance, h_{12}—voltage transfer ratio, h_{21}—current transfer ratio, and h_{22}—admittance.

Notice that the parameters with the same subscript, for example, y_{11} and y_{22}, relate change in voltage to the change in current at the same pair of terminals. The other two parameters, for example, y_{12} and y_{21}, relate the change in voltage at one pair of terminals to the change in current at the other, and therefore they are termed transadmittances or transconductances. For example, y_{21} relates the change in output current to the change in input voltage, for the case when v_o is zero, that is, shorted.

Consider, for illustration, the set of admittance equations, given as Eq. (16.7). Note that each of the currents, i_i and i_o, are composed of two components represented by the two terms on the right side of each equation. The term which relates current and voltage at the same port can be modeled by an admittance, while the remaining term is modeled by a dependent current source with a magnitude expressed by that y-parameter term. Thus, an equivalent circuit for the transistor can be constructed in which the input and output are each represented by an admittance in parallel with a dependent current source as shown in Fig. 16.2(b). A similar circuit model could be developed for each of the other parameter sets.

Figure 16.3 illustrates all the possible circuit configurations for the MOSFET and the BJT, and a two-port network analysis approach may be applied to each of these. For each circuit configuration and for every dc bias condition, a different set of small-signal parameters must be obtained and used for analysis. The process of determining the small-

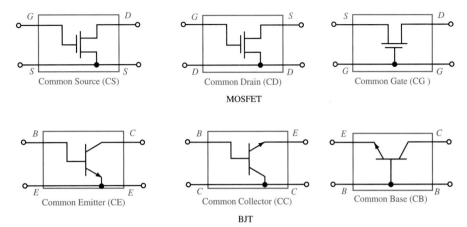

Figure 16.3 Two-port configurations for MOSFET and BJT transistor circuits.

signal parameters can be cumbersome. In actual practice these parameters are often measured directly by computer-controlled instrumentation that applies a dc bias to the transistor and then superimposes upon it small-signal variations in order to measure the small-signal ac parameters. One can also calculate these small-signal parameters from the large-signal output curves. This approach will be illustrated later. In the following example, the small-signal parameters are assumed to be known.

EXAMPLE 16.2

A MOSFET is placed in a common source amplifier circuit with an equivalent load resistance of $R_L = 10\ \text{k}\Omega$. Let us determine the small-signal voltage gain of the circuit if the y parameters of the MOSFET are known to be $y_{11} = 0\ \text{S}$, $y_{12} = 0\ \text{S}$, $y_{21} = 1 \times 10^{-3}\ \text{S}$, and $y_{22} = 0\ \text{S}$, where S is the unit of Siemens.

This network configuration can be represented by the circuit shown in Fig. 16.4. Since $y_{22} = 0$, the second part of Eq. (16.7) reduces to

$$i_o = y_{21}\, v_i$$

In addition, we note from Fig. 16.4 that

$$v_o = -i_o\, R_L$$

Therefore, the small-signal voltage gain is

$$A_v = \frac{v_o}{v_i} = -\frac{i_o\, R_L}{\dfrac{i_o}{y_{21}}} = -y_{21}\, R_L = -10$$ ■

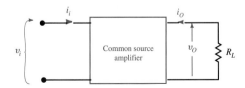

Figure 16.4   Circuit used in Example 16.2.

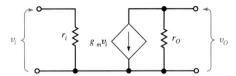

Figure 16.5   Simplified small-signal transistor model.

It is common practice to use two-port network theory to describe small-signal transistor characteristics. However, as the previous example indicates, it is often not necessary to use all four parameters to obtain reasonably accurate transistor modeling. For example, in the mid-band frequency region, the input impedance of a MOSFET is typically so large that z_{11} can generally be assumed to be infinite. For the same reason, we may often assume that $y_{11} = 0$. The parameter y_{12}, which models the signal coupling from the output back to the input, is generally negligible. Therefore, small-signal linear circuit models that simplify the parameter measurement and analysis process are very useful, especially at low and moderate frequencies.

In the mid-band frequency range, where device capacitances and inductances are ignored, the admittance and impedance parameters are reduced to their "real" components, or simply, conductance and resistance.

The general small-signal equivalent circuit of such a model is shown in Fig. 16.5 where

$$r_i = \text{input resistance} = v_i/i_i$$
$$g_m = \text{transconductance} = i_o/v_i \,|\, \text{output shorted; } v_o = 0$$
$$r_o = \text{output resistance} = v_o/i_o \,|\, \text{input shorted; } v_i = 0$$

These parameters can be related to any other set of parameters through algebraic manipulation of the parameter equations.

In the following sections, we will apply this kind of simplified small-signal model to the MOSFET and the BJT.

16.3

The MOSFET Small-Signal Model

Typical output characteristic curves for a common source MOSFET are shown in Fig. 16.6. The small-signal model is developed to describe the device's behavior for *small excursions of voltage and current about the* Q-*point* in the active region where the output

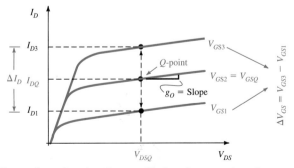

Figure 16.6   Illustration of estimating small-signal parameters from common source output curves.

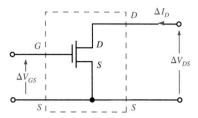

Figure 16.7 Common source amplifier configuration.

curves are almost horizontal. A dc bias is applied to the transistor by a bias circuit which establishes its Q-point at the point defined by V_{DSQ} and I_{DQ}.

The common source amplifier configuration is illustrated in Fig. 16.7. The input voltage is applied from gate to source, and the output is taken from drain to source. An examination of the output curve about the Q-point reveals that only slight changes in the output current I_D occur as V_{DS} is changed. I_D is primarily affected by changes in input voltage, V_{GS}.

Therefore, a small-signal model representing the *output* of the device consists of both a dependent current source and a conductance, as shown in Fig. 16.8.

The dependent current source has a value of $g_m v_{gs}$, where g_m is the transconductance of the device and is defined as:

$$g_m = \left.\frac{\partial I_D}{\partial V_{GS}}\right|_{V_{DS} = V_{DSQ}} \approx \left.\frac{\Delta I_D}{\Delta V_{GS}}\right|_{V_{DS} = constant = V_{DSQ}} \tag{16.10}$$

Note that the parameter g_m is the same as the parameter y_{21} in the two-port admittance model. Recall that when determining y_{21} or equivalently the transconductance g_m, the output is ac shorted, that is, $v_{ds} = 0$ or $V_{DS} = $ constant. The appropriate small-signal model parameter is determined at the Q-point.

The value of g_m can be estimated from the output curves of the device by drawing a vertical line ($\Delta V_{DS} = 0$) through the Q-point which intersects V_{GS} curves above and below the Q-point. Assuming that a change in V_{GS} is the difference between these neighboring curves, and calculating the ratio of that change to the resulting change in output current, yields an approximate value for g_m. With reference to Fig. 16.6, this calculation is

$$g_m \approx \frac{\Delta I_D}{\Delta V_{GS}} = \left.\frac{I_{D3} - I_{D1}}{V_{GS3} - V_{GS1}}\right|_{V_{DS} = constant = V_{DSQ}} \tag{16.11}$$

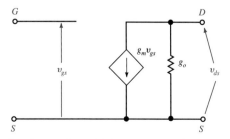

Figure 16.8 Small-signal model for a common source configuration.

We have purposely exaggerated the variations in voltages and currents in this illustration. In general, small-signal variations about the Q-point are quite small.

The output conductance, g_o, is included in the model to provide the small change in output current resulting from changes in output voltage. This conductance accounts for the small but finite slope of the output curves in the active region.

$$g_o = \left. \frac{\partial I_D}{\partial V_{DS}} \right|_{\text{at the Q-point}} \approx \left. \frac{\Delta I_D}{\Delta V_{DS}} \right|_{\text{at the Q-point}} = \textit{slope of curve at Q-point} \quad (16.12)$$

It is interesting to note that g_o is the same as y_{22} in the two-port admittance model.

As illustrated in Fig. 16.7, the *input* to the gate of a MOSFET is electrically insulated from the remainder of the device. Furthermore, since device capacitances are small and need not be considered unless we are performing high-frequency analyses, the MOSFET circuit model for the input is simply an open circuit as shown in Fig. 16.8. Once again, note that this assumption is equivalent to setting y_{12} and y_{11} to zero in the two-port admittance model. The resulting model is quite useful for solving circuit problems.

EXAMPLE 16.3

Consider a MOSFET with output curves as shown in Fig. 16.9. Using the procedures described above, let us determine the small-signal parameters, g_m and g_o, and then use them to calculate the small-signal voltage gain of a common source circuit utilizing this device with a total external load resistance of 8 kΩ.

Figure 16.9 indicates that for V_{GS}, changes of $\pm$ one-half volt ($\Delta V_{GS} = 1$ V), the current change ΔI_D is 1.75 mA. Therefore, using Eq. (16.11), g_m becomes:

$$g_m \approx \frac{\Delta I_D}{\Delta V_{GS}} = \frac{1.75 \times 10^{-3}}{1} = 1.75 \text{ mS}$$

From the figure, we also note that the output conductance, g_o, is

$$g_o \approx \frac{\Delta I_D}{\Delta V_{DS}} = \frac{0.25 \times 10^{-3}}{3} = 83.3 \ \mu\text{S}$$

In order to calculate the voltage gain of the CS circuit represented by this model, we must first determine the total equivalent load resistance, R_{Leq}. This equivalent load is the parallel combination of g_o and the external load resistance of 8 kΩ as illustrated in Fig. 16.10.

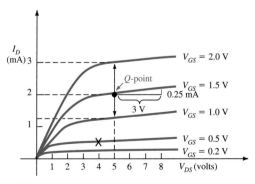

Figure 16.9 MOSFET output curves used in Example 16.3.

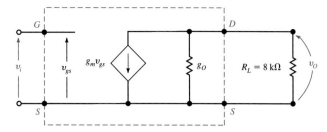

Figure 16.10   MOSFET small-signal model used in Example 16.3.

Hence

$$R_{Leq} = \frac{1}{g_o} \| R_L = \left[\frac{1}{83.3 \ \mu S} \right] \| 8 \ k = 4.8 \ k\Omega$$

And therefore the voltage gain, A_v, is

$$A_v = \frac{v_o}{v_i} = -\frac{v_{gs} g_m R_{Leq}}{v_{gs}} = -g_m R_{Leq} = -(1.75 \times 10^{-3})(4.8 \ k) = -8.4 \qquad \blacksquare$$

The expression

$$A_v = -g_m R_{Leq} \qquad (16.13)$$

generally applies for determining the small-signal voltage gain of a CS amplifier.

16.4
MOSFET Amplifier Circuits—Small-Signal Analysis

The Common Source Amplifier

In this section, a complete amplifier circuit will be analyzed using small-signal analysis. First, the dc circuit is analyzed to determine the dc bias condition. At this specific bias point, the small-signal parameters appropriate for the particular bias condition are determined. Then a small-signal circuit model of the transistor and the circuit in which it operates are constructed. And finally, small-signal calculations are performed to determine the ac mid-band circuit performance.

The principles inherent in the analysis will be presented through the use of several examples.

EXAMPLE 16.4
The complete schematic diagram of a common source amplifier utilizing an *n*-channel MOSFET is shown in Fig. 16.11(a). The MOSFET has the output curves illustrated in Fig. 16.9. Let us determine the dc bias condition (*Q*-point) and the small-signal parameters, g_m and g_o, for this device at that *Q*-point.

We will first determine the dc drain current, I_D, and the drain-to-source voltage, V_{DS}. Since a capacitor is a dc open circuit, the coupling capacitors, C_1 and C_2, block dc current and therefore nothing to the left of C_1 or to the right of C_2 in the circuit need be considered for the dc calculations.

In the active region, the gate-to-source voltage will set the drain current. Therefore,

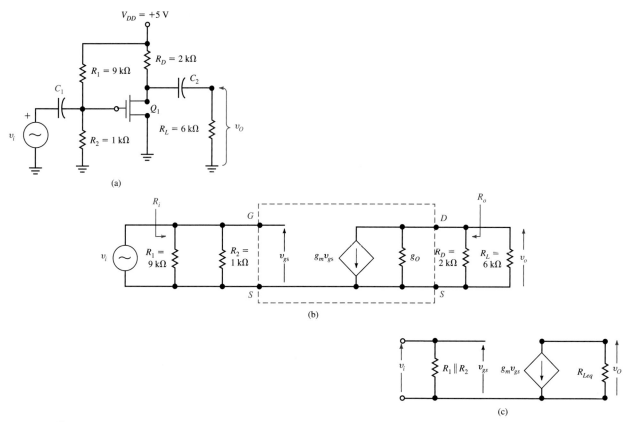

Figure 16.11 A common source amplifier schematic circuit and its small-signal equivalent circuits.

we must first determine V_{GS}. The source voltage is

$$V_s = 0$$

The gate voltage determined by the voltage divider is

$$V_G = V_{DD}\left[\frac{R_2}{R_1 + R_2}\right] = 5[0.1] = 0.5 \text{ V}$$

Therefore

$$V_{GS} = V_G - V_S = 0.5 - 0 = 0.5 \text{ V}$$

We now know that the Q-point will lie somewhere along the curve corresponding to $V_{GS} = 0.5$ Volts in Fig. 16.9. This curve is at a level of approximately $I_D = 0.5$ mA throughout the active region.

If $I_D = 0.5$mA, then V_{DS} can be determined as follows.

$$V_D = V_{DD} - R_D I_D = 5 - (2\text{ k}) (0.5 \times 10^{-3}) = 4 \text{ V}$$

And hence

$$V_{DS} = V_D - V_S = 4 - 0 = 4 \text{ V}$$

Therefore, the Q-point is located on the output curves at the point defined by $I_D = 0.5$ mA and $V_{DS} = 4$ V and marked with an x.

The small-signal transconductance, g_m, will now be determined at the specified Q-point. From Fig. 16.9, note that a variation in V_{GS} of 0.8 volts ($+0.5$, -0.3) yields a drain current variation of approximately 1.0 mA. Therefore

$$g_m \approx \frac{\Delta I_D}{\Delta V_{GS}} = \frac{1.0 \times 10^{-3}}{0.8} = 1.25 \times 10^{-3} \text{ S}$$

The output conductance is often difficult to determine graphically when the slopes are small. However, for this device g_o can be approximated as

$$g_o \approx \frac{\Delta I_D}{\Delta V_{DS}} \approx \frac{0.1 \times 10^{-3}}{4} = 2.5 \times 10^{-5} \text{ S}$$

It is often useful to convert the output conductance to an equivalent resistance value, and hence for this case the output resistance of the transistor is

$$r_o = \frac{1}{g_o} = 40 \text{ k}\Omega \tag{16.14}$$

This value, which we have computed to illustrate the procedure, is relatively low. Typical output resistance values for MOSFETs in the active region are normally much higher. ∎

In general, a small-signal equivalent circuit for an amplifier operating in the mid-band frequency region can be constructed from a schematic drawing of a circuit using the following procedure.

1. Consider all bypass and coupling capacitors as ac shorts.
2. Consider all independent dc voltage sources as ac shorts. If the instantaneous voltage between two nodes is not allowed to vary, then there is no ac voltage component possible and it acts as if there were an ac short. Thus, the supply (V_{DD}) becomes the same node as ground.
3. Consider all independent dc current sources as ac open circuits.
4. Insert the small-signal transistor model into the circuit in place of the transistor.

This procedure is illustrated by the following example.

EXAMPLE 16.5

Once again we consider the amplifier circuit shown in Fig. 16.11(a). For this network, we wish to (a) construct the small-signal equivalent circuit, assuming the amplifier is operating in the mid-band frequency region, and (b) use this equivalent circuit to calculate the small-signal voltage gain.

(a) Using the procedure outlined above, the schematic circuit for this amplifier is reduced to the small-signal equivalent circuit shown in Fig. 16.11(b). Both capacitors act as ac shorts and since the supply node V_{DD} is also an ac ground, R_1 and R_2 are in parallel.

The small-signal MOSFET model is then substituted for the transistor symbol in the circuit drawing. Note that the model representing the transistor is outlined with a box.

Finally, since C_2 is an ac short, the output of this circuit consists of the dc load resistance, R_D in parallel with the external load, R_L.

The ac gain can now be calculated from this small-signal model of the amplifier. From this equivalent circuit, we note that the input voltage, v_i, is the same as v_{gs}.

The three resistors in the output of this circuit are all in parallel and can be reduced to an equivalent load resistance, R_{Leq}. Thus, the circuit can be simplified to that shown in Fig. 16.11(c), where

$$R_{Leq} = r_o \| R_D \| R_L = \frac{1}{\dfrac{1}{r_o} + \dfrac{1}{R_D} + \dfrac{1}{R_L}} = 1.45 \text{ k}\Omega \qquad (16.15)$$

The output voltage can be expressed as

$$v_o = -g_m v_{gs} R_{Leq} = -g_m v_i R_{Leq}$$

And therefore, the voltage gain, A_V

$$A_v = \frac{v_o}{v_i} = -g_m R_{Leq} = -(1.25 \times 10^{-3})(1.45 \text{ k}) = -1.81 \qquad \blacksquare$$

Two other commonly used terms can be defined with the aid of the previous example: (1) amplifier input resistance, R_i, and (2) amplifier output resistance, R_o.

1. The *input resistance of an amplifier*, R_i, is the total equivalent resistance seen looking into the input of the amplifier. In the previous example

$$R_i = R_1 \| R_2 = 1 \text{ k} \| 9 \text{ k} = 900 \ \Omega$$

2. The *output resistance of the amplifier*, R_o, is the total resistance seen looking back into the output terminals of the amplifier, not including any external load resistance.

In the previous example

$$R_o = r_o \| R_D = 40 \text{ k} \| 2 \text{ k} = 1.90 \text{ k}\Omega$$

The arrows in Fig. 16.11(b) provide a reference for measuring both R_i and R_o.

The Common Gate Amplifier

The dc bias techniques for the common gate amplifier are similar to those for the common source. However, in this case the gate is connected to ac ground, the input is applied from source to gate, and the output is taken from drain to gate.

EXAMPLE 16.6

Consider the common gate MOSFET circuit shown in Fig. 16.12(a). If the dc drain current is measured to be 1 mA, let us determine the value of V_{GG} required to establish V_{GS} at +1 volt.

The path of the dc drain current includes V_{DD}, R_D, the transistor (D-to-S), and R_S. Therefore

$$V_S = I_D R_S = (1 \times 10^{-3})(1.5 \text{ k}) = 1.5 \text{ V}$$

And since V_{GS} must be 1 V

$$V_{GG} = V_G = V_{GS} + V_S = 1 + 1.5 = 2.5 \text{ V} \qquad \blacksquare$$

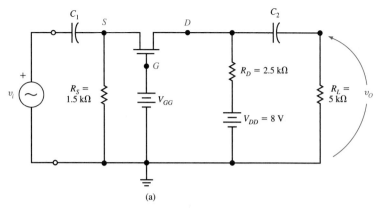

(a)

Figure 16.12(a)   Common gate MOSFET amplifier schematic circuit.

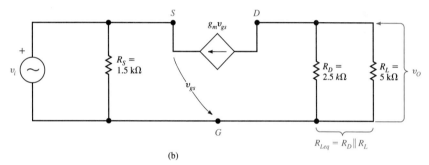

(b)

Figure 16.12(b)   Common gate MOSFET amplifier small-signal equivalent circuit.

EXAMPLE 16.7

Let us determine (a) the small-signal equivalent circuit for the amplifier shown in Fig. 16.12(a), and (b) the small-signal voltage gain of this common gate circuit, assuming the transistor parameters are

$$g_o = 0$$
$$g_m = 5 \times 10^{-3} \text{ S}$$

(a) The small-signal equivalent circuit is shown in Fig. 16.12(b). Since $g_o = 0$, it does not appear in the equivalent circuit.

(b) Note that the only ac current through R_{Leq} is $g_m v_{gs}$; therefore, with reference to Fig. 16.12(b)

$$v_i = -v_{gs}$$

$$v_o = -g_m v_{gs} R_{Leq}$$

and therefore the voltage gain is

$$A_v = \frac{v_o}{v_i} = g_m R_{Leq} = 8.33 \tag{16.16}$$

∎

EXAMPLE 16.8

Let us determine the input resistance, R_i, of the amplifier described in Example 16.7.

The input signal to the transistor is applied at the source terminal, NOT the gate. Therefore, in this case, the resistance looking into the transistor is not infinite.

The equivalent circuit of Fig. 16.12(b) indicates that because of the presence of the dependent source, a change in the input voltage $(-v_{gs})$ at the input terminals of the transistor will result in a change in input current. The input resistance of the transistor is the ratio of the change in input voltage to the change in input current at the terminals S–G. Therefore

$$r_{i(source)} = -\frac{v_{gs}}{-g_m v_{gs}} = \frac{1}{g_m} \tag{16.17}$$

The input resistance of the amplifier is then the parallel combination of R_S and $1/g_m$

$$R_i = R_S \left\| \frac{1}{g_m} = (1.5 \text{ k}) \right\| (200) = 176 \ \Omega \qquad \blacksquare$$

Several general comments can be made about the common gate circuit. The input resistance to this circuit is not high as it is in both the common source and the common drain configurations. The voltage gain is positive, that is, the circuit does NOT invert the signal, and the magnitude of the voltage gain can be quite large. The circuit is also useful for matching a low impedance input to a high impedance output.

It is important to note that Eq. (16.17) can be used to express the resistance looking into the source for any properly biased MOSFET.

The Common Drain Amplifier

A schematic drawing for a common drain circuit using an *n*-channel enhancement-mode transistor was shown in Fig. D15.1. As discussed in the previous chapter, the voltage gain of this circuit is just under unity.

The exact expression for this amplifier's small-signal voltage gain, A_v, can be obtained by inserting the MOSFET model into the small-signal equivalent circuit—a procedure described in this chapter. The result obtained is

$$A_v = \frac{v_o}{v_i} = \frac{g_m R_{Leq}}{1 + g_m R_{Leq}} \tag{16.18}$$

where R_{Leq} is defined as the parallel combination of R_S and any other external load resistance. This amplifier configuration has a very high input resistance, low output resistance, and does not invert the signal.

Summary of Important MOSFET Relationships

Table 16.1 summarizes the important relationships for small-signal circuit analyses of MOSFETs in the mid-band frequency region.

An analytic expression for g_m derived from Eq. (9.1) is $g_m = 2K \, (V_{GS} - V_T)$; similarly, for depletion-mode devices, from Eq. (9.5) $g_m = 2 \, I_{DSS} \, V_P^{-2} \, (V_{GS} - V_P)$.

Table 16.1 Important MOSFET Relationships for Small-Signal Analysis

CONFIGURATION	NOTES	VOLTAGE GAIN, A_v	INPUT RESISTANCE		OUTPUT RESISTANCE	
			To Transistor	To Circuit	Transistor	Circuit
COMMON SOURCE	$R_{Leq} = R_D \| R_L \| r_o$ R_G = equivalent resistance of gate bias resistance (voltage signal inverted)	$A_v = \dfrac{v_o}{v_i}$ $A_v = -g_m R_{Leq}$	$r_i \approx \infty$	$R_i \approx R_G$	$r_o \approx \dfrac{1}{g_o}$	$R_o \approx R_D \| \dfrac{1}{g_o}$
COMMON GATE	$R_{Leq} = R_D \| R_L$ R_G (as above) Assume $r_o = \infty$ (voltage signal not inverted)	$A_v = g_m R_{Leq}$	$r_i \approx \dfrac{1}{g_m}$	$R_i \approx R_S \| \dfrac{1}{g_m}$	$r_o > \dfrac{1}{g_o}$	$R_o \approx R_D$
COMMON DRAIN	$R_{Leq} = R_S \| R_L$ R_G (as above) (voltage signal not inverted)	$A_v = \dfrac{g_m R_{Leq}}{1 + g_m R_{Leq}}$	$r_i \approx \infty$	$R_i \approx R_G$	$r_o \approx \dfrac{1}{g_m}$	$R_o \approx R_S \| \dfrac{1}{g_m}$

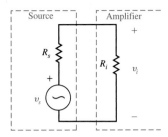

Figure 16.13 Equivalent circuit of the input of an amplifier.

Voltage Gain Loss at the Input Due to Finite Source Resistance

In our calculations of small-signal voltage gain thus far, we have only considered the gain of the transistor stage, that is, v_o/v_i, where v_i is the signal voltage at the INPUT of the transistor. If the transistor stage is driven by a voltage source, v_s, with finite internal resistance, R_s, some gain will be lost in the input circuit. The input circuit of an amplifier stage can be represented by the circuit shown in Fig. 16.13, where R_i represents the input resistance to the amplifier stage. From this circuit

$$v_i = v_s \left[\frac{R_i}{R_i + R_s} \right] \tag{16.19}$$

Normally, R_i is designed to be large compared to R_s, so that

$$v_i \approx v_s \tag{16.20}$$

We often speak of the voltage gain of the input circuit, which is always < 1, as, A_{vs}, where

$$A_{vs} = \frac{v_i}{v_s} = \frac{R_i}{R_i + R_s} \tag{16.21}$$

The gain of the entire circuit from v_s to v_o is the product of the two gains

$$\frac{v_o}{v_s} = \left[\frac{v_i}{v_s} \right]\left[\frac{v_o}{v_i} \right] = A_{vs} A_v \tag{16.22}$$

These same input circuit considerations also apply to the bipolar transistor circuits described in the next section.

16.5

The BJT Small-Signal Circuit Model

There are two major differences between the BJT and the MOSFET that affect small-signal modeling: (1) the input connection to the BJT is always attached to one end of a forward biased base emitter junction, which is not an extremely large impedance as was assumed for the MOSFET gate, and (2) since output current for the BJT is most conveniently modeled as if it were controlled by input current, the output curves for a BJT are constructed by plotting a separate curve for different input currents, NOT voltage, as was the case for a MOSFET.

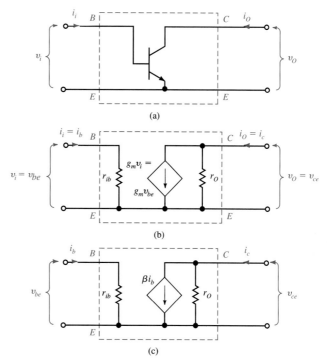

Figure 16.14 (a) Final BJT common emitter circuit configuration; (b) BJT common emitter small-signal model; (c) final BJT common emitter small-signal model.

The Common Emitter Model

We will consider first the most common bipolar circuit configuration, the common emitter. Figure 16.14(a) shows the two-port representation of this configuration, with the input applied between base and emitter and output taken from collector to emitter.

Applying the small-signal model of Fig. 16.5 to this configuration, we obtain the circuit of Fig. 16.14(b). Note the introduction of the term r_{ib}, which represents the input resistance of the device looking into the base terminal. In a similar manner, we will later define another term, r_{ie}, which we use to designate the input resistance looking into the emitter terminal for the common base circuit.

The magnitude of r_{ib} can be easily calculated. This resistance represents the small-signal resistance of the forward biased base emitter diode seen from the base side of the device. In the first section of this chapter, we reviewed the expression for determining the small-signal resistance of a diode and found that it is inversely related to the dc bias current, which in this case is the dc base current, I_B. Therefore, r_{ib} can be expressed as

$$r_{ib} = \frac{kT}{qI_B} \qquad (16.23)$$

EXAMPLE 16.9

We wish to determine the input resistance, r_{ib}, for a transistor which is biased in the common emitter configuration, assuming it has a dc collector current of 1 mA and a beta of 50 at room temperature.

In order to determine r_{ib}, we must first determine I_B. Recall that β relates I_C to I_B in the active region, that is

$$\beta = \frac{I_C}{I_B}$$

From the data we obtain

$$I_B = \frac{I_C}{\beta} = \frac{1 \times 10^{-3}}{50} = 20 \; \mu A$$

Therefore, at room temperature

$$r_{ib} = \frac{0.026}{I_B} = \frac{0.026}{20 \times 10^{-6}} = 1.3 \; k\Omega \qquad (16.24)$$

■

The output section of this *CE* model has a dependent current source $g_m v_i$ and an output resistance, r_o. The "short-circuit output current" is the ac current that flows through a short placed across the output from C to E. For this case all the current from the current source flows in the short and

$$i_o = g_m \, v_i$$

or equivalently

$$i_c = g_m v_{be} \qquad (16.25)$$

Furthermore, we can assume, for purposes of this discussion, that beta represents the ratio of the collector current to base current, for both dc current and ac current. Therefore

$$i_c = \beta \, i_b \qquad (16.26)$$

And hence

$$g_m v_{be} = \beta \, i_b \qquad (16.27)$$

This relationship permits us to replace the current source in the output of the small-signal model with a dependent source proportional to the input current, i_b, as shown in Fig. 16.14(c). While the models shown in Figs. 16.14(b) and (c) are equivalent and either can be used for circuit calculations, the model in Fig. 16.14(c) has found widespread acceptance due to its ease of use, and is called a simplified hybrid -π model.

EXAMPLE 16.10

A BJT is biased in the common emitter configuration with a dc base current of 10 microamperes and has an external equivalent load resistance of 10 kΩ. Let us determine the small-signal voltage gain of this circuit assuming the transistor has a beta of 100 and an output resistance, r_o, of 30 kΩ at the Q-point.

The small-signal model for this configuration is shown in Fig. 16.15. From this model we note that the output voltage is

$$v_o = -(\beta i_b) \, (r_o \, \| \, R_{Leq})$$

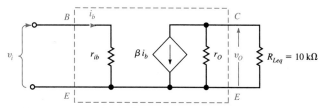

Figure 16.15 Equivalent circuit used in Example 16.10.

and the input current i_b is

$$i_b = \frac{v_i}{r_{ib}}$$

Therefore

$$A_v = \frac{v_o}{v_i} = -\frac{\beta \left(r_o \parallel R_{Leq} \right)}{r_{ib}} \tag{16.28}$$

Recall that at room temperature

$$r_{ib} = \frac{0.026}{I_B} = \frac{0.026}{10 \times 10^{-6}} = 2.6 \text{ k}\Omega$$

and therefore from Eq. (16.28)

$$A_v = -\frac{(100)(7.5 \text{ k})}{2.6 \text{ k}} = -288 \qquad \blacksquare$$

In many practical circuits we find that $r_o \gg R_{Leq}$, and hence r_o is often ignored in the calculation of A_v.

The values of the BJT small-signal parameters can be determined by direct measurement on an actual device, or from calculations using the output curves. The input resistance, r_{ib}, can be calculated using Eq. (16.23) in which only the dc base current is a variable. The parameters on the output side of the model, β and r_o, can be determined from the output curves, as illustrated in the following example.

EXAMPLE 16.11

Consider the typical output curves for a common emitter device as shown in Fig. 16.16. If the transistor is biased with the Q-point at $I_C = 2$ mA and $V_{CE} = 6$ volts, let us calculate approximate values for β and r_o at this Q-point.

In order to determine β, we construct a vertical line through the Q-point that intersects the curve above and below the Q-point. These intersections determine the change in collector current that results from the given change in base current. Performing this analysis for the curves in Fig. 16.16 yields

$$\beta \approx \frac{\Delta I_C}{\Delta I_B} = \frac{(3.2 - 1)10^{-3}}{(15 - 5)10^{-6}} = 220$$

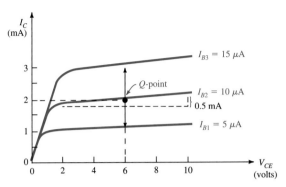

Figure 16.16   Common emitter output curves used in Example 16.11.

The output resistance, r_o, is determined by computing the inverse of the slope of the I_C curve at the Q-point, that is

$$r_o \approx \frac{1}{\dfrac{\Delta I_C}{\Delta V_{CE}}} = \frac{1}{\dfrac{0.5 \times 10^{-3}}{10 - 2}} = 16\ \text{k}\Omega \qquad \blacksquare$$

The Common Base Model

For small-signal analysis of BJT circuits in the common base configuration, the previously developed small-signal model can be used directly, although the calculations are somewhat awkward. However, a more useful *common base small-signal model* can be readily developed directly from the common emitter model. Figure 16.17(a) shows a simplified common emitter small-signal model with the output resistance omitted. The common base configuration has a very high output resistance, and thus ignoring the output resistance is often easily justified in practice.

Figure 16.17(b) shows the same circuit model redrawn so that the base is the common node in the circuit. In this model, the input is from emitter to base, and the output from collector to base. The use of this model complicates the analysis because there is now a current source appearing directly from output to input. However, this circuit can be reconfigured into an equivalent circuit similar to the common emitter circuit shown in Fig. 16.17(c) by changing the circuit parameters as follows.

The input resistance r_{ie} for the common base model in Fig. 16.17(b) is

$$r_{ie} = \frac{v_{eb}}{i_e} = \frac{-v_{be}}{-i_b - \beta i_b} = \frac{1}{\beta + 1} \frac{v_{be}}{i_b} \tag{16.29}$$

Now since

$$\frac{v_{be}}{i_b} = r_{ib}$$

$$r_{ib} = \frac{kT}{qI_B}$$

and

$$(\beta + 1) I_B = I_E$$

Then

$$r_{ie} = \frac{1}{\beta + 1} \, r_{ib} = \frac{1}{\beta + 1} \frac{kT}{qI_B} = \frac{kT}{qI_E} \tag{16.30}$$

Thus, the input resistance seen looking into the emitter is a factor of $1/(\beta + 1)$ less than that seen looking into the base. Similarly, just as the dc currents between emitter and base are related by the factor $(\beta + 1)$, so are the ac currents, and so are the ac resistances seen looking into the base and emitter terminals.

From Fig. 16.17(b), the value of ac current at the collector is given by

$$i_c = \beta \, i_b = \beta \left[\frac{-i_e}{\beta + 1} \right] = \left[\frac{\beta}{\beta + 1} \right] (-i_e) = -\alpha i_e \tag{16.31}$$

Where

$$\alpha = \frac{\beta}{\beta + 1} \tag{16.32}$$

is the *common base current gain.*

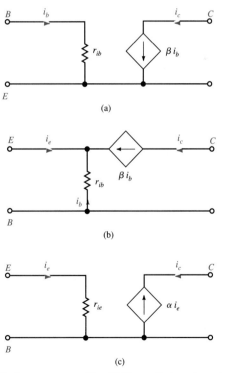

(a)

(b)

(c)

Figure 16.17 (a) Simplified common emitter BJT small-signal model; (b) simplified common base BJT small-signal model; (c) final simplified common base BJT small-signal model.

The result of Eq. 16.31 can be made positive by reversing the polarity of the dependent current source as shown in Fig. 16.17(c).

16.6
BJT Amplifier Circuits—Small-Signal Analysis

As was our approach with MOSFETs, the small-signal analysis of BJT circuits will be illustrated with a series of examples.

The Common Emitter Amplifier

Consider the following example of a common emitter amplifier.

EXAMPLE 16.12

Consider the common emitter amplifier circuit shown in Fig. 16.18(a). The transistor is an NPN BJT described by the output curves shown in Fig. 16.16. Assuming that the dc voltage across the forward biased base emitter junction, V_{BE}, is 0.7 volts, let us determine the values of R_E and R_C that are required to place the Q-point at the position shown in Fig. 16.16.

Recall from Chapters 9 and 15 that when biasing BJT circuits, the coupling capacitors C_1 and C_2 block dc current and nothing to the left of C_1 or to the right of C_2 need be considered in calculating the dc bias. In addition, bias circuits of this type set the base voltage by the resistor divider, R_1 and R_2. The dc emitter voltage is then fixed at 0.7 volts below the dc base voltage, and this resultant voltage across R_E determines the dc emitter current. It is important to note that once any one current for a BJT in the active region is known, the others can be easily found by the following simple relations:

$$\text{If } I_B = x, \text{ then } I_C = (\beta) \, x, \text{ and } I_E = (\beta + 1) \, x$$

Recall that if the current I_B is negligible compared to the current flowing down through the voltage divider composed of R_1 and R_2, then the dc base voltage is set by the resistors R_1 and R_2. From Example 16.11 we know that at the required Q-point, $I_c = 2$ mA and $\beta = 220$.

Therefore, the dc base current is

$$I_B = \frac{I_C}{\beta} = \frac{2 \times 10^{-3}}{220} = 9.1 \ \mu A$$

Ignoring I_B, the current in the voltage divider R_1 and R_2 is

$$I_{R1} \approx I_{R2} \approx \frac{V_{CC}}{R_1 + R_2} = \frac{20}{7 \, k + 3 \, k} = 2 \text{ mA}$$

Clearly

$$I_{R1} \approx I_{R2} >> I_B$$

and therefore I_B can be ignored in calculating the voltage, V_B. V_B is derived from the expression

$$V_B = 20 \left[\frac{R_2}{R_1 + R_2} \right] = 6 \text{ V}$$

and

$$V_E = V_B - V_{BE} = 6 - 0.7 = 5.3 \text{ V}$$

Since I_C must be set at 2 mA, then

$$I_E = I_C + I_B = 2 \times 10^{-3} + 9.1 \times 10^{-6} \approx 2 \text{ mA}$$

And hence

$$R_E = \frac{V_E}{I_E} = \frac{5.3}{2 \times 10^{-3}} = 2.65 \text{ k}\Omega$$

Finally, we must select the proper value of R_C to set the Q-point at $V_{CE} = 6$ volts. Since

$$V_C = V_E + V_{CE} = 5.3 + 6 = 11.3 \text{ V}$$

Then

$$R_C = \frac{V_{CC} - V_C}{I_C} = \frac{20 - 11.3}{2 \times 10^{-3}} = 4.35 \text{ k}\Omega$$

The resistors R_E and R_C have now been determined to set the Q-point to satisfy the specified bias conditions. ∎

EXAMPLE 16.13

Given the circuit shown in Fig. 16.18(a), we wish to construct a small-signal equivalent circuit for the amplifier, compute values for all necessary small-signal parameters, and finally, calculate the small-signal voltage gain and the input resistance of the circuit. Use the transistor from Example 16.11.

Note that the bypass capacitor, C_E, provides an ac ground connection to the emitter effectively shorting R_E. Therefore, the ac equivalent circuit is shown in Fig. 16.18(b).

Every parameter in this circuit is known except the parameter r_{ib}, which can be easily calculated as

$$r_{ib} = \frac{0.026}{I_B} = \frac{0.026}{9.1 \times 10^{-6}} = 2.86 \text{ k}\Omega$$

By combining parallel resistances at both the input and output, this circuit can be simplified to that shown in Fig. 16.18(c).

From this equivalent circuit, we note that

$$v_o = -\beta i_b R_{Leq}$$

and

$$v_i = i_b r_{ib}$$

From these two equations we obtain the result previously expressed in Eq. (16.28).

$$A_v = \frac{v_o}{v_i} = -\frac{\beta R_{Leq}}{r_{ib}} \tag{16.33}$$

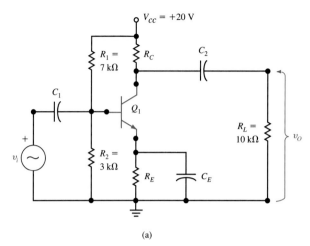

(a)

Figure 16.18(a)   BJT common emitter amplifier schematic circuit.

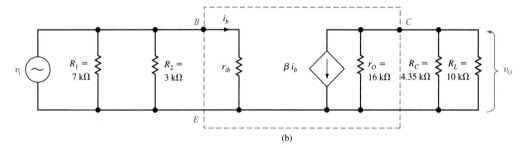

(b)

Figure 16.18(b)   BJT common emitter amplifier small-signal equivalent circuit.

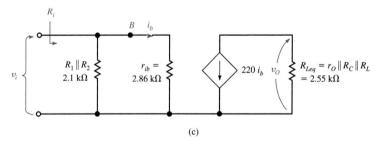

(c)

Figure 16.18(c)   Simplified BJT common emitter amplifier small-signal equivalent circuit.

This equation specifies the gain for any common emitter circuit in which the emitter is grounded. Using the values for this example

$$A_v = -\frac{(220)\,2.55\text{ k}}{2.86\text{ k}} = -196$$

The input resistance of the circuit, R_i, is given by

$$R_i = [R_1\|R_2]\|r_{ib} = 2.1\text{ k}\|2.86\text{ k} = 1.21\text{ k}\Omega$$

■

The Common Collector Amplifier

The common collector amplifier configuration is often called the *emitter follower* and either term refers to the circuit shown in Fig. 16.19(a). In this circuit, the input is applied to the base, the output is taken from the emitter, and the collector is at ac ground potential. This circuit is analogous to the common drain FET circuit, and thus has a voltage gain less than unity.

EXAMPLE 16.14

Let us determine the general expression for the voltage gain of the *CC* circuit shown in Fig. 16.19(a). The small-signal equivalent circuit for the amplifier is shown in Fig. 16.19(b). If we define the parallel combination of R_E and R_L as R_{Leq}, the loop equation for the input loop is

$$v_i = i_b(r_{ib}) + (\beta + 1)\, i_b\, R_{Leq}$$

and the expression for the output voltage is

$$v_o = (\beta + 1)\, i_b\, R_{Leq}$$

Dividing these two equations, we obtain the voltage gain A_v as

$$A_v = \frac{v_o}{v_i} = \frac{R_{Leq}}{R_{Leq} + \dfrac{r_{ib}}{\beta + 1}} = \frac{R_{Leq}}{R_{Leq} + r_{ie}} \tag{16.34}$$

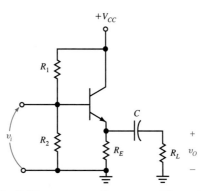

Figure 16.19(a)   BJT common collector amplifier schematic circuit.

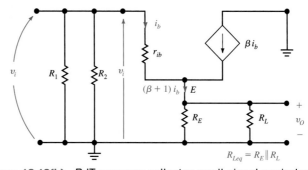

Figure 16.19(b)   BJT common collector small-signal equivalent circuit.

This value is always less than unity; however, since R_{Leq} is generally large compared to r_{ie}, the voltage gain of this circuit is often assumed to be approximately one. Recall r_{ie} is defined by Eq. 16.30. ∎

DRILL EXERCISE

D16.1. If the dc emitter current in the CC circuit is 2 mA and R_{Leq} is given as 1 kΩ, what is the value of r_{ie} and the voltage gain?

Ans: $r_{ie} = 13\ \Omega$ and $A_v = \dfrac{1000}{1013} = 0.99$

The Common Base Amplifier

The analysis of a common base amplifier is illustrated in the following example.

EXAMPLE 16.15

Assuming the PNP transistor has a beta of 50 and a V_{BE} of -0.7 V, let us determine (a) the dc collector current and the dc voltages on the emitter and collector, (b) the small-signal equivalent circuit, and (c) the small-signal voltage gain and the input resistance for the amplifier shown in Fig. 16.20 (a).

(a) Since the capacitor C_1 blocks dc current, the dc loop equation around the emitter base circuit shown in Fig. 16.20(a) is

$$V_{EE} = I_E R_E + V_{EB}$$

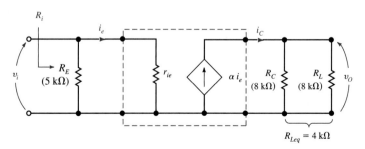

Figure 16.20(a) BJT common base amplifier schematic circuit.

Figure 16.20(b) BJT common base amplifier small-signal equivalent circuit.

Using the known parameter values and solving for I_E yields

$$I_E = 0.46 \text{ mA}$$

Since $\beta = 50$

$$\alpha = \frac{\beta}{\beta + 1} = \frac{50}{51} = 0.98$$

and then

$$I_C = \alpha I_E = 0.98 \,(0.46 \times 10^{-3}) = 0.45 \text{ mA}$$

Since the capacitor C_2 also blocks dc current, the collector voltage can be found from the equation

$$V_C = V_{CC} - R_C I_C = -10 - 8 \text{ k}(-0.45 \times 10^{-3}) = -6.4 \text{ V}$$

And, of course, the voltage on the emitter, V_E, is 0.7 V.

(b) The small-signal equivalent circuit for the amplifier is shown in Fig. 16.20(b), where the value of r_{ie} is

$$r_{ie} = \frac{0.026}{I_E} = \frac{0.026}{0.46 \times 10^{-3}} = 57 \ \Omega$$

(c) The model indicates that the equations for the input and output are

$$v_i = i_e \, r_{ie}; \qquad v_o = \alpha i_e \, R_{Leq}$$

Dividing these two equations yields the voltage gain

$$A_v = \frac{\alpha \, R_{Leq}}{r_{ie}} = \frac{(0.98)\,(4 \text{ k})}{57} = 69 \qquad\qquad (16.35)$$

The input resistance to the amplifier, R_i, is dominated by r_{ie}

$$R_i = R_E \| r_{ie} = 5000 \| 57 \approx 57 \ \Omega \qquad\qquad (16.36)$$

Eq. (16.35) may be used in general for *CB* circuits.   ∎

16.7
Summary of Important BJT Relationships

Table 16.2 summarizes the important relationships for the small-signal analysis of BJT circuits in the mid-band frequency region.

16.8
Summary

- A small-signal model is used when the signal magnitude is sufficiently small that non-linearities in the circuit can be approximated by linear elements.
- The mid-band frequency region is a range of frequencies high enough that all coupling and bypass capacitors may be considered as ac shorts, and NOT high enough for small device and parasitic capacitances to be a factor, that is, they are considered open.

Table 16.2 Important BJT Relationships for Small-Signal Analysis

CONFIGURATION	NOTES	CURRENT GAIN, A_i	VOLTAGE GAIN, A_V	INPUT RESISTANCE — To Transistor	INPUT RESISTANCE — To Circuit	OUTPUT RESISTANCE — Transistor	OUTPUT RESISTANCE — Circuit
COMMON EMITTER	$R_{Leq} = R_C \, \| \, R_L \, \| \, r_o$ R_B = equivalent resistance of base bias resistors	$A_i = \dfrac{i_c}{i_b} = \beta$	$A_v = \dfrac{v_o}{v_i}$ $A_v = -\dfrac{\beta R_{Leq}}{r_{ib}}$	$r_i = r_{ib} = \dfrac{kT}{qI_B} = \dfrac{0.026}{I_B}$ (room temp)	$R_i = R_B \, \| \, r_{ib}$	$r_o = \dfrac{1}{g_o}$	$R_o = R_C \, \| \, \dfrac{1}{g_o}$
COMMON BASE	$R_{Leq} = R_C \, \| \, R_L$ Assume $r_o \approx \infty$	$A_i = \dfrac{i_c}{i_e} = \alpha$ $\alpha = \dfrac{\beta}{\beta+1}$	$A_v = \dfrac{\alpha R_{Leq}}{r_{ie}}$	$r_i = r_{ie} = \dfrac{kT}{qI_E} = \dfrac{0.026}{I_E}$ (room temp) Also $r_{ie} = \dfrac{r_{ib}}{(\beta+1)}$	$R_i = R_E \, \| \, r_{ie}$	$r_o \approx \infty$	$R_o \approx R_C$
COMMON COLLECTOR	$R_{Leq} = R_E \, \| \, R_L$	$A_i = \dfrac{i_e}{i_b} =$ $A_i = (\beta+1)$	$A_v = \dfrac{R_{Leq}}{R_{Leq} + r_{ie}}$	$r_i = r_{ib} + (\beta+1)R_{Leq}$	$R_i = R_B \, \| \, r_i$	$r_o = r_{ie} + \dfrac{R_B \| R_S^*}{\beta+1}$	$R_o \approx R_E \, \| \, r_o$

$^*R_S$ = resistance associated with source, v_i

556

- The ac gain is generally constant and at its maximum over the mid-band frequency range.
- A *p–n* diode has a dynamic or ac resistance, r_d; it is the inverse of the slope of a line tangent to the *Q*-point on the diode curve.
- At room temperature

$$r_d = \frac{0.026}{I_D} \ \Omega$$

- Two-port linear networks may be used to model small-signal amplifiers. The voltage and current for both the input and output are related by a set of two equations and four parameters. There are several parameter sets commonly used, for example, *y* parameters, *z* parameters, and *h* parameters.
- A simplified small-signal model can be used effectively for modeling transistors in the mid-band range.
- The MOSFET small-signal model at mid-band frequencies has two dominant parameters, g_m and g_o.
- The small-signal voltage gain of a common source MOSFET amplifier is $A_v = -g_m R_{Leq}$, where R_{Leq} is the total effective load resistance including g_o.
- Small-signal analysis begins with a dc bias analysis in order to calculate values for the transistor's small-signal parameters; then a small-signal equivalent circuit is developed and the parameter values are used in the model.
- The small-signal voltage gains and other amplifier characteristics for all three MOSFET circuit configurations, that is, common source, common gate, and common drain, are developed and summarized in Table 16.1.
- If an amplifier has input resistance R_i and is driven from a voltage source with an internal resistance R_s, then if $R_i >> R_s$ we avoid significant loss of voltage gain at the input.
- Two small-signal models for the BJT are developed, one more appropriate for common emitter and common collector configurations, and the other for common base configurations.
- The BJT small-signal model for mid-band frequencies contains three parameters, input resistance, current gain, and output resistance. Output resistance can be ignored when it is high compared to the external load resistances.
- The small-signal voltage gains and other amplifier characteristics for all three BJT circuit configurations, that is, common emitter, common base, and common collector, are developed and summarized in Table 16.2.

PROBLEMS

16.1. Define the mid-band frequency region. In this region what can be said about an amplifier's voltage gain?

16.2. A $p-n$ junction diode has a forward bias applied of 0.68 V and is conducting 3 mA. (a) Determine I_o. (b) Determine the static forward resistance. (c) Determine the ac small-signal resistance of the diode at room temperature.

16.3. Determine the ac resistance of a silicon diode at room temperature which is conducting 3 μA.

16.4. Two silicon diodes are connected in series and are conducting a dc current of 1 mA; what is the ac resistance of the combination?

16.5. A two-port network as illustrated in Fig. 16.2 has the following parameters: $y_{11} = 0$; $y_{12} = 0$; $y_{21} = 3.5 \times 10^{-3}$ S; $y_{22} = 1 \times 10^{-4}$ S. If an external load resistance of 10 kΩ is connected to the output, what is the voltage gain of this network?

16.6. A set of h parameters are provided as follows: $h_{11} = 20$ kΩ; $h_{12} = 0$; $h_{21} = 100$; and $h_{22} = 1 \times 10^{-5}$. If the output is connected to an external load resistance of 50 kΩ, determine: (a) the network's input resistance, (b) the network's current gain, (c) the network's voltage gain, (d) the network's output resistance.

16.7. A MOSFET in the common source configuration is modeled as a two-port network as illustrated in Fig. 16.4. If $R_L = 8$ kΩ, and $y_{11} = 0$; $y_{12} = 0$; $y_{21} = 2 \times 10^{-3}$ S; and $y_{22} = 0$, determine the small-signal voltage gain of the amplifier.

16.8. Assume the simplified small-signal model for transistors shown in Fig. 16.5 has $r_i = \infty$, $g_m = 1.8 \times 10^{-3}$ S, and $r_o = 100$ kΩ. This device is placed in the circuit as shown in Fig. P16.8. Calculate the small-signal voltage gain.

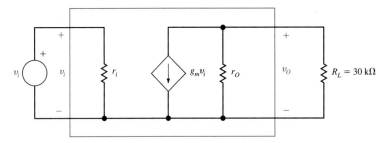

Figure P16.8

16.9. For the n-channel enhancement-mode MOSFET with output curves of Fig. P16.9, determine an approximate value for g_m at (1) Q-point Q_{P1}, and (2) Q-point Q_{P2}.

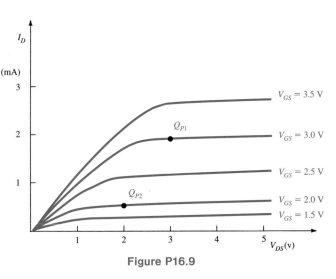

Figure P16.9

16.10. For the MOSFET with output curves in Fig. P16.9 (a) Determine the approximate value of g_o for each Q-point. (b) Redraw the output curves of Fig. P16.9 for a device with approximately the same g_m values, but $g_o = 1 \times 10^{-4}$ S.

16.11. Consider the common source circuit shown in Fig. 16.11(a) with a transistor which has output curves of Fig. P16.9, and (a) determine a new value for R_2 that will place the Q-point on the $V_{GS} = 2.5$ V curve. (b) Plot the dc load line to locate the Q-point. (c) Compute an approximate value for g_m at this Q-point. (d) Draw a small-signal equivalent circuit for this amplifier. (e) Using the equivalent circuit, calculate the small-signal voltage gain.

16.12. The common gate n-channel enhancement mode MOSFET circuit shown in Fig. 16.12(a) has new component values as follows: $R_S = 1$ kΩ, $R_D = 3$ kΩ, and g_m for the transistor is determined to be 2×10^{-3} S; assume r_o is infinite. (a) Draw a small-signal equivalent circuit. (b) Compute the small-signal voltage gain. (c) Determine the amplifier input resistance, R_i.

16.13. The common drain amplifier circuit shown in Fig. P16.13 utilizes a transistor with $g_m = 3.0$ mS and $R_S = 3$ kΩ. (a) Compute the small-signal voltage gain. (b) Compute the voltage gain if a new transistor is used with $g_m = 30$ mS.

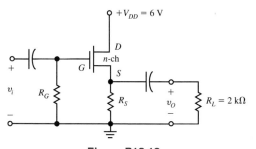

Figure P16.13

16.14. If a transistor amplifier is driven by a voltage source, v_s, with internal resistance, $R_s = 50$ kΩ, what is the gain of the input circuit, A_{vs}, for the following: (note: since $A_{vs} < 1$, this is a loss in signal gain) $R_i = 50$ kΩ, $R_i = 100$ kΩ, $R_i = 500$ kΩ.

16.15. The common emitter small-signal model for the BJT is given in Fig. P16.15. This transistor is placed in the active region by a dc bias circuit where the measured dc currents are $I_C = 2$ mA and $I_B = 16$ μA; when V_{CE} was increased 4 V, I_C increased 0.2 mA. (a) Determine the small-signal model parameters for this model. (b) Construct a small-signal equivalent circuit for the amplifier, assuming the total external load resistance is 2 kΩ. (c) Calculate the small-signal voltage gain.

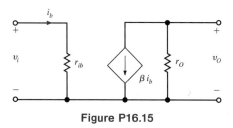

Figure P16.15

16.16. Repeat problem 16.15 if $I_B = 8$ μA.

16.17. In a common emitter amplifier circuit (emitter grounded), the dc base current I_B is 10 μA, and the transistor has $\beta = 100$ and $r_o = 80$ kΩ. (a) Draw an ac small-signal equivalent circuit of the amplifier if the external load resistance is 15 kΩ; the parallel combination of the base bias resistors is 4 kΩ. (b) Calculate the small-signal voltage gain of the circuit. (c) Determine the circuit's input and output resistance.

16.18. Given the circuit in Fig. P16.18, (a) determine the dc bias conditions; assume $\beta = 120$, and $V_{BE} = 0.7$ V. (b) Construct a mid-band ac equivalent circuit. (b) Determine the small-signal voltage gain.

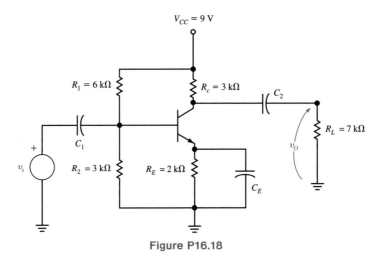

Figure P16.18

16.19. A common emitter amplifier circuit has a transistor with $\beta = 50$ and $r_{ib} = 2$ kΩ. If the circuit provides $R_{Leq} = 6$ kΩ, what is the small-signal voltage gain?

16.20. A transistor with $\beta = 30$ is placed in a common base amplifier circuit with $I_E = 1$ mA and $R_{Leq} = 5$ kΩ. (a) What is the small-signal voltage gain? (b) Does this circuit invert the signal?

16.21. Sketch the small-signal common base BJT model, assuming the output resistance is high enough to be ignored. If $r_{ie} = 45$ Ω, and $\alpha = 0.98$, (a) determine the dc emitter current, I_E, in the bias circuit. (b) Find the dc base current, I_B. (c) Construct a small-signal equivalent circuit for an amplifier using this model and with v_s as the voltage input source, $R_s = 50$ Ω, $R_{Leq} = 2$ kΩ, and assume $R_i = r_{ie}$. (d) Determine the small-signal voltage gain, including any loss in the input circuit; that is, find v_o/v_s.

16.22. A transistor is placed in a common collector circuit with $I_E = 2$ mA and $R_{Leq} = 500$ Ω. Find the small-signal voltage gain.

16.23. The circuit of Fig. P16.23 uses a transistor with $\beta = 120$ and $V_{BE} = -0.6$ V. Find the small-signal voltage gain.

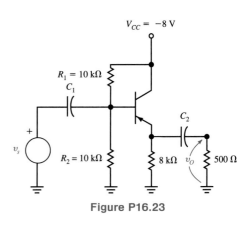

Figure P16.23

16.24. A transistor with a measured $\beta = 60$ is placed in the circuit of Fig. P16.24; assume $V_{BE} = 0.7$ V. (a) Construct an ac equivalent circuit. (b) Determine the small-signal voltage gain.

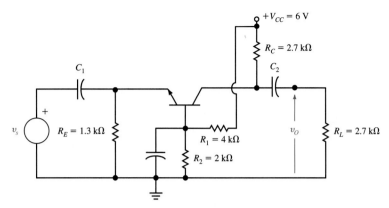

Figure P16.24

16.25. A common base amplifier uses a transistor with $\alpha = 0.97$. If I_E is measured to be 1.3 mA, and the total effective load resistance $R_{Leq} = 2$ kΩ, what is the small-signal voltage gain?

Advanced Analog Circuit Concepts and Applications

17.1

Introduction

There are additional concepts that are important for the design and application of analog systems. There is often the requirement for more gain than is possible from a single-stage amplifier, such as those described in the previous chapters. A single-stage amplifier is one which utilizes a single transistor for gain. Therefore, most analog systems incorporate multi-stage amplifiers in which several single-stage amplifiers are connected together to achieve the required performance.

There is also the issue of frequency response of an analog system. Usually an amplifier has maximum gain over the range of frequencies defined as the mid-band frequency region. There is a frequency below which the gain decreases, and there is also a high-frequency limitation on the gain. In this chapter, the low and high corner frequencies, also called half-power points, will be defined; they describe the frequency limits of the mid-band region.

The differential amplifier (DA) is discussed in detail, and differential amplifier parameters and specifications are given with various illustrative circuits. The comparator circuit is described as a combination of a DA input and a digital output. Various oscillator circuits are presented which utilize an amplifier for gain combined with circuitry to provide positive feedback.

Finally, there are special applications in which a relatively large amount of power must be delivered to the output load. The special output circuits designed for delivering such power will also be described.

17.2
Multi-Stage Amplifiers

A *stage* is defined as an amplifier circuit utilizing one transistor, or other gain element such as a vacuum tube, for amplification. If the system requires more gain than that possible from a single-stage amplifier, a multi-stage amplifier must be used. A *multi-stage amplifier* is one in which the signal is passed sequentially through more than one stage, and each stage, in turn, increases the signal gain.

Figure 17.1 shows as an illustration a three-stage amplifier circuit where each of the three stages is represented by the amplifier symbols, A_1, A_2, and A_3. If the mid-band voltage gain of each of these stages is given as follows:

$$A_{v1} = 20$$
$$A_{v2} = 30$$
$$A_{v3} = 40$$

then the mid-band voltage gain of the entire circuit is the product of the voltage gains of each of the stages, and hence

$$A_{v(total)} = A_{v1}A_{v2}A_{v3} = (20)(30)(40) = 24,000$$

This relationship is easily derived since

$$A_{v(total)} = \frac{v_{out}}{v_{in}} = \frac{v_{out}}{v_b}\frac{v_b}{v_a}\frac{v_a}{v_{in}} = A_{v3}A_{v2}A_{v1}$$

In general, the total voltage gain of a multi-stage amplifier is the product of the individual voltage gains of each of the stages. Therefore

$$A_{v(total)} = A_{v1}A_{v2}\ldots\ldots\ldots\ldots A_{vN} \qquad (17.1)$$

where N is the total number of stages in the multi-stage amplifier. An analogous relation applies for current gain, A_i.

EXAMPLE 17.1

A two-stage voltage amplifier is designed to have a total voltage gain of 1,000. Let us determine the gain of the second stage if the first stage has a gain of 25.

The total voltage gain of 1,000 is the product of 25 and the unknown gain of the second stage. Therefore

$$1,000 = 25\,A_{v2}$$

Solving for A_{v2} we obtain

$$A_{v2} = 40 \qquad\qquad \blacksquare$$

EXAMPLE 17.2

Consider the circuit shown in Fig. 17.2. This is a circuit schematic of a two-stage amplifier consisting of a common base transistor stage followed by a common emitter stage. If

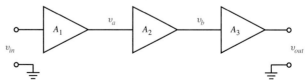

Figure 17.1 A three-stage amplifier.

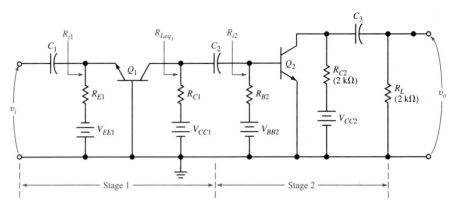

Figure 17.2 A two-stage BJT amplifier schematic circuit (CB-CE).

the voltage gain of the first stage is 50 and that of the second stage is -30, let us determine the voltage gain of the circuit.

The voltage gain of the circuit is

$$A_{v\ (total)} = (50)(-30) = -1500 \qquad ■$$

In Example 17.2, the values for the voltage gain of each stage were given. However, if only circuit component values are known, then we would calculate the gain of each stage, using the procedures described in the previous chapter. In solving such problems, it is generally easier to start at the output of the circuit and work backwards toward the input. The equivalent load resistance for the second stage is $R_{Leq2} = R_L \| R_{C2}$ and therefore, A_{v2} can be determined. However, the equivalent load resistance for the first stage is $R_{Leq1} = R_{i2} \| R_{C1}$, where R_{i2} is the input resistance to the second stage.

DRILL EXERCISES

D17.1. If the voltage gain of the first stage of a three-stage amplifier is doubled and that of the third stage is tripled, by what factor is the total voltage gain of the amplifier changed?

Ans: 6.

D.17.2. In Fig. 17.2, if R_L and R_{C2} are each 2 kΩ, what is the value of dc emitter current (I_{E2}) through Q_2 required to make the mid-band gain of the second stage -30?

Ans: 0.78 mA.

The gain of amplifiers is often a large number, especially in multistage amplifiers. Therefore, it becomes convenient to express the voltage gain in *decibels* (db), using the following relation.

$$gain\ in\ decibels = 20 \log \left[\frac{v_o}{v_i} \right] = 20 \log [A_v] \qquad (17.2)$$

This expression provides a logarithmic measure of amplifier gain. Thus, for multi-stage amplifiers, the gain of the entire amplifier is the sum (in db) of the gains of each stage.

Current gain can also be expressed in decibels using an analogous equation.

EXAMPLE 17.3

If a two-stage amplifier has a first-stage voltage gain of 100 and a second-stage gain of 1,000, let us express the gains of each stage in decibels and find the total gain.

$$A_{v1} = 100; \ 20 \log(100) = 40 \ \text{db}$$
$$A_{v2} = 1,000; \ 20 \log(1,000) = 60 \ \text{db}$$

And hence the gain of the entire amplifier is the SUM of the gains expressed in db:

$$A_{v(total)} = 40 + 60 = 100 \ \text{db}$$

As a check, let us calculate the total gain as the PRODUCT of the gains expressed as voltage ratios:

$$A_{v(total)} = (100)(1,000) = 100,000$$

Note that 100,000 expressed in db is 100 db, as expected. ■

Another advantage of expressing gains in decibels is that plots of amplifier gain (in db) versus the log of frequency result in very convenient graphs for analyzing amplifer performance. Since the gains in db are additive, plots of individual stages can be added graphically to obtain composite performance. It has been shown that the frequency response of RC networks results in straight-line segments on such plots. In the following sections we will make use of such plots to show the frequency dependence of the gain of analog amplifier stages.

17.3
Frequency Response

The gain of an amplifier in its mid-band frequency region is at its maximum magnitude. Because of reactances, typically capacitances, present in the amplifier circuit, the gain decreases or falls off below a frequency designated as the *lower corner frequency, f_{LO}*. Similarly, there is also an *upper corner frequency, f_{HI}*, above which the gain also falls off. These *corner frequencies,* also known as *break frequencies, cutoff frequencies, −3 db frequencies,* or *half-power points,* are defined as the frequencies at which the gain has decreased 3 db, or a factor of $1/\sqrt{2} = 0.707$, from its maximum or mid-band value.

In summary, we will use the designations:

$$f_{LO} = \text{lower corner frequency}$$
$$f_{HI} = \text{upper corner frequency}$$

Therefore, the typical frequency response for the voltage gain of an amplifier circuit appears as shown in Fig. 17.3(a). Note that the voltage gain is plotted on a logarithmic scale, that is, linear in decibels or db. The actual amplifier response follows the smooth curve which asymptotically approaches the straight-line segments with breaks at the corner frequencies. The voltage gain in the mid-band frequency region is defined as A_v.

The relative phase angle of the output signal compared to the input signal is also often plotted on the same graph. Over the central part of the mid-band frequency region, the phase shift is either zero or 180 degrees, depending on whether the amplifier is noninverting or inverting. At the edges of the mid-band range, RC networks begin to alter both the gain and phase of the output signal.

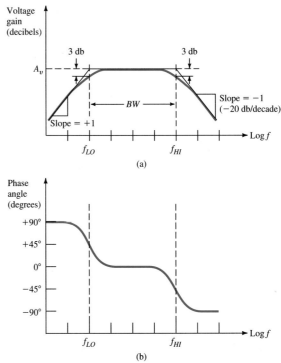

Figure 17.3   Frequency response of the (a) gain, and (b) phase for an amplifier (a Bode plot).

At the corner frequencies, the phase angle will be changed by 45 degrees from its mid-band value, as shown in Fig. 17.3(b). A plot of gain (in db) and phase angle (in degrees) on the same graph (versus log of frequency) is called a *Bode plot*.

The lower corner frequency is generally a result of the frequency dependence of *RC* networks involving circuit resistances and coupling or bypass capacitors.

The upper corner frequency is generally a result of the frequency dependence of *RC* networks involving circuit resistances and device junction and parasitic capacitances.

The *bandwidth, BW,* is defined as the difference between the two corner frequencies. Therefore

$$BW = f_{HI} - f_{LO} \qquad (17.3)$$

In the following sections, calculation of these corner frequencies, which define the normal frequency bounds of useful operation, will be illustrated.

17.4
Amplifier Low-Frequency Response

The low-frequency response of an amplifier stage is determined by evaluating the effect of various *RC* networks that generally are present at the input, output, and other circuit nodes. Thus, the total frequency response is affected by each coupling or bypass capacitor. Each of these networks generally acts as single-pole high-pass filters and therefore

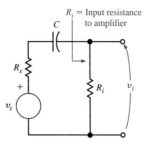

Figure 17.4   Amplifier input circuit.

limits the low-frequency response of the circuit. Coupling capacitors normally link the signal to the input and output of an amplifier. Furthermore, in a multi-stage amplifier, they couple the signal from the output of one stage to the input of the next.

An input circuit with a coupling capacitor is shown in Fig. 17.4. Standard ac circuit analysis can be applied in analyzing such a circuit. The magnitude of the voltage gain of this input stage is given by

$$\left| \frac{v_i}{v_s} \right| = \frac{R_i}{\sqrt{(R_s + R_i)^2 + \left(\frac{1}{\omega C} \right)^2}} \tag{17.4}$$

The magnitude of this function is reduced to 0.707 of its maximum value, that is, a decrease of 3 db when

$$\omega = \omega_{LO} = \frac{1}{C(R_s + R_i)} \tag{17.5}$$

or

$$f_{LO} = \frac{1}{2\pi} \frac{1}{C(R_s + R_i)} \tag{17.6}$$

This is the frequency at which the response of the network is decreased by 3 db at the lower end of the mid-band range.

In general, the corner frequencies of coupling and bypass networks are determined by first calculating an RC time constant, τ_{LO}, defined as

$$\tau_{LO} = R_{eq}C \tag{17.7}$$

where C is the capacitor under consideration, and R_{eq} is the equivalent or total effective resistance across the capacitor.

The corner frequency associated with each τ_{LO} is given by

$$f_{LO} = \frac{1}{2\pi \, \tau_{LO}} \tag{17.8}$$

A typical transistor amplifier stage may have two coupling capacitors, one at the input and one at the output, and a bypass capacitor. A lower corner frequency is computed for each of these capacitors in the circuit and the HIGHEST lower corner frequency is the one that dominates in determining the overall amplifier response.

EXAMPLE 17.4

Given the two-stage amplifier circuit shown in Fig. 17.2, we wish to write an expression for the lower corner frequencies associated with each of the coupling capacitors, C_1, C_2, and C_3.

If we ignore the output resistance of the transistors because it is so large, and assume that v_{in} is a voltage source with zero internal resistance, then

$$\tau_{LO(C1)} = C_1 R_{i1} = C_1[R_{E1} \| r_{ie1}] \tag{17.9}$$

where

$$r_{ie1} = \frac{0.026}{I_{E1}} \tag{17.10}$$

$$\tau_{LO\,(C2)} = C_2[R_{C1} + R_{i2}] = C_2[R_{C1} + (R_{B2} \| r_{ib2})] \tag{17.11}$$

where

$$r_{ib2} = \frac{0.026}{I_{B2}} \tag{17.12}$$

and

$$\tau_{LO\,(C3)} = C_3[R_{C2} + R_L] \tag{17.13}$$

From these time constants, the corresponding corner frequencies can be calculated as

$$f_{LO(C1)} = \frac{1}{2\pi\ \tau_{LO(C1)}} \tag{17.14}$$

$$f_{LO(C2)} = \frac{1}{2\pi\ \tau_{LO(C2)}} \tag{17.15}$$

$$f_{LO(C3)} = \frac{1}{2\pi\ \tau_{LO(C3)}} \tag{17.16}$$

The highest of the three frequencies above would dominate in determining the amplifier's performance. ∎

EXAMPLE 17.5

Let us determine the lower corner frequency associated with C_3 in the circuit of Fig. 17.2, if $C_3 = 2\ \mu F$, and $R_{C2} = R_L = 2\ k\Omega$. The time constant associated with this capacitor is

$$\tau_{LO(C3)} = C_3[R_{C2} + R_L] = [2 \times 10^{-6}][4\ k] = 8 \times 10^{-3}\ \text{sec}$$

and therefore

$$f_{LO(C3)} = \frac{1}{2\pi\tau_{LO(C3)}} = \frac{1}{2(3.14)(8 \times 10^{-3})} = 19.9\ \text{Hz} \qquad \blacksquare$$

EXAMPLE 17.6

We wish to find the lower corner frequency for the common emitter circuit in Fig. 17.5, assuming that beta is 200 and the circuit is driven by a source with a resistance of 5 kΩ.

There are two capacitors that will affect the low-frequency performance, coupling capacitor C_1 and bypass capacitor C_E. We will determine the time constants associated with each and then the corresponding corner frequency. The highest f_{LO} will dominate the circuit response. A dc bias calculation will show that $I_E = 1.5$ mA.

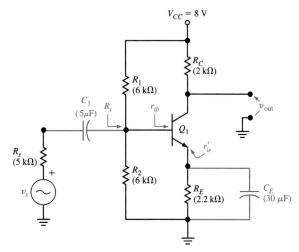

Figure 17.5 Common emitter amplifier schematic circuit.

When calculating the time constant associated with any particular capacitor, we assume the other capacitors act as ac shorts.

For C_1

$$\tau_{LO(C1)} = C_1[R_s + R_i]$$

or

$$\tau_{LO(C1)} = C_1[R_s + [(R_1\|R_2)\|r_{ib}]$$

$$I_B = \frac{I_E}{(\beta + 1)} = \frac{1.5 \times 10^{-3}}{201} = 7.46 \ \mu A$$

and therefore

$$r_{ib} = \frac{0.026}{I_B} = \frac{0.026}{7.46 \times 10^{-6}} = 3.48 \ k\Omega$$

$\tau_{LO(C1)}$ is then

$$\tau_{LO(C1)} = (5 \times 10^{-6})[5 \ k + (3 \ k\|3.48 \ k)] = 3.3 \times 10^{-2} \ \text{sec}$$

and therefore the corner frequency, $f_{LO(C1)}$, is

$$f_{LO(C1)} = \frac{1}{2\pi\tau_{LO(C1)}} = \frac{1}{2\pi \ (3.3 \times 10^{-2})} = 4.8 \ \text{Hz}$$

For C_E

$$\tau_{LO(CE)} = C_E[R_E\|r'_{ie}]$$

where r'_{ie} is the ac resistance seen looking into the emitter of the device. As shown in the figure, r'_{ie} is the sum of two terms: (1) the ac resistance of the emitter junction (r_{ie}), plus (2) the effective resistance from the base terminal to ground divided by $\beta + 1$.

All resistances at the base, when viewed from the emitter, are divided by $\beta + 1$ because of the fixed current ratio forced by the transistor. Correspondingly, any resistance from the emitter to ground seen from the base will be multiplied by $\beta + 1$.

Therefore

$$r'_{ie} = r_{ie} + \frac{(R_1 \| R_2 \| R_s)}{\beta + 1}$$

where

$$r_{ie} = \frac{0.026}{I_E}$$

These two equations, with the corresponding parameter values, yield

$$r_{ie} = 17.3 \ \Omega$$

and

$$r'_{ie} = 17.3 + 9.3 = 26.6 \ \Omega$$

and therefore

$$\tau_{LO(CE)} = (30 \times 10^{-6})(26.6 \| 2.2 \ \text{k}) = (30 \times 10^{-6})(26.3) = 7.89 \times 10^{-4} \ \text{sec}$$

and

$$f_{LO(CE)} = 202 \ \text{Hz}$$

The lower corner frequency of the circuit is dominated by the highest of the lower corner frequencies; since $f_{LO(CE)} \gg f_{LO(C1)}$, the corner frequency of the circuit is approximately 202 Hz.   ∎

Analogous methods may be used in low-frequency analysis of MOSFET circuits.

17.5
The Miller Effect

The *Miller effect* is a very important concept in the ac analysis of inverting amplifiers. It will be used in the next section for determining the high-frequency response of amplifiers, but is generally applicable to any situation where a feedback capacitor is connected from the output back to the input of an inverting amplifier. Such a situation will be shown to produce the equivalent effect of a much larger capacitance at the input of the amplifier.

Figure 17.6 illustrates an inverting amplifier with voltage gain, $-A_v$, across which a feedback capacitor, C_f, has been connected between the output and the inverting input. The effective capacitance looking into the input of this amplifier will be termed the *input Miller capacitance, C_{mi}*. Let us now illustrate how to determine the value of this capacitance.

From Fig. 17.6 we note that

$$v_o = A_v v_i \tag{17.17}$$

and

$$i_i = i_c \tag{17.18}$$

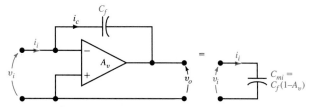

Figure 17.6   Inverting amplifier and equivalent Miller input capacitance.

where A_v is negative because we are using an inverting amplifier. In addition, the voltage across the capacitor C_f is

$$v_c = \frac{1}{C_f} \int i_c dt = v_i - v_o \qquad (17.19)$$

Combining these equations, and solving for v_i, we obtain

$$v_i = \frac{1}{C_f(1 - A_v)} \int i_i dt \qquad (17.20)$$

This expression indicates that at the input, the relationship between voltage and current is determined by a larger equivalent capacitor, C_{mi}, where

$$C_{mi} = C_f(1 - A_v) \qquad (17.21)$$

If A_v is a large negative number, as is the case for a high gain inverting amplifier, the C_{mi} can become quite large.

In the circuit of Fig. 17.6 there is also an equivalent *Miller output capacitance, C_{mo}*; this is the effective capacitance seen looking into the output of the circuit. A similar analysis will show that, for $A_v \gg 1$

$$C_{mo} = C_f \left[\frac{1}{A_v} + 1 \right] \approx C_f \qquad (17.22)$$

EXAMPLE 17.7

In the network in Fig. 17.6, if the feedback capacitor, C_f, is 25 pF, and the magnitude of the inverting voltage gain, A_v, is 1,000, that is, $A_v = -1,000$, let us determine (a) the equivalent input capacitance, and (b) the equivalent capacitance at the output of this circuit.

(a) The input capacitance is

$$C_i = C_{mi} = C_f(1 - A_v) = (25 \times 10^{-12})(1001) \approx 25 \times 10^{-9} = 0.025 \ \mu F$$

(b) The capacitance at the output of this circuit is

$$C_{mo} = C_f \left[\frac{1}{1001} + 1 \right] \approx C_f = 25 \ pF \qquad \blacksquare$$

17.6
Amplifier High-Frequency Response

At mid-band frequencies (and higher), the coupling and bypass capacitors are assumed to have reactances that are negligibly small and are therefore considered as ac shorts. However, as frequency is increased, other capacitances begin to have an effect. The capacitances that limit high-frequency operation are the small device capacitances, that is, those between adjoining structures in the transistor, and other small parasitic capacitances. These small device capacitances are generally measured in picofarads (pF), while the coupling and bypass capacitors are generally in the range of microfarads (μF).

We have not yet included these small device and parasitic capacitances in our transistor model because at both low frequencies and the mid-band region it isn't necessary. At these frequencies the reactance of these capacitors is considered infinitely high, or an ac open circuit. In this section we will modify our transistor models to include the device capacitances and then utilize these new models to analyze the high-frequency behavior of transistor circuits.

The procedure to be followed here is very similar to that used previously for low-frequency analysis:

1. We will construct a high-frequency equivalent circuit which contains the relevant *RC* networks by considering all coupling and bypass capacitors as ac shorts, and voltage supply nodes as ac grounds.
2. For each node in the signal path we will determine an *RC* time constant, τ_{HI}, and its corresponding upper corner frequency, f_{HI}.
3. The LOWEST upper corner frequency will dominate in determining the high-frequency limitation of the circuit.

High-Frequency Model For MOSFETs

The high-frequency circuit model for a MOSFET is shown in Fig. 17.7(a). Note that the transistor model is modified from that previously used by adding two capacitors, one from gate to source (C_{gs}) and another from gate to drain (C_{gd}). The capacitance from drain to

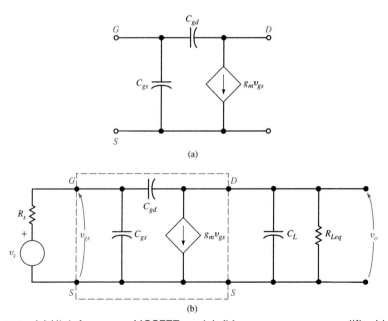

Figure 17.7 (a) High-frequency MOSFET model; (b) common source amplifier high-frequency equivalent circuit.

source is, in general, negligibly small and is not included in the model. The output resistance of the transistor is assumed here to be high enough to be ignored.

Common Source Amplifier

In Fig. 17.7(b) the high-frequency model is used to represent the MOSFET in a common source amplifier circuit. Recall from Chapter 16 that the mid-band voltage gain of this circuit is

$$A_v = -g_m R_{Leq} \tag{17.23}$$

Using this relation, and the Miller effect, we can redraw the high-frequency equivalent circuit as shown in Fig. 17.8, where

$$C_{mi} = C_{gd}(1 + g_m R_{Leq}) \tag{17.24}$$

and

$$C_{mo} \approx C_{gd} \tag{17.25}$$

The high-frequency circuit model also contains a load capacitance, C_L, in parallel with R_{Leq}. Most real loads have a capacitive component, even if it results only from the capacitance of the wire or cable from the amplifier outuput to the load.

From Fig. 17.8, we can now calculate the RC time constants associated with the input (Gate) and output (Drain) nodes. At the gate, the capacitances C_{gs} and C_{mi} are in parallel and are simply added together. R_s is the resistance across this capacitance, and therefore

$$\tau_{HI(G)} = (C_{gs} + C_{mi})(R_s) \tag{17.26}$$

and

$$f_{HI(G)} = \frac{1}{2\pi\tau_{HI(G)}} \tag{17.27}$$

For the drain

$$\tau_{HI(D)} = (C_L + C_{mo})(R_{Leq}) \tag{17.28}$$

and

$$f_{HI(D)} = \frac{1}{2\pi\tau_{HI(D)}} \tag{17.29}$$

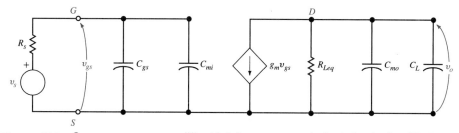

Figure 17.8 Common source amplifier high-frequency equivalent circuit simplified with Miller capacitances.

The lowest of these upper corner frequencies will dominate the circuit's high-frequency response.

EXAMPLE 17.8

A MOSFET with the following parameters: $g_m = 6 \times 10^{-3}$ S, $C_{gs} = 1.5$ pF, and $C_{gd} = 1.5$ pF is placed in a common source amplifier circuit like that of Fig. 17.8, in which $R_{Leq} = 5.2$ kΩ, the load capacitance $C_L = 3$ pF, and the source resistance $R_s = 3$ kΩ.

Let us determine the upper corner frequency, f_{HI}, for the circuit.

In order to determine the upper corner frequency, we first determine the Miller capacitances

$$C_{mi} = C_{gd}(1 + g_m R_{Leq}) = 1.5 \times 10^{-12}(1 + 31.2) = 48 \text{ pF}$$

and

$$C_{mo} \approx C_{gd} = 1.5 \text{ pF}$$

Now the time constants for both gate and drain are

$$\tau_{HI(G)} = (C_{gs} + C_{mi})R_s = (1.5 + 48.3)(10^{-12})(3 \text{ k}) = 1.49 \times 10^{-7} \text{ sec}$$

and

$$\tau_{HI(D)} = (C_{mo} + C_L)(R_{Leq}) = (1.5 + 3)(10^{-12})(5.2 \text{ k}) = 2.34 \times 10^{-8} \text{ sec}$$

Finally, the high-frequency corners are

$$f_{HI(G)} = \frac{1}{2\pi(1.49 \times 10^{-7})} = 1.07 \text{ MHz}$$

and

$$f_{HI(D)} = \frac{1}{2\pi(2.34 \times 10^{-8})} = 6.8 \text{ MHz}$$

The circuit performance is limited by the lowest of these corner frequencies, $f_{HI(G)} = 1.07$ MHz. ∎

Common Gate Amplifier

EXAMPLE 17.9

The high-frequency equivalent circuit of a common gate MOSFET circuit is shown in Fig. 17.9(a). Assume the transistor used is that in Example 17.8 and the circuit has the same source and load characteristics. Let us determine the lower and upper corner frequencies of this amplifier, and compare its voltage gain to that of the amplifier in the previous example.

In the common gate configuration, there is no capacitance in the feedback position and therefore no Miller capacitance. Hence, as shown in Fig. 17.9(b), for the source

$$\tau_{HI(S)} = C_{gs}\left[R_s \| \frac{1}{g_m}\right] = (1.5 \times 10^{-12})(158) = 2.37 \times 10^{-10} \text{ sec}$$

and

$$f_{HI(S)} = \frac{1}{2\pi(2.37 \times 10^{-10})} = 671 \text{ MHz}$$

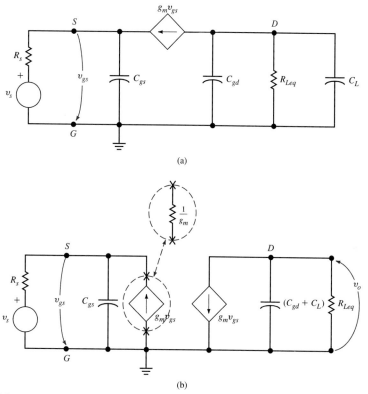

Figure 17.9 (a) High-frequency equivalent circuit for a common gate MOSFET amplifier; (b) simplified high-frequency equivalent circuit for a common gate MOSFET amplifier.

For the drain

$$\tau_{HI(D)} = (C_{gd} + C_L)(R_{Leq}) = (1.5 + 3)(10^{-12})(5.2\ k) = 2.34 \times 10^{-8}\ sec$$

Comparing time constants, we note that the upper corner frequency of this circuit, f_{HI}, is limited by the time constant, $\tau_{HI(D)}$, since $\tau_{HI(D)} \gg \tau_{HI(S)}$. Therefore

$$f_{HI} = f_{HI(D)} = \frac{1}{2\pi(2.34 \times 10^{-8})} = 6.8\ \text{MHz}$$

Note that the bandwidth of this amplifier is larger than that of the common source amplifier analyzed in Example 17.8. ∎

High-Frequency Model For BJTs

The high-frequency model for the BJT is shown in Fig. 17.10(a). Note that this BJT model is basically the mid-band model which has been modified by the addition of two capacitors, one representing the capacitance of the base emitter junction, (C_{be}), and the other representing the capacitance of the base collector junction, (C_{bc}). The capacitance from emitter to collector is small and usually ignored. This model may also include r_o and series base resistance and is called the hybrid-π model.

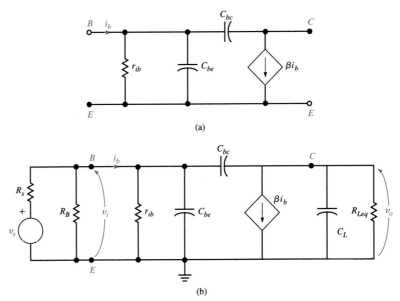

Figure 17.10 (a) High-frequency BJT small-signal model; (b) high-frequency equivalent circuit of common emitter BJT amplifier.

Common Emitter Amplifier

In Fig. 17.10(b), this new BJT model is placed in a high-frequency equivalent circuit for a common emitter BJT amplifier.

The common emitter amplifier is an inverting amplifier, and the capacitor, C_{bc}, is in the feedback position. Therefore, the Miller capacitances can be calculated and the circuit simplified to the form shown in Fig. 17.11. In this circuit the input (Base) and output (Collector) nodes are separated and a high-frequency time constant can be calculated for each. First, C_{mi} and C_{mo} will be calculated.

Recall that the mid-band voltage gain for a common emitter amplifier is given by Eq. (16.33)

$$A_v = -\frac{\beta R_{Leq}}{r_{ib}} \tag{17.30}$$

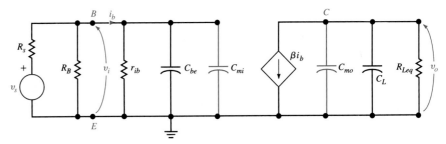

Figure 17.11 Equivalent circuit of common emitter amplifier with Miller capacitances.

Therefore

$$C_{mi} = \left(1 + \frac{\beta R_{Leq}}{r_{ib}}\right) C_{bc} \tag{17.31}$$

and

$$C_{mo} \approx C_{bc} \tag{17.32}$$

The high-frequency time constants are easily calculated as follows. For the base

$$\tau_{HI(B)} = (C_{be} + C_{mi}) [r_{ib}\|R_s\|R_B] \tag{17.33}$$

and for the collector

$$\tau_{HI(C)} = (C_L + C_{mo})R_{Leq} \tag{17.34}$$

EXAMPLE 17.10

A BJT whose parameters are $\beta = 120$, $r_{ib} = 2\ k\Omega$, $C_{be} = 30\ pF$, and $C_{bc} = 4\ pF$ is used in the amplifier configuration shown in Fig. 17.11 with $R_s = 3.0\ k\Omega$, $R_B = 3.0\ k\Omega$, $C_L = 25\ pF$, and $R_{Leq} = 5\ k\Omega$. Let us calculate the mid-band voltage gain, and the high corner frequency. The mid-band voltage gain is

$$A_v = \frac{v_o}{v_i} = -\frac{\beta R_{Leq}}{r_{ib}} = -\frac{(120)(5\ k)}{2\ k} = -300$$

The Miller capacitances are

$$C_{mi} = (1 + 300)\,4 \times 10^{-12} = 1204\ pF$$

and

$$C_{mo} \approx C_{bc} = 4\ pF$$

The time constants for the base and collector are

$$\tau_{HI(B)} = [(30 + 1204)10^{-12}]\,(2\ k\|3.0\ k\|3.0\ k) = 1.06 \times 10^{-6}\ \text{sec}$$

and

$$\tau_{HI(C)} = (25 + 4)(10^{-12})(5\ k) = 1.45 \times 10^{-7}\ \text{sec}$$

The high corner frequency of the circuit is dominated by the LOWEST upper corner frequency, that is, this is the one associated with the largest time constant. Therefore

$$f_{HI} = \frac{1}{2\pi\tau_{HI(B)}} = \frac{1}{2\pi(1.06 \times 10^{-6})} = 1.5 \times 10^5\ \text{Hz} = 150\ \text{kHz} \qquad \blacksquare$$

The transistor parameter C_{bc} is dominated by the junction capacitance of the reverse biased base collector junction. This is relatively easy to measure and is often specified by transistor manufacturers, although it is called by various names, such as C_{ob}.

On the other hand, C_{be} is almost never given in specifications. It models both the base emitter depletion capacitance and the diffusion capacitance, which results from the finite time required for charge carriers to move across the base.

It is, however, possible to estimate a value for C_{be} based on parameters usually provided by manufacturers and the bias conditions. The specifications for a transistor will usually include, in addition to β, one of the following: f_β = the *beta cutoff frequency*, or f_T = the *unity gain frequency*.

Frequency Dependent Beta

High-frequency modeling of the BJT often uses the concept of a *frequency-dependent beta*. That is, beta is assumed to be constant for low and mid-band frequencies; however, at high frequencies, the value of beta rolls off like the frequency response of a low-pass filter described in Chapter 6. If we let $\beta(f)$ represent the frequency-dependent beta, then its magnitude is given by

$$\beta(f) = \frac{\beta}{\sqrt{1 + \left(\dfrac{f}{f_\beta}\right)^2}} \tag{17.35}$$

A typical beta versus frequency plot is shown in Fig. 17.12. Note that for low values of f, $\beta(f)$ is constant. At the frequency, f_β, the value of beta is reduced by a factor of $1/\sqrt{2}$ or 0.707 of its low-frequency value.

Also notice that at very high frequencies ($f \gg f_\beta$), the above equation reduces to

$$\beta(f) \approx \frac{\beta f_\beta}{f} \tag{17.36}$$

In other words, in this range, beta is inversely related to frequency, and there is a particular frequency where beta is reduced to unity. This frequency is defined as f_T. Setting $\beta(f) = 1$ in Eq. (17.36) yields

$$f_T = \beta f_\beta \tag{17.37}$$

The parameter, f_T, is also known as the *current gain-bandwidth product*.

Recall that beta is, by definition, the ratio of ac collector current (i_c) to base current (i_b) with the output shorted ($v_{ce} = 0$), that is

$$\beta = \left.\frac{i_c}{i_b}\right|_{v_{ce} = 0} \tag{17.38}$$

Since $v_{ce} = 0$ shorts the output voltage, $A_v = 0$, the input Miller capacitance is reduced to the feedback capacitance, and therefore the circuit appears as that shown in Fig. 17.13. The roll-off of beta is modeled by the RC time constant of the input circuit and thus the

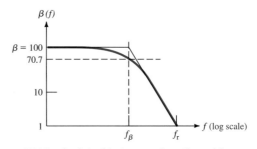

Figure 17.12 A plot of beta as a function of frequency.

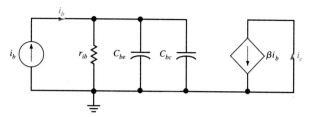

Figure 17.13 Common emitter amplifier circuit for relating f_β to circuit capacitances.

corner frequency, f_β, is

$$f_\beta = \frac{1}{2\pi(r_{ib})(C_{be} + C_{bc})} \tag{17.39}$$

Since r_{ib} and C_{bc} are known, if we know either f_T or f_β, then C_{be} can be calculated from the equation

$$C_{be} = \frac{1}{2\pi f_\beta r_{ib}} - C_{bc} \tag{17.40}$$

EXAMPLE 17.11

Let us calculate the f_T of a transistor with $\beta = 300$ and $f_\beta = 10$ MHz.
Using Eq. (17.37), we find that

$$f_T = \beta f_\beta = (300)(10)^7 = 3.0 \times 10^9 \text{ Hz} = 3 \text{ GHz} \qquad \blacksquare$$

EXAMPLE 17.12

If the transistor employed in Example 17.11 is biased with a dc base current of 25 μA, and has a C_{bc} value of 2.8 pF, let us find the value of C_{be} that should be used in the high-frequency model. In order to compute the capacitance, we first find that

$$r_{ib} = \frac{0.026}{I_B} = \frac{0.026}{25 \times 10^{-6}} = 1.04 \text{ k}\Omega$$

and then

$$C_{be} = \frac{1}{2\pi(10 \times 10^6)(1.04 \text{ k})} - 2.8 \times 10^{-12} = 12.5 \text{ pF} \qquad \blacksquare$$

DRILL EXERCISE

D17.3. If a transistor has an f_T of 220 MHz, and a low-frequency beta of 100, what is f_β, and the value of $\beta(f)$ at $f = 110$ MHz?

Ans: $f_\beta = 2.2$ MHz; at 110 MHz, β (110 MHz) = 2.

EXAMPLE 17.13

A common emitter amplifier driven by a 4 kΩ source and with $R_B = 4$ kΩ, utilizes the transistor in the previous drill problem. The transistor is biased such that $r_{ib} = 2$ kΩ. If

$R_{Leq} = 3$ kΩ and $C_{bc} = 4$ pF, we wish to find the upper corner frequency associated with the input (base) node.

The capacitance, C_{be}, can be determined using Eq. (17.40):

$$C_{be} = \frac{1}{2\pi(2.2 \times 10^6)(2 \text{ k})} - 4 \times 10^{-12} = 32.2 \text{ pF}$$

The amplifier voltage gain, obtained using Eq. (17.30), is

$$A_v = -\frac{\beta \, R_{Leq}}{r_{ib}} = -\frac{(100)(3 \text{ k})}{2 \text{ k}} = -150$$

Therefore, the Miller input capacitance, C_{mi}, is

$$C_{mi} = C_{bc} \, (1 + 150) = 604 \text{ pF}$$

and

$$\tau_{HI(B)} = (C_{be} + C_{mi})(r_{ib}\|R_s\|R_B) = (32.2 + 604)(10^{-12})(2 \text{ k}\|4 \text{ k}\|4 \text{ k}) = 6.36 \times 10^{-7} \text{s}$$

Therefore, the upper corner frequency is

$$f_{HI(B)} = \frac{1}{2\pi(6.36 \times 10^{-7})} = 250 \text{ kHz} \qquad \blacksquare$$

Common Base Amplifier

The small-signal frequency response of the common base BJT amplifier configuration often extends to higher frequencies than that of the same device in the common emitter configuration. It is analogous to the common gate MOSFET amplifier previously discussed. The common base amplifier is noninverting, and there is no device capacitance in the feedback position; therefore, there is no multiplication of the feedback capacitance via the Miller effect. In addition, the input resistance to the *CB* amplifier is relatively low, that is, r_{ie} is lower than r_{ib} by a factor of $\beta + 1$. Thus, the input time constant is small, and the corner frequency is high.

Figure 17.14 shows a small-signal high-frequency equivalent circuit of a common base BJT amplifier. The time constants and corresponding upper corner frequencies can be calculated using the same approach already demonstrated.

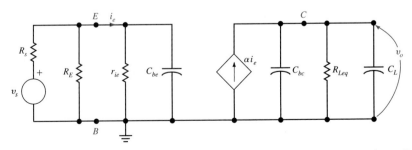

Figure 17.14 Small-signal, high-frequency equivalent circuit for a common base BJT amplifier.

D17.4. A transistor has an $f_T = 1,000$ MHz and an $f_\beta = 30$ MHz. What is α, the common base current gain? If this transistor is placed in a CB circuit with a load capacitance $C_L = 28$ pF, $R_{Leq} = 8$ kΩ, and $C_{bc} = 5.5$ pF, what is the upper corner frequency associated with the output (collector) node?

Ans: $\alpha = 0.97; f_{HI} = 594$ KHz.

D.17.5. If the transistor in problem D17.4 has a dc emitter current of 1 mA, what is the mid-band voltage gain of the circuit?

Ans: $A_v = 299$.

Voltage Follower Circuits: Common Collector and Common Drain

The high-frequency analysis of the *low-output resistance amplifier configurations,* which are sometimes called *voltage follower circuits* since their voltage gain is approximately unity, can be performed by following the procedure already outlined: (1) substitute the high-frequency transistor model into a high-frequency equivalent circuit; (2) develop equivalent capacitances at the input and output nodes; and (3) solve for the time constants and corresponding upper corner frequencies.

In general, because the output resistance of these circuits is so low, the output node will have a very high corner frequency approximating the frequency limit of the transistor, f_T. Therefore, it is unlikely that this node will be the amplifier's frequency limiting factor. The input node may have a high resistance, but input capacitances are low and therefore this node normally allows good high-frequency performance. Calculation of f_{HI} may be performed using the procedures described earlier; these circuit configurations, in general, do not limit the high-frequency performance of multi-stage amplifier circuits.

17.7

The Differential Amplifier

The *differential amplifier* is a very important circuit configuration. It is the amplifier circuit used at the input of op amps and in many other applications such as instrumentation amplifiers and comparators. Basically a differential amplifier has two inputs, one called the inverting ($-$) input, and the other the noninverting ($+$) input. There is generally a single output. The object of a differential amplifier is to amplify "differences" in voltage between the two inputs, and to be unresponsive to voltage changes that appear simultaneously on both inputs.

The block diagram of a general differential amplifier as shown in Fig. 17.15 looks very much like that of the symbol for an op amp. In fact, an op amp is a special type of differential amplifier.

For this amplifier, we define the *differential-mode input signal* as the difference between v_1 and v_2.

$$v_d = v_1 - v_2 \tag{17.41}$$

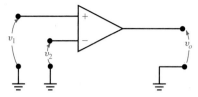

Figure 17.15   A differential amplifier symbol.

The *common mode input signal* is defined as the average value of the two input signals, or

$$v_c = \frac{v_1 + v_2}{2} \tag{17.42}$$

If the two input voltages move exactly in opposite directions, that is, antisymetrically, there is a differential mode signal, but the common mode signal is zero. On the other hand, if the voltages v_1 and v_2 move in such a way that they are always equal, there is no differential mode signal, but the common mode signal is large.

Table 17.1 describes the amplifier's operation under three different input signal conditions.

The output voltage of the differential amplifier, v_o, is related to both the differential mode input signal and the common mode input signal; thus

$$v_o = v_d A_d + v_c A_c \tag{17.43}$$

where A_d is defined as the differential mode voltage gain and A_c is defined as the common mode voltage gain.

A measure of an amplifier's ability to distinguish between differential mode and common mode signals is the *common mode rejection ratio*, CMRR, where

$$CMRR = \frac{|A_d|}{|A_c|} \tag{17.44}$$

This factor is often expressed in decibels as

$$CMRR = 20 \log \frac{|A_d|}{|A_c|} \text{ db} \tag{17.45}$$

A high-quality differential amplifier has a large A_d, a very low A_c, and therefore a high CMRR.

One of the practical values of a differential amplifier is its rejection of unwanted signals or noise. If two wires bring a differential mode signal to the diff amp (an expression for differential amplifier), stray interference from unwanted sources will be picked up on

Table 17.1 Differential and Common-Mode Amplifier Input
Signals

v_1	v_2	v_d	v_c
1 mV	−1 mV	2 mV	0 mV
61 mV	59 mV	2 mV	60 mV
3 mV	0 mV	3 mV	1.5 mV

BOTH input wires and appear as a common mode signal, which will be rejected since A_c is low. The desired signal, the difference in voltage between the input wires, will be amplified by the gain A_d.

EXAMPLE 17.14

If the specifications on a differential amplifier state that it has a differential mode gain of 1,000 and a CMRR of 70 db, let us find the common mode gain of this amplifier.

From Eq. (17.45) we find that

$$A_c = A_d 10^{-\frac{CMRR(db)}{20}} = 1,000 \ (10^{-3.5}) = 0.316$$

From a different perspective, note that the differential mode can be expressed as 60 db, which is equivalent to a factor of 1,000. Recall from Eq. (17.2)

$$A_d (in \ db) = 20 \log A_d$$

and hence

$$60 \ db = 20 \log 1,000$$

The CMRR (db) can be shown to be

$$CMRR \ (db) = A_d(db) - A_c(db)$$

Therefore, in this case

$$A_c(db) = A_d \ (db) - CMRR(db)$$
$$= 60 \ db - 70 \ db$$
$$= -10 \ db$$

which, as expected, is equivalent to a factor of 0.316. ∎

dc Analysis

The analysis of differential amplifier circuits follows the same general procedure as that for other amplifier circuits. A dc analysis is done first to determine the bias point. This analysis will be illustrated with NPN BJTs. However, a similar analysis could be performed with MOSFETs.

The basic differential amplifier circuit is shown in Fig. 17.16. The circuit is symmetric about a vertical line through the center, and we will use this symmetry to simplify the analysis.

If there is no signal applied, v_1 and v_2 are zero, and the bases of both transistors are at zero volts. Therefore, the emitters of both are at a dc voltage of -0.7 V. The current through R_E can be easily found as

$$I_{RE} = \frac{-0.7 - (-V_{EE})}{R_E} \qquad (17.46)$$

If the transistors are matched and have identical inputs, this current down the "tail" of the circuit is evenly divided between the emitters of Q_1 and Q_2 so that

$$I_{E1} = I_{E2} = \frac{1}{2} I_{RE} \qquad (17.47)$$

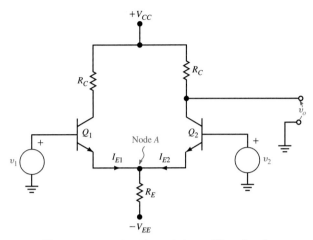

Figure 17.16 A differential amplifier circuit.

This illustrates an important point in understanding the diff amp. The circuitry in the tail at the bottom of the circuit, which in this case is R_E and V_{EE}, sets the dc bias current, and the relative voltages at the inputs v_1 and v_2 determine how this current will be split between the two sides.

Differential Mode Gain

If we consider the common mode signal to be zero and only apply a differential mode signal, v_d, then $v_1 = v_d/2$ and $v_2 = -v_d/2$. When a completely antisymmetric signal is applied, one input always goes up in voltage by the same amount the other one goes down. In this case, the ac voltage at the node in the middle (node A) never changes, and, in fact becomes a VIRTUAL ac ground. Using this fact, we can simplify the analysis by examining only half of the circuit, which with node A grounded, becomes a simple common emitter amplifier which we have analyzed before. The simplified circuit appears as shown in Fig. 17.17.

For this circuit, the voltage gain is

$$A_d = \frac{v_o}{v_d} \tag{17.48}$$

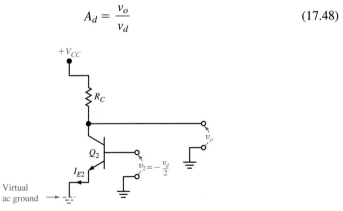

Figure 17.17 Common emitter amplifier representing half of a differential amplifier.

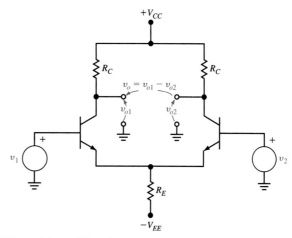

Figure 17.18   Differential amplifier circuit with the output between collectors (differential output).

and

$$v_2 = -\frac{v_d}{2} \tag{17.49}$$

From our analysis of the small-signal gain of common emitter amplifiers in Chapter 16, we know that

$$\frac{v_o}{v_2} = -\frac{\beta\, R_c}{r_{ib}} = -\frac{\beta\, R_c}{(\beta + 1)r_{ie}} \approx -\frac{R_c}{r_{ie}} \tag{17.50}$$

where

$$r_{ie} = \frac{0.026}{I_{E2}} = \frac{0.026}{\frac{1}{2}I_{RE}} \tag{17.51}$$

Combining Eq. (17.48) through (17.50) yields

$$A_d \approx \frac{1}{2}\frac{R_c}{r_{ie}} \tag{17.52}$$

Note that the output in this case is measured from one side of the network to ground, as shown in Fig. 17.16. However, sometimes the output is connected between the two collectors, as shown in Fig. 17.18. This circuit provides a differential output voltage, and since the voltage at one collector swings an equal and OPPOSITE amount from that at the other, the output voltage, defined as $v_o = v_{o1} - v_{o2}$, is twice that of the previous case. Therefore, for this circuit

$$A_d \approx \frac{R_c}{r_{ie}} \tag{17.53}$$

Common Mode Gain

We will again make use of the symmetry of the circuit in Fig. 17.16 to analyze the common mode gain. If we redraw the circuit but split R_E into two parallel resistors, we ob-

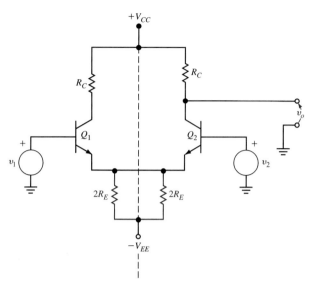

Figure 17.19 Differential amplifier circuit for calculating common mode gain.

tain the symmetrical circuit in Fig. 17.19. Each tail resistor is $2\,R_E$ so that the parallel combination will be R_E.

Again, by analyzing half this circuit using the common emitter analysis techniques, and assuming that R_E is much larger than r_{ie}, we obtain

$$A_c \approx \frac{R_c}{2\,R_E} \tag{17.54}$$

Therefore, in order to get the CMRR as high possible, the common mode gain A_c should be as small as possible. This requires a very large value for R_E. In many modern applications a transistor circuit configured as a current source (current sink in this case) is used which establishes the dc bias current and also has a very large ac resistance. The current mirror described in Chapter 15 works well for this purpose. A differential amplifier with an active current sink is shown in Fig. 17.20.

In addition, we note from Eq. (17.52) that the differential mode gain can be increased if R_C is increased. This can be accomplished by substituting for the resistors R_C the high ac resistance of a current mirror. Such a circuit, implemented with PNP transistors, is shown in Fig. 17.21, and this circuit is often found in the input of a modern integrated differential amplifier or op amp, such as the popular 741 op amp.

Input Offset Voltage

On the data sheet for an op amp or other typical differential amplifier, we will find several other parameters not yet discussed which may be important in a detailed analysis. The *input offset voltage*, V_{OS}, is the dc voltage required to be placed between the two inputs to bring the output to zero volts. Ideally, V_{OS} should be zero; however, due to manufacturing irregularities, it is impossible to make a circuit which is EXACTLY balanced, so there is some finite offset voltage required to balance the amplifier, or bring the out-

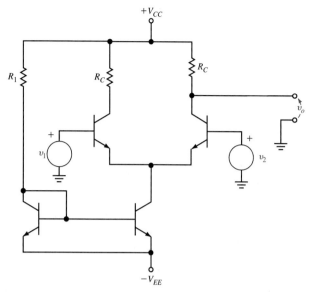

Figure 17.20 Differential amplifier with an active current sink.

put voltage to zero. This offset and its drift, or change with time or temperature, is an important source of error in op amp circuits. The specification on V_{OS} must be small enough that it does not cause a problem in the particular application. Some circuits provide connections to points in the circuit so that an external adjustment of the offset can be made. Integrated op amps may have offset voltages measured in fractions of a millivolt.

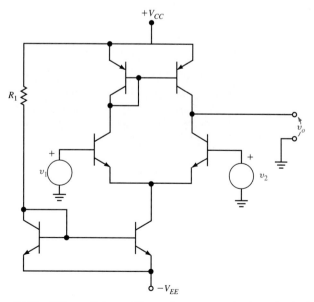

Figure 17.21 Differential amplifier using a current mirror (active) load.

Input Bias Current and Bias Current Offset

Bipolar differential amplifiers require a dc path to the base of both input transistors in order to supply base current. This base current is called the *input bias current* and is usually defined as the average of the two base currents. This is a viable situation if the amplifier is driven from a voltage source. However, if the driving sources have high impedances, then differences in base current can cause different voltage offsets at the inputs. For this reason, the *bias current offset*, I_{OS}, is defined as the difference in base currents

$$I_{OS} = I_{B1} - I_{B2} \qquad (17.55)$$

Slew Rate

The ac frequency response of an amplifier describes its ability to respond to rapidly changing SMALL-SIGNAL inputs. Due to the finite availability of bias currents in the amplifier and other limitations, there is a limit to how fast the voltage at the output of an amplifier can move over large voltage swings. This is defined by the *slew rate, SR,* and is usually expressed in volts per microsecond.

EXAMPLE 17.15

If an op amp has a slew rate (SR) of 10 V/μs and a step increase in voltage is applied, let us find the approximate time it will take the output to move from 0 to 6 volts.

The output change in voltage is $\Delta V = 6$ V, and therefore

$$time\ required = \frac{6}{10} = 0.6\ \mu s \qquad \blacksquare$$

Comparators

A *comparator* is a special type of circuit that has analog inputs and a digital output. It is used to compare the relative value of two analog input voltages and register a digital output of 1 or 0 depending on which input voltage is higher.

A block diagram of a typical comparator circuit is shown in Fig. 17.22. For such a circuit, if $v_1 > v_2$, then the output is 1 and if $v_1 < v_2$, then the output is 0.

A comparator has an input circuit that is typical of a differential amplifier, and its output circuit is typical of a logic gate. Separate power supplies are often used for the analog and digital portions of the network. In the circuit shown, the output is supplied by 5 V, which is the common supply voltage for digital TTL circuits and the input section has + V and −V supply voltages.

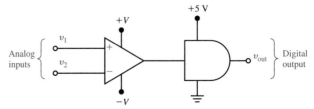

Figure 17.22   A comparator circuit.

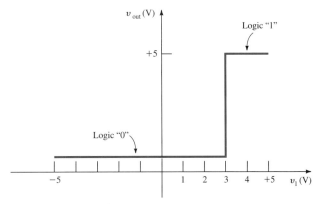

Figure 17.23   Plot used in Example 17.16.

Comparators are very useful circuits in interfacing analog signals to digital circuitry. They are often configured such that one of the inputs is connected to a reference voltage and the incoming signal is applied to the other input. The output of the comparator is then a digital signal that indicates whether the incoming signal is above or below the reference.

EXAMPLE 17.16

If the comparator circuit of Fig. 17.22 has a dc voltage reference of $+3$ V connected to v_2, let us plot the output voltage as a function of v_1 for the range $-5 < v_1 < +5$ V.
This plot is shown in Fig. 17.23.   ■

17.8
Oscillators

An *oscillator* is an electronic circuit designed to create or generate a periodic ac electronic signal. dc power is applied to this circuit, and an ac signal emerges. This circuit is composed of two essential elements: (1) there must be gain in the circuit and (2) there must be some form of positive feedback in which some of the output signal is fed back to the input. In general, the block diagram of an oscillator circuit appears as shown in Fig. 17.24. In this figure **A** is the amplifier gain, and **H** is the gain of the feedback network. Either or both of these networks may have gain and phase responses which are frequency dependent. Oscillations will be sustained when (1) the loop gain, **AH,** is equal to unity, and (2) the phase shift around the loop is 0 or 360°. If the loop gain is greater than unity, the oscillations will grow until some circuit limitation arrests the growth.

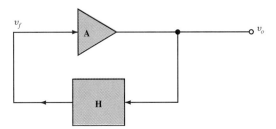

Figure 17.24   General oscillator block diagram.

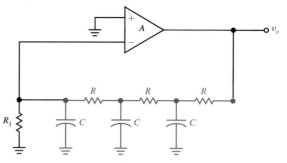

Figure 17.25   Phase shift oscillator.

Phase Shift Oscillator

A common oscillator circuit which illustrates the oscillation principles is the *phase shift oscillator* shown in Fig. 17.25. The amplifier provides a gain of A and since the feedback signal is returned to the inverting input, a phase shift of 180° is obtained in the amplifier. The remaining 180° phase shift required for oscillation is obtained by three RC networks. The frequency of oscillation is the frequency at which this network provides the additional 180° required. It can be shown that the frequency of oscillation of such a circuit is

$$f_o = \frac{1}{2\pi RC\sqrt{6}} \tag{17.56}$$

Resonant Circuit Oscillator

The oscillator circuit of Fig. 17.26 is like a tuned amplifier with a frequency selective "tank" circuit in the output. This parallel LC circuit has a resonant frequency at which its

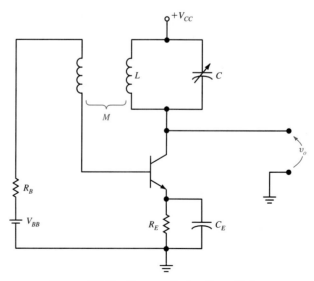

Figure 17.26   Resonant circuit oscillator.

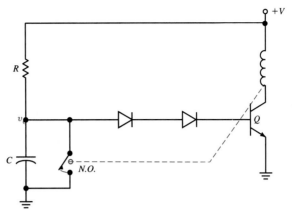

Figure 17.27 Relaxation oscillator.

impedance is very large, resulting in a large R_{Leq}, and therefore the gain of the circuit is very high at that frequency. There is some signal coupled from the output back to the input via mutual inductance to another coil. If the coupling is sufficient, the loop gain will exceed unity and this circuit will produce a sinusoidal output signal at the frequency of the tuned tank circuit. Recall from Chapter 6 that this is

$$f_o = \frac{1}{2\pi\sqrt{LC}} \text{ Hz} \tag{17.57}$$

Relaxation Oscillator

In a *relaxation oscillator,* the voltage on a key node is allowed to grow over time, and when it reaches some threshold, the circuit acts to reduce the voltage and let the charging process begin again. A simple relaxation oscillator is shown in Fig. 17.27. In this circuit, the voltage across C gradually increases as it is charged by current flowing through R. The RC time constant determines the rate of charging. When the voltage across C reaches a critical level, in this case, 3 diode drops = $3 \times 0.7 = 2.1$ V, the transistor Q turns on, and current flows through the relay coil and closes the relay contacts across C. This action discharges C, the voltage falls, and the transistor turns off. The process then begins again, creating an ongoing repetition of charging and discharging. The voltage across C, v_C, appears as shown in Fig. 17.28. There are many variations of this type of

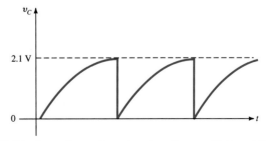

Figure 17.28 Capacitor voltage in relaxation oscillator of Fig. 17.27.

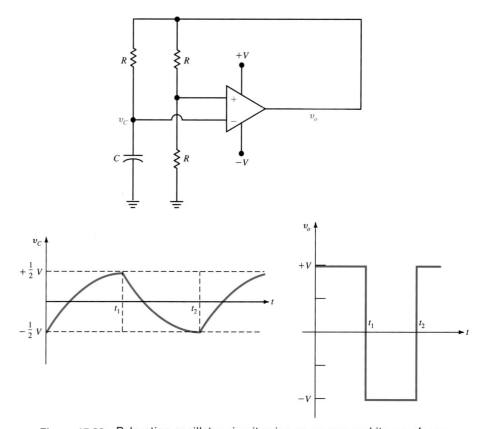

Figure 17.29   Relaxation oscillator circuit using an op amp and its waveforms.

circuit, and a modern circuit would normally use a transistor or op amp instead of a re-lay. One such circuit is shown in Fig. 17.29. The op amp output voltage can be assumed to swing between either $+V$ or $-V$ depending on whether the $(+)$ input is above or below the $(-)$ input. This circuit oscillates with C being alternately charged and discharged, and v_C swings from $+V/2$ to $-V/2$.

17.9
Power Amplifiers And Output Stages

Power amplifiers are used to deliver relatively large amounts of power to a load. While there are a wide variety of approaches to power amplification, most power amplifiers are multi-stage amplifiers which provide significant voltage and current gain. Typically, the early stages may have small signals, be amenable to small-signal analysis, and be used to create voltage gain. The output stage of a high-power multi-stage voltage amplifier has special requirements. In such an amplifier, the final or output stage must: (1) be able to deliver as much power as possible to the load, that is, deliver high-output current at relatively high voltages, and (2) have a low-output resistance which simulates an ideal voltage source with zero internal resistance.

The transistor circuit configurations that provide high current gain and low output re-

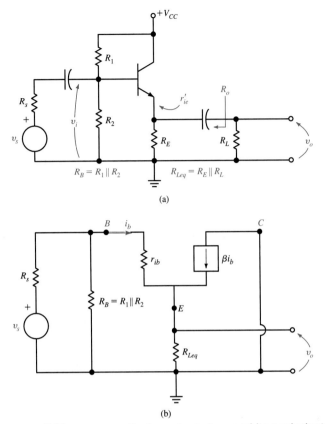

Figure 17.30 BJT common collector output stage and its equivalent circuit.

sistance are *voltage followers,* which are either common collector or common drain configurations. The designation of emitter follower (for BJTs) or source follower (for FETs) arises because the output voltage of these configurations *follows* the input almost exactly, and they both have voltage gains very close to unity. The value of these circuits lies primarily in the current gain and impedance transformation they provide.

These circuits were introduced in Chapter 16. Once again consider, however, the basic emitter follower circuit shown in Fig. 17.30(a). Substituting the BJT model for the transistor, we obtain the circuit of Fig. 17.30(b).

This circuit can be used to calculate the voltage gain, $A_v = v_o/v_i$. The result is

$$A_v = \frac{v_o}{v_i} = \frac{R_{Leq}}{R_{Leq} + r_{ie}} \tag{17.58}$$

which can never be greater than 1.

The current gain for this circuit is large and equal to $(\beta + 1)$.

The resistance r'_{ie}, which is the resistance looking back into the emitter not including R_{Leq}, is given by

$$r'_{ie} = r_{ie} + \frac{R_s \| R_B}{\beta + 1} \tag{17.59}$$

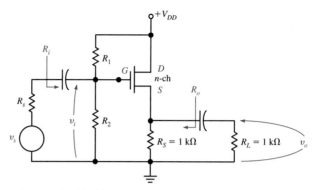

Figure 17.31 MOSFET common drain output stage.

where $R_B = R_1 \| R_2$. R_o for the amplifier is $r'_{ie} \| R_E$. Note that the output resistance is low, and has a component determined by the effective resistance seen in the base circuit DIVIDED by $(\beta + 1)$. The availability of high current and low impedence at the output makes this circuit an excellent output stage for a voltage amplifier.

The source follower or common drain circuit shown in Fig. 17.31 has similar characteristics.

The voltage gain of the source follower was shown to be

$$A_v = \frac{v_o}{v_i} = \frac{g_m R_{Leq}}{1 + g_m R_{Leq}} \tag{17.60}$$

In Chapter 16 we also derived the resistance looking into the source of a MOSFET while studying the input resistance for the common gate circuit. We found that this resistance was $1/g_m$. Therefore, the output resistance, R_o, for a source follower is

$$R_o = \left[\frac{1}{g_m} \| R_s \right] \tag{17.61}$$

This value is typically small and therefore this configuration is also useful as the output stage of a voltage amplifier.

Most op amps and many other analog amplifiers, particularly power amplifiers, require the output voltage to swing both positive and negative. The availability of comple-

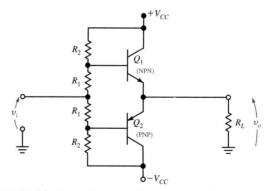

Figure 17.32 Output stage with complementary symmetry.

mentary transistor types, such as NPN and PNP, makes it possible to merge two emitter followers in the same amplifier output stage. This is termed a *complementary symmetry* output stage and is shown in Fig. 17.32. Only the top half or the bottom half of the network is active at any one time. If the output voltage is swinging positive, Q_1 is active and Q_2 is off, and the analysis is similar to that performed above for a single emitter follower. If the output voltage swings negative, the opposite situation exists.

The analysis of power amplifiers often requires the use of the large-signal techniques described in Chapter 15 because of the large voltage and current excursions that may occur in these type circuits.

17.10
Summary

- The voltage gain of a multi-stage amplifier is the product of the voltage gains of each of the stages, that is

$$A_{v(total)} = A_{v1}A_{v2}A_{v3} \ldots A_{vN}$$

- Voltage gain expressed in decibels is

$$A_v \ (in \ db) = 20 \ \log[A_v]$$

- Analogous expressions exist for current gain.
- An amplifier has maximum gain in the mid-band frequency region. The lower corner frequency, f_{LO}, is the lower frequency at which the mid-band gain falls to 0.707 of its mid-band value. The upper corner frequency, f_{HI}, is the high frequency at which the gain is reduced to 0.707 of its mid-band value. The corner frequencies are also called cutoff frequencies or half-power points.
- The phase shift of the amplifier is 0° or 180° over the mid-band frequency region; it deviates 45° from its mid-band value at the corner frequencies, as shown in Fig. 17.3.
- Determining f_{LO} is accomplished by evaluating the RC time constants associated with the coupling and bypass capacitors in the amplifier; $\tau_{LO} = R_{eq}C$ where R_{eq} is the equivalent resistance across C. There is a low corner frequency associated with each such RC network:

$$f_{LO} = \frac{1}{2\pi\tau_{LO}}$$

The highest f_{LO} in the circuit dominates in determining the corner frequency of the amplifier.

- Determining f_{HI} is accomplished by evaluating the RC time constants associated with the transistor's device capacitances and parasitic capacitances; $\tau_{HI} = R_{eq}C$ where R_{eq} is the equivalent resistance across C. There is a high corner frequency associated with each such RC network:

$$f_{HI} = \frac{1}{2\pi\tau_{HI}}$$

The lowest f_{HI} in the circuit dominates in determining the corner frequency of the amplifier.

- If C_f is the capacitance in the feedback of an inverting amplifier, the Miller effect: (1) creates an effective capacitance at the amplifier's input of $(1 - A_v)C_f$; this can be large if A_v has a large negative value. (2) There is an effective capacitance at the amplifier's output approximately equal to C_f.
- High-frequency models for the MOSFET are developed and used to create ac equivalent circuits for the complete amplifier. From this model the frequency performance of the amplifier can be determined.
- High-frequency models for the BJT are developed and used to create ac equivalent circuits for the complete amplifier. Determining BJT device capacitance for the emitter base junction may require use of the frequency-dependent beta. From the ac model of the amplifier, its frequency performance can be determined.
- The differential amplifier responds to differential mode signals and seeks to reject common mode signals; that is

$$v_o = v_d A_d + v_c A_c \text{ where } A_d > A_c$$

- Common mode rejection ratio is expressed in *db* as

$$CMRR = 20 \log \frac{|A_d|}{|A_c|} \ db$$

- A comparator is a circuit with a differential amplifier input circuit, and a digital output; it provides a logic 1 if the " + " input is at a higher voltage than the "−" input and a logic 0 at the output otherwise.
- Oscillators are electronic circuits which provide a continuous periodic signal.
- Power amplifiers often use output stages with high current gain, for example, common drain or common collector circuits.

PROBLEMS

17.1. Determine the composite voltage gain of three voltage amplifiers cascaded together as shown in Fig. 17.1 and with gains as follows: $A_{v1} = 40$, $A_{v2} = 50$, and $A_{v3} = -20$.

17.2. Repeat problem 17.1 if the voltage gains are: $A_{v1} = 25$, $A_{v2} = 80$, and $A_{v3} = 0.9$.

17.3. Determine the current gain of the multi-stage current amplifier comprised of two stages, $A_{i1} = 30$, $A_{i2} = -15$.

17.4. (a) Express the voltage gains of each of the stages in problem 17.1 in decibels. (b) Compute the gain for the multi-stage amplifier in db.

17.5. The voltage gain of a multi-stage amplifier is measured to be 100 db. The gain of the first stage was calcu-

lated as $A_{v1} = 8$. Determine the total gain of the remaining stages.

17.6. The input stage of an amplifier is shown in Fig. P17.6, where R_i represents the input resistance to the amplifier. Calculate f_{LO} associated with C.

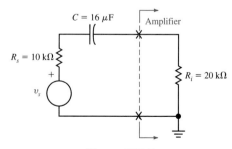

Figure P17.6

17.7. For the circuit in Fig. P17.7, assume the emitter current, $I_E = 2$ mA, $\beta = 200$, r_o is very high, and $V_{BE} = 0.7$ V. (a) Find f_{LO} associated with C_1. (b) Find f_{LO} associated with C_2. (c) Find f_{LO} associated with C_E. (d) Comment on which one(s) will dominate in determining the amplifier's lower corner frequency.

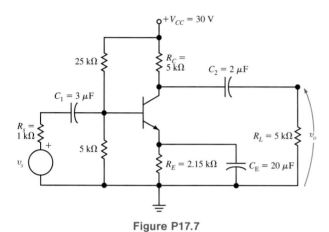

Figure P17.7

17.8. Figure P17.8 shows an ac equivalent circuit of a common base amplifier; assume $I_E = 1$ mA and the transistor's output resistance is very high. (a) Find f_{LO} associated with C_1. (b) Find f_{LO} associated with C_2. (c) Comment on which f_{LO} would dominate in setting the amplifier's low-frequency corner.

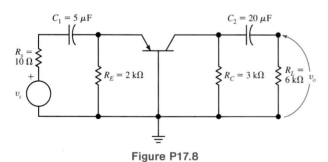

Figure P17.8

17.9. Figure P17.9 illustrates a circuit for a common source MOSFET amplifier; assume g_m for the transistor is 2.5 mS and r_o can be neglected. (a) Find f_{LO} associated with C_1. (b) Find f_{LO} associated with C_2. (c) Find f_{LO} associated with C_S. (Hint: recall the resistance "looking into" the source of an FET biased in the active region is $1/g_m$.)

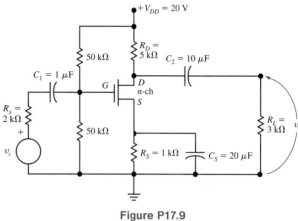

Figure P17.9

17.10. An inverting amplifier has a gain of $A_v = -2500$, and has a capacitance of 10 pF connected from output to input. (a) Find the Miller input capacitance, C_{mi}. (b) Find the Miller output capacitance, C_{mo}.

17.11. Assume the high-frequency circuit model for the MOSFET shown in Fig. 17.7(a) is placed in a common source bias circuit with a load resistance of $R_L = 20$ kΩ, a gate bias resistor of $R_G = 800$ kΩ and $R_s = 60$ kΩ; for the transistor, $g_m = 4$ mS, $C_{gs} = 1.2$ pF, and $C_{gd} = 0.9$ pF. (a) Find the mid-band voltage gain. (b) Construct an ac equivalent circuit and determine the Miller capacitances. (c) Determine the total capacitance at the input. (d) Determine the total output capacitance. (e) Find f_{HI} for the input node. (f) Find f_{HI} for the output node. (g) Comment on which of these corner frequencies will dominate in determining the amplifier's upper corner frequency.

17.12. The n-channel MOSFET common source amplifier circuit of Fig. P17.9 is reduced to a high-frequency equivalent circuit in Fig. P17.12. The transistor has $g_m = 5$ mS, $C_{gs} = 1.7$ pF, $C_{gd} = 1.6$ pF. (a) Determine the mid-band voltage gain, v_o/v_i. (b) Draw the ac equivalent circuit simplified by using the Miller capacitances. (c) Find f_{HI} for the RC network at the amplifier's input. (d) Repeat (c) for the output node. (e) Predict the upper corner frequency of this amplifier.

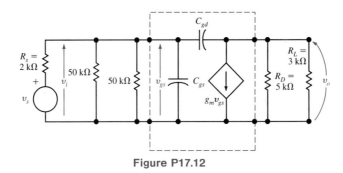

Figure P17.12

17.13. A MOSFET with the same parameters as the device in problem 17.12 is placed in a common gate bias circuit as shown in Fig. P17.13. (a) Draw the small-signal equivalent circuit. (Hint: Be sure to include the resistance seen "looking into" the source.) (b) Find f_{HI} for the input node. (c) Find f_{HI} for the output node. (d) Predict the upper corner frequency for this amplifier.

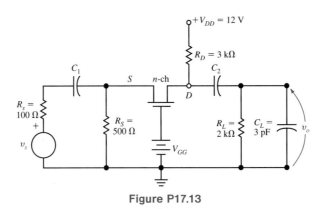

Figure P17.13

17.14. Figure P17.14 is the small-signal equivalent circuit of the common emitter amplifier shown in Fig. P17.7, and using a BJT small-signal model like that of Fig. 17.10(a). The BJT small-signal model parameters are given as: $\beta = 100$, $C_{be} = 22$ pF, $C_{bc} = 1.9$ pF. We are also given that $I_B = 20\ \mu A$; this could be calculated from Fig. P17.7. (a) Construct a simplified small-signal equivalent circuit and find values for all the elements. (b) Find f_{HI} for the input node. (c) Find f_{HI} for the output node. (d) Predict the upper corner frequency for this amplifier.

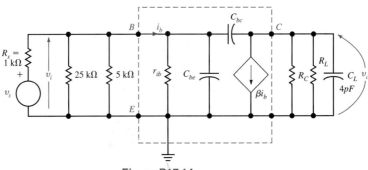

Figure P17.14

17.15. (a) Construct the small-signal equivalent circuit for the common base amplifier shown in Fig. P17.8. Assume the BJT has the same parameters as given in problem 17.14 and that the dc emitter current $I_E = 0.75$ mA. There is also a load capacitance, $C_L = 2.5$ pF in parallel with R_L. (b) Find f_{HI} for the input node. (c) Find f_{HI} for the output node. (d) Predict the upper corner frequency for this amplifier.

17.16. (a) For a BJT if $f_T = 500$ MHz and the low frequency $\beta = 250$, plot β as a function of frequency on a log-log scale. (b) Determine f_β.

17.17. A BJT has $C_{bc} = 1.2$ pF, $f_\beta = 20$ MHz, and is biased with $I_B = 50$ μA. Determine C_{be}.

17.18. Repeat problem 17.17 if $f_\beta = 10$ MHz.

17.19. A BJT has a low frequency beta of 50 and $f_\beta = 8$ MHz. Find f_T.

17.20. A differential amplifier has two inputs identified as v_1 and v_2. The following sets of voltages are applied referenced to ground. For each set find the differential mode voltage and the common mode voltage applied.

v_1	v_2	v_d	v_c
+2 mV	−2 mV		
+3 V	+3 V		
+20 mV	−10 mV		
+32 mV	+12 mV		

17.21. A differential amplifier has a differential mode gain of 60 db and a common mode gain of −20 db. Find the common mode rejection ratio (in db).

17.22. A differential amplifier has a common mode gain of 0.1 and a common mode rejection ration 10^4. Find the differential mode gain.

17.23. The circuit drawing for a basic differential amplifier is shown in Fig. P17.23. The transistors have $V_{BE} = 0.7$ V and $\beta = 200$. The dc current in R_E is 0.5 mA. (a) Find the differential mode gain. (b) Find the common mode gain. (c) Find the CMRR. (d) Determine the suppy voltage, $-V_{EE}$.

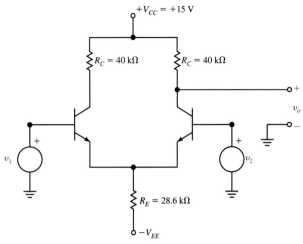

Figure P17.23

17.24. Determine the input bias current for the differential amplifier of problem 17.23.

17.25. For the op amp circuit shown in Fig. 17.21, if $R_1 = 50$ kΩ, the β's = 250, and V_{CC} and $V_{EE} = +15$ and $−15$ V, respectively, find the input bias current.

17.26. An op amp with a slew rate of 30 V/μs has a step change in the input voltage. Approximately how long will it take the output voltage to move from $−3$ to $+3$ V?

17.27. A comparator is used in an automobile warning circuit to alert the car's computer if the electrical system is failing. The circuit of Fig. P17.27 will provide a logic 1 if the main battery voltage drops below 9 volts. Otherwise it provides a logic 0. Construct a similar circuit to warn against an "overvoltage condition." Your circuit should provide a logic 1 output only when the battery voltage exceeds 15 volts.

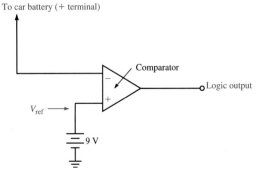

Figure P17.27

17.28. List and explain the conditions required to achieve oscillation in an electronic circuit.

17.29. A phase shift oscillator circuit is shown in Fig. 17.25. If $R = 5$ kΩ, find the required value for C for the circuit to produce an output frequency of 440 Hz.

17.30. The oscillator circuit of Fig. 17.26. has an inductor $L = 0.1$ mH. Find the value of C that will cause oscillations at 1 MHz.

17.31. (a) Derive an expression for the frequency of oscillation of the circuit in Fig. 17.29. (b) If $R = 4$ kΩ and $C = 0.5$ μF, determine the frequency of oscillation.

17.32. The circuit in Fig. P17.32 has $I_E = 3.5$ mA. The BJT has $\beta = 120$. (a) Find the voltage gain, v_o/v_i. (b) Find the circuit's output resistance, R_o, as indicated in the figure. (c) Verify that R_i, the input resistance to the circuit, is 7.97 kΩ. (d) From the results of (a) and (c) determine the total voltage gain, v_o/v_s. (e) Determine the current gain of this circuit.

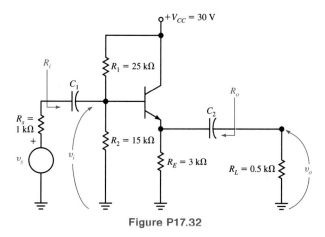

Figure P17.32

17.33. The MOSFET in Fig. 17.31 has $g_m = 4.5$ mS. (a) Determine the voltage gain, v_o/v_i. (b) Determine the input resistance, R_i. (c) Determine v_o/v_s. (d) Verify that the output resistance of the circuit, $R_o = 182$ Ω. $R_1 = R_2 = 60$ kΩ; $R_s = 10$ kΩ.

ELECTROMECHANICAL SYSTEMS AND SAFETY

Magnetically Coupled Circuits and Transformers

18.1
Introduction

In this chapter some of the basic concepts of magnetic fields are introduced. It is shown that these concepts form the basis for the development of magnetic circuits, which are fundamental building blocks in transformers and electric machinery. An analogy between electric circuits and magnetic circuits is also presented.

In order to provide a basis for understanding the operation of a transformer, we first examine two inductors that are placed in close proximity to one another. Because they are physically located close to one another, they share a common magnetic flux. The magnetic coupling between the coils is the vehicle by which energy is transmitted.

Once the coupled inductors have been introduced, the concepts are extended to the development of transformers. First, linear transformers are presented, and then coils that are coupled with ferromagnetic material are examined and used to derive an approximation for ideal coupling, called the ideal transformer.

Finally, some of the practical considerations of the use of transformers in modern technology is presented and discussed.

18.2
Magnetic Circuits

Magnetic circuits are a fundamental component of the study of electrical engineering. Their importance stems from the fact that they form the basis for the operation of transformers, generators, and motors. In this chapter, we will introduce magnetic circuits and illustrate their use in transformers. In the following chapters, we will show that an elec-

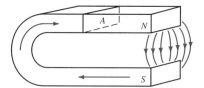

Figure 18.1 Permanent magnet.

tromagnetic field is the basic mechanism by which generators and motors transform mechanical energy into electrical energy and vice versa.

We begin our discussion by considering the permanent magnet shown in Fig. 18.1. This device creates a magnetic field between the north (N) and south (S) poles. A *magnetic flux,* or simply flux, emanates from the north pole and returns through the south pole in a closed path. The flux is measured in webers (Wb) and named after the German physicist Wilhelm Weber (1804–1891). The existence of the flux lines is dramatically illustrated in Fig. 18.2, where a bar magnet is placed under a sheet of paper and iron filings are sprinkled on the paper. A compass needle placed near the iron filings will align itself tangential to the flux lines just as a compass used for direction would align itself with the magnetic field lines of the earth.

If we examine a cross-sectional area A of the magnet shown in Fig. 18.1 and assume that the flux is uniformly distributed over the area, the *magnetic flux density, B,* is defined

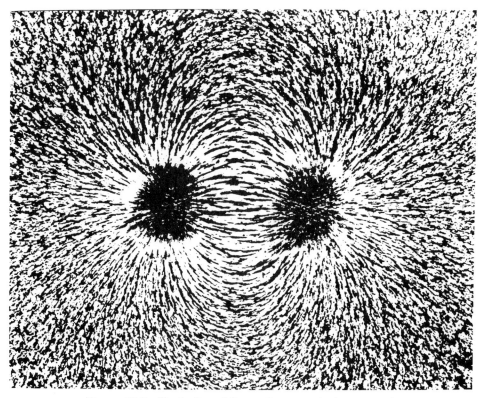

Figure 18.2 Illustration of flux paths around a bar magnet.

by the equation

$$B = \phi/A \tag{18.1}$$

where the units of flux density are Wb/m² or Tesla after Nikola Tesla (1856–1943).

In a more general sense, the flux density is related to the magnetic flux through a surface by the expression

$$\phi = \int_s \mathbf{B} \cdot \mathbf{ds} \tag{18.2}$$

Note that **B** and **ds** are vectors; however, in the case where **B** is uniform over the surface area A and perpendicular to that surface area, Eq. (18.2) is reduced to

$$\phi = BA \tag{18.3}$$

Furthermore, it is important to note that the flux will pass through anything, including air. However, ferromagnetic materials such as iron and steel have flux densities that are typically one thousand times greater than that obtained in air. In essence, ferromagnetic materials form a superhighway for the flux.

A magnetic field can also be created by a current flowing through a conductor as shown in Fig. 18.3. The flux lines encircle the current-carrying conductor and the direction of the magnetic field is determined by a right-hand rule which states that if the thumb on the right hand points in the direction of positive current, then the fingers of the right hand curl in the direction of the magnetic field.

The current carried by the conductor is related to the *magnetic field intensity, H,* by *Ampere's law,* which is

$$\oint \mathbf{H} \cdot \mathbf{dl} = I_{Enclosed} \tag{18.4}$$

where the field intensity **H** is related to the flux density **B** by the relationship

$$\mathbf{B} = \mu\mathbf{H} \tag{18.5}$$

where μ is the *permeability* of the material which carries the flux. If μ is constant, the relative material is said to be "magnetically linear." It is important to note that while the magnetic field intensity at any point depends on the current in the conductor and the distance from the point to the conductor, the flux density is directly proportional to the permeability of the material through which the flux passes.

In general, the permeability μ is written in the following form:

$$\mu = \mu_o\mu_r \tag{18.6}$$

where μ_o is the permeability of free space, that is, $4\pi \times 10^{-7}$ Henry/m and μ_r is the

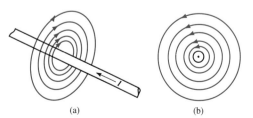

(a)   (b)

Figure 18.3 (a) Magnetic field surrounding a current-carrying conductor; (b) end view of current-carrying conductor. Current is coming out of page.

Table 18.1 Relative Permeabilities of Some Common Materials

Material	Relative Permeability
air	1.0000004
wood	0.9999995
water	0.999991
cast iron	60
purified iron	200,000
typical ferrite	1000

relative permeability of the material containing the flux. Since our objective in a magnetic circuit is a large confined flux, the materials employed typically have a large μ_r, such as iron and steel, rather than materials with a small μ_r, such as wood. Table 18.1 provides a short list of the relative permeabilities of some common materials of interest.

It is interesting to note that the creation of a magnetic field by a current-carrying conductor is the reason people are concerned about the biological effects of using electronic devices such as cellular telephones as well as living or working near high-voltage lines.

Consider now the magnetic circuit shown in Fig. 18.4(a). It consists of an iron core and an electrical winding of N turns carrying a current of I amperes. As Fig. 18.3 demonstrates, the current will produce a magnetic flux which circulates in the core. While there may be some small leakage flux which circulates around the conductor and through the surrounding air, almost all of the flux will be confined to the ferromagnetic core. It follows from Ampere's law that the flux is not only dependent upon the current but the number N of turns of the wire around the core. Thus, the driving force, which pushes the flux around the magnetic circuit, is the number of ampere-turns NI just as the voltage source is the driving mechanism which forces the current in an electrical circuit. This magnetic driving force is called *magnetomotive force* (mmf), $\mathcal{F}$, and defined as

$$\mathcal{F} = NI \tag{18.7}$$

where I is the current in amperes in an N-turn coil, and therefore $\mathcal{F}$ is measured in ampere-turns.

The relationship between the flux and the mmf for the magnetic circuit shown in Fig. 18.4(a) is demonstrated by the curve in Fig. 18.4(b). As indicated in the figure, for small

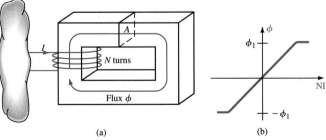

Figure 18.4 (a) Magnetic circuit consisting of an iron core excited by *NI* Amp-turns; (b) relationship between flux and amp-turns. ϕ_1 is the saturation flux.

Table 18.2 Electric/Magnetic Circuits Analogies

Electric Circuit	Magnetic Circuit
$V = IR$	$\mathcal{F} = \phi \mathcal{R}$
V = voltage in volts	$\mathcal{F} = NI$ = mmf, in ampere-turns
I = current in amperes	ϕ = flux in webers
R = resistance in ohms	$\mathcal{R}$ = reluctance in amp-turns/Wb
G = conductance in siemens	$\mathcal{P} = \dfrac{1}{\mathcal{R}}$ = permeance in Wb/amp-turns

values of ϕ, doubling either the current I or the number of turns N will double ϕ. In other words, the relationship between ϕ and NI is linear until the iron begins to saturate, at which point further increases in the driving force NI yield little or no increase in the flux ϕ.

Although the mmf produces a flux which penetrates some medium such as the iron core, this flux is limited by the magnetic resistance, which is called *reluctance*. Reluctance is defined by the equation

$$\mathcal{R} = \frac{\mathcal{F}}{\phi} \tag{18.8}$$

and the units for reluctance are ampere-turns/wb. Similarly, *magnetic permeance* is defined as:

$$\mathcal{P} = \frac{1}{\mathcal{R}} \tag{18.9}$$

As expected, materials with high reluctance, or magnetic resistance, are things like free space, teflon, and wood. Furthermore, materials such as iron and steel have low reluctance. Therefore, we find that reluctance is inversely proportional to permeability. While reluctance is a useful concept, its applicability is limited by the fact that the ϕ vs NI curve is nonlinear as shown in Fig. 18.4(b) and therefore the reluctance is not constant.

Equation (18.8) is often called Ohm's law for the magnetic circuit. Table 18.2 lists analogous relationships between electric and magnetic circuits.

EXAMPLE 18.1

Consider the ferrite core shown in Fig. 18.5. The relative permeability of the core is $\mu_r = 1000$ and the distances around the core are

$$\ell_1 = \ell_3 = 50 \text{ cm}$$
$$\ell_2 = \ell_4 = 100 \text{ cm}$$

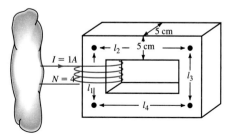

Figure 18.5 Ferrite core used in Example 18.1.

The B field is assumed to be uniform over the cross-sectional area. We wish to find the flux ϕ in the core.

If the B field is uniform over an area A, and the area is uniform over a length ℓ, then in general the reluctance of the flux path can be determined by the equation

$$\mathcal{R} = \frac{\ell}{\mu_o \mu_r A} \tag{18.10}$$

Hence, the reluctance for the path ℓ_1 is

$$\mathcal{R}_1 = \mathcal{R}_3 = \frac{(50 \text{ cm})(100 \text{ cm/m})}{(1000)(4\pi \times 10^{-7} \text{H/m})(25 \text{ cm}^2)}$$

$$= 159.2 \times 10^3 \text{H}^{-1}$$

or

$$\mathcal{R}_1 = \mathcal{R}_3 = 159.2 \times 10^3 \text{ amp-turns/Wb}$$

In a similar manner, $\mathcal{R}_2$ and $\mathcal{R}_4$ are found to be

$$\mathcal{R}_2 = \mathcal{R}_4 = 318.4 \times 10^3 \text{ amp-turns/Wb}$$

The total reluctance for the path is then

$$\mathcal{R}_T = \mathcal{R}_1 + \mathcal{R}_2 + \mathcal{R}_3 + \mathcal{R}_4$$

$$= 954.9 \times 10^3 \text{ amp-turns/Wb}$$

The flux ϕ is then obtained from the equation

$$\phi = NI/\mathcal{R}_T$$

$$= \frac{4}{954.9 \times 10^3}$$

$$= 4.19 \times 10^{-6} \text{ Wb}$$

Note the resemblance between these calculations and those required for a single-loop circuit containing one voltage source and four resistors. ∎

DRILL EXERCISE

D18.1. The magnetic field intensity H is measured at some point to be 26.530 A/m. Calculate the magnetic flux density B at that point if the point is in (a) water, (b) cast iron, and (c) purified iron.

Ans: (a) 33.3 μWb/m^2, (b) 2 mWb/m^2 and (c) 6.668 Wb/m^2.

Having described some of the basic concepts of magnetic circuits, we now turn our attention to their application. We will demonstrate that when two current-carrying conductors are in close proximity, their magnetic fields interact. It is through the interactive coupling of the magnetic fields that we transfer energy.

18.3

Mutual Inductance

Consider the situation illustrated in Fig. 18.6. To simplify our discussion, consider all conductor resistance to be negligible. Now recall *Faraday's law,* which can be stated as follows: The induced voltage in a coil is proportional to the time rate of change of *flux linkage* ($\lambda = N\phi$), where N is the number of turns in the coil. Two coupled coils are shown in Fig. 18.6 together with the following flux components.

ϕ_{L1}	The flux in coil 1, which does not link coil 2, that is produced by the current in coil 1.
ϕ_{L2}	The flux in coil 2, which does not link coil 1, that is produced by the current in coil 2.
ϕ_{12}	The flux in coil 1 produced by the current in coil 2.
ϕ_{21}	The flux in coil 2 produced by the current in coil 1.
$\phi_{11} = \phi_{L1} + \phi_{21}$	The flux in coil 1 produced by the current in coil 1.
$\phi_{22} = \phi_{L2} + \phi_{12}$	The flux in coil 2 produced by the current in coil 2.
ϕ_1	The total flux in coil 1.
ϕ_2	The total flux in coil 2.

In order to write the equations that describe the coupled coils, we define the voltages and currents, using the passive sign convention, at each pair of terminals as shown in Fig. 18.6.

Mathematically, Faraday's law can be written as

$$v_1(t) = N_1 \frac{d\phi_1}{dt} \tag{18.11}$$

The flux ϕ_1 will be equal to the algebraic sum of ϕ_{11}, the flux in coil 1 caused by current coil 1, and ϕ_{12}, the flux in coil 1 caused by current in coil 2; that is

$$\phi_1 = \phi_{11} + \phi_{12} \tag{18.12}$$

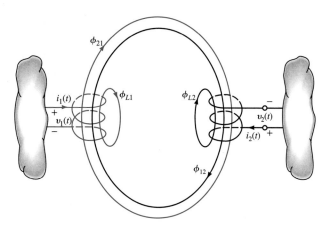

Figure 18.6 Flux relationships for mutually coupled coils.

The equation for the voltage can be written

$$v_1(t) = N_1 \frac{d\phi_1}{dt}$$

$$= N_1 \frac{d\phi_{11}}{dt} + N_1 \frac{d\phi_{12}}{dt} \tag{18.13}$$

where N_1 is the number of turns in coil 1. For magnetically linear (constant permeability) situations:

$$\phi_{11} = N_1 i_1 \mathscr{P}_{11}$$

$$\phi_{12} = N_2 i_2 \mathscr{P}_{12} \tag{18.14}$$

where the $\mathscr{P}$'s are constant permeances defined in Eq. (18.9) and depend on the magnetic paths taken by the flux components. The voltage equation can be written

$$v_1(t) = N_1^2 \mathscr{P}_{11} \frac{di_1}{dt} + N_1 N_2 \mathscr{P}_{12} \frac{di_2}{dt} \tag{18.15}$$

The constant $N_1^2 \mathscr{P}_{11} = L_{11}$ (the same L that we used before) is now called the *self-inductance,* and the constant $N_1 N_2 \mathscr{P}_{12} = L_{12}$ is called the *mutual inductance.* Therefore

$$v_1(t) = L_{11} \frac{di_1}{dt} + L_{12} \frac{di_2}{dt} \tag{18.16a}$$

Using the same technique, we can write

$$v_2(t) = N_2^2 \mathscr{P}_{22} \frac{di_2}{dt} + N_1 N_2 \mathscr{P}_{21} \frac{di_1}{dt}$$

$$= L_{21} \frac{di_1}{dt} + L_{22} \frac{di_2}{dt} \tag{18.16b}$$

If the media through which the magnetic flux passes its linear, then $\mathscr{P}_{12} = \mathscr{P}_{21}$. Hence, $L_{12} = L_{21} = M$. For convenience, let us define $L_1 = L_{11}$ and $L_2 = L_{22}$. Thus

$$v_1(t) = L_1 \frac{di_1}{dt} + M \frac{di_2}{dt}$$

$$v_2(t) = L_2 \frac{di_2}{dt} + M \frac{di_1}{dt} \tag{18.17}$$

In order to indicate the physical relationship of the coils, we employ what is commonly called the *dot convention.* Dots are placed beside each coil so that if currents are entering both dotted terminals or leaving both dotted terminals, the mutual fluxes produced by these currents will add. In order to place the dots on a pair of coupled coils, we arbitrarily select one terminal of either coil and place a dot there. Using the right-hand rule, we determine the direction of the flux produced by this coil current when entering the dotted terminal. We then examine the other coil to determine which terminal the current would have to enter to produce a flux that would be in the same direction as the flux produced by the first coil. Place a dot on this terminal. The dots have been correctly placed on the two coupled circuits in Fig. 18.7(a). The corresponding circuit symbol is shown in Fig. 18.7(b).

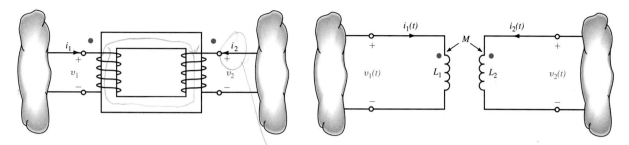

Figure 18.7(a) Physical situation.   **Figure 18.7(b)** Corresponding circuit diagram.

Figure 18.7 Two strongly coupled coils.

Note carefully that the mathematical equations are written for the circuit diagram in Fig. 18.7(b), and hence changing a voltage polarity or current direction will reverse the sign of the appropriate variable in the equations. It is interesting to note that this procedure is directly analogous to the use of Ohm's law. For sinusoidal steady state analysis, the following transformations are appropriate:

$$v \rightarrow \mathbf{V}$$
$$i \rightarrow \mathbf{I}$$
$$L\frac{d}{dt}(\ ) \rightarrow j\omega \mathbf{L}$$

Thus, Eq. 18.17 becomes

$$\mathbf{V}_1 = j\omega L_1 \mathbf{I}_1 + j\omega M \mathbf{I}_2$$
$$\mathbf{V}_2 = j\omega L_2 \mathbf{I}_2 + j\omega M \mathbf{I}_1 \tag{18.18}$$

The symbology of the coupled circuit in the frequency domain is identical to that in the time domain except for the way the elements and variables are labeled.

EXAMPLE 18.2

We wish to determine the output voltage $\mathbf{V}_o$ in the circuit in Fig. 18.8.

The two KVL equations for the network are

$$+ 24\underline{/30°} - 2\mathbf{I}_1 - j4\mathbf{I}_1 + j2\mathbf{I}_2 = 0$$
$$-j6\mathbf{I}_2 + j2\mathbf{I}_1 - (-j2)\mathbf{I}_2 - 2\mathbf{I}_2 = 0$$

or

$$(2 + j4)\mathbf{I}_1 - j2\mathbf{I}_2 = 24\underline{/30°}$$
$$-j2\mathbf{I}_1 + (2 + j6 - j2)\mathbf{I}_2 = 0$$

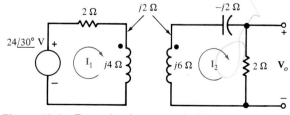

Figure 18.8 Example of a magnetically coupled circuit.

Solving the equations yields

$$\mathbf{I}_2 = 2.68 \underline{/\ 3.43°} \text{A}$$

Therefore

$$\mathbf{V}_0 = 2\mathbf{I}_2$$

DRILL EXERCISE

D18.2. Find the currents $\mathbf{I}_1$ and $\mathbf{I}_2$ and the output voltage $\mathbf{V}_0$ in the network in Fig. D18.2.

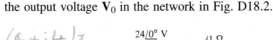

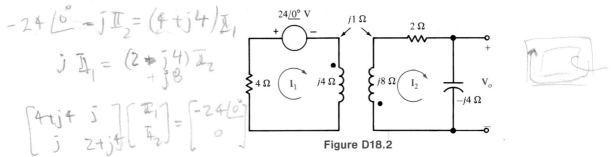

Figure D18.2

Ans: $\mathbf{I}_1 = -4.29 \underline{/-42.8°}$ A, $\mathbf{I}_2 = 0.96 \underline{/-16.26°}$ A, $\mathbf{V}_0 = 3.84 \underline{/-106.26°}$ V.

We define the *coefficient of coupling* between two coils as

$$k = \frac{M}{\sqrt{L_1 L_2}} \tag{18.19}$$

and its range is

$$O \le k \le 1 \tag{18.20}$$

This coefficient is an indication of how tightly the coils are magnetically coupled; that is, if all the flux in one coil links the other coil, then we have 100% coupling and $k = 1$. The previous equations indicate that the value for the mutual inductance is confined to the range

$$O \le M \le \sqrt{L_1 L_2} \tag{18.21}$$

Recall that the energy stored in the magnetic field of an inductor is

$$w_L = \frac{1}{2} L i^2$$

For the two-coil situation of Fig. 18.7, the energy is:

$$w_L = \frac{1}{2} L_1 i_1^2 + \frac{1}{2} L_2 i_2^2 + L_{12} i_1 i_2$$

EXAMPLE 18.3
The coupled circuit in Fig. 18.9a has a coefficient of coupling of 1 (i.e., $k = 1$). We wish to determine the energy stored in the mutually coupled inductors at time $t = 5$ ms. $L_1 = 2.653$ mH and $L_2 = 10.61$ mH.

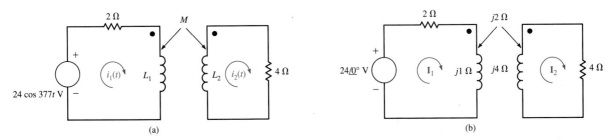

Figure 18.9   Circuit used in Example 18.3.

From the data the mutual inductance is

$$M = \sqrt{L_1 L_2} = 5.31 \text{ mH}$$

The frequency-domain equivalent circuit is shown in Fig. 18.9b, where the impedance values for X_{L_1}, X_{L_2}, and X_M are 1, 4, and 2, respectively. The mesh equations for the network are then

$$(2 + j1)\mathbf{I}_1 - 2j\mathbf{I}_2 = 24\angle 0°$$

$$- j2\mathbf{I}_1 + (4 + 4j)\mathbf{I}_2 = 0$$

Solving these equations for the mesh currents yields

$$\mathbf{I}_1 = 9.41 \angle -11.31° \text{ A} \quad \text{and} \quad \mathbf{I}_2 = 3.33 \angle +33.69° \text{ A}$$

and therefore,

$$i_1(t) = 9.41 \cos (377t - 11.31°) \text{ A}$$

$$i_2(t) = 3.33 \cos (377t + 33.69°) \text{ A}$$

At $t = 5$ ms, $377t = 1.885$ rad or $108°$, and therefore,

$$i_1(t = 5 \text{ ms}) = 9.41 \cos (108° - 11.31°) = -1.10 \text{ A}$$

$$i_2(t = 5 \text{ ms}) = 3.33 \cos (108° + 33.69°) = -2.61 \text{ A}$$

Therefore, the energy stored in the coupled inductors at $t = 5$ ms is

$$w(t)|_{t=0.005 \text{ sec}} = 0.5(2.653)(10^{-3})(-1.10)^2 + 0.5(10.61)(10^{-3})(-2.61)^2$$

$$- (5.31)(10^{-3})(-1.10)(-2.61)$$

$$= (1.55)(10^{-3}) + (36.14)(10^{-3}) - (15.25)(10^{-3})$$

$$= 22.44 \text{ mJ}$$ ■

8.4
The Linear Transformer

A transformer is a device that exploits the properties of two magnetically coupled coils. A typical transformer network is shown in Fig. 18.10. The source is connected to what is called the *primary* of the transformer, and the load is connected to the *secondary*. Thus, R_1 and L_1 refer to the resistance and self-inductance of the primary, and R_2 and L_2 refer to the secondary's resistance and self-inductance. The transformer is said to be *linear* if the magnetic permeability (μ) of the paths through which the fluxes pass is constant as discussed in the previous section. Without the use of high μ material, the coefficient of coupling, k, may be very small. Transformers of this type find wide application in such products as radio and TV receivers.

With reference to Fig. 18.10, let us compute the input impedance to the transformer as seen by the source. The network equations are

$$\mathbf{V}_s = \mathbf{I}_1(R_1 + j\omega L_1) - j\omega M\mathbf{I}_2$$
$$0 = -j\omega M\mathbf{I}_1 + (R_2 + j\omega L_2 + \mathbf{Z}_L)\mathbf{I}_2$$

Solving the second equation for $\mathbf{I}_2$ and substituting it into the first equation yields

$$\mathbf{V}_s = \left(R_1 + j\omega L_1 + \frac{\omega^2 M^2}{R_2 + j\omega L_2 + \mathbf{Z}_L}\right)\mathbf{I}_1$$

Therefore, the *input impedance* is

$$\mathbf{Z}_i = \frac{\mathbf{V}_s}{\mathbf{I}_1} = R_1 + j\omega L_1 + \frac{\omega^2 M^2}{R_2 + j\omega L_2 + \mathbf{Z}_L} \tag{18.22}$$

As we look from the source into the network to determine $\mathbf{Z}_i$, we see the impedance of the primary (i.e., $R_1 + j\omega L_1$) plus an impedance that the secondary of the transformer reflects, due to mutual coupling, into the primary. This *reflected impedance* is

$$\mathbf{Z}_R = \frac{\omega^2 M^2}{R_2 + j\omega L_2 + \mathbf{Z}_L} \tag{18.23}$$

Note that this reflected impedance is independent of the dot locations.

If $\mathbf{Z}_L$ in Eq. (18.23) is written as

$$\mathbf{Z}_L = R_L + jX_L \tag{18.24}$$

then

$$\mathbf{Z}_R = \frac{\omega^2 M^2}{R_2 + R_L + j(\omega L_2 + X_L)}$$

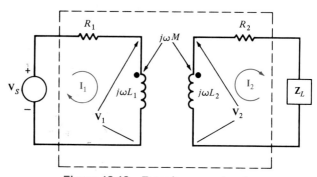

Figure 18.10 Transformer network.

which can be written as

$$\mathbf{Z}_R = \frac{\omega^2 M^2[(R_2 + R_L) - j(\omega L_2 + X_L)]}{(R_2 + R_L)^2 + (\omega L_2 + X_L)^2} \tag{18.25}$$

This equation illustrates that if X_L is an inductive reactance, or if X_L is a capacitive reactance with $\omega L_2 > X_L$, then the reflected reactance is capacitive. If $\omega L_2 + X_L = 0$ (i.e., the secondary is in resonance), $\mathbf{Z}_R$ is purely resistive and

$$\mathbf{Z}_R = \frac{\omega^2 M^2}{R_2 + R_L} \tag{18.26}$$

EXAMPLE 18.4

For the network shown in Fig. 18.11, we wish to determine the input impedance. Following the development that led to Eq. (18.22), we find that the input impedance is

$$\mathbf{Z}_i = 12 + j10 + \frac{(1)^2}{16 + j8 - j4 + 4 + j6}$$

$$= 12.04 + j9.98$$

$$= 15.64 \underline{/39.65°} \ \Omega \qquad\blacksquare$$

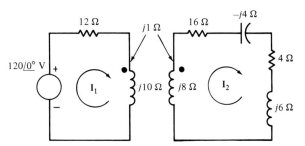

Figure 18.11 Example transformer circuit.

D18.3. Given the network in Fig. D18.3, find the input impedance of the network and the current in the voltage source.

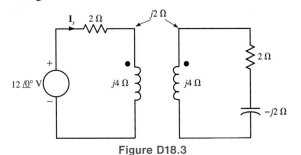

Figure D18.3

Ans: $\mathbf{Z}_i = 3 + j3 \ \Omega$, $\mathbf{I}_s = 2 - j2$ A.

A *perfect transformer* is a linear transformer for which $R_1 = R_2 = 0$, and $k =$ unity.

18.5

The Ideal Transformer

An *ideal transformer* is a linear transformer for which $L_1 \to \infty$, $L_2 \to \infty$, $k \to 1$, $R_1 \to 0$, and $R_2 \to 0$. Consider the situation illustrated in Fig. 18.12, showing two coils of wire wound on a single closed magnetic core. The magnetic core flux links all the turns of both coils. Therefore

$$v_1 = N_1 \frac{d\phi}{dt}$$

$$v_2 = N_2 \frac{d\phi}{dt}$$

and therefore

$$\frac{v_1}{v_2} = \frac{N_1}{N_2} \frac{\dfrac{d\phi}{dt}}{\dfrac{d\phi}{dt}} = \frac{N_1}{N_2} \tag{18.27}$$

Another relationship can be developed between the currents $i_1(t)$ and $i_2(t)$ and the number of turns in each coil. To develop this relationship, we employ Ampere's law. Note in Fig. 18.12 that $i_2(t)$ is defined positive out of the dot. Under this condition, Ampere's law, Eq. (18.4) can be written as

$$\oint \mathbf{H} \cdot dl = i_{enclosed} = N_1 i_1 + N_2(-i_2) \tag{18.28}$$

where $\mathbf{H}$ is the magnetic field intensity and the integral is over the closed path traveled by the flux around the transformer core. For the ideal core material, $\mu = \infty$, and $\mathbf{H} = 0$. Therefore

$$N_1 i_1 - N_2 i_2 = 0 \tag{18.29}$$

or

$$\frac{i_1}{i_2} = \frac{N_2}{N_1} \tag{18.30}$$

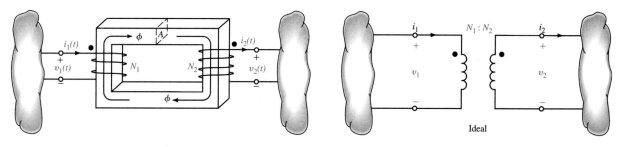

Figure 18.12(a)   Situation resulting in an ideal transformer.   **Figure 18.12(b)**   Circuit symbol.

Figure 18.12   The ideal transformer.

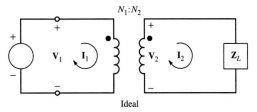

Figure 18.13 Ideal transformer circuit used to illustrate input impedance.

Note that if we divide Eq. (18.29) by N_1 and multiply it by v_1 we obtain

$$v_1 i_1 - \frac{N_2}{N_1} v_1 i_2 = 0$$

However, since $v_1 = (N_1/N_2)v_2$, then

$$v_1 i_1 = v_2 i_2$$

and hence the power *into* the device equals the *output* power, which means that an ideal transformer is *lossless*.

Consider now the circuit shown in Fig. 18.13. The phasor voltages $\mathbf{V}_1$ and $\mathbf{V}_2$ are related by the expression

$$\frac{\mathbf{V}_1}{\mathbf{V}_2} = \frac{N_1}{N_2}$$

and the phasor currents, from Eq. (18.30), are related by

$$\frac{\mathbf{I}_1}{\mathbf{I}_2} = \frac{N_2}{N_1}$$

These two equations above can be rewritten as

$$\mathbf{V}_1 = \frac{N_1}{N_2} \mathbf{V}_2$$

$$\mathbf{I}_1 = \frac{N_2}{N_1} \mathbf{I}_2$$

Also note that

$$\mathbf{S}_1 = \mathbf{V}_1 \mathbf{I}_1^* = \left(\frac{N_1}{N_2} \mathbf{V}_2\right)\left(\frac{N_2}{N_1} \mathbf{I}_2\right)^*$$

$$= \mathbf{V}_2 \mathbf{I}_2^* = \mathbf{S}_2$$

From the figure we note that $\mathbf{Z}_L = \mathbf{V}_2/\mathbf{I}_2$, and therefore the input impedance

$$\mathbf{Z}_i = \frac{\mathbf{V}_1}{\mathbf{I}_1} = \left(\frac{N_1}{N_2}\right)^2 \mathbf{Z}_L \tag{18.31}$$

If we now define the turns ratio as

$$n = \frac{N_2}{N_1} \tag{18.32}$$

then the defining equations for the *ideal transformer* are

$$\mathbf{V}_1 = \frac{\mathbf{V}_2}{n}$$

$$\mathbf{I}_1 = n\mathbf{I}_2$$

$$\mathbf{S}_1 = \mathbf{S}_2 \qquad (18.33)$$

$$\mathbf{Z}_i = \frac{\mathbf{Z}_L}{n^2}$$

Equations (18.33) define the important relationships for an ideal transformer. The signs on the voltages and currents are dependent on the assigned references defined in Fig. 18.12.

EXAMPLE 18.5

Given the circuit shown in Fig. 18.14, we wish to determine all indicated voltages and currents.

Because of the relationships between the dots and the currents and voltages, the transformer equations are

$$\mathbf{V}_1 = -\frac{\mathbf{V}_2}{n} \text{ and } \mathbf{I}_1 = -n\mathbf{I}_2$$

where $n = 1/4$. The reflected impedance at the input to the transformer is

$$\mathbf{Z}_i = 16(2 + j1) = 32 + j16 \ \Omega$$

Therefore, the current in the source is

$$\mathbf{I}_1 = \frac{120\angle 0°}{18 - j4 + 32 + j16} = 2.23\angle -13.5° \text{ A}$$

The voltage across the input to the transformer is then

$$\mathbf{V}_1 = \mathbf{I}_1 \mathbf{Z}_1$$
$$= (2.33\angle -13.5°)(32 + j16)$$
$$= 83.50\angle 13.07° \text{ V}$$

Hence, $\mathbf{V}_2$ is

$$\mathbf{V}_2 = -n\mathbf{V}_1$$
$$= -(1/4)(83.50\angle 13.07°)$$
$$= 20.88\angle 193.07° \text{ V}$$

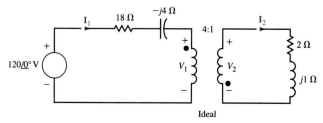

Figure 18.14 Ideal transformer circuit.

The current $\mathbf{I}_2$ is

$$\mathbf{I}_2 = -\frac{\mathbf{I}_1}{n}$$

$$= -4(2.33 \angle -13.5°)$$

$$= 9.32 \angle 166.50° \text{ A} \qquad \blacksquare$$

EXAMPLE 18.6

In the transformer network shown in Fig. 18.15, the commercial load is 1 kW at 0.85 pf lagging. We wish to determine the required source voltage $\mathbf{V}_s$.

Using the relationship $P = |\mathbf{V}||\mathbf{I}|\cos \theta$, we can compute the magnitude of $\mathbf{I}_2$ as

$$I_2 = \frac{1000}{(120)(0.85)}$$

$$= 9.8 \text{ A}$$

The phase angle is

$$\theta = \cos^{-1} 0.85$$

$$= 31.79°$$

Hence

$$\mathbf{I}_2 = 9.8 \angle -31.79° \text{ A}$$

Therefore

$$\mathbf{V}_2 = 2\mathbf{I}_2 + 120 \angle 0°$$

$$= 19.6 \angle -31.79° + 120 \angle 0°$$

$$= 137.05 \angle -4.32° \text{ V}$$

Given the transformer ratio $n = 2$ and the dot locations, we obtain

$$\mathbf{V}_1 = \frac{-\mathbf{V}_2}{n} = -68.52 \angle -4.32° \text{ V}$$

and

$$\mathbf{I}_1 = -n\mathbf{I}_2 = -19.6 \angle -31.79° \text{ A}$$

then

$$\mathbf{V}_s = 6\mathbf{I}_1 + \mathbf{V}_1$$

$$= 6(-19.6 \angle 31.79°) + (-68.52 \angle -4.32°)$$

$$= 181.18 \angle 158.26° \text{ V}$$

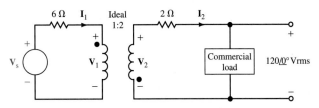

Figure 18.15   Ideal transformer feeding a commercial load.

D18.4. Compute the current $\mathbf{I}_1$ in the network in Fig. D18.4.

Ans: $\mathbf{I}_1 = 3.07 \underline{/39.81°}$ A.

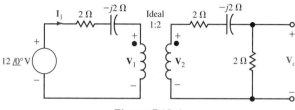

Figure D18.4

D18.5. Find $\mathbf{V}_0$ in the network in Fig. D18.4.

Ans: $\mathbf{V}_0 = 3.07 \underline{/39.82°}$ V.

18.6

Transformer Applications

As we have noted, transformers can be used in applications which require the conversion of transient or ac voltage from one level to another. Practical transformers have "ratings," that is, values at which they are designed to operate. Typically ratings are interpreted as upper limits of operation, although this is not true in general, since sometimes it is potentially harmful to operate the device below its ratings. The two basic ratings of transformers involve rms voltage and apparent power, that is, the voltage level(s) at which the transformer was designed to operate, and the power it was designed to process. A very common engineering problem is to determine appropriate transformer ratings for a specific application.

There are two broad categories of use: "electronic" transformers, which operate at the microwatt, milliwatt, or watt level, and "power" transformers, which process kilowatts or megawatts. We shall discuss two common applications of electronic and power transformers.

Electronic Transformers. Consumer electronic equipment includes radios, television sets, VCRs, compact disk players, personal computers, and dozens of other devices. These devices contain electronic circuits designed for many functions, such as amplification, signal processing, encoding, and decoding. All such circuits require a source of electrical energy to properly operate, typically in the dc form. If the power requirements are very small, that is, in the micro or milliwatt range, batteries may serve this function. However, at larger power levels, the appropriate batteries may be too large, heavy, and/or expensive; in addition, batteries need to be replaced periodically.

Another option is to connect the device to the residential ac 120 V system. There are two obvious problems: first, most electronic circuits operate at a much lower voltage level, and second, the required source is typically dc. A circuit designed to accept 120 V ac as input and convert power to different voltage levels is called an electronic "power supply," as was first introduced in Chapter 8. The transformer frequently is used to solve the voltage conversion problem.

Consider the circuit of Fig. 18.16. Suppose that electronic loads require 5 volts dc and a maximum of 17.3 watts to operate properly. The filter capacitor will smooth the volt-

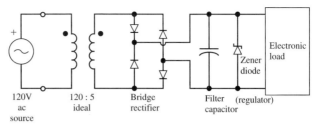

Figure 18.16   A typical electronic power supply employing a transformer.

age to the peak secondary value, and the zener diode will "regulate," or clamp the voltage to 5 volts, as long as the capacitor voltage is ≥ 5 V (with the zener disconnected). It is good design to provide a little more than 5 volts to accommodate low ac input voltage, the internal transformer voltage drop, and the forward voltage drop in the rectifier diodes. Suppose the total forward voltage drop, including compensation for low voltage, is estimated at 2.0 V. This would produce a maximum transformer secondary voltage of $5.0 + 2.0 = 7.0$ volts for a 120 V ac rms transformer primary input voltage. Thus, V_2 rms $= 7.00/\sqrt{2} = 4.950$ V. An appropriate transformer rating would be 120 V: 5 V; 20 VA.

Consider the engineering issues even in this simple situation. If we undersize the transformer, it will get too hot, damaging itself and nearby components. If we oversize it, it will be too big, too heavy, and too expensive. If we are confused about conversion between rms and maximum values we will produce the wrong output voltage, possibly destroying the zener diode and damaging the circuits it supplies. Finally we had to consider that the transformer was approximately a power-invariant device. In a practical design we would need more precise data about low voltage operation and transformer series voltage drop.

Power Transformers. Bulk electrical power is optimally generated at "medium" ac voltage levels of from about 7 to 25 kV; whereas it is transmitted from generation to load points at "high" ac voltage levels, ranging from 115 to 800 kV. Thus, it is required to convert large amounts of power from medium voltage to high voltage at high efficiency. The power transformer is admirably suited for this purpose. Power at this level is always processed in the balanced polyphase form. Typically, but not always, the transformer itself is a three-phase device; for our purposes we shall use three single-phase transformers.

EXAMPLE 18.7

The specific application we shall consider in this example is connecting a 17 kV, 800 MVA, 0.866 power factor (PF) lagging, three-phase generator into a 500 kV transmission system. Our job is to determine appropriate ratings and compute the currents everywhere. See Fig. 18.17. Note that we have elected to use a delta-wye transformer connection. We shall model the transformers as ideal.

Operating at rated conditions, the magnitude of the generator voltages are:

$$V_{AB} = V_{BA} = V_{BC} + V_{CA} = 17 \text{ kV}$$

See the phasor diagram for phase angles for all quantities. The generator currents are

$$I_A = I_B = I_C = \frac{800 \times 10^6}{\sqrt{3} \times 17 \text{ k}} = 27.17 \text{ kA}$$

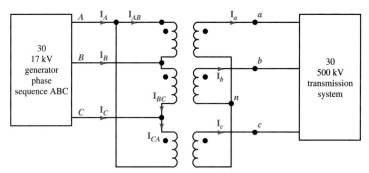

Figure 18.17(a)   Circuit diagrams.

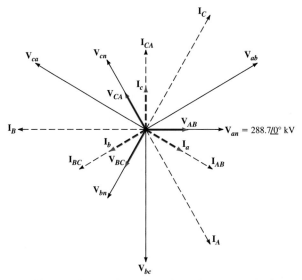

Figure 18.17(b)   Phasor diagram in which all angles between phasors are 30°.

Figure 18.17   Diagrams used in Example 18.7.

Also

$$I_{AB} = I_{BC} = I_{CA} = 27.17 \text{ k}/\sqrt{3} = 15.69 \text{ kA}$$

It is desired to make

$$V_{ab} = V_{bc} = V_{ca} = 500 \text{ kV}$$

Therefore

$$V_{an} = V_{bn} = V_{cn} = 50 \text{ k}/\sqrt{3} = 288.7 \text{ kV}$$

The high-side currents will be

$$I_A = I_B = I_C = 800 \times 10^6/(\sqrt{3} \cdot 500 \text{ k}) = 923.8 \text{ A}$$

Thus, an appropriate transformer bank would be rated as

Three single-phase 266.7 MVA 17 kV:289 kV transformers connected
low voltage delta; high voltage wye.   ∎

The foregoing examples should give the reader an appreciation for the dozens of applications of transformers in electrical systems. Advantages of transformers include simplicity, high efficiency, and electrical isolation between primary and secondary. Disadvantages include bulk, weight, and fairly high cost. Advances in power electronics have enabled electronic circuits to replace transformers in some cases; however, the transformer remains an important component in many practical applications.

18.7

Summary

- Magnetic circuits form the basis for understanding the operation of transformers, generators, and motors.
- Magnetic fields are produced by permanent magnets and current-carrying conductors.
- There is a direct analogy between an electric circuit and a magnetic circuit.
- Linear transformers, that is, those in which the magnetic permeability of the paths through which the fluxes pass is constant, are widely used in radio and TV receivers.
- Ideal transformers are lossless devices in which the flux path is on infinitely permeable magnetic material.
- Transformers are used in electronic and power applications.

PROBLEMS

18.1. At a particular point in an unknown material, the magnetic flux density is $B = 0.6$ Wb/m^2 and the magnetic field intensity is $H = 20,000$ A/m. Compute the relative permeability of the material.

18.2. The magnetic flux density $B = 0.20$ Wb/m^2 at a point in a material with a relative permeability of 80. Find the magnetic field intensity at the point.

18.3. The core in Fig. P18.3 is made of cast iron. If the core is excited by 200 amp-turns, determine the flux in the core.

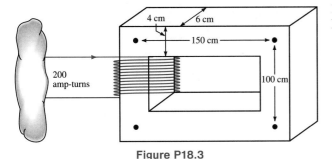

Figure P18.3

18.4. The core in Fig. P18.4 is made of cast iron. The desired flux in the core is 0.0015 Wb. Determine the number of amp-turns required to generate this flux.

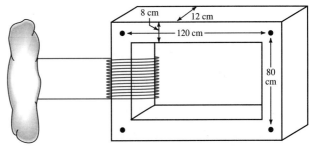

Figure P18.4

18.5. Write the mesh equations for the circuit shown in Fig. P18.5.

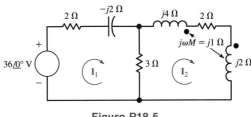

Figure P18.5

18.6. Write the equations necessary to find $\mathbf{V}_o$ in the network shown in Fig. P18.6.

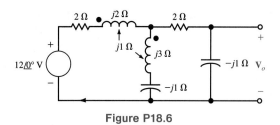

Figure P18.6

18.7. Write the circuit equations necessary to find I_2 in the network shown in Fig. P18.7.

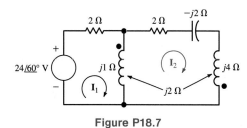

Figure P18.7

18.8. Find I_2 in the network in Fig. P18.8.

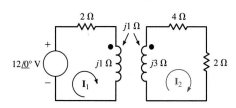

Figure P18.8

18.9. Find the current I_2 in the circuit in Fig. P18.9.

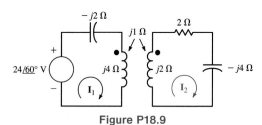

Figure P18.9

18.10. Find the current I_1 in the circuit in Fig. P18.10.

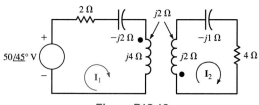

Figure P18.10

18.11. Find the voltage V_o in the network shown in Fig. P18.11.

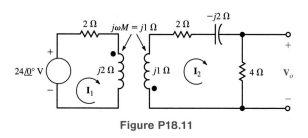

Figure P18.11

18.12. Find the voltage gain V_o/V_s of the network shown in Fig. P18.12.

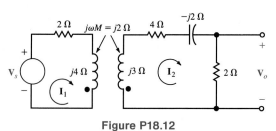

Figure P18.12

18.13. Find V_A in the network shown in Fig. P18.13.

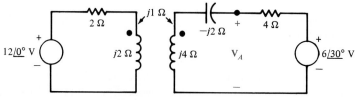

Figure P18.13

18.14. Find $\mathbf{V}_o$ in the network in Fig. P18.14.

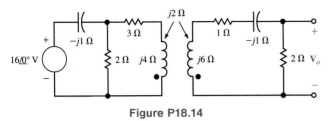

Figure P18.14

18.15. Find $\mathbf{I}_A$ in the network in Fig. P18.15.

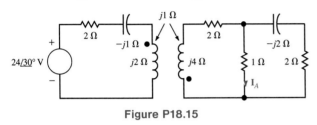

Figure P18.15

18.16. Find the voltage gain $\mathbf{V}_o/\mathbf{V}_s$ for the network shown in Fig. P18.16.

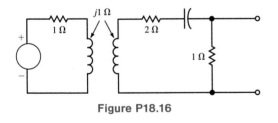

Figure P18.16

18.17. Determine the impedance seen by the source in the network shown in Fig. P18.17.

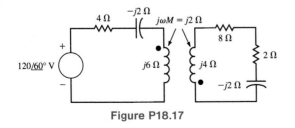

Figure P18.17

18.18. Determine the impedance seen by the source in the network in Fig. P18.18.

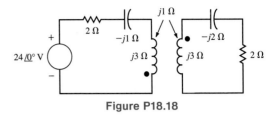

Figure P18.18

18.19. Given the network shown in Fig. P18.19, determine the value of the capacitor C that will cause the reflected impedance to the primary to be purely resistive.

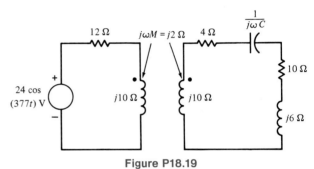

Figure P18.19

18.20. Analyze the network in Fig. P18.20 and determine if a value of X_C can be found such that the output voltage is equal to twice the input voltage.

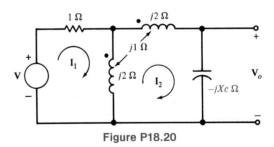

Figure P18.20

18.21. Calculate all currents and voltages in the circuit shown in Fig. P18.21.

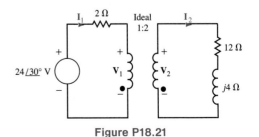

Figure P18.21

18.22. Given that $\mathbf{V}_o = 48 \angle 30°$ V in the circuit shown in Fig. P18.22, determine $\mathbf{V}_s$.

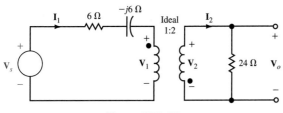

Figure P18.22

18.23. If the voltage source $\mathbf{V}_s$ in the circuit of problem 18.22 is $50 \angle 0°$ V, determine $\mathbf{V}_o$.

18.24. In the network in Fig. P18.24 the voltage $\mathbf{V}_o = 10 \angle 0°$ V. Find the input voltage $\mathbf{V}_s$.

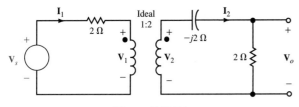

Figure P18.24

18.25. Determine the input impedance seen by the source in the network shown in Fig. P18.25.

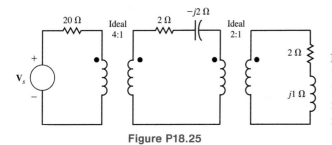

Figure P18.25

18.26. Determine the average power delivered to each resistor in the network shown in Fig. P18.26.

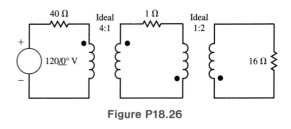

Figure P18.26

18.27. Given the network in Fig. P18.27, show that $\mathbf{V}_2$ can be written as

$$\mathbf{V}_2 = \frac{n\mathbf{Z}_L}{n^2\mathbf{Z}_s + \mathbf{Z}_L} \, \mathbf{V}_s$$

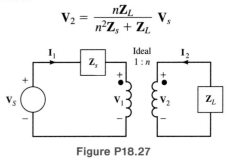

Figure P18.27

18.28. In the network shown in Fig. P18.28, determine the value of the load impedance for maximum power transfer.

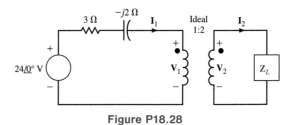

Figure P18.28

18.29. The output stage of an amplifier is to be matched to the impedance of a speaker as shown in Fig. P18.29. If the impedance of the speaker is 8 Ω and the amplifier requires a load impedance of 3.2 kΩ, determine the turns ratio of the ideal transformer.

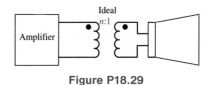

Figure P18.29

18.30. If the load voltage in the network shown in Fig. P18.30 is $100 \angle 0°$ V rms, determine the input voltage $\mathbf{V}_s$.

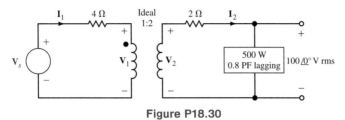

Figure P18.30

18.31. In the circuit shown in Fig. P18.31, if $\mathbf{V}_2 = 120 \angle 0°$ V rms and the load $\mathbf{Z}_L$ absorbs 400 W at 0.9 PF lagging, determine the wattmeter reading.

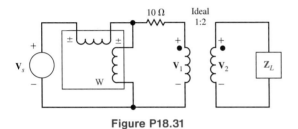

Figure P18.31

18.32. In the network shown in Fig. P18.32, if $\mathbf{V}_L = 120 \angle 0°$ V rms and the load $\mathbf{Z}_L$ absorbs 500 W at 0.85 PF lagging, compute the voltage $\mathbf{V}_s$.

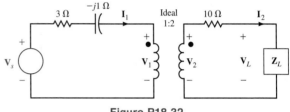

Figure P18.32

19.1
Introduction

In this chapter we will discuss the basic fundamentals and some of the applications of dc machines. First, we will describe the basic physical principles which govern their operation. We will then show how these concepts can be employed to construct a dc machine. The equivalent circuits for dc motors and generators will be presented and analyzed to exhibit some of their operating characteristics. Finally, we will describe some applications, which demonstrate the use of dc machines in industry.

19.2
Fundamental Concepts

Perhaps the easiest method of describing the fundamentals which govern the operation of a dc machine is to first examine what is called a linear machine. A simple linear machine is shown in Fig. 19.1. In this figure, a voltage source is connected through a switch to a pair of electrical conducting rails. A bar which makes electrical contact with the rails is free to move along the rails in the x direction. The rails and bar are present in a constant magnetic $\mathbf{B}$ field that is directed into the page, that is, in the negative z direction. Let us now examine the operation of this machine in three separate time intervals, that is, $t < t_0$, $t = t_0$, and $t > t_0$. Prior to the time $t = t_0$, no current exists in the rails.

According to the Lorentz force equation, the force which is exerted upon the sliding bar is

$$\mathbf{F} = (\mathbf{i} \times \mathbf{B})\,\ell \qquad (19.1a)$$

where $\times$ represents the *vector cross product,* $\mathbf{F}$ is measured in the positive x direction,

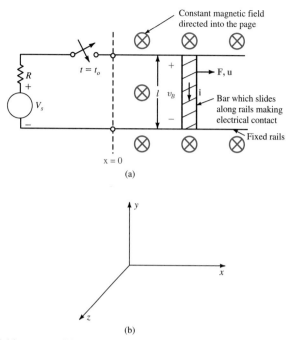

(a)

(b)

Figure 19.1 (a) Linear machine; (b) coordinates used in discussing the linear machine.

and the vector **i** is measured in the direction of the positive current flow. Since there is no force acting on the bar, the velocity **u** of the bar is zero. Therefore, in view of Faraday's law, we find that

$$v_B = |(\mathbf{u} \times \mathbf{B})\,\ell|$$ (19.2a)

$$= 0$$

and hence the voltage developed across the bar is zero.

At time t_0, the magnitude of the current is $i\,(t_0) = v_s/R$. This current produces a force (**F**) on the bar. The direction of the force defined by the vector cross product is obtained using the *right-hand rule,* as illustrated in Fig. 19.2. If we rotate the vector **i** into the vector **B,** the force is in the direction as shown. Note that this is the direction of motion of a right-hand screw. The force on the bar causes the bar to accelerate in the $+x$ direction. This is actually the principle behind the rail gun. As the bar gains speed, the voltage across

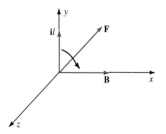

Figure 19.2 Relationships for the Lorentz force equation.

the bar begins to increase in accordance with the expression in Eq. (19.2). The polarity of the induced voltage v_B is obtained by the right-hand rule in which the vector cross product points to the positive terminal.

Since **F, B,** and **i** (and correspondingly, **u,** and **B**) are perpendicular, the scalar versions of Eqs. (19.1a) and (19.2a) are

$$F = iB\ell \tag{19.1b}$$

and

$$v_B = uB\ell \tag{19.2b}$$

For time $t > t_0$, the magnitude of the current is given by the expression

$$i(t) = [v_s - v_B(t)]/R \tag{19.3}$$

and the bar continues to accelerate until

$$v_B(t) = uB\ell = v_s$$

At this point the current is zero, the force is then zero, the bar moves at constant speed, and the machine is in an equilibrium state. Conversion of energy from electrical to mechanical form is the action of a *motor.*

Now suppose that by some means we increase the speed of the bar beyond the equilibrium speed. This increase in velocity causes the induced voltage $v_B(t)$ to become greater than the applied voltage v_s, which, in turn, reverses the direction of the current. Using the Lorentz force equation, we find that reversing the direction of the current also reverses the direction of the force on the bar. Since the bar tries to move in the negative x direction, we must apply force in the positive x direction in order to keep the bar moving in this direction at a constant speed, which is greater than the equilibrium speed. The application of this external force produces a voltage at the input terminals, and therefore, in this mode the linear machine acts like a *generator.*

Note that the key factor which determines whether the machine operates as a motor or generator is the velocity of the bar with respect to the equilibrium speed.

DRILL EXERCISE

D19.1. Consider the system in Fig. 19.1. Suppose the **B** field into the page is 2 Wb/m², $R = 5\ \Omega$, $\ell = 1$ m, $v_s = 150$ V, and the switch has been closed "forever." At what speed (u) must the bar move for the current (i) to be (a) zero, (b) +10 A, or (c) −10 A?

Ans: (a) $u = +75$ m/s, (b) $u = +50$ m/s (motoring), and (c) $u = +100$ m/s (generating).

19.3
A Simple Rotating Machine

Consider the rotating loop of wire shown in Fig. 19.3(a). The loop rotates in a counterclockwise direction around the z axis, and is immersed in a constant magnetic **B** field that is directed along the negative x axis. θ is used to measure the angular position of the loop and is positive in the counterclockwise direction. In addition, we define $\theta = 0$ when the rotating loop lies in the xz plane.

With reference to Fig. 19.3(a), let us consider the conductor ab, that is, the top side of the rectangular loop $abcd$. A two-dimensional "free body diagram" of the conductor immersed in the **B** field is shown in Fig. 19.3(b). The magnitude of the tangential velocity of the conductor $\mathbf{u}_{tan}$ is $r\omega$ where r is the radius of the loop and ω is the angular velocity. The induced voltage along the conductor ab, which we will call e_{ab}, is then

$$e_{\mathbf{ab}} = |(\mathbf{u}_{tan} \times \mathbf{B})\,\ell| \qquad (19.4)$$

The velocity $\mathbf{u}_{tan}$ can be split into two components; one tangential to the **B** field and one perpendicular to it. The cross product of the component of $\mathbf{u}_{tan}$ along the **B** field, with the

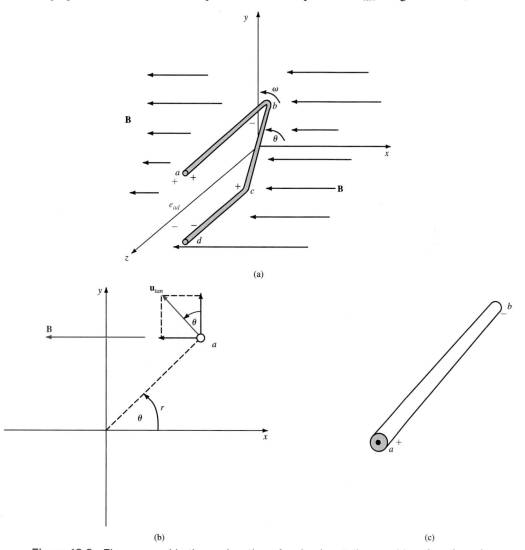

(a)

(b)   (c)

Figure 19.3 Figures used in the explanation of a simple rotating machine. (continues)

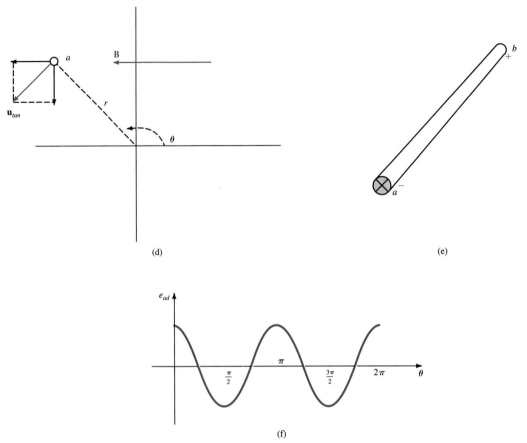

(d)

(e)

(f)

Figure 19.3 (cont.)

B field, is zero. However, the cross product of the velocity component perpendicular to the **B** field with the **B** field produces an induced voltage (out of the page) as shown in Fig. 19.3(c). If the **B** field is constant, then as shown in Fig. 19.3(b), Eq. (19.4) can be expanded as

$$e_{ab} = (r\omega \cos\theta)\, B\ell \tag{19.5}$$

Suppose now that the conductor ab is in the position shown in Fig. 19.3(d). Once again, only the velocity component perpendicular to the **B** field yields a nonzero cross product. However, in this case the cross product produces an induced voltage (into the page) as shown in Fig. 19.3(e). A similar argument follows for the conductor cd and hence

$$e_{ad} = e_{ab} + e_{cd}$$

$$= 2Br\ell\omega\cos\theta \tag{19.6}$$

Note that the voltage e_{ad} is a maximum at $\theta = 0°$. Furthermore, the analysis illustrates that the induced voltage as a function of the angle θ is of the form shown in Fig. 19.3(f).

DRILL EXERCISE

D19.2. Consider the U-shaped coil shown in Fig. 19.3(a). Its dimensions are $ab = cd = 50$ cm and $bc = 20$ cm. $B = 2$ Wb/m² and $\omega = 100$ rad/s. Evaluate the voltage e_{ad} at the following coil positions: (a) $\theta = 0°$, (b) $\theta = 45°$, (c) $\theta = 90°$, (d) $\theta = 135°$, (e) $\theta = 180°$, (f) $\theta = 225°$, (g) $\theta = 270°$, and (h) $\theta = 315°$.

Ans: (a) +20 V, (b) +14.14 V, (c) 0 V, (d) −14.14 V, (e) −20 V, (f) −14.14 V, (g) 0 V, (h) +14.14 V.

Now suppose that our loop of wire *abcd* in Fig. 19.3(a) is connected to what is called a segmented ring, which is composed of two semicircular pieces of metal as shown in Fig. 19.4. These segments are in constant contact with *brushes,* which are the connection points

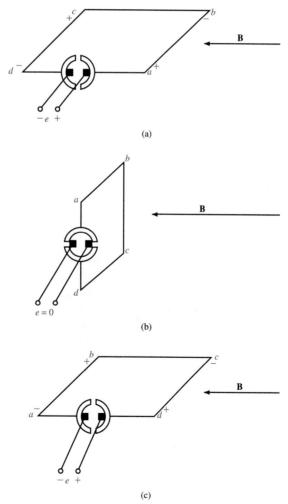

(a)

(b)

(c)

Figure 19.4   Demonstration of commutator action for a single loop.

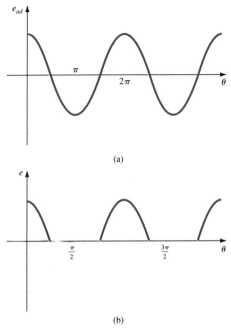

(a)

(b)

Figure 19.5   (a) Loop voltage, and (b) Brush voltage for the rotating wire in Fig. 19.4.

for the loop voltage. As the loop rotates, the segments slide under the brushes. This mechanism is called a *commutator,* which is essentially a mechanical rectifier.

When the loop is in the position shown in Fig. 19.4(a), the voltage e is positive and decreasing as the loop rotates to the position shown in Fig. 19.4(b). When the loop is in the position shown in Fig. 19.4(b), note that the loop voltage is zero. As the loop rotates beyond the position shown in Fig. 19.4(b), the voltage e is positive and increasing as the loop rotates to the position shown in Fig. 19.4(c). Therefore, while the voltage between the commutator segments, that is, e_{ad}, is as shown in Fig. 19.5(a), the voltage at the brushes, that is, e, is as shown in Fig. 19.5(b). Note that this voltage shown in Fig. 19.5(b) is a rectified version of the voltage shown in Fig. 19.5(a).

The previous analysis illustrates the basic principles of a rotating machine. However, in an actual machine, there are numerous coils of wire and commutator sections as shown in Fig. 19.6. Both the loops of wire and the commutator are built on a shaft (referred to

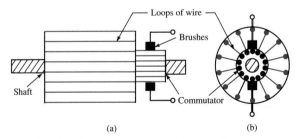

(a)   (b)

Figure 19.6   Multiple loops and commutator sections built on a shaft.

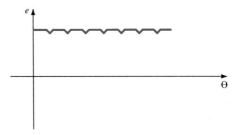

Figure 19.7   Terminal voltage of a rotating machine with multiple loops and commutator sections.

as the *armature* or *rotor*) as shown in Fig. 19.6. The brushes are mounted on the frame which encloses this rotating shaft and contains the materials necessary to construct the **B** field. When multiple loops and commutator sections are used, the terminal voltage will resemble that shown in Fig. 19.7, which when smoothed will yield a constant dc voltage. It is important to note that while the geometry of the coils and commutator are represented as shown in Fig. 19.6, the actual electrical connections are more complex.

19.4
The Basic dc Machine

At this point let us try to relate the concepts we have just discussed to an actual machine. The diagram in Fig. 19.8(a) illustrates the basic design of a *dc machine.* For simplicity, we assume that the rotor has a single coil (loop) as shown in both Figs. 19.8(a) and (b). The rotor and the two field poles which surround it are made of iron. The voltage v_f establishes the current i_f. Because of Ampere's rule, this current creates a constant magnetic field as shown in Fig. 19.8(a). Figure 19.9 indicates that flux lines of the magnetic field are perpendicular to the rotor at every point except in the two small gaps where the rotor is not covered by a field pole. Like current, which takes the path of least resistance, the

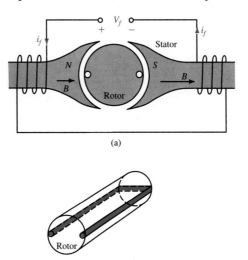

(a)

(b)

Figure 19.8   (a) A basic dc machine; (b) rotor displaying a single coil.

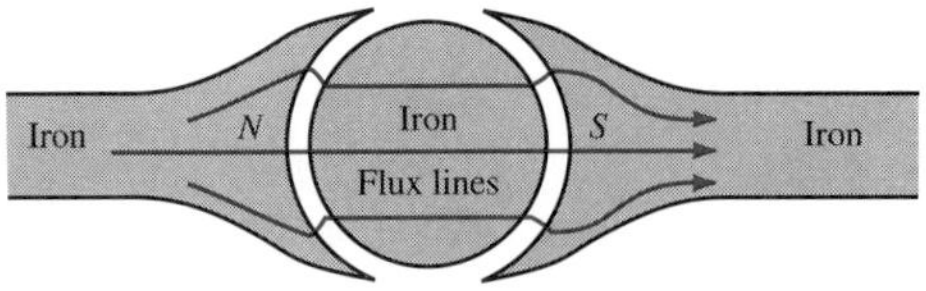

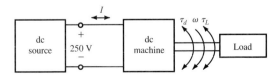

Figure 19.9 Illustration of the flux paths in a dc machine.

flux will follow the path of least reluctance (essentially magnetic resistance). This path is the shortest possible distance from one piece of iron to another through the least amount of air. Therefore, the *B* field is essentially always perpendicular to the velocity vectors of the loop of wire except in the interpolar regions.

Note that the basic dc machine we have described satisfies the conditions that the conducting loop rotates in the constant magnetic field in such a way that the conductor's velocity vector is always perpendicular to the field except in the interpolar regions. The use of multiple conducting loops and commutator sections will then generate a voltage at the brushes which is of the form shown in Fig. 19.7.

Thus far we have described the basic principles of a *dc machine*. However, the machine can be operated either as a motor or a generator. In the *motor mode* the input is electrical power and the output is mechanical power. When operated in the *generator mode*, mechanical power is the input and electrical power is the output. The following example illustrates the dual modes of a dc machine.

EXAMPLE 19.1

A permanent magnet field dc machine is connected as shown in Fig. 19.10, and may be assumed to be 100% efficient, that is, $P_{in} = P_{out} = 40$ kW. The speed is $\omega = 100$ rad/s. We wish to find the directions and magnitude of the current and torques if the dc machine is operating in (a) the motor mode, and (b) the generator mode.

(a) In the motor operating mode the current I shown in Fig. 19.10 is

$$I = \frac{P_{in}}{250}$$
$$= 160 \text{ A}$$

and is directed *into* the machine.

The developed torque, τ_d, is given by the fundamental relationship

$$\tau_d = \frac{P_{out}}{\omega}$$
$$= \frac{40000}{100}$$
$$= 400 \text{ Nm}$$

Figure 19.10 Schematic diagram used in Example 19.1.

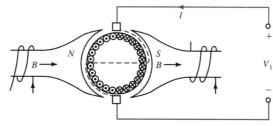

Figure 19.11   Illustration of the current path in a dc motor.

and is *in* the direction of rotation. τ_L, the load torque, is always equal and opposite to τ_d for steady state operation.

(b) In the generator mode the current is $I = 160$ A as calculated in part (a); however, the current is directed *out of* the machine. In addition, the torque $\tau_d = 400$ Nm is in the direction *opposite to* that of rotation.   ■

In the more general case we can illustrate the motoring action of a dc machine via Fig. 19.11. The voltage V_1 causes a current I to exist in the loops which are embedded in the rotor. The current enters the upper brush and through commutator action, is directed into the loops under the S pole. The current goes around the loops on the rotor and out of the loops under the N pole. The dotted line in the figure illustrates the current path for a single loop. Applying the Lorentz force equation to the single loop, we find that for the portion of the single loop under the S pole, $F = (B\ell I)$ yields a downward force producing a clockwise rotation. For the portion of the loop under the N pole, $F = (B\ell I)$ yields a force in the upward direction, which also produces a clockwise rotation. The result of the force produced by all loops is the creation of a torque. This torque is proportional to the force, that is,

$$\tau \propto F\ell$$
$$\propto I\ell B$$

Thus, we find that the current produced by V_1 produces a torque which causes the rotor to rotate. The rotor speed will continue to increase until the induced voltage, that is, $(u \times B)\,\ell$ is equal to the applied voltage V_1. This equality defines the stable operating speed of the motor.

Our previous discussion has employed a boldface notation for the variables because our description of the physical phenomena required the use of vectors. At this point in our analysis, however, we recognize that the machine is designed in such a way that all quantities are mutually perpendicular and hence the same equations hold for scalar magnitudes. Thus, we can simplify the analysis by the subsequent use of scalar notation.

19.5
Equivalent Circuits and Analysis

The schematic diagram for a dc machine operating in the motor mode is shown in Fig. 19.12(a). The input is electrical power and the output is mechanical power. τ_d and P_d represent the *electromagnetic developed torque* and *power,* respectively. τ_L represents the *load torque,* which opposes the direction of rotation. J is the moment of inertia.

The equivalent circuits for the motor are shown in Fig. 19.12(b). R_f represents the re-

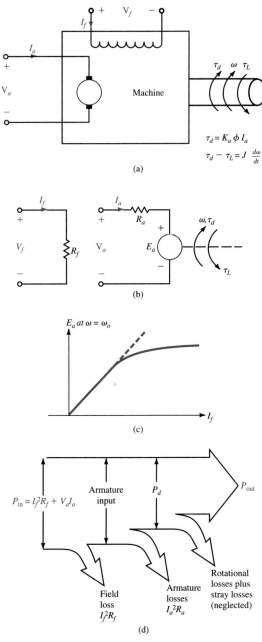

Figure 19.12   dc machine characteristics: motor sign conventions. (a) Schematic diagram; (b) equivalent circuits; (c) magnetization curve; (d) power flow in a motor mode.

sistance of the separately excited field windings, and the field (stator) circuit establishes the magnetic **B** field within the machine. In the armature (rotor) circuit V_a is the voltage at the commutator terminals, R_a is the equivalent resistance of the rotating windings, I_a is the armature current, and E_a is the total voltage induced between brushes.

The circuit equations for the equivalent circuits are derived from KVL as

$$V_a = I_a R_a + E_a$$
$$V_f = I_f R_f \tag{19.7}$$

In addition, from Faraday's law we know that

$$E_a = K_a \phi \omega \tag{19.8}$$

where K_a is a constant and ϕ the flux per pole. In separately excited or permanent magnet rotors, one may consider ϕ constant and express the equation as

$$E_a = K_\phi \omega \tag{19.9}$$

where $K_\phi = K_a \phi$. Information on $K_a \phi$ versus I_f is available from the manufacturer in the form of a *magnetization curve* such as that shown in Fig. 19.12(c). The curve is generated by running the machine at some reference rated speed (ω_0) under open circuit armature conditions ($I_a = 0$), and measuring $V_a = E_a$ versus I_f. The relationship exhibited in the figure occurs because ϕ is proportional to the **B** field and, in turn, the **B** field is created by $N_f I_f$. Therefore, E_a is "caused by" I_f.

It is desirable to operate close to the knee of the magnetization curve for maximum utilization of the iron. Note that machine weight depends on the size and density of the field poles, which determine the **B** field.

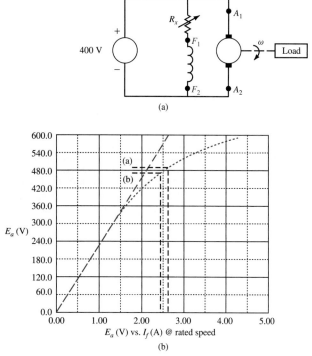

(a)

E_a (V) vs. I_f (A) @ rated speed

(b)

Figure 19.13 (a) Schematic diagram and (b) Magnetization curve employed in Example 19.2.

The power converted from electrical to mechanical form, that is, the developed power, as shown in Fig. 19.12(d), is

$$P_d = E_a I_a$$
$$= K_a \phi \omega I_a \tag{19.10}$$

Consequently, the developed torque is

$$\tau_d = \frac{P_d}{\omega}$$
$$= \frac{K_a \phi \omega I_a}{\omega} \tag{19.11}$$
$$= K_a \phi I_a$$

And if ϕ is constant

$$\tau_d = K_\phi I_a \tag{19.12}$$

Equations (19.7) to (19.12) are the basic equations which govern fundamental dc machine analysis.

EXAMPLE 19.2

A dc machine with the following parameters has the schematic diagram shown in Fig. 19.13(a) and the magnetization curve shown in Fig. 19.13(b).

dc Machine Parameters

Ratings

Arm. Voltage	=	400.0 V;	Horsepower	=	60.0 hp
Arm. Current	=	119.7 A;	No. of poles	=	4
Field Current	=	4.27 A;	Rated Speed	=	1800 rpm = 188.5 rad/s
Base Speed	=	1200 rpm	J Motor	=	0.581 kg-m sq.

Equivalent Circuit Values

Ra = 0.0669 ohms;	Rf = 93.60 ohms

We wish to find the field current, the armature current, the operating mode, and the developed torque and power if the shaft speed is 1500 rpm and the field rheostat (variable resistor) is set to (a) 60.4 Ω and (b) 68.4 Ω.

(a) If the field rheostat is set to 60.4 Ω, then

$$I_f = \frac{400}{93.6 + 60.4}$$
$$= 2.597 \text{ A}$$

From the magnetization curve

$$K_a \phi = \frac{488.2}{188.5}$$
$$= 2.590 \text{ Wb}$$

Since

$$\omega = \frac{2\pi \times 1500}{60}$$
$$= 157.1 \text{ rad/s}$$

Then

$$E_a = K_a\phi\omega$$
$$= 406.9 \text{ V}$$

Since $E_a > V_a$, current is directed *out of* the machine

$$I_a = \frac{E_a - V_a}{R_a}$$
$$= 103.5 \text{ A}$$

and the machine operates as a *generator.* The developed torque and power are then

$$\tau_d = K_a\phi I_a$$
$$= 268.1 \text{ Nm}$$

and

$$P_d = \tau_d\omega$$
$$= 42.12 \text{ kW}$$

and since 746 watts = 1 hp

$$P_d = 56.46 \text{ hp}$$

(b) If the field rheostat is set to 68.4 Ω, then

$$I_f = \frac{400}{93.6 + 68.4}$$
$$= 2.469 \text{ A}$$

For the magnetization curve we find that

$$K_a\phi = \frac{475.6}{188.5}$$
$$= 2.523 \text{ Wb}$$

And since $\omega = 157.1$ rad/s

$$E_a = K_a\phi\omega$$
$$= 396.3 \text{ V}$$

In this case $E_a < V_a$ and hence current is directed *into* the machine and

$$I_a = \frac{V_a - E_a}{R_a}$$
$$= 54.7 \text{ A}$$

and the machine operates as a *motor.*

The developed torque and power are

$$\tau_d = K_a \phi I_a$$
$$= 137.9 \text{ Nm}$$

and

$$P_d = \tau_d \omega$$
$$= 21.66 \text{ kW}$$
$$= 29.03 \text{ hp}$$ ∎

At this point let us investigate speed control under varying load conditions. In order to do this, we need a relationship between the output torque and the shaft speed. From Eq. (19.12)

$$\tau_d = K_\phi I_a$$

Recall that the circuit equation for the armature is

$$V_a = I_a R_a + E_a$$

Therefore, the armature current can be written as

$$I_a = (V_a - E_a)/R_a$$

and since

$$E_a = K_\phi \omega$$

I_a can be expressed as

$$I_a = [V_a - K_\phi \omega]/R_a$$

and then the developed torque is

$$\tau_d = \frac{V_a K_\phi}{R_a} - \left(\frac{K_\phi^2}{R_a}\right)\omega$$

The torque/speed curve represented by this equation is shown in Fig. 19.14. Note that V_a and R_a are known constants, and K_ϕ for the separately excited machine is dependent only upon the field current. Note that this torque/speed characteristic permits us to determine the changes which will occur in speed as a result of varying load conditions, that is, changes in torque.

Machine manufacturers may provide multiple windings which can be employed to cre-

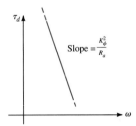

Figure 19.14   Torque/speed curve for the dc machine.

ate the constant magnetic field. The different methods in which these windings are used lead to machines operating as motors or generators with different torque-speed characteristics. Two types of field windings are *shunt* and *series*. The shunt winding can be either *self-excited* or *separately excited,* that is, the terminals of the winding can be connected across the input/output voltage terminals or fed from a separate voltage source. The series winding is normally in series with the armature and therefore the series field current is the armature current.

If both series and shunt windings are used, the machine is said to be compounded. If the situation is such that the mmf's add, this connection is called *additive* or *cumulative compounding.* If the fields are connected so that the mmf's oppose one another, then the connection is called *subtractive* or *differential compounding.*

The steady state performance of a dc machine can be modeled with reasonable accuracy using the simple circuits, graphs, and equations summarized in Fig. 19.12. Hence, at least in part, dc machine analysis reduces to dc circuit analysis. The key point to note is that the machine is an electrical-magnetic-mechanical device, and its operation can only be understood when the interaction between these EMM phenomena is properly accounted for.

In addition to the machines described above, there also exists a *brushless dc machine.* However, this device is discussed in the following chapter on ac machines since it employs what is traditionally called a three-phase synchronous machine.

19.6 Applications of dc Machines

dc machines are widely used in industrial and commercial applications. Examples of small machine applications include automobile power windows, food blenders, hand power tools, and record player turntables. Examples of large machine applications include elevators, hoists, and heavy metal rolling mills.

Our final example will deal with a specific real world use of dc machines. This *example* is more of a *case study* since the analysis will consider a wide variety of technical aspects and force us to blend dc machine concepts with those of mechanics, dynamics, and basic physics. Because of the many facets which are addressed in this example, we will dispense with our usual practice of presenting all data "up-front" and provide the necessary information on a "just-in-time" basis.

EXAMPLE 19.3

Let us consider a mythical electrical railway system, which we will call the "Metro." Similar systems are common in many American and European cities. The Metro has twenty stations, spaced 3 km apart. The trains stay in each station for 45 seconds; accelerate at a maximum rate to full speed and brake at about the same rate to a stop. The trains operate from 6 AM to midnight daily; the trains reverse at the end of the line and run "backwards." Each train makes an integral number of round-trip runs per day.

A train consists of a string of six coupled identical "cars," each of which is powered by eight on-board dc motors (four per truck, one per wheel). Electrical power is supplied to each car from a third "hot" rail at a dc voltage of 750 V relative to the two-rail track, which is grounded. The motors apply torque to their respective shafts, which, through gearing, apply torque to the car wheels. The car wheels convert torque to thrust which

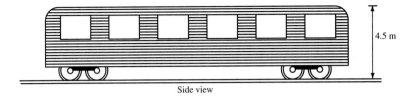

Side view

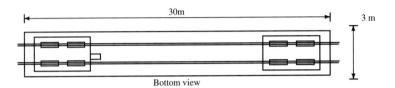

Bottom view

Passenger seats 72 @ 75 kg each
Standing 72 @ 75 kg each

Empty car: 19,200 kg

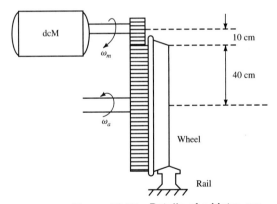

Figure 19.15 Details of a Metro car.

moves the cars. Opposing motion is a drag force, composed of all frictional forces, including aerodynamics. This drag force, acting through the wheels and gearing, appear as a counter-torque on each motor shaft opposing rotation. The details of a metro car are shown in Fig. 19.15.

The motors will be armature-controlled, that is, the field is constant and we shall control the applied armature voltage. The dc machine circuit model is shown in Fig. 19.16. We assume that all motors share the load equally.

To study this system we shall consider a series of specific questions and problems that

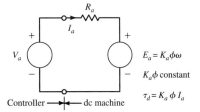

Figure 19.16   dc machine circuit model.

deal with engineering issues relevant to the dc machines as they relate to the overall design of the Metro. We shall use numerical data to enable us to generate specific numerical results. Also, we shall assume certain data are available to us as we progress through the exercise. Our purpose is to provide the reader with a "feel" for some of the considerations involved in practical design; it is obviously not to provide a comprehensive design analysis, since many issues are beyond the scope of this book. We begin with a consideration of the requisite motor speed.

A system design specification defines the normal maximum system speed. The issues involved are the distance between stations, the track and wheel tolerances, and the maximum comfortable acceleration limits. Suppose for our example system the maximum speed is specified to be 23 m/s. The train speed is converted through the wheels to wheel angular velocity:

$$\omega_{wheel} = v/r$$
$$= 23/0.4$$
$$= 57.5 \text{ rad/s}$$

Each motor shaft is connected through gearing to a wheel. Therefore

$$\omega_{motor} = \omega_{wheel} \, (r_{wheel}/r_{motorgear})$$
$$= 57.5 \, (4/1)$$
$$= 230 \text{ rad/s}$$
$$= 2196 \text{ rpm}$$

Thus, we would select a motor base speed of 2400 rpm, which is the nearest standard speed rating equal to or greater than the requisite value of 2196 rpm.

Let us now determine the motor power ratings. Analysis and testing shows that the total drag force on a car at 23 m/s is 7652 Nm, which may be assumed to vary linearly with speed. The greatest uphill grade on the system is 4%. From the data in Fig. 19.15, the mass of a fully loaded car is

$$M = 19200 + (72 + 72) \times 75 \text{ kg/person}$$
$$= 30{,}000 \text{ kg}$$

Hence, the total ("worst case") force opposing motion is

$$F = 7652 + Mg \sin (\tan^{-1} (0.04))$$
$$= 7652 + 30000 \, (9.8) \, (0.04)$$
$$= 19412 \text{ Nm}$$

where g is the acceleration due to gravity. The corresponding power is

$$P = Fv$$
$$= 19.142 \ (23)$$
$$= 446.5 \text{ kW}$$

Since there are eight wheels, thus eight motors, per car

$$P_{\text{motor}} = 446.5/8$$
$$= 55.81 \text{ kW}$$
$$= 74.8 \text{ hp}$$

A 75 hp motor is a standard size. This represents a worst case steady state analysis. However, system dynamic performance must also be considered. How fast should the train accelerate? The answer is as fast as possible, without causing harm or discomfort to the passengers. Suppose this value is specified at 0.18 $g = 1.764$ m/s². On a level surface, and ignoring drag:

$$F = Ma = 30000 \times 1.764$$
$$= 52920 \text{ Nm}$$

or 272.6% of the greatest steady state thrust. Car thrust translates to motor torque. Recall that

$$\tau_d = K_a\phi I_a$$

This means that for fixed $K_a\phi$, I_a would need to be 272.6% of rated value during the starting period, which is excessive. If we doubled the size of the motor to 150 hp, the starting current would be 272.6/2 = 136.3% of rated, which is deemed acceptable. Note that most of the time the motor current is far below this value. The time interval from zero speed to running speed is

$$T_{\text{start}} \cong v/a$$
$$= 23/1.764$$
$$= 13 \text{ seconds}$$

which is reasonable.

Thus, based upon our analysis, we select a 750 V, 150 hp, 2400 rpm (base speed) 4-pole—because it is a commonly available—dc machine. Reasonable parameters for such a motor are: I_a (rated) = 160 A; R_a = 0.14 Ω; $K_a\phi$ = 3.00 Wb (constant); Rotational loss = 2.2 kW @ running speeds and negligible at low speed. Full load efficiency = 93%.

In "real life," this information would be obtained from the motor manufacturer, or supplier. We intend to use armature control, so $K_a\phi$ may be considered constant, that is, I_f = constant, or we could use a permanent magnet field motor which would also provide a constant field. Now that a specific dc motor has been selected, let us determine the armature voltage and current for maximum speed on a level surface for the fully loaded train.

We intend to use armature control, which simply means that we will apply adjustable voltage ($-750 \leq V_a \leq + 750$ V) to the armature. This variable-voltage source, or controller, which is connected directly to the armature, is called a "dc-dc solid state electronic converter." We need not concern ourselves with the converter details, only that it is available. Thus, we need to determine certain V_a levels that correspond to certain system operating conditions for proper converter settings. The voltage E_a is

$$E_a = K_a\phi\omega$$
$$= (3)(230)$$
$$= 690 \text{ V}$$

Now recall that

$$F_{drag}@ 23 \text{ m/s} = 7652 \text{ Nm}$$

Therefore

$$P_{drag} = 7652(23)$$
$$= 176.0 \text{ kW}$$

Then for each motor

$$P_{out} = 176/8$$
$$= 22 \text{ kW}$$

The developed power is

$$P_d = 22 + 2.2$$
$$= 24.2 \text{ kW}$$

However

$$P_d = E_a I_a$$
$$= (690)I_a$$

Therefore

$$I_a = 24200/690$$
$$= 35.07\text{A}$$

And hence

$$V_a = R_a I_a + E_a$$
$$= 35.07(0.14) + 690$$
$$= 694.9 \text{ V}$$

Therefore, the controller design would require that for a motor speed of 230 rad/s, V_a must be set to $V_a = 694.9$ V.

Let us now determine each motor's V_a, I_a, and τ_d at starting on a level surface for the fully loaded train. The armature controller (converter) is designed to be inherently current limiting; that is, the voltage collapses if the current tries to exceed a user-defined maximum setting. In our previous work we had decided to limit the current to 136% of rated value. Since we ignored drag, let's use 140% of rated value to partially compensate. Hence

$$I_a = 1.4 \times 160$$
$$= 224 \text{ A}$$

Therefore

$$\tau_d = K_a \phi I_a$$
$$= 3\ (224)$$
$$= 672\ \text{Nm}$$

However, since $\omega = 0$

$$E_a = 0$$

Thus

$$V_a = I_a R_a$$
$$= 224\ (0.14)$$
$$= 31.36\ \text{V}$$

We now estimate the time for a station-to-station run and the electrical energy required. We previously estimated 13 seconds for the time required to go from a dead stop to running speed. Assuming constant acceleration, the average velocity is $23/2 = 11.5$ m/s, and the distance traversed $= 11.5 \times 13 = 149.5$ m. An equal time and distance will be required for stopping. Thus, the time (T_1) required for running at 23 m/s between stations is

$$T_1 = \text{distance/velocity}$$
$$= (3000 - 2 \times 149.5)/23$$
$$= 208\ \text{s}$$

The total station-to-station trip time (T_2) is

$$T_2 = 208 + 45 + 13 \times 2$$
$$= 279\ \text{s}$$

We previously determined that at 23 m/s each motor was loaded to 22 kW. Thus, the energy required to run at 23 m/s for 208 seconds (w_1) is

$$w_1 = 22\ \text{kW} \times 208\ \text{s} \times 8\ \text{motors/car} \times 6\ \text{cars}$$
$$= 219.6\ \text{MJ}$$

During starting, the average speed was about $23/2 = 11.5$ m/s, and the drag force is $7652/2 = 3826$ N. Hence, the energy expended in starting and stopping due to drag (w_2) is

$$w_2 = 3826\ \text{N} \times 11.5\ \text{m/s} \times 13\ \text{s} \times 6\ \text{cars}$$
$$= 3.43\ \text{MJ}$$

Neglect the stopping energy, assuming mechanical braking. The kinetic energy of the train moving at 23 m/s (w_3) is

$$w_3 = 0.5\ M\ v^2$$
$$= 0.5(6 \times 30000)\ (23)^2$$
$$= 47.61\ \text{MJ}$$

Thus, the total energy required for a station-to-station run is

$$w_{mech} = w_1 + w_2 + w_3$$
$$= 219.6 + 3.43 + 47.61$$
$$= 270.6 \text{ MJ}$$

If we assume that the converter is 93% efficient, and the motor is 94% efficient, the combined converter-motor efficiency is $0.93 \times 0.94 = 0.8742$, or 87.42%. Hence

$$w_{elec} = 270.6/0.8742$$
$$= 309.5 \text{ MJ}$$
$$= 85.98 \text{ kW-hrs}$$

We ignored uphill, downhill operation because they will approximately cancel each other as the train makes a round trip. ■

This example illustrates the wide variety of technical issues that come into play when considering real-world type problems.

Finally, it is important to note that historically, dc machines were used as the generator part of "m-g" (motor-generator) sets where ac to dc power conversion was needed. This application has all but disappeared, since modern solid state electronic devices can convert ac to dc efficiently at essentially any power level. Remaining dc motor applications exist where there are critical speed and/or position control requirements. As ac motor control technology continues to improve both technically and economically, there will likely be fewer applications for dc machines because of maintenance problems associated with its brushes and commutator.

19.7
Summary

- In the basic dc machine an interrelationship exists between a conducting loop of wire rotating in a constant magnetic field and the loop's terminal voltage.
- A dc machine can be operated as a generator or a motor. In a motor, the input is electrical and the output is mechanical. In a generator, the reverse is true.
- Equivalent circuits exist for dc machines. These circuits, together with the machine's magnetization curve, are used to analyze these machines.
- dc machine field windings may be either shunt or series. If both windings are used, the machine is said to be compounded in which case it may be either cumulative or differential.

PROBLEMS

19.1. A dc shunt motor has an armature current and armature resistance of 200 A and 0.15 Ω, respectively. If the induced voltage, E_a, is 370 V, find the voltage at the terminals of the motor.

19.2. Find the induced voltage, E_a, in a shunt-connected dc motor given that $I_a = 50$ A, $V_a = 250$ V, and $R_a = 0.25$ Ω.

19.3. A shunt-connected dc motor has the following parameters: $V_a = 400$ V, $E_a = 350$ V, $R_a = 0.08$ Ω, and $R_f = 125$ Ω. Find the armature and field currents in the motor.

19.4. Find the armature current in a dc shunt motor if $E_a = 300$ V and the motor develops 4 kW of power.

19.5. If $K_a\phi = 1.1$ Wb, determine the speed of the motor in problem 19.4.

19.6. Determine the torque that is developed by the motor described in problems 19.4 and 19.5.

19.7. A dc shunt motor develops 13.41 hp. Under this condition, $E_a = 250$ V and it is known that $R_a = 0.1$ Ω. Find the terminal voltage V_a.

19.8. A dc shunt motor develops 22 Nm of torque with $K_a \phi = 1.6$ Wb. If $R_a = 0.2$ Ω and $V_a = 300$ V, find E_a.

19.9. Find the speed of the motor in problem 19.8.

19.10. A dc shunt motor develops 30 hp with $V_a = 350$ V and $E_a = 340$ V. Find the value of the armature resistance R_a.

19.11. If the motor in problem 19.10 operates at 1423 rpm, find the values of $K_a\phi$ and τ_{dev}.

19.12. A separately excited dc shunt motor is operated at constant terminal voltage and constant developed power. Assuming the armature resistance is zero, if the field current (and therefore $K_a\phi$) is decreased, will the speed of the machine increase or decrease, and why?

19.13. Given the motor and operating conditions described in problem 19.12, will the armature current increase, decrease, or remain the same, and why?

19.14. Find the terminal current I_a in a dc shunt motor if $V_a = 250$ V, $R_f = 100$ Ω, $R_a = 0.1$ Ω, $K_a\phi = 1.25$ Wb, and the shaft speed is 1800 rpm.

19.15. A dc shunt motor has the following characteristics: $V_a = 400$ V, $R_f = 110$ Ω, and $R_a = 0.3$ Ω. The magnetization curve for the machine running at a speed of $\omega = 3476$ rpm is shown in Fig. P19.15. Find the machines armature current and developed shaft torque at $\omega = 3476$ rpm.

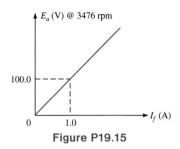

Figure P19.15

19.16. Determine the input power that must be supplied to the machine in problem 19.15.

19.17. If shaft rotational losses are neglected, determine the efficiency of the dc motor described in problem 19.15.

19.18. A dc shunt motor has the following characteristics: $V_a = 250$ V, $R_a = 0.2$ Ω, $R_f = 100$ Ω, and $\omega = 1800$ rpm. Use the magnetization curve in Fig. P19.18 to find the torque developed by the machine.

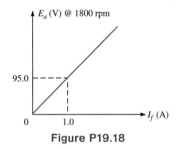

Figure P19.18

19.19. Given all the data for the machine in problem 19.18, derive a torque-speed curve for this motor and use it to find the developed torque if the shaft speed drops to 1200 rpm.

19.20. A shunt-connected dc motor with a terminal voltage of 600 V is used to drive the wheels of a subway train. Find the torque available for starting the train if a 25Ω starting resistor is connected in series with the armature and the motor characteristics are $R_a = 0.5$ Ω, $R_f = 100$ Ω, and $K_a\phi = 0.504$ I_f.

19.21. A separately excited dc shunt generator is modeled using the circuits shown in Fig. P19.21. The generator supplies 120 kW at 600 V. If the armature resistance is 0.2 Ω, find the value of E_a.

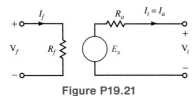

Figure P19.21

19.22. If the field current for the generator in problem 19.21 is fixed to provide a $K_a \phi = 1.3$ Wb, determine the shaft speed of the generator in rpm.

19.23. Determine the mechanical torque required to turn the generator described in problems 19.21 and 19.22. Neglect all losses.

19.24. If the magnetization curve shown in Fig. P19.24 applies to the machine described in problems 19.21, 19.22, and 19.23, find the value of the field resistance if the field voltage, V_f is 300 V.

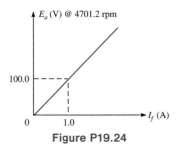

Figure P19.24

19.25. A separately excited dc generator develops 100 Nm of torque. Find the terminal voltage V_a if $E_a = 320$ V, $R_a = 0.3$ Ω, and $K_a \phi = 1.5$ Wb.

19.26. A separately excited dc generator develops 90 Nm of torque and 18 kW of power. Compute the generator speed.

19.27. If the generator in problem 19.26 has an armature resistance of 0.3 Ω and $K_a \phi = 1.4$ Wb, calculate the terminal voltage V_a.

19.28. If the generator described in problem 19.26 is operated as a self-excited generator, determine the field resistance R_f if the field current $I_f = 3.0$ A.

19.29. If the generator in problem 19.26 has the following characteristics: $R_a = 0.1$ Ω, $K_a \phi = 1.6$ Wb, $\tau_{dev} = 100$ Nm, and $P_{dev} = 20$ kW, calculate the terminal voltage.

19.30. A separately excited dc shunt generator produces a terminal voltage of 400 V at no load with the shaft rotating at 2000 rpm. If the field current and armature current remain unchanged, find the new terminal voltage if the shaft speed drops to 1800 rpm.

19.31. A 300 V self-excited dc generator is used to supply power to a 300 V dc power distribution grid in an underground coal mine. Note that this application requires a constant terminal voltage of 300 V. If the torque developed by the machine is 100.4 Nm, $R_a = 1$ Ω, $R_f = 85$ Ω, and $\omega = 1800$ rpm, find the machine's terminal output power.

19.32. A dc motor is used to drive a dc generator. If we assume that all losses can be neglected, what is the electrical power input to the motor if the electrical power output of the generator is 50 kW?

19.33. A separately excited dc motor is used to drive a separately excited dc generator as shown in Fig. P19.33(a).

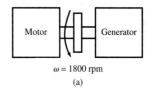

$\omega = 1800$ rpm

(a)

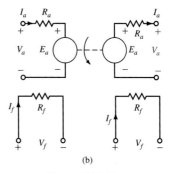

(b)

Figure P19.33

The electrical models for the machines are shown in Fig. P19.33(b). Given the characteristics for each machine as tabulated below, find the terminal voltage of the generator.

Parameter	Motor	Generator
V_a	300 V	?
R_a	0.5 Ω	0.7 Ω
V_f	300 V	300 V
R_f	100 Ω	100 Ω
I_a	50 A	40 A

19.34. Determine the efficiency of the motor-generator (m-g) set described in problem 19.33.

19.35. The motor-generator set in problem 19.33 is used to supply power to a 300 V dc power grid. The machine parameters are listed below.

Parameter	Motor	Generator
V_a	300 V	300 V
R_a	0.5 Ω	0.7 Ω
V_f	300 V	300 V
R_f	100 Ω	100 Ω

The field voltages and resistances for both machines are set to produce a generator terminal voltage of 300 V when the generator terminal current is zero. Assuming linear magnetics, find the new terminal voltage of the motor if the generator must supply 50 kW to the power grid. Recall that although the no-load speed is 1800 rpm, this speed may change under load. The field voltages and the generator terminal voltage remain constant. In addition, neglect all shaft rotational losses.

ac Polyphase Machines

20.1
Introduction

In this chapter we will examine two important types of ac polyphase machines: induction and synchronous. In each case we will examine the basic characteristics, present the equivalent circuits for the machines, and discuss the operational modes through a number of examples.

20.2
The Revolving Magnetic Field

Consider a stator structure on which are mounted two identical N-turn sinusoidally distributed windings (a-a' and b-b'), as illustrated in Fig. 20.1(a). Suppose we supply two balanced currents to the windings, so that

$$i_a = I_m \cos(\omega t)$$
$$i_b = I_m \sin(\omega t) \tag{20.1}$$

and

$$\omega = 2\pi f$$

which is the stator radian frequency, in rad/s. These currents are plotted in Fig. 20.1(b). Note the spatial angle θ, measured positive CCW, referenced from the winding a-a' magnetic axis, which is coincident with the positive x axis. The mmf's produced by these windings are

$$\mathscr{F}_a(\theta,t) = N\, i_a \cos(\theta) = N\, I_m \cos(\omega t)\, \cos(\theta)$$
$$\mathscr{F}_b(\theta,t) = N\, i_b \sin(\theta) = N\, I_m \sin(\omega t)\, \sin(\theta) \tag{20.2}$$

653

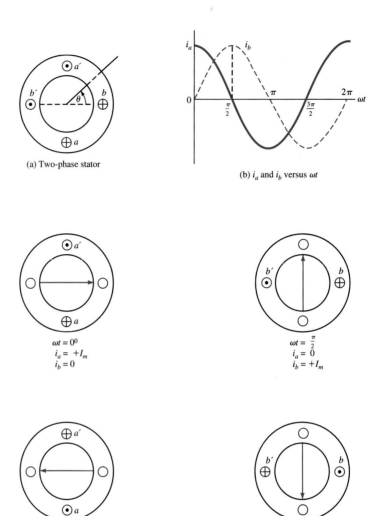

Figure 20.1   Figures used to describe a revolving magnetic field.

Hence, the total mmf, $\mathscr{F}(\theta,t)$, inside the stator is

$$\mathscr{F}(\theta,t) = \mathscr{F}_a(\theta,t) + \mathscr{F}_b(\theta,t)$$
$$= NI_m \left[\cos (\omega t) \cos (\theta) + \sin (\omega t) \sin (\theta)\right]$$
$$= NI_m \cos (\omega t - \theta) \tag{20.3}$$

This mmf varies sinusoidally in space and time and may be visualized as follows. Suppose we draw an arrow in the direction of maximum field intensity with magnetic flux lines flowing in the direction of the arrow. From Eq. (20.3), this orients the arrow such that

$$\omega t - \theta = 0$$

or equivalently

$$\theta = \omega t$$

Figure 20.1(c) shows the spatial orientation of the mmf at four different ωt values (0, $\pi/2$, π, $3\pi/2$). The overall effect is that $\mathcal{F}(\theta,t)$ appears to be revolving, or spatially rotating, at angular velocity ω, *which is also* the radian frequency of the currents. This angular velocity is called the *synchronous speed* and given the symbol ω_s.

We have investigated the mmf $\mathcal{F}(\theta,t)$ produced by two balanced ac currents, 90° apart in *phase,* in two sinusoidally distributed windings, positioned 90° apart in *space.* In general, "n" balanced ac currents, 360°/n apart in phase, which exist in n sinusoidally distributed windings, positioned 360°/n apart in space, will also produce a rotating mmf $\mathcal{F}(\theta,t)$, for all integers $n \geq 3$. For this more general case

$$\mathcal{F}(\theta,t) = (nNI_m/2) \cos(\omega t - \theta) \tag{20.4}$$

In machine terminology, these "windings" are also called "phases" and collectively the "n-windings" configuration, the "n-phase" case. Commonly $n = 3$, producing a "three-phase" stator winding configuration, and we shall limit our discussion to this case. Also, the foregoing discussion is relevant to a so-called "two-pole" field geometry. The winding locations may be modified to form, in general, a "P-pole" field, where P is any even integer. To deal with the P-pole situation, it is convenient to define two different units of angular measure. We define:

$$\theta_e = \text{electrical angle} \tag{20.5}$$

where electrical radian is equal to one electrical cycle/2π and electrical degree is equal to one electrical cycle/360.
Then

$$\theta_m = \text{mechanical angle} \tag{20.6}$$

where 2π mechanical radians are equivalent to one mechanical revolution or 360 mechanical degrees. Thus,

$$\theta_m = \frac{2}{P}\theta_e \tag{20.7}$$

Given this definition, if we now differentiate Eq. (20.7), we obtain

$$\omega_m = \frac{2}{P}\omega_e \tag{20.8}$$

And thus the synchronous speed in mechanical rad/s is

$$\omega_s = \frac{2}{P}(2\pi f) \tag{20.9}$$

Finally, three balanced ac currents, 120 electrical degrees apart in phase, in three sinusoidally distributed P-pole windings, will produce a rotating mmf $\mathcal{F}(\theta_m,t)$ of the form

$$\mathcal{F}(\theta_m,t) = (3NI_m/2) \cos(\omega_s t - \theta_m) \tag{20.10}$$

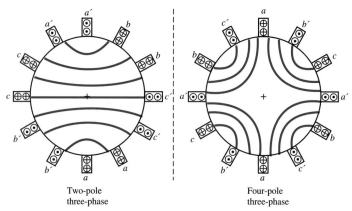

Figure 20.2 Field patterns in ac polyphase machine stators.

This revolving magnetic field, produced by balanced three-phase stator currents, is called the *stator field*, and is fundamental to the operation of all ac three-phase machines. The two- and four-pole cases are illustrated in Fig. 20.2(a) and 20.2(b), respectively.

DRILL EXERCISE

D20.1. Consider the two-phase stator shown in Fig. 20.1. Note that the direction of field rotation is counterclockwise (ccw). Determine the direction of field rotation if (a) $i_a = +I_m \cos\omega t$ and $i_b = -I_m \sin\omega t$, (b) $i_a = -I_m \cos\omega t$ and $i_b = +I_m \sin\omega t$, (c) $i_a = -I_m \cos\omega t$ and $i_b = -I_m \sin\omega t$, and (d) $i_a = +I_m \cos\omega t$ and $i_b = +I_m \sin\omega t$.

Ans: (a) cw, (b) cw, (c) ccw, and (d) ccw.

20.3
The Polyphase Induction Machine: Balanced Operation

Basic Principles of Operation

We shall next discuss the first of the two basic types of ac machines, namely the *three-phase induction machine* operating at constant speed under balanced three-phase conditions. The stator is designed as discussed in previous sections, and produces a sinusoidally distributed P-pole revolving field, revolving at speed ω_s. To simplify things, all further diagrams and discussion will be restricted to the two-pole case; however, the mathematics will be valid for the P-pole case unless specifically stated otherwise.

Consider a cylindrical rotor structure on which are mounted three sinusoidally distributed windings (A-A', B-B', C-C'), separated 120° (electrical) apart, which for the moment we shall consider to be open-circuited. We further consider this *wound rotor* to be rotating at speed ω_r, which we shall assume is less than ω_s. The situation is illustrated in Fig. 20.3(a). From the rotor's perspective, the *stator field* sweeps over the rotor surface at a speed of $(\omega_s - \omega_r)$. If we define $s = slip$ as follows:

$$s = slip = \frac{\omega_s - \omega_r}{\omega_s} \tag{20.11}$$

then the stator field rotates at speed $s\omega_s$ (in **mechanical** rad/s) relative to the rotor. This

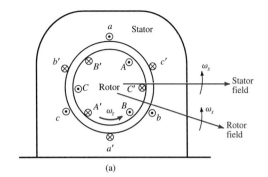

(a)

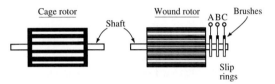

Both behave as "P" pole 3-phase wye connected ac windings.

(b)

Figure 20.3   Induction machine details.

field will induce balanced three-phase voltages in the rotor windings at radian frequency $s\omega_s$ (in **electrical** rad/s); likewise, the rotor cyclic frequency in Hz is $f_r = sf$, where $f =$ stator frequency, in Hz.

If we now connect the rotor windings in a balanced three-phase passive termination, a balanced pattern of ac three-phase currents will flow in the rotor with frequency f_r. Therefore, the rotor currents will produce a second sinusoidally distributed P-pole revolving field, revolving at speed $s\omega_s = (\omega_s - \omega_r)$. This *rotor field* rotates at $(\omega_s - \omega_r) + \omega_r = \omega_s$, and thus turns synchronously with the stator field. It is the interaction of these two synchronized fields that is the mechanism of torque production, and thus power conversion, in the polyphase induction machine.

The above discussion is valid for all conceivable balanced rotor terminations, including a short circuit. In fact, a short circuit is the best possible termination for many operating conditions, since it maximizes the rotor current, and hence the rotor field and torque. Since the number of rotor winding turns is irrelevant, consider a one-turn case. A short circuit would, in effect, short out each rotor conductor. For this case, we don't need a rotor winding at all, simply provision for appropriate conducting paths on the rotor structure. Such a rotor design is called a *squirrel cage* or *cage rotor,* because of the conductor arrangement resemblance to the exercise cage used for pet gerbils, hamsters, or other small rodents. The two rotor designs are illustrated in Figure 20.3(b).

The Equivalent Circuit

Balanced constant speed operation of the induction machine may be modeled by the ac per-phase wye equivalent circuit shown in Fig. 20.4. The parameters shown on the circuit are defined as follows:

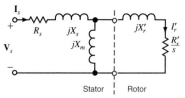

Figure 20.4 Per-phase Wye equivalent circuit for an induction machine.

$\mathbf{V}_s$ = Line-to-neutral stator phase-*a* phasor voltage applied at the terminals, usually used as phase reference in volts.

$\mathbf{I}_s$ = Stator phase-*a* current in amperes.

R_s = Phase-*a* winding resistance of the stator in ohms.

X_s = Leakage reactance associated with the stator flux that does not link rotor windings in ohms.

X_m = Magnetizing reactance in ohms.

I'_r = Rotor phase-*a* current reflected into the stator in amperes.

X'_r = Leakage rotor reactance, reflected into the stator, associated with the rotor flux that does not link stator windings in ohms.

R'_r = Rotor resistance reflected into stator, in ohms.

s = Refers to the slip which is a measure of rotor speed. It is defined as

$$s = \frac{\omega_s - \omega_r}{\omega_s}$$

It is important to note that the rotor components are reflected to the stator circuit in the same manner in which the secondary circuit of a transformer is reflected to the primary, and are indicated by primes.

The power flow diagram for an induction motor is shown in Fig. 20.5(a). The per-phase circuit diagram can be redrawn as shown in Fig. 20.5(b) in order to account for some of these terms. In this new circuit, R_s represents the element which accounts for the stator copper losses, R_r the rotor copper losses, and $R'_r(1 - s)/s$ accounts for the mechanical power developed. Other losses noted in Fig. 20.5(a) will be neglected in our calculations.

Since the circuit in Fig. 20.5(b) is a per-phase equivalent circuit, and the induction motor is a 3ϕ machine, the 3ϕ power developed is

$$P_d = 3(I'_r)^2 R'_r \frac{(1 - s)}{s} \tag{20.12}$$

where $I'_r = |\mathbf{I}'_r|$. The torque developed is

$$\tau_d = \frac{P_d}{\omega_r} \tag{20.13}$$

which after some algebraic manipulation can be expressed as

$$\tau_d = \frac{3(I'_r)^2 (R'_r/s)}{\omega_s} \tag{20.14}$$

Given the equivalent circuit and the equations which have been developed, many induction motor problems may be reduced to circuit problems.

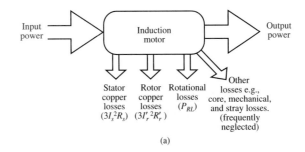

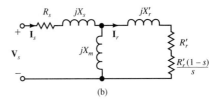

(b)

Figure 20.5   Induction motor diagrams.

Table 20.1

# of Poles	ω_s
4	$377/2 = 188.5$ rad/s
6	$377/3 = 125.67$ rad/s
8	$377/4 = 94.25$ rad/s

EXAMPLE 20.1

An induction motor is to be used to drive a conveyer belt by turning the main drive roller at an angular velocity of 120 rad/s. We wish to determine (a) the maximum number of pole-pairs that the motor can possess, and (b) the slip at which the motor is running when operating under the conditions in (a).

Table 20.1, calculated from Eq. (20.9), indicates the relationship between the number of poles and the speed of the stator magnetic field where 377 rad/s corresponds to the standard power frequency of 60 Hz. The table indicates that the machine should have 6 poles. Note that 8 poles would result in a speed that is too slow. Since $\omega_s = 125.67$ rad/s and $\omega_r = 120$ rad/s, the slip is

$$s = \frac{125.67 - 120}{125.67}$$

$$= 0.0451$$ ∎

EXAMPLE 20.2

Consider the three-phase wound rotor induction machine described by Table 20.2.

For a shorted rotor termination, and a slip of 0.015, compute: rotor speed in rpm and rad/s; rotor frequency; currents (I_s, I'_r); developed, rotational, and output torques; stator copper, rotor copper, and rotational losses; input and output powers; power factor, and efficiency.

Table 20.2 Three-Phase Induction Motor Data

Ratings	
Line Voltage = 460 Volts;	Horsepower = 75 hp
Stator Frequency = 60 Hz	No. of Poles = 4
Rotor Type: Wound;	Synchronous Speed = 1800.0 rpm

*** Equivalent Circuit Values (in stator ohms) ***

$R_s = 0.0564;$ $X_s = 0.2539; X_m = 9.875$
$R_r' = 0.0564;$ $X_r' = 0.2539$
Rotational Loss Torque = $\tau_{RL} = 0.03149\ \omega_r$ Nm

The synchronous speed in mechanical rad/s is

$$\omega_s = 2\pi(60)/2$$
$$= 188.5 \text{ rad/s}$$
$$= 1800 \text{ rpm}$$

The rotor speed in mechanical rad/s is

$$\omega_r = (1 - s)1800$$
$$= (0.985)188.5$$
$$= 185.7 \text{ rad/s}$$
$$= 1773.0 \text{ rpm}$$

The a-phase equivalent circuit is shown in Fig. 20.6. The mesh currents for the circuit are

$$\mathbf{V}_s = (0.056 + j10.129)\mathbf{I}_s + (0.000 - j9.875)\mathbf{I}_r'$$
$$0 = (0.000 - j9.875)\mathbf{I}_s + (3.762 + j10.129)\mathbf{I}_r'$$

Solving these equations yields

$$\mathbf{I}_s = 73.62 \angle -27.55°\text{A}$$
$$\mathbf{I}_r' = 67.28 \angle -7.17°\text{A}$$

The developed torque is

$$\tau_d = 3(I_r')^2 R_r'/s)/\omega_s$$
$$= \frac{3(67.28)^2(0.0564/0.015)}{188.5}$$
$$= 271.0 \text{ Nm}$$

The rotational loss torque is

$$\tau_{RL} = 0.03149(185.7)$$
$$= 5.85 \text{ Nm}$$

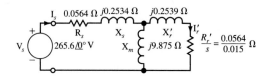

Figure 20.6   Circuit for the machine in Example 20.2.

And the load torque is

$$\tau_{load} = \tau_d - \tau_{RL}$$
$$= 271.0 - 5.8$$
$$= 265.2 \text{ Nm}$$

The input power to the machine is

$$P_{in} = 3V_s I_s \cos\theta$$
$$= 3(265.6)\,(73.62)\cos(-27.55°)$$
$$= 52.00 \text{ kW}$$

The stator copper loss (SCL) is

$$SCL = 3I_s^2 R_s$$
$$= 0.917 \text{ kW}$$

The rotor copper loss (RCL) is

$$RCL = 3(I_r')^2 R_r'$$
$$= 0.766 \text{ kW}$$

The power developed by the machine is

$$P_d = P_{in} - SCL - RCL$$
$$= 50.32 \text{ kW}$$

And the rotational losses are

$$P_{RL} = \tau_{RL}\omega_r$$
$$= 5.85(185.7)$$
$$= 1.086 \text{ kW}$$

The total losses are then

$$Total\ Loss = SCL + RCL + P_{RL}$$
$$= 2.769 \text{ kW}$$

And hence the output power is

$$P_{out} = P_{in} - Total\ Loss$$
$$= 49.24 \text{ kW}$$
$$= 66.00 \text{ hp}$$

The machine's power factor is

$$Power\ Factor = \cos(-27.55°)$$
$$= 0.8866 \text{ } lagging$$

And hence the efficiency is

$$Efficiency = P_{out}/P_{in}$$
$$= 94.67\%$$

D20.2. A three-phase, four-pole, 60Hz cage rotor induction motor runs at 1746 rpm, drawing a rotor current $I_r' = 100$ A. The rotational loss of the machine is 4 kW. Find the developed power, rotor copper losses, and output power if $R_r' = 0.09\ \Omega$.

Ans: $P_d = 87.3$ kW, $RCL = 2.7$ kW, and $P_{out} = 83.3$ kW.

20.4
The Polyphase Synchronous Machine: Balanced Operation

Basic Principles of Operation

The second major type of ac machine is the *three-phase synchronous machine.* The first point to note is that the stator structure is essentially identical to the three-phase induction machine, that is, a cylindrical ferromagnetic structure inside which are mounted balanced three-phase P-pole windings. Like the induction device, the fundamental purpose of the stator is to produce a revolving P-pole magnetic field, revolving at speed ω_s. Recall that

$$\omega_s = \frac{2\pi f}{P/2}\ mechanical\ rad/s$$

Like the three-phase induction machine, the synchronous machine can operate in both the generator and motor modes and has important applications when used both ways.

The difference between synchronous and induction machines is in the design of the rotor. There are two types: *salient* and *non-salient* as illustrated in Fig. 20.7. The number of stator and rotor poles must always be the same, that is, a P-pole stator always requires a P-pole rotor for proper operation. Another basic difference is that the synchronous machine rotor is always synchronized with the stator field, in the steady state. That is

$$\omega_r = \omega_s$$

And hence

$$s = 0$$

Because there is no relative motion between the rotor and stator field, there are no induced rotor voltages and currents, and hence the rotor field cannot be produced by in-

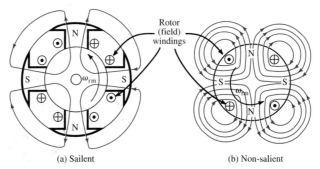

(a) Sailent   (b) Non-salient

Figure 20.7   Two four-pole synchronous machine rotor types.

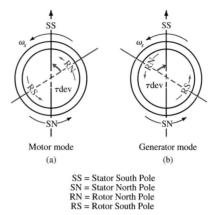

SS = Stator South Pole
SN = Stator North Pole
RN = Rotor North Pole
RS = Rotor South Pole

Figure 20.8   Basic synchronous machine operation.

duction processes. The rotor field is produced by dc currents in windings provided for that purpose, called the "field" windings, as shown in Fig. 20.7.

The mechanism of torque production is illustrated in Fig. 20.8. In Fig. 20.8(a), note that the stator field south pole ("SS"), would attract the rotor north pole ("RN"), producing a torque in the direction of rotor rotation. In a similar manner, "SN" attracts "RS." We observed earlier that developed electromagnetic torque on the rotor in the direction of motion was a positive indication of *motor* operation. Now consider the situation in Fig. 20.8(b). Again, the stator field south pole ("SS"), would attract the rotor north pole ("RN"), and likewise, "SN" attracts "RS." However, this time the torque on the rotor **opposes** rotation, a sure sign of *generator* operation. Thus, it appears that the relative angular **position** of the stator and rotor fields determines the machine operating mode. Remember that both fields rotate at the same speed $\omega_r = \omega_s$.

The Nonsalient Synchronous Machine Equivalent Circuit

Balanced three-phase ac constant speed nonsalient synchronous machine operation can be predicted with reasonable accuracy using the equivalent circuits shown in Fig. 20.9. Consider the rotor field circuit illustrated in Fig. 20.9(a). The adjustable dc source, called the *exciter*, provides the dc field current I_f needed to create the rotor field. The rotor field circuit may be modeled electrically by its resistance R_f. The ac stator voltage E_f is created by the rotor field sweeping over the stator conductors at $\omega_r = \omega_s$, which also determines the frequency of E_f. Since E_f is directly proportional to the rotor field, E_f is functionally related to I_f. This function is called the *magnetization characteristic* of the machine, and is shown in Fig. 20.9(b). It will be sufficiently accurate for our purposes to use a linearized approximation to the magnetization characteristic in our calculations. The key equations are

$$I_f = V_{ex}/R_f$$
$$E_f = K_{ag}I_f$$

where K_{ag} is the slope of the magnetization characteristic. The per-phase stator ac wye-equivalent circuit is shown in Fig. 20.9(c). The source E_f models rotor field effects. The

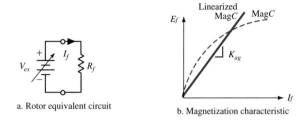

a. Rotor equivalent circuit

b. Magnetization characteristic

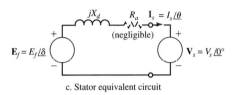

c. Stator equivalent circuit

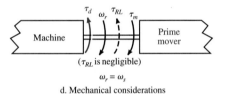

(τ_{RL} is negligible)

$\omega_r = \omega_s$

d. Mechanical considerations

Figure 20.9 Synchronous machine equivalent circuits *generator* convention.

reactance X_d accounts for two basic phenomena: "armature reaction," that is, the voltages induced into the stator windings caused by the rotating stator magnetic field which is produced by the balanced three-phase stator currents, and the "leakage field," that is, that portion of the stator field which does not cross the air gap. The circuit also has resistance R_a, which we shall neglect since $X_d \gg R_a$. We shall limit our study to the case where the machine is always connected to a large external balanced three-phase ac system, modeled with an ideal voltage source $\mathbf{V}_s$, which shall always be our phase reference; that is, $\mathbf{V}_s = |\mathbf{V}_s| \angle 0°$. The stator current $\mathbf{I}_s$ represents the a-phase line current. The angle δ is the phase angle of $\mathbf{E}_f$ relative to $\mathbf{V}_s$; the angle θ is the phase angle of I_s, relative to V_s, and hence, θ is the power factor angle. The key equations are

$$\mathbf{E}_f = \mathbf{V}_s + jX_d\mathbf{I}_s \qquad (20.15)$$

$$P_{1\phi} = V_sI_s\cos(\theta) = E_fI_s\cos(\delta - \theta) \qquad (20.16)$$

$$Power\ Factor = \mathrm{PF} = \cos(\theta) \qquad (20.17)$$

Assume the machine is operating at a rated voltage of 440 V and draws a current of 50 A.

An alternate expression for $P_{1\phi}$ is

$$P_{1\phi} = (E_f V_s / X_d) \sin(\delta) \tag{20.18}$$

See problem 20.30 for a discussion of this equation. Furthermore, remember that

$$P_{3\phi} = 3P_{1\phi} \tag{20.19}$$

Finally, consider the mechanical issues at the shaft, as shown in Fig. 20.9(d). Neglecting rotational losses, and for constant speed operation, the torque, τ_m, of the prime mover (for example, steam turbine), is equal to the electromagnetic torque, that is

$$\tau_m = \tau_d \tag{20.20}$$

Likewise, the prime mover power is equal to the electromagnetic power, that is

$$P_m = P_d$$

or

$$\begin{aligned} P_m &= \tau_m \omega_r \\ &= \tau_{dev} \omega_r \\ &= P_d \end{aligned} \tag{20.21}$$

and

$$\omega_r = \omega_s \tag{20.22}$$

Consider the following example.

EXAMPLE 20.3

A 3ϕ 2300 V 4-pole 1000 kVA 60 Hz synchronous machine has $X_d = 5 \ \Omega$, $R_f = 10 \ \Omega$, and $K_{ag} = 200 \ \Omega$; it is to be used as a generator connected to a balanced three-phase ac system.

We wish to find (a) the rated stator current, (b) the exciter setting, V_{ex}, for operation at rated conditions for a power factor of 0.866 lagging, (c) V_{ex} in part (b) if the power factor is 0.866 leading, (d) V_{ex} in part (b) for unity power factor and the same real power output as in parts (b) and (c), and (e) the complex power delivered by the generator to the system for parts (b) and (c).

(a) The rated stator current is found from the complex power as follows:

$$\begin{aligned} S_{1\phi rated} &= 1000 \ k/3 \\ &= 333.3 \ kVA \end{aligned}$$

and

$$\begin{aligned} V_{srated} &= 2.3 \ k/\sqrt{3} \\ &= 1.328 \ kV \end{aligned}$$

then

$$\begin{aligned} I_{srated} &= S_{1\phi rated} / V_{srated} \\ &= 333.3 \ k/1.328 \ k \\ &= 251.0 \ A \end{aligned}$$

(b) In order to find the exciter setting V_{ex} for operation at rated conditions for a power factor of 0.8666 lagging, we first determine

$$\theta = \cos^{-1}(0.866)$$
$$= -30°$$

then

$$\mathbf{E}_f = 1328 \underline{/0°} + j5(251\underline{/-30°})$$
$$= 2495 \underline{/14.6°} \text{ V}$$

and since

$$I_f = E_f/K_{ag}$$
$$= 2495/200$$
$$= 12.48 \text{ A}$$

Therefore

$$V_{ex} = I_f R_f$$
$$= 12.48(10)$$
$$= 124.8 \text{ V}$$

(c) For a power factor of 0.8666 leading we find that

$$\theta = \cos^{-1}(0.866)$$
$$= +30°$$

Then

$$\mathbf{E}_f = 1328\underline{/0°} + j5(251\underline{/+30°})$$
$$= 1293\underline{/57.2°} \text{ V}$$

Since

$$I_f = E_f/K_{ag}$$
$$= 1293/200$$
$$= 6.465 \text{ A}$$

Then

$$V_{ex} = I_f R_f$$
$$= 6.465(10)$$
$$= 64.65 \text{ V}$$

(d) V_{ex} for operation at rated voltage unity power factor, for the same real power output as in parts (b) and (c) is derived as follows:
From parts (b) and (c)

$$P_{1\phi} = V_s I_s \cos(\theta)$$
$$= 1328(251)(0.866)$$
$$= 288.7 \text{ kW}$$

Therefore, at unity PF

$$I_s = P_{1\phi}/V_s$$
$$= 288.7 \text{ k}/1.328 \text{ k}$$
$$= 217.4 \text{ A}$$

Since

$$\theta = \cos^{-1}(1.000)$$
$$= 0°$$

Then

$$\mathbf{E}_f = 1328 \angle 0° + j5(217.4 \angle 0°)$$
$$= 1716 \angle 39.3° \text{ V}$$

And

$$I_f = E_f/K_{ag}$$
$$= 1716/200$$
$$= 8.580 \text{ A}$$

Therefore

$$V_{ex} = I_f R_f$$
$$= 8.580(10)$$
$$= 85.80 \text{ V}$$

(e) For parts (b), (c), and (d), the complex power delivered by the generator to the system is derived from the equation

$$\mathbf{S}_{3\phi} = 3\mathbf{V}_s\mathbf{I}_s^*$$

Thus, for part (b)

$$\mathbf{S}_{3\phi} = 3(1.328)(251 \angle +30°)$$
$$= 866 \text{ kW} + j500 \text{ kvar}$$

For part (c)

$$\mathbf{S}_{3\phi} = 3(1.328)(251 \angle -30°)$$
$$= 866 \text{ kW} - j500 \text{ kvar}$$

And for part (d)

$$\mathbf{S}_{3\phi} = 3(1.328 \text{ k})(217.4)$$
$$= 866 \text{ kW} + j0 \text{ kvar} \qquad \blacksquare$$

Example 20.3 demonstrates some general points about synchronous generator operation. Lagging, unity, and leading generator power factor operation is associated with high, medium, and low excitation levels called "over, normal, and under" excitation. Note that the *generator lagging* mode is associated with reactive power Q **delivery,** and *generator leading* operation means that Q is **absorbed** by the machine. The point is that field control can be used to control Q flow into and out of the machine regardless of motor or generator operation.

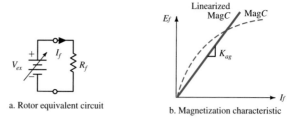

a. Rotor equivalent circuit

b. Magnetization characteristic

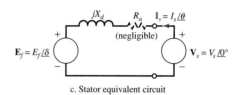

c. Stator equivalent circuit

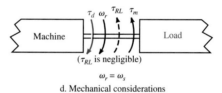

$(\tau_{RL}$ is negligible)

$\omega_r = \omega_s$

d. Mechanical considerations

Figure 20.10   Synchronous machine equivalent circuits *motor* convention.

For the machine operating in the motor mode, the current $\mathbf{I}_s$ would be phase-positioned in the first or fourth quadrants. Likewise, the powers $P_{1\phi}$ and $P_{3\phi}$ would be negative. To avoid this correct, but awkward, situation, we reverse the positive definition of $\mathbf{I}_s$ and δ, as shown in Fig. 20.10, that is, $\mathbf{I}_s$ now is defined positive **into** the machine, and δ is defined positive when $\mathbf{E}_f$ **lags** $\mathbf{V}_s$. The *motor convention* machine equations are

$$\mathbf{V}_s = \mathbf{E}_f + jX_d\,\mathbf{I}_s \tag{20.23}$$

Power flow into the machine is

$$P_{1\phi} = V_sI_s\cos(\theta)$$
$$= E_fI_s\cos(\delta - \theta) \tag{20.24}$$

The power factor is

$$PF = \cos(\theta) \tag{20.25}$$

And

$$P_{1\phi} = (E_fV_s/X_d)\sin(\delta) \tag{20.26}$$

EXAMPLE 20.4

A 3ϕ 2300 V 4-pole 1000 kVA 60 Hz synchronous machine has $X_d = 5\ \Omega$; $R_f = 10\ \Omega$; and $K_{ag} = 200\ \Omega$, that is, the machine of Example 20.3, and is to be used as a motor.

We wish to find (a) the exciter setting, V_{ex}, for operation at rated conditions for a power factor of 0.866 lagging, (b) V_{ex} in part (a) if the power factor is 0.866 leading, (c) V_{ex} in part (a) for unity power factor and the same real power as in parts (a) and (b), and (d) the complex power absorbed by the machine in parts (a), (b), and (c).

(a) In order to find the exciter setting V_{ex} for operation at rated conditions for a power factor of 0.866 lagging, we first determine

$$\theta = \cos^{-1}(0.866) = -30°$$

Then

$$\mathbf{E}_f = 1328\underline{/0°} - j5(251\underline{/-30°})$$
$$= 1293\underline{/-57.2°}\ \text{V}$$

And since

$$I_f = E_f / K_{ag}$$
$$= 1293/200$$
$$= 6.465\ \text{A}$$

Therefore

$$V_{ex} = I_f R_f$$
$$= 6.465(10)$$
$$= 64.65\ \text{V}$$

(b) For a power factor of 0.866 leading, we find that

$$\theta = \cos^{-1}(0.866)$$
$$= +30°$$

Then

$$\mathbf{E}_f = 1328\underline{/0°} - j5(251\underline{/+30°})$$
$$= 2237\underline{/-29.1°}\ \text{V}$$

Since

$$I_f = E_f / K_{ag}$$
$$= 2237/200$$
$$= 11.19\ \text{A}$$

Hence

$$V_{ex} = I_f R_f$$
$$= 11.19(10)$$
$$= 111.9\ \text{V}$$

(c) V_{ex} for operation at rated voltage, unity power factor, and the same real power input as in parts (a) and (b) is derived as follows:

From parts (a) and (b)

$$P_{1\phi} = V_s I_s \cos(\theta)$$
$$= 1328(251)(0.866)$$
$$= 288.7 \text{ kW}$$

Therefore, at unity PF

$$I_s = P_{1\phi}/V_s$$
$$= 288.7 \text{ k}/1.328 \text{ k}$$
$$= 217.4 \text{ A}$$

Since

$$\theta = \cos^{-1}(1.000) = 0°$$

Then

$$\mathbf{E}_f = 1328\angle 0° - j5(217.4\angle 0°)$$
$$= 1716\angle -39.3° \text{ V}$$

And

$$I_f = E_f/K_{ag}$$
$$= 1716/200$$
$$= 8.580 \text{ A}$$

Therefore

$$V_{ex} = I_f R_f$$
$$= 8.580(10)$$
$$= 85.80 \text{ V}$$

(d) For parts (a), (b), and (c), the complex power absorbed by the machine is as follows:
For part (a)

$$\mathbf{S}_{3\phi} = 3(1.328 \text{ k})(251\angle +30°)$$
$$= 866 \text{ kW} + j500 \text{ kvar}$$

For part (b)

$$\mathbf{S}_{3\phi} = 3(1.328 \text{ k})(251\angle -30°)$$
$$= 866 \text{ kW} - j500 \text{ kvar}$$

And for part (c)

$$\mathbf{S}_{3\phi} = 3(1.328 \text{ k})(217.4)$$
$$= 866 \text{ kW} + j0 \text{ kvar} \qquad \blacksquare$$

Example 20.4 demonstrates some general points about synchronous motor operation. Leading, unity, and lagging motor power factor operation is associated with high, medium, and low excitation levels called "over, normal, and under" excitation. Note that the *motor leading* mode is associated with reactive power Q flow from motor to system; *motor lagging* operation means that Q is absorbed by the machine.

D20.3. A medium head hydroelectric power plant in Europe employs a hydraulic turbine that produces peak efficiency at 635 rpm. If the European power frequency is 50 Hz, select an appropriate number of poles for the generator.

Ans: $P = 10$.

20.5
ac Motor Applications

The overwhelming majority of industrial motor applications utilize ac three-phase cage rotor induction motors, particularly where speed control is noncritical. Included are pumps, fans, compressors, and drives for industrial processes. Where speed control is important, both induction and synchronous machines, with electronic controllers, called *drives*, compete with dc machines. The speed is controlled by varying stator applied voltage magnitude and frequency such that $V/f = $ constant. As ac drive technology advances, dc drives are becoming less common.

Applications which utilize ac three-phase wound induction motors are less common because of their much greater cost; however, a significant number are still being used in situations which demand unusually large starting torque and moderate speed control. Very few induction devices are used as generators.

The overwhelming majority of bulk electric energy production in the world utilizes the ac three-phase synchronous machine as the generator. The two main types of electric energy-producing plants, that is, *power plants*, are thermal and hydro, where the former accounts for better than 80% of the U.S. production, and almost all of the balance is produced by the latter. Thermal plants use either fossil fuels such as coal, oil, gas, and biomass, or nuclear fuel such as enriched uranium. The nonsalient rotor design is typically used for thermal plants, whereas the salient pole type is used in slower speed hydro applications. For applications which require constant speed and few starts, synchronous motors are ideal. As we have observed, the ac three-phase synchronous machine also provides the capability of reactive power control in both motor and generator modes.

Another machine which oddly enough falls into the category of machines described in this chapter is the *brushless dc machine*. Recall that ac three-phase synchronous machines operate at a speed determined by the applied stator voltage frequency; indeed, the term *synchronous* means that the stator and rotor permanent magnet fields, and the rotor structure, are synchronized, that is, turn at the same speed. Hence, if you control the frequency, you control the speed. Suppose we start from a constant voltage constant frequency balanced three-phase ac source. This 3ϕ ac source can be rectified to dc, and inverted back to variable voltage magnitude and frequency, balanced three-phase ac, which serves as the input stator voltage to a three-phase synchronous machine. This integrated system-rectifier, inverter, synchronous machine, is called a "brushless dc" machine, since its speed controllability is comparable to the dc machine.

Finally, the most common type of electric motor is the ac single-phase type, which, although more complicated, can be analyzed using the same principles. These are the motors found in household appliances, including mixers, dryers, washing machines, refrigerators, blenders, rotisseries, garage door openers, and many other low-power applications.

20.6
Summary

- Induction and synchronous machines are two important types of polyphase machines.
- In threephase machines, balanced three-phase currents produce a revolving magnetic field.
- Both the induction and synchronous machines can be operated in either the generator or motor mode.
- Balanced constant speed operation of an induction machine can be analyzed using a per-phase equivalent circuit.
- Synchronous machines are distinguished by the type of rotor employed: salient or non-salient.
- Balanced three-phase constant speed nonsalient synchronous machine operation can be predicted using equivalent circuits.

PROBLEMS

20.1. A 3ϕ, 440 V, 60 Hz, 8-pole, Y-connected induction motor has the following parameters: $R_s = 0.29\ \Omega$, $X_s = 1.25\ \Omega$, $X'_r = 1.25\ \Omega$, $R'_r = 0.1\ \Omega$, and $X_m = 18.5\ \Omega$. Find the terminal current for a slip of 10%.

20.2. Given the description in problem 20.1, find the power factor at the motor terminals for a slip of 5%.

20.3. For the induction motor described in problem 20.1, find the power developed for a slip of 0.01.

20.4. Given the induction motor described in problem 20.1, determine the torque developed for a slip of 0.08.

20.5. A 3ϕ, 6-pole, 60 Hz, Y-connected induction motor has the following parameters: $R_s = 0.21\ \Omega$, $X_s = 1.01\ \Omega$, $X'_r = 1.01\ \Omega$, $R'_r = 0.11\ \Omega$, and $X_m = \infty$. The motor is used to drive a fan. The fan must turn at 1145 rpm and 10 hp of shaft power is required to run the fan at this speed. Find the magnitude of the terminal voltage required to achieve this operating condition.

20.6. A 3ϕ, 4-pole, 60 Hz, Y-connected induction motor rated at 440 V and 1750 rpm has the following per-phase model parameters: $R_s = 0.35\ \Omega$, $X_s = 1.1\ \Omega$. $X'_r = 1.1\ \Omega$, $R'_r = 0.15\ \Omega$, and $X_m \gg |R'_r/s + jX'_r|$ and can therefore be neglected. If the motor is operated at rated voltage and shaft speed, find the magnitude of the line current if the motor is used in a hoist that requires a torque of 146.29 Nm to drive a load.

20.7. An induction motor is operating at rated speed and terminal voltage. The motor slip remains constant. Find the developed torque if the terminal voltage is reduced by 1/2. Express the answer in terms of percent of the rated torque. Neglect X_m.

20.8. A 3ϕ, 4-pole, 440 V, 60 Hz, Y-connected induction motor has the following parameters: $R_s = 0.22\ \Omega$, $X_s = 1.3\ \Omega$, $X'_r = 1.0\ \Omega$, $R'_r = 0.1\ \Omega$, and it is assumed that $X_m = \infty$. Find the torque developed at starting if rated voltage is applied to the machine terminals.

20.9. A 3ϕ, 4-pole, 60 Hz, 550 V, Y-connected induction motor has the following parameters: $R_s = 1.0\ \Omega$, $X_s = 3.0\ \Omega$, $X'_r = 4.5\ \Omega$, $R'_r = 0.9\ \Omega$, and $X_m = 35\ \Omega$. When the motor is running at rated speed, the frequency of the rotor currents is measured to be 5 Hz. (a) Find rated speed. (b) Find the torque developed.

20.10. A 3ϕ, 2-pole, 60 Hz, 550 V, Y-connected induction motor has the following parameters: $R_s = 0.3\ \Omega$, $X_s = 0.75\ \Omega$, $X'_r = 0.75\ \Omega$, $R'_r = 0.17\ \Omega$, and $X_m = \infty$. The motor is used to drive a centrifugal pump. When the pump is primed and running, the motor slip is 0.1. Neglecting rotational and stray losses, find the efficiency of the motor at this slip.

20.11. A common method of controlling the starting current in induction machines is the use of the external resistance which is inserted into the rotor windings by connecting the windings to slip rings and brushes. The slip rings and brushes allow the rotor windings to be terminated into any external resistance, thereby changing the value of R'_r in the model. The modified model can be represented as shown in Fig. P20.11.

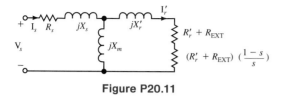

Figure P20.11

If the induction motor is a 3ϕ, 4-pole, 440 V, 60 Hz, Y-connected machine with parameters: $R_s = 0.2\ \Omega$, $X_s = 0.95\ \Omega$, $X_r' = 0.95\ \Omega$, $R_r' = 0.1\ \Omega$, and $X_m = \infty$, find the value of the external resistance that must be used to limit the starting current to 50 amperes if the full rated voltage is applied at starting.

20.12. A 3ϕ, 8-pole, 60 Hz, 440 V, Y-connected induction machine has the following parameters: $R_s = 0.5\ \Omega$, $X_s = 1.0\ \Omega$, $X_r' = 1.0\ \Omega$, $R_r' = 0.4\ \Omega$, and $X_m = \infty$. If the motor draws a line current of 29.09 A, find the shaft speed.

20.13. A 3ϕ, 4-pole, 60 Hz, 220 V, Y-connected induction machine has the following parameters: $R_s = 0.2\ \Omega$, $X_s = 0.5\ \Omega$, $X_r' = 0.5\ \Omega$, $R_r' = 0.15\ \Omega$, and $X_m = 10\ \Omega$, and has a rotor speed of 195 rad/s. Find the real power output of this machine. Note: induction generation occurs when the slip is negative, that is, the rotor is turned at a speed greater than synchronous speed.

20.14. A local power company charges $0.08 per kW-hr. of energy used and pays $0.02 kW-hr that a customer supplies back to the power company's system. A 3ϕ, 4-pole, 60 Hz, 440 V, Y-connected motor with the following parameters: $R_s = 0.8\ \Omega$, $X_s = 1.9\ \Omega$, $X_r' = 1.9\ \Omega$, $R_r' = 0.45\ \Omega$, and $X_m = \infty$, is employed at a local construction site to raise and lower a hoist. The motor speed in the "raise" mode is 1600 rpm and the speed in the "lower" mode is 2000 rpm, and the shaft always turns in the same direction because of a mechanical gearbox. If it takes 10 seconds to raise the loaded hoist and 5 seconds to lower the hoist, find the total cost to operate the hoist through one complete raise-and-lower cycle. Neglect rotational losses.

20.15. A 3ϕ, 4-pole, Y-connected synchronous motor is rated at 440 V. When used to turn a fan, the motor draws 50 amperes of current and the power factor is 0.89 lagging. If the synchronous reactance is known to be 0.254 Ω, calculate $\mathbf{I}_s$ and $\mathbf{E}_f$ and draw the phasor diagram illustrating $\mathbf{V}_s$, $\mathbf{I}_s$, and $\mathbf{E}_f$.

20.16. A 3ϕ, 4-pole Y-connected synchronous motor is rated at 440 V and has a synchronous reactance of $X_d = 3.0\ \Omega$. If the line current is measured to be 65 A and the power factor at the motor terminals is 0.9 leading, find $\mathbf{E}_f$.

20.17. A 3ϕ, 6-pole, Y-connected synchronous motor is known to be operating in an overexcited state. Is this machine supplying or absorbing reactive power?

20.18. An industrial plant requires a 3ϕ 2080 V synchronous motor to supply 300 kvars of reactive power. Find the induced voltage E_f if the shaft power output is 400 kW. $X_s = 7\Omega$.

20.19. A 3ϕ, 6-pole, Y-connected synchronous machine is operating as a motor. The machine is rated at 2080 volts and $X_s = 4.0\ \Omega$. If the magnitude of the induced voltage E_f is 1500 volts, find the maximum three-phase power that the motor can develop without losing synchronism.

20.20. A 3ϕ, 4-pole, Y-connected synchronous motor is operating at a rated voltage of 550 V and the input current is $\mathbf{I}_s = 45\angle -25°A$. Find the three-phase power developed by the machine.

20.21. A 3ϕ, 8-pole, 440 V, Y-connected synchronous motor is used to supply reactive power to a power system. Find the combination of E_f and δ that will allow the motor to supply 100 kvars if the motor is operated at a rated voltage and $X_d = 1.5\ \Omega$, such that $P_{3\phi dev} = P_{3\phi input} = 0$.

20.22. A 3ϕ, 6-pole, Y-connected synchronous generator is rated at 550 volts and has a synchronous reactance of $X_d = 2.0\ \Omega$. If the generator supplies 50 kVA at a rated voltage and a power factor of 0.95 lagging, find $\mathbf{I}_s$ and $\mathbf{E}_f$ and sketch the phasor diagram for $\mathbf{V}_s$, $\mathbf{I}_s$, and $\mathbf{E}_f$.

20.23. A 3ϕ, 6-pole, Y-connected synchronous generator with a synchronous reactance of $X_d = 12\ \Omega$ is rated at 4160 V. If the dc field current is adjusted to produce an induced voltage of 5000 V and the rotor angle delta is known to be 35°, find the three-phase complex power output at the generator terminals.

20.24. For the machine in problem 20.23, if the three-phase real power output does not change, find the new rotor angle if the field current is adjusted to produce an $E_f = 4000$ V while the terminal voltage is held constant.

20.25. A 3ϕ, 10-pole synchronous machine is used as a generator in a power plant. If the frequency of the power system is 50 Hz, find the speed of the synchronous machine rotor.

20.26. A 3ϕ, 8-pole, 4160 V, Y-connected synchronous generator is operated at a rated terminal voltage to supply a terminal current of 100 A at 0.9 power factor lagging. If $X_d = 12.0 \ \Omega$ and the dc field current in the rotor is related to E_f by the equation $I_f = 0.15 \ E_f$, find the dc field current required to operate the generator.

20.27. Repeat problem 20.26 if the machine is operated as a motor and draws a current of $75.0 \underline{/\ 15°}$A.

20.28. A 3ϕ, 6-pole, 2080 V, Y-connected synchronous machine is operated as a generator. The generator delivers 300 kVA at a power factor of 0.75 lagging. Find the induced voltage, E_f, and rotor angle, δ, if the per-phase stator impedance is $\mathbf{Z}_s = 0.1 + j8 \ \Omega$. Note that in this case we are simply including the effects of stator winding resistance.

20.29. Given that the power output of a synchronous generator is given by the equation

$$P_{3\phi} = \frac{3V_sE_f}{X_d}\sin\delta$$

(a) Sketch the power output-vs-delta curve for $0 \le \delta \le \pi$.
(b) If the line-to-neutral voltage $\mathbf{V}_s = 254\underline{/\ 0°}$ V, $E_f = 300$ V, and $X_d = 4.0 \ \Omega$, find the two values of δ that result in $P_{3\phi} = 28.575$ kW. Which value is most desirable and why?

20.30. Consider the synchronous machine in the generator mode as modeled in Fig. 20.9. Show that the real power flow from the machine is

$$P_{1\phi} = \frac{E_fV_s}{X_d}\sin\delta$$

where $\mathbf{E}_f = E_f\underline{/\delta}$ and $\mathbf{V}_s = V_s\underline{/0°}$. Hint: Start with the equation $P_{1\phi} = Re[\mathbf{V}_s\mathbf{I}_s^*]$.

Electrical Safety

21.1 Introduction

Electrical safety is a very broad and diverse topic that would require several volumes for a comprehensive treatment. Instead, we will limit our discussion to a few introductory concepts and illustrate them with examples.

It would be difficult to imagine that anyone in our society could have reached adolescence without having experienced some form of electrical shock. Whether that shock was from a harmless electrostatic discharge or from accidental contact with an energized electrical circuit, the response was probably the same—an immediate and involuntary muscular reaction. In either case, the cause of the reaction is current flowing through the body. The severity of the shock depends on several factors, the most important of which are the magnitude, the duration, and the pathway of the current through the body.

21.2 Human Response To Electric Shock

The effect of *electrical shock* varies widely from person to person. Figure 21.1 shows the general reactions that occur as a result of 60 Hz ac current flow through the body from hand to hand, with the heart in the conduction pathway. Observe that there is an intermediate range of current, from about 0.1 to 0.2 A, which is most likely to be fatal. Current levels in this range are apt to produce *ventricular fibrillation,* a disruption of the orderly contractions of the heart muscle. Recovery of the heartbeat generally does not occur without immediate medical intervention. Current levels above that fatal range tend to cause the heart muscle to contract severely, and if the shock is removed soon enough, the heart may resume beating on its own.

675

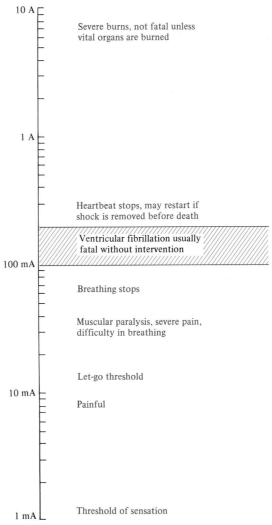

Figure 21.1 Effects of electrical shock (from C. F. Dalziel and W. R. Lee, "Lethal Electric Currents," *IEEE Spectrum*, Feb. 1969, pp. 44–50 and C. F. Dalziel, "Electric Shock Hazard," *IEEE Spectrum*, Feb. 1972, pp. 41–45).

The voltage required to produce a given current depends on the quality of the contact to the body and the impedance of the body between the points of contact. The electrostatic voltage such as might be produced by sliding across a car seat on a dry winter day may be on the order of 20,000 to 40,000 V, and the current surge upon touching the door handle, on the order of 40 A. However, the pathway for the current flow is mainly over the body surface and its duration is for only a few microseconds. Although that shock could be disastrous for some electronic components, it causes nothing more than mild discomfort and aggravation to a human being.

Electrical appliances found about the home typically require 120 or 240 V rms for operation. Although the voltage level is small compared with that of the electrostatic shock, the potential for harm to the individual and to property is much greater. Accidental contact is more apt to result in current flow either from hand to hand or from hand to foot—either of which will subject the heart to shock. Moreover, the relatively slowly changing (low-frequency) 60 Hz current tends to penetrate more deeply into the body as opposed to remaining on the surface as a rapidly changing (high-frequency) current would tend to do. Additionally, the energy source has the capability of sustaining a current flow without depletion. Thus, subsequent discussion will concentrate primarily on hazards associated with the 60 Hz ac power system.

21.3
Ground-Fault Interrupter

The single-phase three-wire system shown in Fig. 21.2 is commonly, although not exclusively, used for electrical power distribution in residences. Two important aspects of this, or any, system that relate to safety are *circuit fusing* and *grounding*.

Each branch circuit, regardless of the type loads it serves, is protected from excessive current flow by circuit breakers or fuses. Receptacle circuits are generally limited to 20 amps and lighting circuits to 15 amps. Clearly, these cannot protect persons from lethal shock. The primary purpose of these current limiting devices is to protect equipment.

The neutral conductor of the power system is connected to ground (earth) at a multitude of points throughout the system and, in particular, at the service entrance to the residence. The connection to earth may be by way of a driven ground rod or by contact to a cold water pipe of a buried metallic water system. The 120 V branch circuits radiating from the distribution panel (fuse or breaker box) generally consist of three conductors rather than only two, as shown in Fig. 21.2. The third conductor is the ground wire, as shown in Fig. 21.3.

The ground conductor may appear to be redundant, since it plays no role in the normal operation of a load that might be connected to the receptacle. Its role is illustrated by the following example.

EXAMPLE 21.1

Joe College has a workshop in his basement where he uses a variety of power tools such as drills, saws, and sanders. The basement floor is concrete and being below ground level, it is usually damp. Damp concrete is a relatively good conductor. Unknown to Joe, the insulation on a wire in his electric drill has been nicked and the wire is in contact with (or shorted to) the metal case of the drill, as shown in Fig. 21.4.

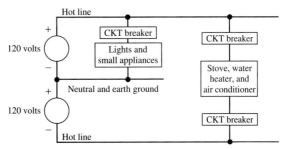

Figure 21.2   Single-phase three-wire system shown without circuit breakers.

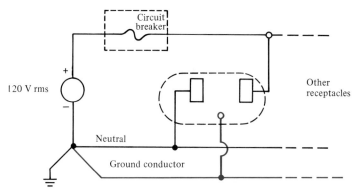

Figure 21.3   A household receptacle.

Without the ground conductor connected to the metal case of the tool, Joe would receive a severe, perhaps fatal shock when he attempts to use the drill. The voltage between
his hand and his feet would be 120 V and the current through his body would be limited
by the resistance of his body and of the concrete floor. Typically, the circuit breakers
would not operate. However, if the ground conductor is present and properly connected
to the drill case, the case will remain at ground potential, the 120 V conductor becomes
shorted to ground, the circuit breaker operates, and Joe lives to drill another hole. ■

It was mentioned that the circuit breaker or fuse cannot provide effective protection
against shock. There is, however, a special type of device called a *ground-fault interrupter*
(GFI) which can provide protection for personnel. This device detects current flow outside the normal circuit. Consider the circuit of Fig. 21.4. In the absence of a fault condition, the current in the neutral conductor must be the same as that in the line conductor.
If a fault occurs, the neutral and line currents will differ by the current flowing to ground
through the fault. The GFI detects that imbalance of currents between the neutral and line
conductor and opens the circuit in response. Its principle of operation is illustrated by
the following example.

EXAMPLE 21.2
Consider the action of the magnetic circuit in Fig. 21.5. Under normal operating conditions, i_1 and i_2 are equal and, if the coils in the neutral and line conductors are identical,
as we learned in Chapter 18, the magnetic flux in the case will be zero. Consequently, no
voltage will be induced in the sensing coil.

If a fault should occur at the load, current will flow in the ground conductor and per-

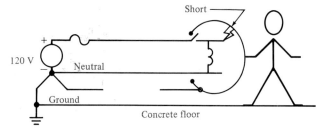

Figure 21.4   Faulty circuit.

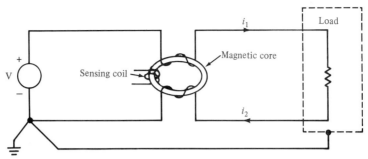

Figure 21.5   Ground–fault interrupter circuit.

haps in the earth, thus i_1 and i_2 will no longer be equal, the magnetic flux will not be zero, and a voltage will be induced in the sensing coil. That voltage can be used to activate a circuit breaker. This is the essence of the GFI device.   ■

Ground-fault interrupters are available in the form of circuit breakers and also as receptacles. They are now required in branch circuits that serve outlets in areas such as bathrooms, basements, garages, and outdoor sites. The devices will operate at ground-fault currents on the order of a few milliamperes. Unfortunately, the GFI is a relatively new device and electrical code requirements are generally not retroactive. Thus, few older residences have them.

Requirements for the installation and maintenance of electrical systems are meticulously defined by various codes that have been established to provide protection of personnel and property. Installation, alteration, or repair of electrical devices and systems should be undertaken only by qualified persons. The subject matter that we study in this book does not provide that qualification.

21.4
Modeling
Techniques

We now provide some modeling techniques which help us quantify the danger involved in the potential misuse of electrical devices and systems.

The following examples illustrate the potential hazards that can be encountered in a variety of everyday situations. We begin by revisiting a situation described in a previous example.

EXAMPLE 21.3

Suppose that a man is working on the roof of a mobile home with a hand drill; the mobile home is tied to ground with metal straps. It is early in the day, the man is barefoot, and dew covers the mobile home. The ground prong on the electrical plug of the drill has been removed. Will the man be shocked if the "hot" electrical line shorts to the case of the drill?

To analyze this problem, we must construct a model that adequately represents the situation described. In his book *Medical Instrumentation* (Houghton Mifflin Company, Boston, 1978), John G. Webster suggests the following values for resistance of the human body:

R_{skin} (dry) = 15 kΩ, R_{skin} (wet) = 150 Ω, R_{limb} (arm or leg) = 100 Ω, and R_{trunk} = 200 Ω.

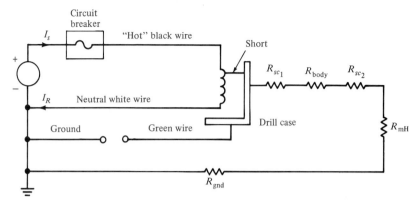

Figure 21.6 Model for Example 21.3.

The network model is shown in Fig. 21.6. Note that since the ground line is open circuited, a closed path exists from the hot wire through the short, the human body, the mobile home, and the ground. For the conditions stated above, we assume that the surface contact resistance R_{sc1} and R_{sc2} are 150 Ω each. The body resistance, R_{body}, consisting of arm, trunk, and leg, is 400 Ω. The mobile home resistance is assumed to be zero, and the ground resistance, R_{gnd}, from the mobile home ground to the actual source ground is assumed to be 1 Ω. Therefore, the magnitude of the current through the body from hand to foot would be

$$I_{body} = \frac{120}{R_{sc1} + R_{body} + R_{sc2} + R_{gnd}}$$
$$= \frac{120}{701}$$
$$= 171 \text{ mA}$$

A current of this magnitude can easily cause heart failure.

It is important to note that additional protection would be provided if the circuit breaker were a ground-fault interrupter. ■

EXAMPLE 21.4

Two boys are playing basketball in their backyard. In order to cool off, they decide to jump into their pool. The pool has a vinyl lining, so the water is electrically insulated from the earth. Unknown to the boys, there is a ground fault in one of the pool lights. One boy jumps in and while standing in the pool with water up to his chest, reaches up to pull in the other boy, who is holding onto a grounded hand rail as shown in Fig. 21.7. What is the impact of this action?

The action in Fig. 21.7(a) is modeled as shown in Fig. 21.7(b). Note that since a ground fault has occurred, there exists a current path through the two boys. Assuming that the fault, pool, and railing resistances are approximately zero, the magnitude of the current

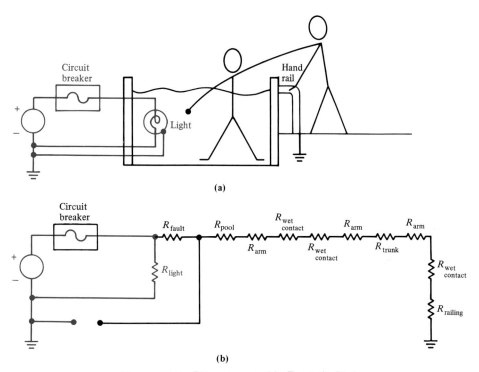

Figure 21.7   Diagrams used in Example 21.4.

through the two boys would be

$$I = \frac{120}{3R_{arm} + 3R_{wetcontact} + R_{trunk}}$$
$$= \frac{120}{950}$$
$$= 126 \text{ mA}$$

This current level would cause severe shock in both boys. The boy outside the pool would be more likely to experience heart failure. ■

EXAMPLE 21.5

A patient in a medical laboratory has a muscle stimulator attached to his left forearm. His heart rate is being monitored by an EKG machine with two differential electrodes over the heart and the ground electrode attached to his right ankle. This activity is illustrated in Fig. 21.8(a). The stimulator acts as a current source that drives 150 mA through the muscle from the active electrode to the passive electrode. If the laboratory technician mistakenly decides to connect the passive (ground) electrode of the stimulator to the ground electrode of the EKG system to achieve a common ground, is there any risk?

When the passive electrode of the stimulator is connected to the ground electrode of the EKG system, the equivalent network in Fig. 21.8(b) illustrates the two paths for the

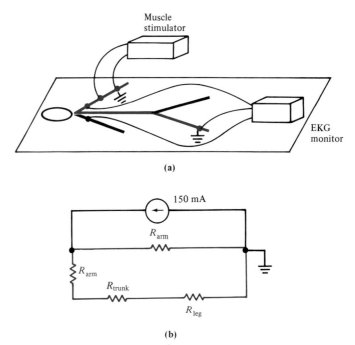

Figure 21.8 Diagrams used in Example 21.5.

stimulator current. Using current division, the magnitude of the body current is

$$I_{body} = \frac{(150)(10^{-3})(100)}{100 + 100 + 200 + 100}$$

$$= 30 \text{ mA}$$

Therefore, a dangerously high level of current will flow from the stimulator through the body to the EKG ground. ■

EXAMPLE 21.6

A cardiac care patient with a pacing electrode has ignored the hospital rules and is listening to a cheap stereo. The stereo has an amplified 60 Hz hum that is very annoying. The patient decides to dismantle the stereo partially in an attempt to eliminate the hum. In the process, while he is holding one of the speaker wires, the other touches the pacing electrode. What are the risks in this situation?

Let us suppose that the patient's skin is damp and that the 60 Hz voltage across the speaker wires is only 10 mV. Then the circuit model in this case would be shown in Fig. 21.9. The magnitude of the current through the heart would be

$$I = \frac{(10)(10^{-3})}{150 + 100 + 200}$$

$$= 22.2 \ \mu A$$

It is known that 10 μA delivered directly to the heart is potentially lethal. ■

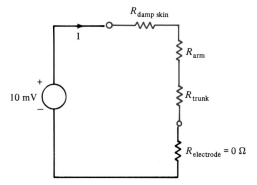

Figure 21.9   Circuit model for Example 21.6.

EXAMPLE 21.7

While maneuvering in a muddy area, a crane operator accidentally touched a high-voltage line with the boom of the crane as illustrated in Fig. 21.10. The line potential was 7200 V. The neutral conductor was grounded at the pole. When the crane operator realized what had happened, he jumped from the crane and walked in the direction of the

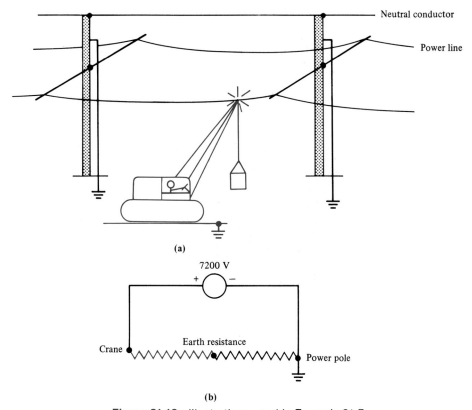

(a)

(b)

Figure 21.10   Illustrations used in Example 21.7.

pole, which was approximately 10 m away. He was electrocuted as he walked. Can we explain this very tragic accident?

The conditions depicted in Fig. 21.10(a) can be modeled as shown in Fig. 21.10(b). The crane was at 7200 V with respect to earth. Therefore, a gradient of 720 V/m existed along the earth between the crane and the power pole. This earth between the crane and the pole is modeled as a resistance. If the man's stride was about 1 m, the difference in potential between his feet was approximately 720 V. A man standing in the same area with his feet together was unharmed. ■

EXAMPLE 21.8

Two adjacent homes, *A* and *B,* are fed from different transformers as shown in Fig. 21.11(a). A surge on the line feeding house *B* has caused the circuit breaker *X–Y* to open. House *B* is now left without power. In an attempt to help his neighbor, the resident of house *A* volunteers to connect a long extension cord between a wall plug in house *A* and wall plug in house *B,* as shown in Fig. 21.11(b). Later, the lineman from the utility company comes to reconnect the circuit breaker. Unaware of the extension cord connection, the lineman

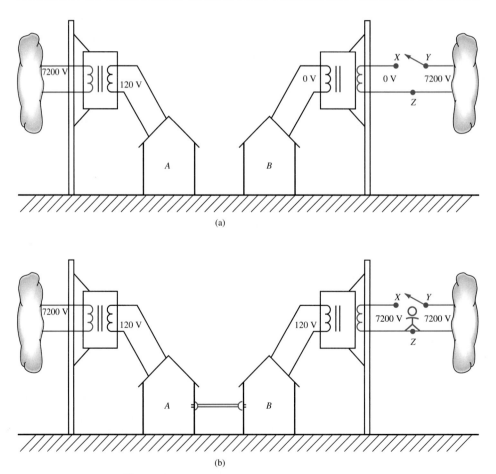

Figure 21.11 Diagrams used in Example 21.8.

believes that there is no voltage between points X and Z. However, because of the electrical connection between the two homes, 7200 V exists between the points and the lineman could be seriously injured or even killed if he comes in contact with this high voltage.   ■

DRILL EXERCISE

D21.1. A young lady is driving her car in a violent rainstorm. While she is waiting at an intersection, a power line falls across her car and makes contact. The power line voltage is 7200 V.

(a) Assuming that the resistance of the car is negligible, what is the potential current through her body if while holding the door handle with a dry hand, she steps out onto the wet ground?

(b) If she remained in the car, what would happen?

Ans: (a) $I = 463$ mA, extremely dangerous; (b) She will probably be safe.

21.5
Some Guidelines for Avoiding Injury

The examples of this section have been provided in an attempt to illustrate some of the potential dangers that exist when working or playing around electric power. In the worst case, failure to prevent an electrical accident can result in death. However, even nonlethal electrical contacts can cause such things as burns or falls. Therefore, we must always be alert to ensure not only our own safety, but also that of others who work and play with us.

The following guidelines will help us minimize the chances of injury.

1. Avoid working on energized electrical systems.
2. Always assume that an electrical system is energized unless you can absolutely verify that it is not.
3. Never make repairs or alterations which are not in compliance with the provisions of the prevailing code; have such repairs made by a licensed professional.
4. Do not work on potentially hazardous electrical systems alone.
5. If another person is "frozen" to an energized electrical circuit, deenergize the circuit if possible. If that cannot be done, use nonconductive material such as dry wooden boards, sticks, belts, and articles of clothing to separate the body from the contact. Act quickly but take care to protect yourself.
6. When handling long metallic equipment such as ladders, antennas, and so on, outdoors, be continuously aware of overhead power lines and avoid any possibility of contact with them.

Safety when working with electric power must always be a primary consideration. Regardless of how efficient or expedient an electrical network is for a particular application, it is worthless if it is also hazardous to human life.

In addition to the numerous deaths that occur each year due to electrical accidents, fire damage that results from improper use of electrical wiring and distribution equipment amounts to millions of dollars per year.

To prevent the loss of life and damage to property, very detailed procedures and specifications have been established for the construction and operation of electrical systems to ensure their safe operation. *The National Electrical Code ANSI C1* (ANSI—American National Standards Institute) is the primary guide. There are other codes, however; for example, *The National Electric Safety Code, ANSI C2*, which deals with safety requirements for public utilities. The Underwriters' Laboratory (UL) tests all types of devices and systems to ensure that they are safe for use by the general public. We find the UL label on all types of electrical equipment that is used in the home, such as appliances and extension cords.

Electric energy plays a very central role in our lives. It is extremely important to our general health and well-being. However, if not properly used, it can be lethal.

21.6
Summary

- Electrical shock can be fatal to human beings.
- Circuit breakers and fuses are designed to protect equipment, not humans.
- Ground-fault interrupters are used to protect humans from electric shock.
- Electrical modeling techniques can be used to help quantify the danger involved in the misuse of electrical systems.
- There are a number of guidelines that help minimize the chances of injury from electrical systems.
- The National Electrical Code is the primary guide for the construction and operation of electrical systems for safe operation.

PROBLEMS

21.1. A man accidentally let his parakeet out the door. The bird flew up to a power line and sat there. Unable to coax the bird down, the man stood on an aluminum ladder and reached for the bird with an aluminum pole. Is there any potential danger in this situation? If so, what, and if not, why not?

21.2. In order to test a light socket, a young lady, while standing on cushions that insulate her from the ground, sticks her finger into the socket, as shown in Fig. P21.2. The tip of her finger makes contact with one side of the line and the side of her finger makes contact with the other side of the line. Assuming that any portion of a limb has

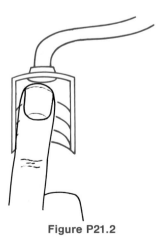

Figure P21.2

a resistance of 100 Ω, is there any current in the body? Is there any current in the vicinity of the heart?

21.3. A young mechanic is installing a 12 V battery in a car. The negative terminal has been connected. He is currently tightening the bolts on the positive terminal. With a tight grip on the wrench, he turns it so that the gold ring on his finger makes contact with the frame of the car. This situation is modeled in Fig. P21.3, where we assume that the resistance of the wrench is negligible and the resistance of the contact is as follows:

$$R_1 = R_{\text{bolt to wrench}} = 0.01 \ \Omega$$
$$R_2 = R_{\text{wrench to ring}} = 0.01 \ \Omega$$
$$R_3 = R_{\text{ring}} = 0.01 \ \Omega$$
$$R_4 = R_{\text{ring to frame}} = 0.01 \ \Omega$$

What power is quickly dissipated in the gold ring, and what is the impact of this power dissipation?

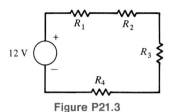

Figure P21.3

21.4. A man and his son are flying a kite. The kite becomes entangled in a 7200 V power line close to a power pole. The man crawls up the pole to remove the kite. While trying to remove the kite, the man accidentally touches the 7200 V line. Assuming the power pole is well grounded, what is the potential current through the man's body?

CONTROLS AND COMMUNICATIONS

CHAPTER

22

Control Systems Analysis

22.1
Introduction

The concept of feedback is extremely important. It is fundamental to control and is an integral part of the majority of industrial processes. We ourselves routinely act as very accurate feedback systems in a variety of functions which are a normal part of our daily lives. For example, consider the act of opening a door. When we reach for the door knob we don't stop before our hand reaches it, nor do we "overshoot" and ram our hand into the knob. Our accuracy in reaching for it correctly every time is a result of proper feedback control. Our eye continuously measures the difference in position between our hand and the knob. The error signal, which is the difference between our hand position and the position of the knob, is fed to the brain which activates the muscles to accurately drive the position of the hand to the position of the knob. To understand the improvement in performance which can result from the use of feedback, we need only attempt the same task with eyes closed, after an initial measurement of the door knob position is obtained.

In this chapter, we will discuss certain basic properties of feedback systems as well as some of the fundamental methods of analysis and design.

22.2
The Laplace Transform

The tool we will employ in our discussion of feedback control systems is the *Laplace transform.*

Our use of the Laplace transform to solve control problems is an extension of using phasors in sinusoidal steady state analysis. Using the Laplace transform we transform the problem from the time domain to the frequency domain, solve the problem using algebra in the frequency domain, and then convert the solution in the complex frequency domain back to the time domain. Therefore, as we shall see, the Laplace transform is an integral transform that converts a set of linear simultaneous integrodifferential equations to a set of simultaneous algebraic equations.

We will first define the Laplace transform, and then present a number of transform pairs which are useful in our analysis of feedback systems.

Definition

The *unilateral* or *"one-sided" Laplace transform* of a function $f(t)$ is defined by the equation

$$\mathcal{L}[f(t)] = \mathbf{F}(s) = \int_0^\infty f(t)e^{-st}\, dt \tag{22.1}$$

where s is the complex frequency

$$s = \sigma + j\omega \tag{22.2}$$

Note that this Laplace transform is unilateral ($0 \le t < \infty$) and hence our analysis of control systems using the Laplace transform will focus on the time interval $t \ge 0$. It is important to note that it is the initial conditions that account for the operation of the system prior to $t = 0$, and therefore our analyses will describe system operation only for $t \ge 0$.

In order for a function $f(t)$ to possess a Laplace transform, it must satisfy the condition

$$\int_0^\infty e^{-\sigma t}|f(t)|dt < \infty \tag{22.3}$$

for some real value of σ. Because of the convergence factor $e^{-\sigma t}$, there are a number of important functions that have Laplace transforms. All of the inputs we will apply to systems possess Laplace transforms. Functions that do not have Laplace transforms (e.g., e^{t^2}) are seldom of interest to us in practical feedback system analysis. The *inverse Laplace transform* is defined by the relationship

$$\mathcal{L}^{-1}[\mathbf{F}(s)] = f(t) = \frac{1}{2\pi j} \int_{\sigma_1 - j\infty}^{\sigma_1 + j\infty} \mathbf{F}(s)e^{st}\, ds \tag{22.4}$$

where σ_1 is real and $\sigma_1 > \sigma$ in the previous equation. The evaluation of this integral is based on complex variable theory; normally we will circumvent its use by developing and using a set of Laplace transform pairs.

Some Useful Transform Pairs

One of the most important inputs which is employed in feedback control systems is the *unit step function* defined by the following mathematical relationship:

$$u(t) = \begin{cases} 0 & t < 0 \\ 1 & t > 0 \end{cases} \tag{22.5}$$

This function, which is dimensionless, is equal to zero for negative values of the argument and unity for positive values of t. It is undefined for a zero argument where the function is discontinuous. A graph of the unit step is shown in Fig. 21.1.

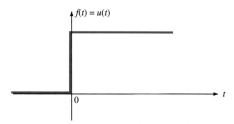

Figure 22.1   A unit step function.

EXAMPLE 21.1

Let us determine the Laplace transform of the unit step function $f(t) = u(t)$

$$\mathbf{F}(s) = \int_0^\infty u(t)e^{-st}dt$$

$$= \int_0^\infty 1e^{-st}dt$$

$$= \left. \frac{-1}{s}e^{-st} \right|_0^\infty$$

$$= \frac{1}{s}$$

Therefore, if $f(t) = u(t)$, then $\mathbf{F}(s) = \dfrac{1}{s}$ is one transform pair.

EXAMPLE 22.2

The Laplace transform of the function $f(t) = t$ is

$$\mathbf{F}(s) = \int_0^\infty te^{-st}dt$$

Using integration by parts we select

$$u = t \qquad dv = e^{-st}dt$$

$$du = dt \qquad v = -\frac{1}{s}e^{-st}$$

then

$$\mathbf{F}(s) = \left. uv \right|_0^\infty - \int_0^\infty vdu$$

$$= \left. \frac{-t}{s}e^{-st} \right|_0^\infty + \frac{1}{s}\int_0^\infty e^{-st}dt$$

$$= \frac{1}{s^2}$$

Hence, if $f(t) = t$, then $\mathbf{F}(s) = \dfrac{1}{s^2}$.

DRILL EXERCISE

D22.1. If $f(t) = \cos\omega t$, find $\mathbf{F}(s)$ using Euler's formula, that is, $\cos\omega t = (e^{j\omega t} + e^{-j\omega t})/2$.

Ans: $\mathbf{F}(s) = \dfrac{s}{s^2 + \omega^2}$.

A short table of commonly used Laplace transform pairs and properties is shown in Table 22.1. A very important property of the Laplace transform is *linearity*. Transforms of linear combinations of functions $f_1(t)$ and $f_2(t)$ are simply linear combinations of $\mathbf{F}_1(s)$ and $\mathbf{F}_2(s)$. In addition, the transform approach only applies to linear time-invariant analysis, so it is used to analyze systems that don't change with time.

Table 22.1 Short Table of Laplace Transform Pairs and Properties

	$f(t)$	$F(s)$
Pairs	$\delta(t)$	1
	$u(t)$	$\dfrac{1}{s}$
	e^{-at}	$\dfrac{1}{s+a}$
	t	$\dfrac{1}{s^2}$
	$\dfrac{t^n}{n!}$	$\dfrac{1}{s^{n+1}}$
	te^{-at}	$\dfrac{1}{(s+a)^2}$
	$\dfrac{t^n e^{-at}}{n!}$	$\dfrac{1}{(s+a)^{n+1}}$
	$\sin bt$	$\dfrac{b}{s^2 + b^2}$
	$\cos bt$	$\dfrac{s}{s^2 + b^2}$
	$e^{-at}\sin bt$	$\dfrac{b}{(s+a)^2 + b^2}$
	$e^{-at}\cos bt$	$\dfrac{s+a}{(s+a)^2 + b^2}$
Properties	$Af(t)$	$A\mathbf{F}(s)$
	$f_1(t) \pm f_2(t)$	$\mathbf{F}_1(s) \pm \mathbf{F}_2(s)$
	$\dfrac{d^n f(t)}{dt^n}$	$s^n\mathbf{F}(s) - s^{n-1}f(0) - s^{n-2}\dfrac{df(0)}{dt} \dots s^0\dfrac{d^{n-1}f(o)}{dt^{n-1}}$

22.3

Transfer Functions

Transfer functions are the Laplace transform representation of the relationship between a process input and output. For the process shown in Fig. 22.2, the transfer function is

$$G(s) = \frac{Y(s)}{U(s)} \tag{22.6}$$

Let us consider now some simple examples involving actual hardware. The circuit shown in Fig. 22.3(a) consists of a fixed voltage source such as a battery and two potentiometers similar to a volume control. The voltage measured at each wiper (V_1, V_2) is proportional to the angle of the pot shafts (θ_1, θ_2), that is

$$V_1 = \frac{\theta_1}{\theta_{max}} V_B$$

$$V_2 = \frac{\theta_2}{\theta_{max}} V_B$$

The difference between V_1 and V_2 can be found by applying Kirchhoff's voltage law to obtain the equation

$$V = V_1 - V_2$$

which can be expressed as

$$V = \frac{V_B}{\theta_{max}} (\theta_1 - \theta_2) \tag{22.7}$$

which is shown in block diagram form in Fig. 22.3(b).

Let us now consider dc motors and generators. It is helpful to describe dc motors and generators as electromechanical devices, which are modeled with electrical and mechanical equations, plus coupling equations that relate electrical quantities to mechanical quantities.

A dc motor with a constant field current is shown in Fig. 22.4(a). In this figure, v_a, i_a, and R_a are the armature voltage, current, and resistance, respectively. J_m is the inertia

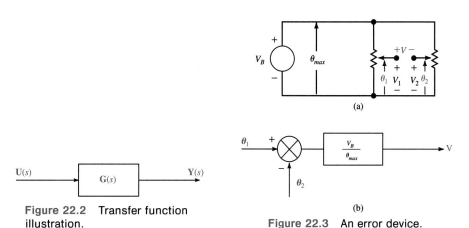

(a)

(b)

U(s) → G(s) → Y(s)

Figure 22.2   Transfer function illustration.

Figure 22.3   An error device.

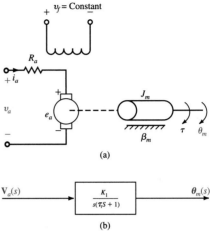

(a)

(b)

Figure 22.4 dc motor characteristics.

constant, β_m is the viscous friction constant, and θ_m is the shaft angle. The electrical dynamics are described by the equation

$$v_a = i_a R_a + e_a \tag{22.8}$$

where e_a is the induced voltage. The mechanical dynamics are described by the equation

$$\tau = J_m \frac{d^2\theta_m}{dt^2} + \beta_m \frac{d\theta_m}{dt} \tag{22.9}$$

The two coupling equations which relate the electrical and mechanical parameters are

$$e_a = K_\phi \frac{d\theta_m}{dt} \tag{22.10}$$

and

$$\tau = K_T i_a \tag{22.11}$$

as described in Chapter 19. Combining Eqs. (22.8) to (22.11) yields

$$v_a = \frac{R_a J_m}{K_T} \frac{d^2\theta_m}{dt^2} + \frac{R_a \beta_m}{K_T} \frac{d\theta_m}{dt} + K_\phi \frac{d\theta_m}{dt}$$

Using the Laplace transformation and assuming zero initial conditions, the equation can be written as

$$\mathbf{V}_a(s) = s \left[\frac{R_a J_m}{K_T} s + \frac{R_a \beta_m + K_T K_\phi}{K_T} \right] \boldsymbol{\theta}_m(s)$$

and thus the transfer function is

$$\frac{\boldsymbol{\theta}_m(s)}{\mathbf{V}_a(s)} = \frac{K_1}{s(\tau_1 s + 1)} \tag{22.12}$$

where

$$K_1 = \frac{K_T}{R_a \beta_m + K_T K_\phi} \tag{22.13}$$

and

$$\tau_1 = \frac{R_a J_m}{R_a \beta_m + K_T K_\phi} \tag{22.14}$$

Thus, in block diagram form the dc motor is represented as shown in Fig. 22.4(b).

A dc generator, operated at constant speed, is shown in Fig. 22.5(a). v_i is the input voltage, and i_f, R_f, and L_f are the field current, resistance, and inductance, respectively. The generator input voltage is described by the equation

$$v_i = i_f R_f + L_f \frac{di_f}{dt} \tag{22.15}$$

The output voltage is linearly related to the field current by the expression

$$K_G = \frac{v_o}{i_f} \tag{22.16}$$

Combining the equations yields

$$v_i = \frac{R_f}{K_G} v_o + \frac{L_f}{K_G} \frac{dv_o}{dt}$$

Once again, employing the Laplace transformation we obtain

$$\frac{\mathbf{V}_o(s)}{\mathbf{V}_i(s)} = \frac{K_2}{\tau_2 s + 1} \tag{22.17}$$

where

$$K_2 = \frac{K_G}{R_f} \tag{22.18}$$

and

$$\tau_2 = \frac{L_f}{R_f} \tag{22.19}$$

Thus, in block diagram form the dc generator is represented as shown in Fig. 22.5(b).

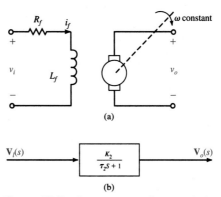

Figure 22.5   dc generator characteristics.

22.4

Block Diagrams

Although feedback theory has been in use since the Roman Empire, where it was employed to control the water level in the reservoir of common bathroom plumbing fixtures, it became a prominent technology during World War II when it was extensively employed in a variety of applications such as fire control. Since World War II feedback theory has been an integral part of a myriad of *automatic control systems.*

Block diagrams are routinely used to describe some very complicated systems, for example, missiles, airplanes, or a wide variety of chemical or mechanical industrial processes.

An elementary *feedback control system* can be illustrated in block diagram form as shown in Fig. 22.6.

In this diagram, $\mathbf{R}(s)$ is the input or desired output. $\mathbf{C}(s)$ is the actual output, and in the simplest case, that is, $\mathbf{H}(s) = 1$, the error signal $\mathbf{E}(s)$ is the output of the comparator which computes the difference between the input and output, that is

$$\mathbf{E}(s) = \mathbf{R}(s) - \mathbf{C}(s)$$

In general, the error signal drives the process in such a way that the error signal is minimized.

Using *block diagram algebra* we can write the equations which describe the feedback system in Fig. 22.6. Note that

$$\mathbf{E}(s) = \mathbf{R}(s) - \mathbf{H}(s)\,\mathbf{C}(s)$$

and

$$\mathbf{C}(s) = \mathbf{E}(s)\,\mathbf{G}(s)$$

Eliminating the error signal between the two equations yields the expression

$$\mathbf{C}(s) = \frac{\mathbf{G}(s)\mathbf{R}(s)}{1 + \mathbf{G}(s)\mathbf{H}(s)} \tag{22.20}$$

Therefore, the *transfer function,* that is, the ratio of the output signal to the input signal, is

$$\frac{\mathbf{C}(s)}{\mathbf{R}(s)} = \frac{\mathbf{G}(s)}{1 + \mathbf{G}(s)\mathbf{H}(s)}$$

It is important to note that this is the *closed loop* transfer function. $\mathbf{G}(s)\mathbf{H}(s)$ is an important and useful quantity which we call the *open loop function.*

This block diagram algebra can be used to obtain the transfer function of more complicated feedback systems.

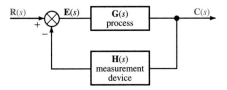

Figure 22.6 Simple feedback system.

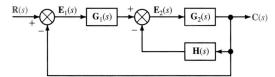

Figure 22.7   An example system.

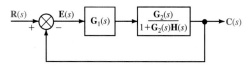

Figure 22.8   System in Fig. 22.7 redrawn.

EXAMPLE 22.3
Let us determine the transfer function of the feedback system shown in Fig. 22.7. The equations which represent the system are

$$\mathbf{E}_1(s) = \mathbf{R}(s) - \mathbf{C}(s)$$
$$\mathbf{E}_2(s) = \mathbf{E}_1(s)\,\mathbf{G}_1(s) - \mathbf{H}(s)\,\mathbf{C}(s)$$
$$\mathbf{C}(s) = \mathbf{E}_2(s)\,\mathbf{G}_2(s)$$

Eliminating $\mathbf{E}_1(s)$ and $\mathbf{E}_2(s)$ in the equations yields the transfer function

$$\frac{\mathbf{C}(s)}{\mathbf{R}(s)} = \frac{\mathbf{G}_1(s)\,\mathbf{G}_2(s)}{1 + \mathbf{G}_1(s)\,\mathbf{G}_2(s) + \mathbf{G}_2(s)\,\mathbf{H}(s)}$$

It is interesting to note that the system in Fig. 22.7 can be redrawn as shown in Fig. 22.8 using the results of the transfer function analysis for Fig. 22.6.
The equations for this system are

$$\mathbf{E}(s) = \mathbf{R}(s) - \mathbf{C}(s)$$
$$\mathbf{C}(s) = \frac{\mathbf{G}_1(s)\,\mathbf{G}_2(s)\,\mathbf{E}(s)}{1 + \mathbf{G}_2(s)\,\mathbf{H}(s)}$$

Once again, eliminating $\mathbf{E}(s)$ in the equations yields the transfer function

$$\frac{\mathbf{C}(s)}{\mathbf{R}(s)} = \frac{\mathbf{G}_1(s)\,\mathbf{G}_2(s)}{1 + \mathbf{G}_1(s)\,\mathbf{G}_2(s) + \mathbf{G}_2(s)\,\mathbf{H}(s)}$$ ■

DRILL EXERCISE

D.22.2. Compute the transfer function for the system shown in Fig. D22.2.

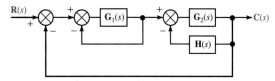

Figure D22.2

Ans: $\dfrac{\mathbf{C}(s)}{\mathbf{R}(s)} = \dfrac{\mathbf{G}_1(s)\,\mathbf{G}_2(s)}{1 + \mathbf{G}_1(s) + \mathbf{G}_2(s)\,\mathbf{H}(s) + \mathbf{G}_1(s)\,\mathbf{G}_2(s)\,\mathbf{H}(s) + \mathbf{G}_1(s)\,\mathbf{G}_2(s)}$

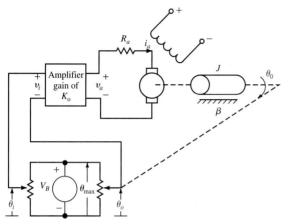

Figure 22.9 A position control system.

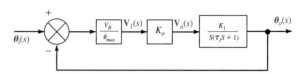

Figure 22.10 Block diagram for system in Fig. 22.9.

Let us now use what we have learned in the development of two feedback systems.

EXAMPLE 22.4

A shaft position control system is shown in Fig. 22.9. Let us develop a block diagram representation for the system and compute the closed loop transfer function $\dfrac{\theta_o(s)}{\theta_i(s)}$.

Based on our previous developments, the block diagram representation of the system is shown in Fig. 22.10.

The closed loop transfer function is then

$$\frac{\theta_o(s)}{\theta_i(s)} = \frac{K_L/s(\tau_1 s + 1)}{1 + K_L/s(\tau_1 s + 1)}$$

$$= \frac{K_L}{s(\tau_1 s + 1) + K_L}$$

where

$$K_L = \frac{V_B K_0 K_1}{\theta_{max}}$$

■

EXAMPLE 22.5

A velocity control system is shown in Fig. 22.11. We wish to develop a block diagram for the system and compute the closed loop transfer function.

Note that in this system the transfer function for the amplifier is K_o. The generator transfer function is $K_2/(\tau_2 s + 1)$. Since the output variable is motor speed ω_m, the equations for the motor are

$$v_a = i_a R_a + e_a$$
$$e_a = K_\phi \omega_m$$

$$\tau = (J_m + J_L) \frac{d\omega_m}{dt} + (\beta_m + \beta_L)\, \omega_m$$

and

$$\tau = K_T i_a$$

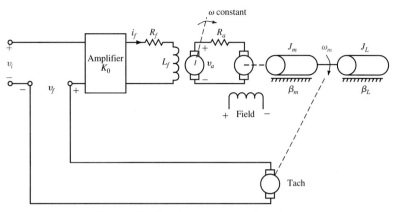

Figure 22.11   A speed control system.

Combining the equations we obtain

$$v_a = i_a R_a + K_\phi \omega_m$$

$$i_a = \frac{(J_m + J_L)}{K_T} \frac{d\omega_m}{dt} + \left(\frac{\beta_m + \beta_L}{K_T}\right)\omega_m$$

and therefore

$$v_a = \frac{R_a(J_m + J_L)}{K_T} \frac{d\omega_m}{dt} + \left[\left(\frac{\beta_m + \beta_L}{K_T}\right) R_a + K_\phi\right]\omega_m$$

Using the Laplace transformation, the transfer function becomes

$$\frac{\omega_m(s)}{V_a(s)} = \frac{K_3}{(\tau_3 s + 1)}$$

where

$$K_3 = \frac{K_T}{(\beta_m + \beta_L) R_a + K_T K_\phi}$$

$$\tau_3 = \frac{R_a(J_m + J_L)}{(\beta_m + \beta_L) R_a + K_T K_\phi}$$

The tachometer produces a voltage that is directly proportional to speed and therefore the transfer function for this element is

$$\frac{V_F(s)}{\omega_m(s)} = K_4$$

The block diagram for the system shown in Fig. 22.11 is illustrated in Fig. 22.12. The closed loop transfer function is

$$\frac{\omega_m(s)}{V_i(s)} = \frac{\dfrac{K_0 K_2 K_3}{(\tau_2 s + 1)(\tau_3 s + 1)}}{1 + \dfrac{K_0 K_2 K_3 K_4}{(\tau_2 s + 1)(\tau_3 s + 1)}}$$

$$= \frac{K_0 K_2 K_3}{(\tau_2 s + 1)(\tau_3 s + 1) + K_0 K_2 K_3 K_4}$$

∎

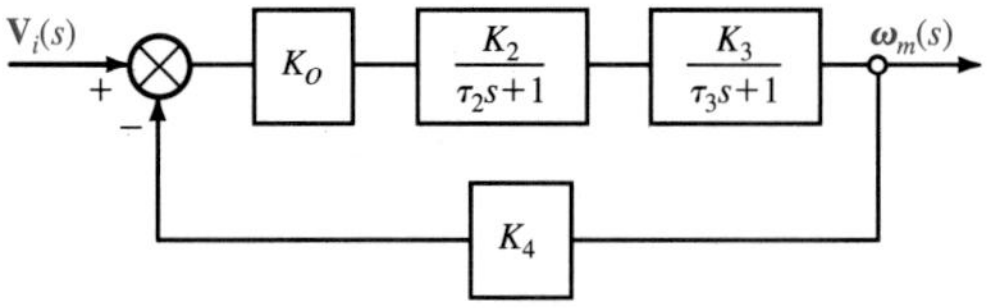

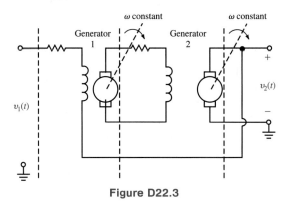

Figure 22.12 Block diagram for system shown in Fig. 22.11.

DRILL EXERCISE

D22.3. The transfer function for each of the generators shown in Fig. D22.3 is $\mathbf{G}(s) = 1/(0.5\ s + 1)$. Compute $\dfrac{\mathbf{V}_2(s)}{\mathbf{V}_1(s)}$ with the loop open and loop closed.

Figure D22.3

Ans: open loop: $\dfrac{\mathbf{V}_2(s)}{\mathbf{V}_1(s)} = 1/(0.5\ s + 1)^2$

closed loop: $\dfrac{\mathbf{V}_2(s)}{\mathbf{V}_1(s)} = 1/(0.25s^2 + 1s + 2)$

22.5

System Stability

The *stability* of a system is an inherent requirement since if the system is not stable, other qualities are of little or no importance. Thus, we now examine the stability of a feedback system.

The general form of the closed loop transfer function of a feedback system can be expressed as the quotient of two polynomials in the Laplace transform variable s.

$$\frac{\mathbf{C}(s)}{\mathbf{R}(s)} = \frac{\mathbf{G}(s)}{1 + \mathbf{G}(s)\ \mathbf{H}(s)}$$

$$= \frac{\mathbf{N}(s)}{\mathbf{D}(s)} = \frac{a_m s^m + a_{m-1} s^{m-1} + \cdots + a_1 s + a_0}{b_n s^n + b_{n-1} s^{n-1} + \cdots + b_1 s + b_0} \tag{22.21}$$

The roots of $\mathbf{D}(s)$, that is, the *poles* of the *closed loop transfer function,* determine system stability. If the roots of $\mathbf{D}(s)$ can be found, that is, if the polynomial can be factored, then the stability of the closed loop system can be determined immediately. For example,

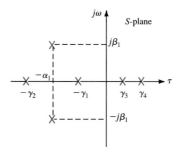

Figure 22.13   *s*-plane plot for **D**(*s*).

suppose that **D**(*s*) has factors of the form

$$\mathbf{D}(s) = (s + \gamma_1)(s + \gamma_2) \cdots (s^2 + \alpha_0^2) \cdots (s + \alpha_1 + j\beta_1)(s + \alpha_1 - j\beta_1) \cdots$$
$$(s - \gamma_3)(s - \gamma_4)$$

If we plot these poles in the *s*-plane where $s = \sigma + j\omega$, then they appear as shown in Fig. 22.13. These poles in the *s*-plane give rise to the following time functions:

$$(s + \gamma_1) \rightarrow e^{-\gamma_1 t}$$
$$(s + \alpha_1 + j\beta_1)(s + \alpha_1 - j\beta_1) \rightarrow e^{-\alpha_1 t} \cos{(\beta_1 t + \theta)}$$
$$(s^2 + \alpha_0^2) \rightarrow \sin\alpha_0 t$$
$$(s - \gamma_3) \rightarrow e^{+\gamma_3 t}$$

Note that the term $e^{-\gamma_1 t}$ approaches 0 as *t* approaches α. The term $e^{-\alpha_1 t} \cos{(\beta_1 t + \theta)}$ performs in a similar manner because of the damping factor α_1. The term $\sin \alpha_0 t$ represents a constant oscillation which is undamped, and the term $e^{+\gamma_3 t}$ grows without bound as *t* approaches ∞. Therefore, systems with poles in the *left half-plane* are *stable,* those with poles on the $j\omega$ axis and those with poles in the *right half-plane* are *unstable.* Hence, if **D**(*s*) can somehow be factored, the stability of the system will be immediately known. Modern hand-held calculators and personal computers can easily solve for the roots of polynomials, and thus the stability of a system can be readily determined.

22.6
Root Loci

The location of the closed loop poles in the *s*-plane tells us a great deal about the feedback system under consideration. These roots of the differential equation that describe the dynamic response of the closed loop system immediately indicate not only stability but the response of the system as well. For example, if all the closed loop poles are in the left half-plane, the system is stable. If we assume that the closed loop system is second order, or that it can be approximated by a second-order system, then the *transient response* of the system is also known.

Suppose that the system transfer function is of the form

$$\frac{\mathbf{C}(s)}{\mathbf{R}(s)} = \frac{K}{s^2 + 2\xi\,\omega_n s + \omega_n^2}$$

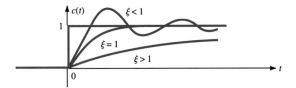

(a) Second order system response as a function of ξ

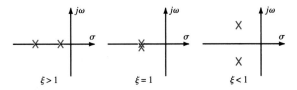

(b) Second order pole location as a function of ξ

Figure 22.14 Relationship between the damping factor ξ, pole locations, and system response.

If the system input is a unit step function, that is, $\mathbf{R}(s) = 1/s$,

$$\mathbf{C}(s) = \frac{K}{s(s^2 + 2\xi\,\omega_n s + \omega_n^2)}$$

As illustrated in Chapter 3, the system response will fall into three separate categories depending upon the value of the *damping constant* ξ. If $\xi > 1$, the roots of $s^s + 2\xi\,\omega_n s + \omega_n^2$ are real and unequal and the system response is said to be *overdamped*. If $\xi = 1$, the roots are real and equal and the response is *critically damped*. If $\xi < 1$, the roots are complex conjugates and the system response is *underdamped*. The second-order response to a unit step input for the three cases is shown in Fig. 22.14.

In a feedback system the location of the closed loop poles is dependent upon the *open loop gain K*. The following simple example demonstrates this fact and illustrates the impact of the gain K on the system response.

EXAMPLE 22.6
Consider the unity feedback system shown in Fig. 22.15. The closed loop transfer function for this system is

$$\frac{\mathbf{C}(s)}{\mathbf{R}(s)} = \frac{\mathbf{G}(s)}{1 + \mathbf{G}(s)}$$

$$= \frac{K}{s^2 + s + K}$$

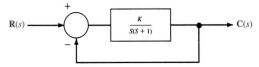

Figure 22.15 A unity feedback system.

Using the quadratic formula, the roots of the characteristic equation, that is, the closed loop poles, are

$$s_1, s_2 = -\frac{1}{2} \pm \frac{\sqrt{1 - 4K}}{2}$$

$$= -\frac{1}{2} \pm \sqrt{\frac{1}{4} - K}$$

Now let's examine the movement of the closed loop poles in the s-plane as the gain K varies. We will graphically illustrate the movement by plotting the pole locations in the s-plane as shown in Fig. 22.16. For $K = 0$, the closed loop poles are located at positions labeled (1), that is, they start at the open loop poles. For K in the range $0 < K < 1/4$, the closed loop poles are located at positions (2), that is, the roots are real and unequal. For $K = 1/4$, both closed loop poles are located at position (3), that is, the roots are real and equal. For $K > 1/4$, the closed loop poles are located at positions (4), that is, the roots are complex conjugates. ■

The heavy line in the s-plane of Fig. 22.16 is called the *root loci* for the feedback system shown in Fig. 22.15. This case was however fairly simple. In general, the characteristic equations may be of higher order, and therefore some guidelines for plotting the root loci are necessary.

It is important to note that the root locus analysis is very convenient because it gives us information about *closed loop* system behavior from knowledge of the *open loop function* $\mathbf{G}(s)\mathbf{H}(s)$. Quite often both $\mathbf{G}(s)$ and $\mathbf{H}(s)$ are easy to determine via modeling, but analyzing $1 + \mathbf{G}(s)\mathbf{H}(s) = 0$ can be quite difficult numerically. In fact, *both* the root locus and Bode plot (to be discussed in the next section) design techniques share the quality of being *graphical* techniques for determining *closed loop* behavior from *open loop* information. It is also important to make the distinction between open loop transfer functions, for example, $\mathbf{G}(s)$ or $\mathbf{H}(s)$, and the open loop function $\mathbf{G}(s)\mathbf{H}(s)$. There may be many of the former, but there is only one of the latter.

Since the root locus is a plot of the roots of the characteristic equation as K varies from 0 to ∞, we are looking for the roots of the equation

$$1 + \mathbf{G}(s)\,\mathbf{H}(s) = 0 \tag{22.22}$$

From this equation we obtain the condition

$$\mathbf{G}(s)\,\mathbf{H}(s) = -1 \tag{22.23}$$

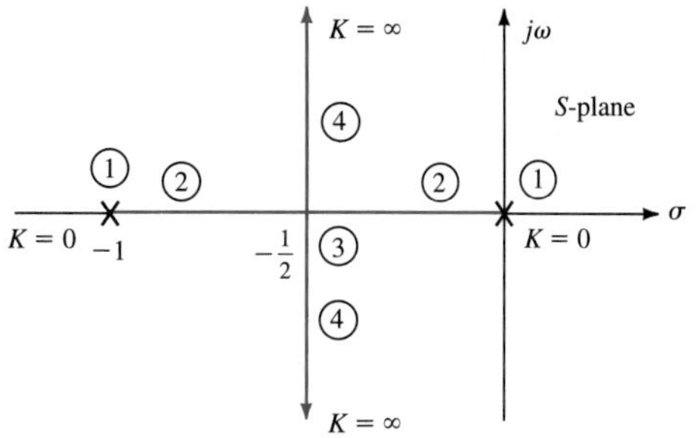

Figure 22.16 Loci of the closed loop poles as the gain K varies from $K = 0$ to $K = \infty$.

or equivalently, the "magnitude" condition

$$|\mathbf{G}(s)\ \mathbf{H}(s)| = 1 \tag{22.24}$$

and the "angle" condition

$$\angle\,\mathbf{G}(s)\ \mathbf{H}(s) = 180° \tag{22.25}$$

From these two conditions we can derive the rules for plotting a root locus. Although we can list many rules for drawing a root locus, we will simply share a few guidelines that will permit us to quickly sketch the loci and then give several examples illustrating their use. Reference to the previous example may help in visualizing the construction process.

First of all, the closed loop poles start ($K = 0$) at the open loop poles and terminate ($K = \infty$) at the open loop zeros or infinity. The number of branches is equal to the order of the characteristic equation. Since the characteristic equation has real coefficients, when the roots appear in the complex plane, they appear in complex conjugates. Every point on the root locus must satisfy the basic condition

$$\mathbf{G}(s)\mathbf{H}(s) = \frac{K(s^m + a_1 s^{m-1} + \ldots)}{(s^{n+m} + b_1 s^{n+m-1} + \ldots)} = -1$$

Note that n = number of poles − number of zeros of $\mathbf{G}(s)\mathbf{H}(s)$. As n loci approach infinity, they do so at angles

$$\alpha_k = \frac{180° + k360°}{n} \qquad k = 0, 1, 2 \cdots n - 1$$

Therefore, if $n = 1$, the asymptote to infinity is at $180°$, if $n = 2$ (the previous example), the asymptotes are at $\pm90°$, if $n = 3$, the asymptotes are at $180°$ and $\pm60°$; if $n = 4$, the asymptotes are at $\pm45°$ and $\pm135°$, etc.

Since $\angle\,\mathbf{G}(s)\mathbf{H}(s) = 180°$, the locus will lie on the real axis at points where the number of poles and zeros to the right is an odd number. Finally, there is symmetry of loci in the s-plane, with respect to the real axis.

EXAMPLE 22.7

We wish to sketch the root loci for the system shown in Fig. 22.17(a).

The loci start at the open loop poles $s = 0, -1$, and -2 and terminate at ∞. The characteristic equation is third order and therefore there are three loci. The asymptotes to in-

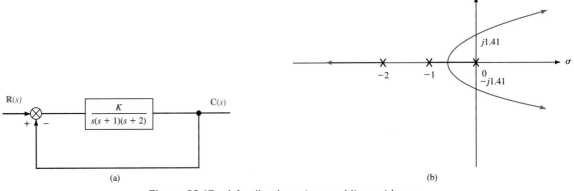

Figure 22.17 A feedback system and its root locus.

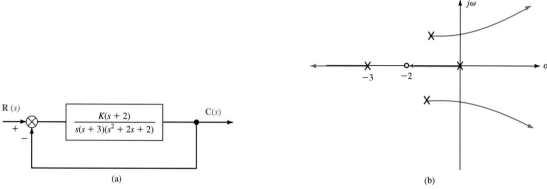

Figure 22.18 A feedback system and its root loci.

finity are at the angles 60°, −60°, and 180°. The roots on the loci at ±60° will be complex conjugates. Loci will appear on the negative real axis between $s = 0$ and $s = -1$ and between $s = -2$ and $s = \infty$.

In view of this analysis we find that the closed loop pole that starts at $s = -2$ follows the loci along the negative real axis to ∞. The closed loop poles that start at $s = 0$ and $s = -1$ come together and split into the complex plane following the asymptotes at ±60° to ∞. A final sketch of the root loci is shown in Fig. 22.17(b). ■

EXAMPLE 22.8

Consider the feedback system shown in Fig. 22.18(a). Loci start at 0, − 3, and −1 ± j1, and one loci terminates at −2, the others terminate at ∞. The characteristic equation is

$$s\,(s + 3)\,(s^2 + 2s + 2) + K\,(s + 2) = 0$$

or

$$s^4 + 5s^3 + 8s^2 + (6 + K)s + 2K = 0$$

And hence there are four loci branches. Since there are four closed loop poles and one open loop zero, the asymptotes to ∞ are at ±60° and 180°. Loci will appear on the negative real axis between $s = 0$ and $s = -2$ and between $s = -3$ and $s = \infty$. Hence, the total root locus is shown in Fig. 22.18(b). ■

22.7
Bode Plots

In Chapters 6 and 17 we illustrated the performance of a network as a function of frequency. If the network characteristics had been plotted on a semi-log scale, that is, a linear scale in decibels for the ordinate and a logarithmic scale for the abscissa, they would have been known as *Bode plots*. This technique can also be used to study feedback system performance as a function of frequency.

As was the case with the root locus, we begin with the system's open loop function. We can express this function as

$$\mathbf{G}(j\omega)\mathbf{H}(j\omega) = M(\omega)e^{j\phi(\omega)}$$

where $M(\omega) = |\mathbf{G}(j\omega)\mathbf{H}(j\omega)|$ and $\phi(\omega)$ is the phase. A plot of these two functions, which are commonly called the *magnitude* and *phase characteristics,* display the manner in which the system response varies with the input frequency ω.

For a feedback system we plot $20 \log_{10}|G(j\omega)H(j\omega)|$ in decibels (dB) and $\angle G(j\omega)H(j\omega)$ in degrees on a semi-log scale, that is, the magnitude in dB and the phase in degrees are on a linear ordinate and the frequency ω is on a logarithmic abscissa.

The following example will be used as a vehicle to demonstrate the analysis procedure. Reference to Example 22.7 will permit us to explore some of the interesting connections which exist between this analysis and the root locus technique.

EXAMPLE 22.9

Once again let us consider the feedback system shown in Fig. 22.17(a). The root locus for this system is shown in Fig. 22.17(b). In our frequency analysis we examine the system performance along the positive $j\omega$ axis in the s-plane. Therefore, if we let $s = j\omega$, the open loop function is

$$G(j\omega) = \frac{K}{j\omega \, (j\omega + 1)(j\omega + 2)}$$

$$= \frac{K_1}{j\omega(j\omega + 1)\left(\dfrac{j\omega}{2} + 1\right)}$$

where $K_1 = \dfrac{K}{2}$.

Figure 22.19 illustrates the Bode plot for the open loop function $G(j\omega)$. Note the changes in both magnitude and phase as a function of frequency.

At this point let us define some important terms that are used in conjunction with Bode plots. The *gain crossover frequency* is the angular frequency ω_1 at which $|G(j\omega_1)H(j\omega_1)| = 0$ dB. At this frequency we measure the phase $\angle G(j\omega_1)H(j\omega_1)$ and the difference between that phase and $-180°$ is the *phase margin*. The *phase crossover frequency* is the angular frequency ω_2 at which $\angle G(j\omega_2)H(j\omega_2) = -180°$. At this frequency we measure the gain $|G(j\omega_2)H(j\omega_2)|$ and the difference between that gain and 0 dB is the *gain margin*. These two quantities (gain margin and phase margin) are illustrated in Fig. 22.20(a),

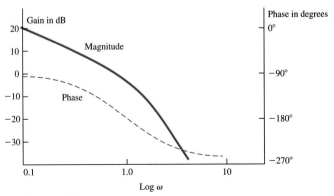

Figure 22.19   Bode plot for the system in Example 22.9.

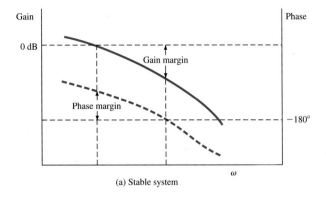

(a) Stable system

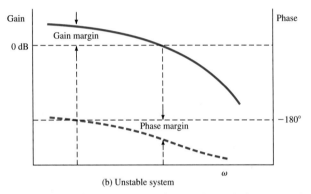

(b) Unstable system

Figure 22.20 Illustration of gain margin and phase margin.

for a stable system and in Fig. 22.20(b), for an unstable system. A system becomes unstable if the composite gain characteristic is above 0 dB when the composite phase characteristic goes below −180°. The shift required for either characteristic to obtain the marginally stable condition in which the gain curve crosses 0 dB when the phase curve crosses −180° is referred to as the margin.

Returning to the Bode plot in Fig. 22.19, we note that the value of ω at which the phase curve crosses the −180° line is $\omega \cong 1.41$ rad/s. The open loop function has a gain of −9.54 dB. Therefore, the gain can be raised by this amount (i.e., gain margin = +9.54 dB) before the system becomes unstable. Hence

$$20 \log_{10} K_1 = 9.54$$
$$K_1 = 3$$

or

$$K = 6$$

Note carefully that in the root locus analysis of this system the loci crossed the $j\omega$ axis at $\omega = 1.41$ rad/s and it can be shown that the gain at that point is $K = 6$. ∎

22.8

Compensation

The original feedback system used to perform a certain task, for example, motor control, may not have the exact performance specification desired. However, it is possible to alter the performance of what could be a very expensive system by using simple and inexpensive *compensating networks* as shown in Fig. 22.21. Perhaps the three most popular networks employed for compensation are the *lead, lag,* and *lead-lag networks* shown in Fig. 22.22. The lead network is typically used to improve the system's transient response;

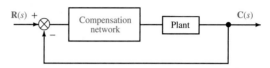

Figure 22.21 Feedback system with a compensator.

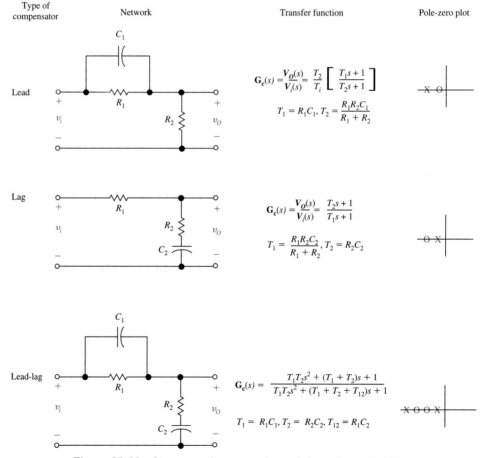

Figure 22.22 Compensation networks and their characteristics.

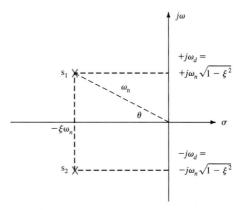

Figure 22.23   The roots of the second-order system for $\xi < 1$.

the lag network is used to improve the low-frequency gain and thus the system's steady state error; and the lead-lag network, as the name implies, combines the properties of the other two.

One important performance criteria is the transient response. For a second-order system the characteristic equation may be written in standard form as

$$1 + \mathbf{G}(s)\,\mathbf{H}(s) = s^2 + 2\xi\omega_n s + \omega_n^2$$

where ω_n is the undamped natural frequency and ξ is the damping ratio. For a good transient response ξ must be less than one. Therefore, the roots of the characteristic equation are

$$= -\xi\omega_n \pm j\omega_d$$
$$s_1,\, s_2 = -\xi\omega_n \pm j\omega_n \sqrt{1 - \xi^2}$$

where ω_d is the *damped frequency*. These poles appear in the s-plane as shown in Fig. 22.23. From the figure we can see that the $\cos\theta = \xi$ or $\xi = \cos^{-1}\theta$. In addition, it can be shown that the system time constant $\tau = 1/\xi\omega_n$. Therefore, a large ω_n for a given ξ will produce a fast response. Hence, for second-order systems or higher order systems that can be approximated by a second-order system, the use of a lead compensator will permit us to adjust the system gain so that the closed loop poles, that is, the roots of the characteristic equation, have selected values of ξ and ω_n which produce some specified transient response as shown in Fig. 22.24.

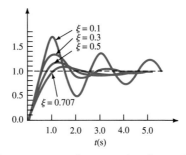

Figure 22.24   The unit step response of a second-order system as a function of ξ with $\omega_n = 1$ rad/sec.

EXAMPLE 22.10

Given the system shown in Fig. 22.25(a), we wish to select a lead compensator and a gain factor K so that the closed loop system will have a damped frequency $\omega_d = 1$ radian per second and a damping factor $\xi = 0.6$.

Since $\omega_d = 1$ rad/s and $\theta = \cos^{-1} \xi = 53°$, two of the closed loop poles should go through the points labeled s_1, s_2 on the s-plane diagram shown in Fig. 22.25(b). Therefore, $s_1, s_2 = -0.75 \pm j1$. If we now arbitrarily select the compensator zero to be at $s = -0.2$, we can calculate the compensator pole so that the root locus shown in Fig. 22.25(c), satisfies the basic condition

$$\angle\, \mathbf{G}(s)\, \mathbf{H}(s) = 180°$$

Therefore, by adding the angles of the vectors from the zeros to point s_1, and subtracting the angles from the poles we find that the compensator pole should be at $s = -1.95$. Finally, the system gain is set to satisfy the condition

$$|\mathbf{G}(s)\, \mathbf{H}(s)| = 1$$

at point s_1. Therefore, solving the equation

$$\frac{K(s_1 + 0.2)}{(s_1)^2(s_1 + 1.95)} = 1$$

yields $K = 1.95$. The value of K can also be found graphically using the vector lengths, that is, multiplying the distances to poles and dividing by the distances to zeros. The compensated system which will perform as specified is shown in Fig. 22.25(d). ■

In design problems, such as the one above, the computer can be used to run test cases quickly. Since the form of the compensator and the characteristic equation are known, root solver routines can also be employed to obtain a solution. In addition, modern con-

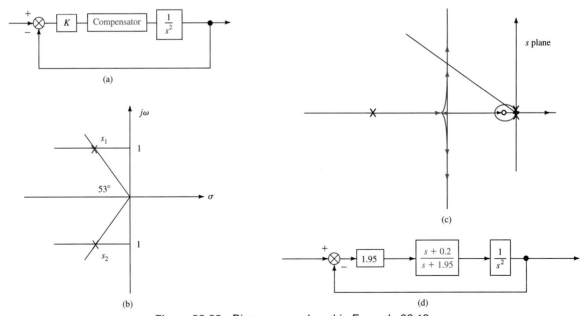

Figure 22.25   Diagrams employed in Example 22.10.

trol software routinely includes functions to generate the graphs for the engineer. While the effect of the lead compensator is easily seen on a root locus, the effect of a lag compensator is more easily visualized using a Bode plot.

22.9
Summary

- The Laplace transform is used to convert a set of linear simultaneous integrodifferential equations to a set of simultaneous algebraic equations.
- The transfer function of a device is the ratio of the Laplace transform of the output to the Laplace transform of the input.
- Block diagrams can be used to describe complicated dynamic systems.
- Two important quantities in the analysis and design of control systems are the closed loop transfer function and the open loop function.
- Block diagram algebra can be used to reduce a complicated system to a single closed loop transfer function.
- A control system is stable as long as its closed loop poles are in the left half of the s-plane.
- A root locus is a plot of the movement of the closed loop poles as a function of the system gain.
- A Bode plot can be used to study both system performance as a function of frequency and stability.
- Lead, lag, and lead-lag compensator networks can be used to adjust system performance.

PROBLEMS

22.1. Find the Laplace transform of the time function $f(t) = e^{-t} + e^{-2t}$.

22.2. Find $\mathbf{F}(s)$ if $f(t) = (1/2)(t - 4e^{-2t})$.

22.3. Find $\mathbf{F}(s)$ if $f(t) = e^{-4t}(t - e^{-t})$.

22.4. Prove that if

(a) $f(t) = e^{-at}$, $\mathbf{F}(s) = \dfrac{1}{s + a}$

(b) $f(t) = te^{-at}$, $\mathbf{F}(s) = \dfrac{1}{(s + a)^2}$

22.5. Prove that if $f(t) = \sin bt$, then $\mathbf{F}(s) = \dfrac{b}{s^2 + b^2}$.

22.6. Compute the transfer function for the block diagram shown in Fig. P22.6.

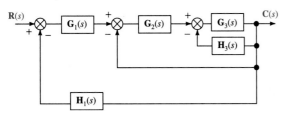

Figure P22.6

22.7. Compute the transfer function for the block diagram shown in Fig. P27.7.

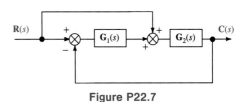

Figure P22.7

22.8. Find the transfer function for the block diagram in Fig. P22.8.

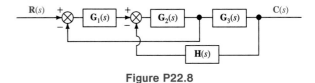

Figure P22.8

22.9. Find the transfer function for the block diagram in Fig. P22.9.

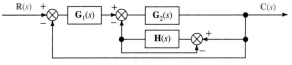

Figure P22.9

22.10. Find the transfer function for the block diagram shown in Fig. P22.10.

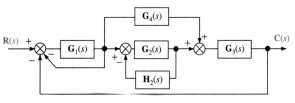

Figure P22.10

22.11. Find the transfer function for the block diagram shown in Fig. P22.11.

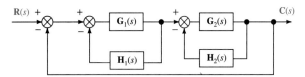

Figure P22.11

22.12. Given the following characteristic equations, determine if the system is overdamped, underdamped, or critically damped.
(a) $1 + G(s) = s^2 + 5s + 4$
(b) $1 + G(s) = s^2 + 4s + 5$

22.13. Repeat problem 22.12 for the following characteristic equations.
(a) $1 + G(s) = s^2 + 6s + 9$
(b) $1 + G(s) = s^2 + 6s + 18$

22.14. Given the system in Fig. P22.14, find the value of K which will produce a critically damped closed loop response.

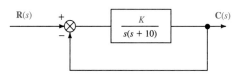

Figure P22.14

22.15. Select a value for the parameter α in the system in Fig. P22.15 so that the closed loop response will have a damping ratio of 0.6.

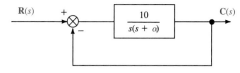

Figure P22.15

22.16. The following open loop functions are employed in a unity feedback system. Sketch the root loci for these systems.

(a) $G(s) = \dfrac{K(s+1)}{s(s+2)}$

(b) $G(s) = \dfrac{K(s+2)}{s(s+1)}$

22.17. Repeat problem 22.16 for the open loop functions:

(a) $G(s) = \dfrac{K(s+5)}{s(s+3)(s+6)}$

(b) $G(s) = \dfrac{K}{s(s+1)(s+3)(s+4)}$

22.18. Repeat problem 22.16 for the following open loop functions:

(a) $G(s) = \dfrac{K(s)(s+2)}{(s+5)(s^2+6s+25)}$

(b) $G(s) = \dfrac{K}{s(s+4)(s^2+6s+13)}$

22.19. Repeat problem 22.16 for the following open loop functions:

(a) $G(s) = \dfrac{K}{s(s+4)(s^2+4s+8)}$

(b) $G(s) = \dfrac{K}{s(s+4)(s^2+2s+5)}$

22.20. Repeat problem 22.16 for the following open loop functions:

(a) $G(s) = \dfrac{K(s+1)(s+2)}{s(s^2+4s+8)}$

(b) $G(s) = \dfrac{K(s+2)}{s(s+1)(s^2+4s+8)}$

22.21. The characteristic equation for a feedback system is

$$1 + \mathbf{G}(s) = s^2 + 3.25\ s + 64$$

Determine the undamped natural frequency and the damping ratio for this system.

22.22. The closed loop transfer function for a system is given by the expression

$$\frac{\mathbf{C}(s)}{\mathbf{R}(s)} = \frac{K_1}{0.04s^2 + 0.16\ s + 1}$$

find the value of ξ and ω_n.

22.23. Sketch the Bode plot for (a) a lead network and (b) a lag network.

22.24. Sketch the root loci for the open loop function

$$\mathbf{G}(s) = \frac{K}{s^2(s + 4)}$$

If $\mathbf{G}(s)$ is used in a unity feedback system, is the system stable?

22.25. Given the unity feedback system described in problem 22.24, can a lead compensator be used to provide a stable operating range? If so, how? If not, why not?

22.26. A lead compensator with $\mathbf{G}_c(s) = \dfrac{s + 2.2}{s + 5}$ is added to the forward path of a unity feedback system with $\mathbf{G}(s) = \dfrac{K}{s(s + 2)(s + 4)}$. Sketch the root loci for this system.

22.27. Repeat problem 22.26 if the compensator is changed to $\mathbf{G}_c(s) = \dfrac{s + 1.8}{s + 3.8}$.

22.28. A lag compensator with $\mathbf{G}_c(s) = \dfrac{s + 0.2}{s + 0.1}$ is added to the forward path of a unity feedback system with $\mathbf{G}(s) = \dfrac{K}{s(s + 1)(s + 2)}$. Sketch the root loci for this system.

23

Communications

23.1
Introduction

The field of *communications* is broad and encompasses all the areas of engineering and science that deal with information and its transfer from one point to another. Most modern communications systems have a large component that is electronic, and this field is one of active growth in electrical engineering.

The rapidly expanding services and capabilities of computer networks, cellular telephone systems, and satellite communications systems, to mention a few applications, are greatly changing our daily lives and the way in which we do our work. In many companies, employees are able to perform significant work away from their office, even at home, through computer communications links with other employees and the office.

Electronic communication systems collect information from an originating source, convert this information into electric currents, fields, or light waves, transmit these signals over electrical networks, optical fibers, or through space to another point, and then reconvert them into a form suitable for interpretation by the receiving station.

In this chapter the fundamental engineering principles of communications systems will be developed and the operational details of several examples of historically important and modern electronic communication systems will be described, including radio, television, telephones, and various computer (data) communications networks.

23.2
Example of Typical Communications System

A typical everyday example of such a system is that of a radio station transmitting its program to a typical listener standing with their cat and transistor radio by the pool, as illustrated in Fig. 23.1. The station announcer speaks into a microphone which converts her

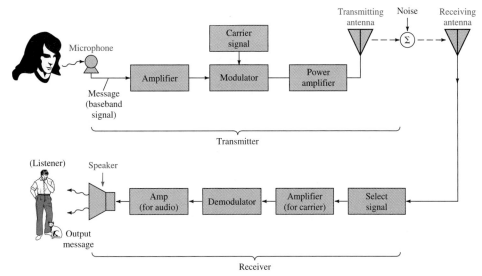

Figure 23.1   Radio communications system.

voice into a weak electronic signal. This signal or message, called the *baseband signal,* is amplified and attached to a higher frequency signal called the *carrier wave* by a process called *modulation;* this new signal (carrier with message attached) is amplified to a high power level and connected to the transmitting antenna. The signal is radiated from this antenna through space and some small portion is captured on the receiving antenna of the radio receiver by the pool; the receiver also receives significant radiation from many other sources not containing the desired information and therefore noise is introduced into the communication channel. The radio receiver "takes in" both the noise and the desired signal; it attempts to reduce the noise and extract the original baseband signal for presentation to the listener. The *signal-to-noise* (S/N) *ratio,* should be made as high as possible to obtain good performance.

The "message" is the original baseband signal to be transmitted; this signal might be the voice signal from a microphone (as in this example) or, for example, the video signal from a television camera. Both of these examples are generally analog signals. If the message is a digital signal, a stream of 1s and 0s from a computer, for example, then the communications system is termed a digital communications system or a data communications system.

The process of attaching the baseband signal to the carrier frequency is termed modulation. In Chapter 7 the basics of *amplitude modulation* (AM), *frequency modulation* (FM), and *phase modulation* were introduced; these concepts will be developed further in this chapter. In general, a modulator converts the baseband signal into a form that can be more readily transmitted over the available *channel* or *medium,* the path connecting the transmitter and the receiver. Generally this consists of attaching the message to a higher frequency signal which will radiate through the atmosphere or propagate more efficiently through the wire or fiber that links transmitter to receiver. At the receiver, the reconversion process in which the baseband signal is extracted from the received signal is called *demodulation.* The noise that appears mixed with the received signal can often be modeled as originating from a single noise source; therefore, the entire communication system described in this example can be modeled, as previously indicated, by Fig. 23.1.

23.3
Types of Communication Systems

Electronic communication systems can be broadly categorized into several types: *direct baseband systems,* or *carrier transmission systems.* In addition, they can also be generally classified as *analog* or *digital transmission systems.* The radio system described in the previous section was an analog, carrier transmission system. An example of a direct baseband system would be an intercom directly wired between two rooms in your home. In this latter case, the audio signal is transmitted directly as an audio frequency signal over the medium (wires) connecting the transmitter to the receiver.

Direct Baseband Systems

Those systems that send and receive the baseband signal directly are called *direct baseband systems.* These are generally the simplest systems and are typical of the original electronic communications systems of historical significance. The later development of carrier systems using modulation and demodulation increased the ease and effectiveness of signal transmission over large distances.

The Telegraph. The first electrical communications system, the *telegraph* which was invented in 1844 by Samuel Morse, made a significant impact on society. It marked the first time that messages could be sent great distances faster than a man or an animal could carry them. Morse, the inventor who launched the modern age of electronic communications, was not known as a scientist or engineer. He was considered one of the foremost artists of his time, and was professor of painting and sculpture at a college in New York. He struggled with the development of the telegraph for decades and with great frustration. Pages from his notebook made in 1832 during a voyage from Europe show early details of his transmitting device and the code that would eventually make him famous. The first intercity transmission between Baltimore and Washington on May 24, 1844, was a successful public demonstration that created momentum for the telegraph's commercial development. It never fell into government hands as Morse had wanted, but was used by several competing companies until 1856, when the various interests were consolidated into the first giant monopoly in American business, Western Union.

It is interesting that this first system was a digital system and used the transmission of a digital binary signal in a code, known as the *Morse code,* shown in Fig. 23.2. The *continental code* was a refinement of the Morse code that eliminated spaces within a single letter. This code represents letters and numbers by a series of pulses of two lengths, a short pulse and a long pulse (often these are called a "dot" and a "dash," or a "dit" and a "dah"). The telegraph was a digital direct baseband system in that the message was carried by direct connection from sender to receiver without any modulation/demodulation or other conversion.

Figure 23.3 shows a circuit drawing of the basic telegraph system; pressing or closing the "key," a simple switch with a spring to keep it "open" unless pressed, completes a circuit and creates a "click" for the listener by the movement of a metal piece over an energized electromagnet.

Continental code	Morse code
Alphabet	

	Continental code		Morse code	
A	.▬		A	.▬
B	▬...		B	▬...
C	▬.▬.		C	.. .
D	▬..		D	▬..
E	.		E	.
F	..▬.		F	.▬.
G	▬▬.		G	▬▬.
H	....		H	....
I	..		I	..
J	.▬▬▬		J	▬.▬.
K	▬.▬		K	▬.▬
L	.▬..		L	▬▬▬
M	▬▬		M	▬▬
N	▬.		N	▬.
O	▬▬▬		O	. .
P	.▬▬.		P	.....
Q	▬▬.▬		Q	..▬.
R	.▬.		R	. ..
S	...		S	...
T	▬		T	▬
U	..▬		U	..▬
V	...▬		V	...▬
W	.▬▬		W	.▬▬
X	▬..▬		X	.▬..
Y	▬.▬▬		Y	.. ..
Z	▬▬..		Z	... .

There are no space letters in the Continental code

C, O, R, Y, Z and & are composed of dots and spaces
T is a short dash
L is a longer dash
Zero (0) is usually abbreviated to T

Numerals	

	Continental code		Morse code	
1	.▬▬▬▬		1	.▬▬.
2	..▬▬▬		2	..▬..
3	...▬▬		3	...▬.
4	....▬		4	....▬
5	.....		5	▬▬▬
6	▬....		6	......
7	▬▬...		7	▬▬...
8	▬▬▬..		8	▬.....
9	▬▬▬▬.		9	▬..▬
0	▬▬▬▬▬		0	▬▬▬▬▬

Figure 23.2   The Morse and Continental codes.

The Telephone. The development and patenting of the *telephone* by Alexander Graham Bell in 1876 extended the capability of baseband communication systems to carry the human voice; this was another significant communications milestone. Bell was not alone in the race to produce a practical instrument for the transmission and reception of human speech. The Bell telephone was similar to that of another American inventor, Elisha Gray. It is interesting that the receiver in the telephones of today is not that dif-

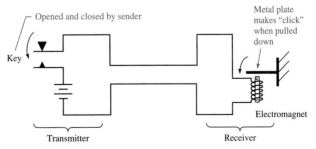

Figure 23.3   Basic telegraph system.

ferent from that in Bell's original telephone. The telephone that Bell developed and produced was an analog baseband system.

Figure 23.4 illustrates a circuit drawing similar to that of the original Bell telephone. The speaker's voice vibrates a diaphragm attached to a compartment of compressed carbon granules. This vibration produces a varying pressure on the carbon granules and the electrical resistance through the carbon is varied in exact proportion to the acoustical pressure waves from the speaker's voice. A dc battery supplies power to the circuit, and the changing resistance in the transmitter creates a changing current in the circuit.

At the receiving end, this changing current varies the intensity of attraction of the electromagnet in the receiving set, which in turn moves the receiving diaphragm over the electromagnet in a vibrating pattern just as that created on the transmitting diaphragm. Therefore, the original sound waves are reproduced at the receiver. Modern telephone systems, of course, are not so simple and often use various forms of carrier transmission as described in the next section.

Carrier Transmission Systems

The discovery that high-frequency electric signals could be connected to an antenna, and that electromagnetic waves would then be produced and radiate through space, led immediately to the notion of carrier transmission systems. That is, if certain frequencies of electromagnetic radiation propagate through space readily, then the message signal could be used to modulate the carrier and thus carry the desired information over great distances at the speed of light without the need for wires connecting sender to receiver. This formed the basis of *wireless* or radio (and later television) communication systems.

Referring to the electromagnetic spectrum of Fig. 1.2, recall that electromagnetic energy radiates as radio frequency waves (RF) from about 10^5 Hz up to approximately 10^{12} Hz; above this range is light radiation, with visible light at about 10^{15} Hz.

There are therefore two types of carrier transmission communication systems, based on the frequency range of the carrier signal:

1. *Radio frequency (RF) systems*
2. *Light wave systems*

We will discover later in this chapter that the carrier frequency must be much higher than the highest frequency in the message or baseband signal to be transmitted; therefore, at higher carrier frequencies more information can be transmitted. This explains in part

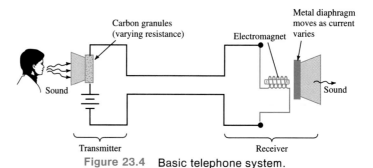

Figure 23.4   Basic telephone system.

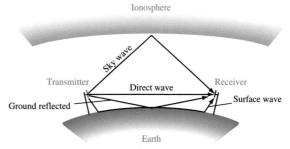

Figure 23.5 Possible radio wave transmission paths.

the growing interest in light wave communications often accomplished with optical fibers. Light wave systems have extremely high information transmission capacity.

Radio Frequency Transmission

Radio frequency or RF electromagnetic waves will propagate through the atmosphere from the transmitting antenna to the receiving antenna of a communication system. The signal received at the receiving antenna is the vector sum of the signals traveling by several different paths: (1) the direct wave, (2) the sky wave, (3) the ground-reflected wave, and (4) the surface wave. These paths are illustrated in Fig. 23.5.

The Direct Wave. The direct wave or "line-of-sight" transmission is the easiest to understand. Radio transmission in free space from a point source (nondirectional) results in a decrease in energy per unit area that follows an inverse square law. There is therefore a limit to the range or distance of successful transmission based on a decrease of signal strength. There is also usually a limit on line-of-sight transmission based on the curvature of the earth and the height above ground of the transmitting and receiving antennas.

EXAMPLE 23.1

Suppose that a transmitting antenna is placed at the top of a tower 1,000 feet tall. If the receiving antenna is also on top of a 1,000-foot tower, let us calculate the maximum possible distance for line-of-sight transmission, as illustrated in Fig. 23.6. Assuming the earth's radius is 4,000 miles and applying geometry to the right triangle shown in Fig. 23.6 gives

$$\phi = \arccos\left[\frac{(4000 \text{ mi})(5280 \text{ ft/mi})}{(4000 \text{ mi})(5280 \text{ ft/mi}) + 1000 \text{ ft}}\right] = 0.558 \text{ degrees}$$

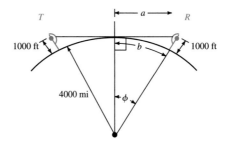

Figure 23.6 Illustration of Example 23.1.

and therefore

$$a = (4000 \text{ mi}) \sin \phi = 39 \text{ miles}$$

Since ϕ is so small, we may assume that $b \approx a$, and therefore the total line-of-sight distance is $2a$, or 78 miles.   ■

The direct waves propagate through the earth's *trophosphere* which contains the earth's air, water, and weather conditions. The air density, temperature, water vapor, and other factors all affect propagation; the extent of the effect of each of these factors depends on the frequency of transmission.

The combination of the direct wave, ground-reflected wave, and surface wave makes communications at lower frequencies (below about 30 MHz) possible over distances greater than that assumed by the direct geometrical calculation above. Therefore, the range of transmission by ground waves increases somewhat at lower frequencies.

The Sky Wave. The sky wave is extremely important and makes possible worldwide shortwave communications. The *ionosphere* surrounds the earth with ionized particles in shells at altitudes above the earth between about 50 km and 600 km. Radio waves in a relatively narrow frequency region, between about 3 MHz and 30 MHz, reflect off these high-altitude "mirrors." The reflected wave can bounce back and forth between an ionospheric reflection and a ground reflection, and thus travel, literally, around the world, as shown in Fig. 23.7.

The charge condition of the ionosphere changes hour by hour, much like the weather is continuously changing on earth. The ionosphere primarily responds to the stream of radiation from the sun (the solar wind) and sun spot activity. Therefore, shortwave radio communications are highly variable. Transmission may be excellent in a certain band of frequencies, a condition known as "the band being open," and then quickly fade away. Meanwhile, communications may be simultaneously improving in some other frequency band. The ionosphere is essentially transparent above 30 MHz and no reflections occur.

Figure 23.8 is qualitative, but gives an idea of the approximate range of communications expected using standard equipment and moderate antenna heights. Several features are worthy of note:

1. The sky wave has a narrow frequency band but is very effective in this region.
2. The ground-wave range increases at lower frequencies.
3. Above 100 MHz, geometrical optics can be used as was done in Example 23.1 for first-order approximations.

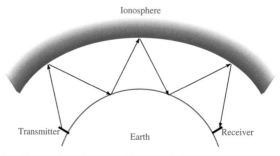

Figure 23.7   Multiple reflections enable worldwide shortwave communications.

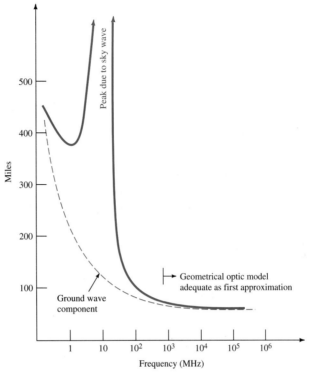

Figure 23.8   Approximate range of transmission using nominal equipment.

Coaxial Cable. Radio frequency signals are often transmitted from point to point by a special cable designed to restrict or contain the RF energy and provide only a small loss in the signal amplitude per unit length of cable. The most common such cable, shown in Fig. 23.9, is called *coaxial cable* and has a cross-section of concentric layers. There is a central copper wire surrounded by an insulating layer or sheath, outside of which is an-

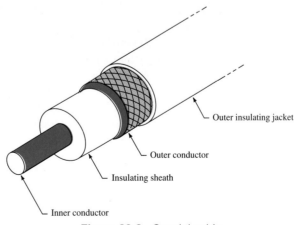

Figure 23.9   Coaxial cable.

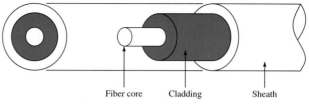

Figure 23.10   Optical fiber.

other concentric layer of conducting material. Finally, the outside layer is another insulating layer or jacket for electrical isolation and protection from the environment.

Coaxial cables are used for the transmission of signals from fractions of a megahertz up to hundreds of megahertz. They are commonly used for the distribution of cable TV signals along streets and into homes.

Light Wave Transmission

Light is electromagnetic radiation with a frequency much higher than radio waves, and therefore as a carrier wave, can contain much more information. In addition, light wave communications has other advantages and some disadvantages.

First the disadvantages will be discussed. Light is not useful for general broadcast over a wide area for many receivers like radio. It is generally used only for information exchange directly between two points, and the light is beamed or directed to the receiver by optical techniques. Light transmission as a directed beam through the atmosphere is possible, but highly susceptible to changing atmospheric conditions. Depending on the frequency, light transmission can be degradated by smoke, fog, haze, rain, snow, etc. Therefore, except for very short special-purpose applications, the use of light as a carrier for reliable information transmission requires that it travel in a protected environment.

The most popular such transmission medium is the *optical fiber*. The optical fiber is a cable designed to carry light waves; it is nonmetallic, and generally has a cross-sectional structure as shown in Fig. 23.10. There is a transparent (light conducting) core, surrounded by a cladding with optical properties that keep the light rays confined to the inner core. Around the cladding is a concentric layer for mechanical strength and an outer cover for protection.

The core material is usually a thin glass strand with an optical index of refraction of n_c. The cladding must have a lower index of refraction, n_{cl}. For this situation, according to Snell's law, the light rays with a low initial angle of incidence entering the cable will suffer "total internal reflection" at the interface between core and cladding; thus, these light rays will bounce back and forth as they travel down the fiber core, as illustrated in Fig. 23.11. This confines the light signal to the core.

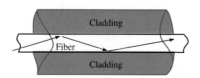

Figure 23.11   Light is confined in the fiber core by "total internal reflections."

There is a loss in signal power as the light travels down the fiber; this is called *attenuation,* and is usually specified in db of loss per kilometer of fiber. Losses are mainly the result of four factors: (1) scattering due to microscopic imperfections in the fiber, (2) absorption due to conversion of some light energy into heat, (3) connection losses at splices and joints in the fiber, and (4) losses at bends in the fiber which cause some light rays to be lost from the core.

Fiber optic cables are lighter and less expensive than copper cables and have very high immunity to interference from outside sources. Fiber optic cables are primarily used to transmit digital data by switching the light source on and off to represent a series of bits (1s and 0s). The primary advantage of fiber optic systems is that the light beam can be pulsed at a very high rate, making high data transmission rates of a Gbps (Gigabit per second) or more possible.

There are two main types of light sources used for optical data transmission: the LED (*light emitting diode*) and the *laser diode.* The laser diode emits coherent laser light at a single frequency. These devices couple efficiently into fiber light guide and are a common light source for optical communication systems. The LED disperses energy over a broader range of frequencies and over a rather large angle, and therefore typically couples much less energy into a given fiber. It is less expensive than a laser diode and also has a longer expected life. Thus, the LED is also commonly used in optical communication systems.

23.4
Signals and Their Spectra

One of the most important concepts in communications is that a signal can be represented as a function of time, or equivalently as a function of frequency and that if you know one, you can determine the other. That is, there is a unique relationship between a signal expressed in the time domain and its expression in the frequency domain.

As a simple example, let's consider a periodic signal, the sinusoidal signal

$$f(t) = A_m \sin(\omega_T t) \tag{23.1}$$

In the time domain, we are familiar with the plot of a sinusoidal signal with an angular frequency of ω_T as a function of time; this is shown in Fig. 23.12(a). The *frequency spectrum* of this signal which is a plot of signal intensity as a function of frequency is shown in Fig. 23.12(b). This plot is simply a vertical line at frequency ω_T. There is no signal energy at any other frequency.

Now, for example, consider a more complex signal

$$f(t) = A_m \cos(\omega_T t) - \frac{A_m}{3} \cos(3\omega_T t) \tag{23.2}$$

The plot of this signal as a function of time is shown in Fig. 23.13(a); the two sinusoidal components (cosine functions) are added graphically to show the plot of $f(t)$. If a frequency spectrum of this signal is constructed as shown in Fig. 23.13(b), it consists of two frequency components. The plot of the magnitude of the spectrum doesn't convey phase relations; the second term in Eq. (23.2) has a phase difference of 180 degrees from that of the first (because of the "−" sign). Information about the phase of each sinusoidal component is also important, and spectra are often plotted in complex or phasor notation to retain the phase information.

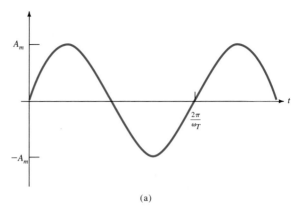

(a)

Figure 23.12(a) Sinusoidal signal with $\omega = \omega_T$ plotted as a function of time.

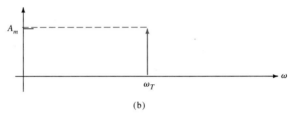

(b)

Figure 23.12(b) Frequency spectrum of sinusoidal signal with $\omega = \omega_T$.

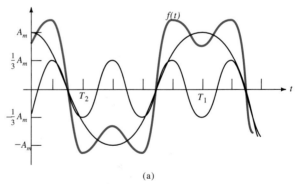

(a)

Figure 23.13(a) Plot of $f(t)$ and its two cosine components.

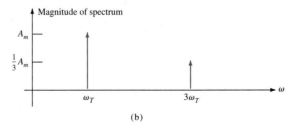

(b)

Figure 23.13(b) Magnitude of spectrum of $f(t)$.

Periodic Signals

A periodic signal is one that repeats in time with a period, T, such as the sinusoidal signals of the previous section. That is

$$f(t) = f(t + nT), \, n = 1,2,3 \ldots \tag{23.3}$$

The value of ω_T in the following equations is the "fundamental" frequency which is simply that associated with the period, T, as

$$\omega_T = 2\pi f_T = \frac{2\pi}{T} \tag{23.4}$$

As illustrated in Chapter 6, any arbitrary periodic signal can be represented by a sum of sine and cosine waves with frequencies which are integer multiples of ω_T; this series is called the *Fourier series*. The Fourier series was introduced in Chapter 4 and is given by the expression

$$f(t) = a_0 + a_1\cos\omega_T t + a_2\cos2\omega_T t + a_3\cos3\omega_T t + \ldots$$
$$+ \, b_1\sin\omega_T t + b_2\sin2\omega_T t + b_3\sin3\omega_T t + \ldots \tag{23.5}$$

where the Fourier coefficients are

$$a_o = \frac{1}{T} \int_{-T/2}^{T/2} f(t) \, dt \tag{23.6a}$$

$$a_n = \frac{2}{T} \int_{-T/2}^{T/2} f(t) \cos n\omega_T t \, dt \qquad \text{for } n = 1, 2, 3, \ldots \tag{23.6b}$$

$$b_n = \frac{2}{T} \int_{-T/2}^{T/2} f(t) \sin n\omega_T t \, dt \qquad \text{for } n = 1, 2, 3, \ldots \tag{23.6c}$$

EXAMPLE 23.2

Given the square wave shown in Fig. 23.14(a), we wish to (a) find the Fourier series, (b) sketch the frequency spectrum, and (c) plot the sum of the first two frequency components.

(a) For this waveform, $T = 2\pi$ and thus $\omega_T = 1$. The term a_o is

$$a_o = \frac{1}{\pi} \int_0^\pi (+V) \, dt + \frac{1}{\pi} \int_\pi^{2\pi} (-V) \, dt = \frac{V}{\pi} (\pi - 0 - 2\pi + \pi) = 0$$

The value of a_o is the average or dc value of the waveform. a_n is then

$$a_n = \frac{1}{\pi} \int_0^\pi (+V)\cos nt \, dt + \frac{1}{\pi} \int_\pi^{2\pi} (-V)\cos nt \, dt$$

$$= \frac{V}{n\pi} (2 \sin n\pi - \sin n \, 2\pi) = 0 \qquad \text{for } n = 1, 2, 3, \ldots$$

and b_n is

$$b_n = \frac{1}{\pi} \int_0^\pi (+V) \sin nt \, dt + \frac{1}{\pi} \int_\pi^{2\pi} (-V) \sin nt \, dt$$

$$= \frac{V}{n\pi} (1 - 2 \cos n\pi + \cos n \, 2\pi)$$

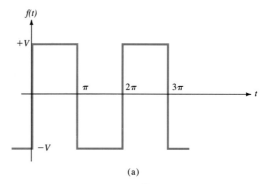

Figure 23.14(a)   Square wave.

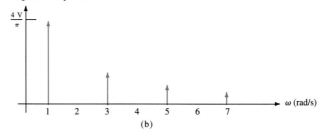

Figure 23.14(b)   Frequency spectrum of square wave.

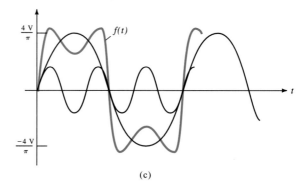

Figure 23.14(c)   Plot of first two terms of Fourier series of $f(t)$.

Therefore

$$b_1 = \frac{4V}{\pi}, \, b_2 = 0, \, b_3 = \frac{4V}{3\pi}, \, b_4 = 0, \, b_5 = \frac{4V}{5\pi}, \, \ldots$$

and

$$f(t) = \frac{4V}{\pi}\left[\sin \omega t + \frac{1}{3}\sin 3\omega t + \frac{1}{5}\sin 5\omega t + \ldots\right] \qquad (23.7)$$

(b) The frequency spectrum shown in Fig. 23.14(b) is a *line spectrum* in which each of the Fourier coefficients represents the magnitude of the signal at frequencies that are integer multiples of the "fundamental" frequency, ω_T. The frequency $3\omega_T$ is termed the third harmonic; $5\omega_T$ is termed the fifth harmonic, etc.

(c) A plot of the first two terms of the series is shown in Fig. 23.14(c); it is similar to the plot previously shown in Fig. 23.13(a) except it is a sine series. Notice that with only the first two terms of the series plotted, the general shape of a square wave is emerging. However, higher harmonics are needed to create sharp corners. We may associate discontinuities and corners with the higher frequency components of a waveform. ■

Note that a periodic waveform will always produce a *comb* or line frequency spectrum consisting of discrete frequencies.

DRILL EXERCISES

D23.1. Determine the Fourier series for the "full-wave rectified" signal shown in Fig. D23.1.

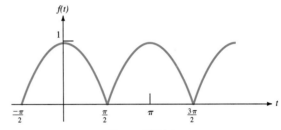

Figure D23.1

Ans: $f(t) = \dfrac{2}{\pi} \left(1 + \dfrac{2}{3} \cos 2\omega t - \dfrac{2}{15} \cos 4\omega t + \dfrac{2}{35} \cos 6\omega t + \ldots \right)$

D23.2. Determine the Fourier series for the series of pulses shown in Fig. D23.2.

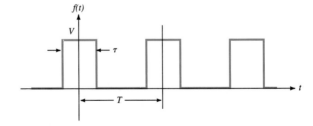

Figure D23.2

Ans: The coefficient a_o = average value = (area per period)/period

$$a_o = V\tau/T \tag{23.8}$$

The coefficient b_n can be shown to be zero for all n.
The coefficient a_n can be calculated and therefore the Fourier series is

$$f(t) = \frac{V\tau}{T} + \sum_{n=1}^{\infty} \frac{2V\tau}{T} \frac{\sin \dfrac{n\pi\tau}{T}}{\dfrac{n\pi\tau}{T}} \cos n\omega_T t \tag{23.9}$$

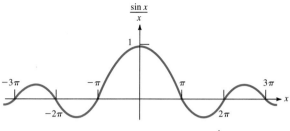

Figure 23.15   Plot of $\text{sinc}(x) = \dfrac{\sin x}{x}$

Notice that the coefficient, a_n, in the previous drill exercise has the functional form of $(\sin x)/x$. This is a very common result in communication studies, and is given the special name, $\text{sinc}(x)$. That is

$$\text{sinc}(x) = \frac{\sin x}{x} \tag{23.10}$$

A plot of $\text{sinc}(x)$ versus x is shown in Fig. 23.15; it has a central peak, and decreasing magnitude as x is increased. The zero crossings occur at integer multiples of π, or at $x = n\pi$.

DRILL EXERCISE

D23.3. In Drill Exercise 23.2, if the special case of a square wave is considered, where $\tau = T/2$, compare the resulting Fourier series with the result obtained in Example 23.2. Note that there is a difference in both the dc level and the position of the wave relative to the $t = 0$ axis.

Nonperiodic Signals

If we determine the sinusoidal frequency components of a nonperiodic signal, that is, one that doesn't repeat in time, we find its frequency spectrum contains a continuous range of frequencies, instead of the discrete integer frequency multiples seen in the Fourier series. One way to view this is that a nonperiodic signal is a limiting case of a periodic signal in which the period, T, approaches infinity. For this case, the fundamental frequency, ω_T, approaches zero, and the frequency harmonics get closer and closer together forming a continuous band.

For signal analysis of nonperiodic signals, the *Fourier transform* is utilized to convert signals from the time domain to the frequency domain and the inverse transform is used to convert back. In Chapter 22, the Laplace transform was introduced and used to convert from the time domain to the complex frequency domain for the solution of systems and controls problems. The Fourier transform is similar and more convenient for most communications problems where the real part of the complex frequency is zero. We define the Fourier transform as

$$\mathbf{F}(f) = \int_{-\infty}^{+\infty} f(t)\, e^{-j2\pi ft}\, dt \tag{23.11}$$

Since the variable t is a dummy variable, after the limits of integration are used, the result is a function of only the frequency f. For convenience of notation, we often let a script F represent the operation on $f(t)$ described in Eq. (23.11); therefore

$$\mathbf{F}(f) = \mathscr{F}[f(t)] \tag{23.12}$$

Likewise, the inverse Fourier transform is defined as

$$f(t) = \int_{-\infty}^{+\infty} \mathbf{F}(f)\, e^{j2\pi ft} dt \tag{23.13}$$

and this is often notated by the shorthand

$$f(t) = \mathscr{F}^{-1}[\mathbf{F}(f)] \tag{23.14}$$

The Fourier transform (spectrum) of an arbitrary nonperiodic signal produces a signal density function; if the signal is voltage, for example, it would represent volts per hertz. It describes the intensity of the signal over the continuous range of frequencies at which signal energy is present.

EXAMPLE 23.3

Let us calculate the Fourier transform or spectrum of a single voltage pulse as shown in Fig. 23.16(a), which has an amplitude of A, and a duration of τ.

The figure indicates that

$$v(t) = A \qquad \text{for } -\tau/2 < t < \tau/2$$
$$= 0 \qquad \text{elsewhere}$$

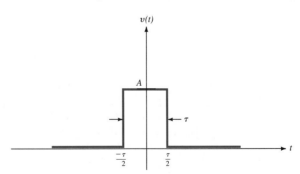

Figure 23.16(a)   Pulse of amplitude A and width τ.

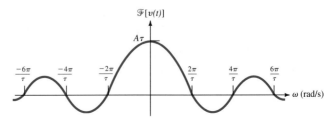

Figure 23.16(b)   Fourier transform of $v(t)$ pulse.

Calculating the Fourier transform

$$\mathbf{F}(f) = \int_{-\tau/2}^{\tau/2} A \, e^{-j2\pi ft} dt$$

$$\mathbf{F}(f) = \frac{A \, e^{-j2\pi ft}}{-j2\pi f} \bigg|_{-\tau/2}^{\tau/2} = A\tau \left[\frac{\sin 2\pi f(\tau/2)}{2\pi f(\tau/2)} \right] = A\tau \, \text{sinc}(\pi f\tau)$$

A plot of this spectrum, shown in Fig. 23.16(b), illustrates the characteristic $\sin(x)/x$ shape; the zero crossings occur when the sin function is zero, which is at integer multiples of π. This occurs in this example at frequencies which are the even integer multiples of $1/\tau$. ■

The frequency spectra representing single and repetitive pulses are very important in the study of digital communications systems.

Bandwidth and Sampling

The *bandwidth* of a signal is of critical importance in determining how this signal might be transmitted via a communications system. The absolute magnitude of the Fourier transform of a baseband signal is usually centered at dc (zero frequency) and might appear as in Fig. 23.17(a). The bandwidth for such a signal is f_{mHI} where f_{mHI} is the positive fre-

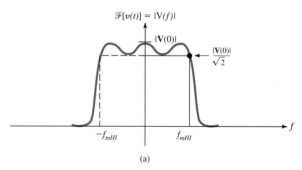

(a)

Figure 23.17(a)   Spectrum of baseband signal.

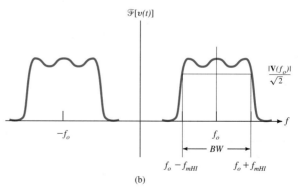

(b)

Figure 23.17(b)   Fourier transform of signal in which the center frequency has been shifted to f_o.

quency at which the magnitude of the Fourier transform of the message signal falls to $1/\sqrt{2}$ of its value at dc.

As we will soon see, sometimes the signal energy is not centered around dc, but rather around some other frequency, f_o. The Fourier transform of such a signal and its bandwidth is shown in Fig. 23.17(b). The bandwidth of this signal extends from $f_o - f_{mHI}$ to $f_o + f_{mHI}$.

Most modern digital systems that input analog data use an A/D converter to sample the analog signal at discrete points in time. A digital representation of the continuously changing analog signal is constructed by storing each of these sampled values in its memory as a digital word. The question naturally arises, how often must one sample and store a value of the analog signal in order to completely represent it in sampled digital form? Of course the answer depends on how fast the analog signal is changing, and that is related to the highest frequency components in the signal.

If the analog signal is changing values rapidly it will have high frequency components and a large bandwidth, and therefore it will require a high *sampling rate*. The converse is also true.

Consider a baseband signal, $f(t)$, which is bandlimited (e.g., passed through a low-pass filter) so that its Fourier transform $\mathbf{F}(f)$ is zero for frequencies outside of the bandwidth, *BW*. If we let f_m be the maximum signal frequency which occurs at the edge of the bandpass, then the signal can be uniquely determined by samples taken at evenly spaced time intervals separated in time by no more than T_s where

$$T_s = \frac{1}{f_s} \le \frac{1}{2f_m} \text{ seconds} \tag{23.15}$$

Stated alternatively, the sampling frequency must be at least twice the highest frequency of the signal being sampled; the *minimum sampling rate* is known as the *Nyquist frequency*, $2f_m$.

$$f_s \ge 2f_m \tag{23.16}$$

EXAMPLE 23.4
If the audio signal in a telephone line has a maximum frequency of 2 kHz, let us determine the rate at which an A/D converter must sample this signal in order to capture an accurate record of it.

The Nyquist frequency is $2f_m$, therefore the sampling rate must be a minimum of $2 \times$ 2 kHz or 4 kHz. ∎

DRILL EXERCISE

D23.4. A signal is bandlimited with a maximum frequency of 25 kHz. It is to be sampled at a rate 50% higher than the minimum allowed rate to ensure margin. These samples will be used by the system to reconstruct the original signal at a later time. Determine the time interval to be used between samples.

Ans: 13.3 μs.

23.5

Modulation

In Chapter 7 the basic principles of signal modulation were introduced and examples given for AM (amplitude modulation), FM (frequency modulation), and phase or angle modulation. In this section these concepts will be developed further including their frequency spectra.

Recall that *modulation* is the process of encoding the information contained in one signal in a second signal by varying a parameter of the second signal such as amplitude, frequency, or phase in a manner related to the information or message to be transmitted. Carrier-based communications systems depend on taking the information contained in the baseband signal (such as audio or video) and modulating a higher frequency carrier signal. This process transforms the baseband information to a higher frequency where it can be easily transmitted, either as RF or as light.

Amplitude Modulation

Amplitude modulation is historically the oldest method for transmitting information on a radio signal. In amplitude modulation (AM) the amplitude of a "carrier wave" is changed in accordance with the information to be transmitted.

Figure 23.18(a) shows a continuous sinusoidal carrier wave of frequency, f_c. Figure 23.18(b) shows the information which we wish to transmit, $v_m(t)$; for illustration, we will consider a sinusoidal signal of frequency, f_m. Finally, Fig. 23.18(c) shows the high-frequency wave, f_c, amplitude modulated with the information at a frequency of f_m.

Mathematically, we let the carrier wave be represented as

$$v_c(t) = V_c \sin (\omega_c t) \tag{23.17}$$

where V_c is the peak amplitude of the carrier signal in volts and ω_c is the carrier frequency in radians/second.

For convenience we will initially assume that the message signal is a single frequency (a tone) expressed as a sinusoid of a much lower frequency; that is, $\omega_m \ll \omega_c$. Hence

$$v_m(t) = V_m \sin (\omega_m t) \tag{23.18}$$

where V_m is the peak amplitude of the modulating signal in volts and ω_m is the angular frequency of the message signal in radians/second.

To construct an expression for amplitude modulation, we let the carrier signal amplitude be changed by that of the message. Thus

$$v_{AM}(t) = [V_c + V_m \sin \omega_m t] \sin (\omega_c t) \tag{23.19}$$

This expression can be written as

$$v_{AM}(t) = V_c \left[1 + \frac{V_m}{V_c} \sin(\omega_m t) \right] \sin \omega_c t \tag{23.20}$$

letting m = the modulation index = V_m/V_c, which should be < 1

$$v_{AM}(t) = V_c \sin\omega_c t + V_c m (\sin\omega_m t) (\sin\omega_c t) \tag{23.21}$$

Using a trigonometric identity for the product of the two sine functions, this equation can

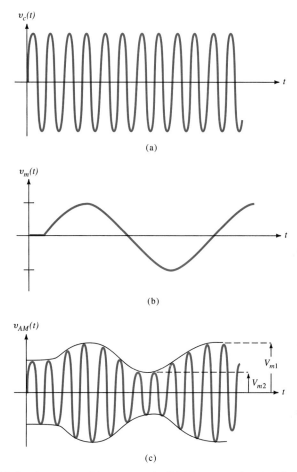

Figure 23.18 (a) Carrier wave of frequency f_c; (b) Message signal of frequency, f_m ($f_m <$ f_c); (c) Amplitude modulated signal.

be written with the last term expressed as a "sum" and a "difference" frequency, as

$$v_{AM}(t) = V_c \, sin\omega_c t + m \, \frac{V_c}{2} \cos(\omega_c - \omega_m)t - m \, \frac{V_c}{2} \cos(\omega_c + \omega_m)t \quad (23.22)$$

This important result shows that an AM modulated signal has a frequency component at the carrier frequency, but, in addition, there are two components with the information, one at a frequency of $(\omega_c - \omega_m)$ and another at $(\omega_c + \omega_m)$. These frequency components are called the *lower* and *upper sidebands,* respectively. A frequency spectrum of this AM modulated signal is shown in Fig. 23.19.

The *modulation index, m,* should ideally be as close to unity as possible, but not exceed it. If m exceeds unity, a condition known as "overmodulation" occurs which creates distortion and loss of signal clarity. The magnitude of the sidebands (relative to the carrier) depends on m.

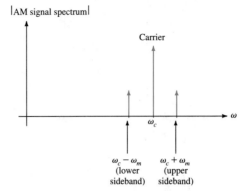

|AM signal spectrum|

Carrier

ω_c

ω

$\omega_c - \omega_m$
(lower
sideband)

$\omega_c + \omega_m$
(upper
sideband)

Figure 23.19   Spectrum of AM signal with sinusoidal message.

EXAMPLE 23.5

Let us calculate the modulation index for the AM modulated signal shown in Fig. 23.18(c).
   Let V_{m1} = the amplitude of the carrier at the maximum, and
      V_{m2} = the amplitude of the carrier at the minimum.

Then

$$\frac{V_{m1} - V_{m2}}{2} = V_m \qquad (23.23)$$

and

$$\frac{V_{m1} + V_{m2}}{2} = V_c \qquad (23.24)$$

Taking the ratio of these two equations yields

$$m = \frac{V_m}{V_c} = \frac{V_{m1} - V_{m2}}{V_{m1} + V_{m2}} = \frac{\dfrac{V_{m1}}{V_{m2}} - 1}{\dfrac{V_{m1}}{V_{m2}} + 1} \qquad (23.25)$$

Therefore, in Fig. 23.18(c), where $V_{m1}/V_{m2} = 3$, we find that $m = 0.5$.   ∎

EXAMPLE 23.6

Let us plot the anticipated frequency spectrum of a carrier signal with an amplitude of unity and frequency $f_c = 10$ MHz that is AM modulated $(m = 1)$ with a signal, $v_m(t)$, where

$$v_m(t) = .8 \sin (2\pi 10000\ t) + .4 \sin (2\pi 22000\ t)$$

   There is a carrier present with a magnitude of unity; each sinusoidal signal produces an upper and a lower sideband at half the magnitude of the message signal; therefore, the frequency spectrum appears as in Fig. 23.20.   ∎

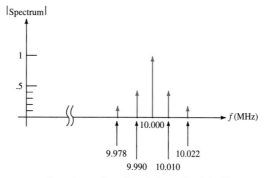

Figure 23.20   Spectrum for signal described in Example 23.6.

D23.5. The carrier frequency of an AM signal is 28.00 MHz; if the upper sideband of this signal has a frequency component at 28.15 MHz and another of the same amplitude at 28.20 MHz, (a) plot the spectrum of the message signal, (b) plot the spectrum of the AM signal including the lower sideband.

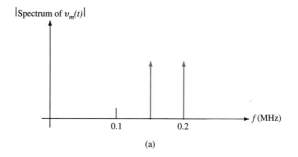

(a)

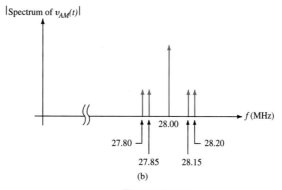

(b)

Figure D23.5

Ans: (a) See Fig. D23.5(a), (b) See Fig. D23.5(b).

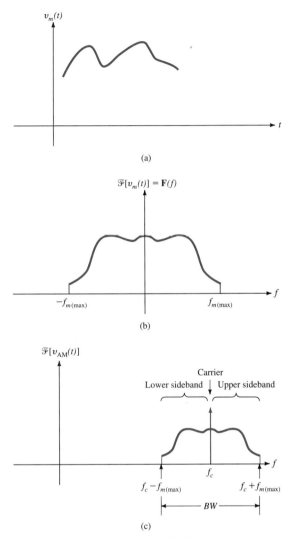

Figure 23.21 (a) Nonperiodic message signal; (b) Fourier transform of nonperiodic signal; (c) Spectrum of AM signal (carrier at f_c) modulated by nonperiodic signal.

In general the *spectrum* of an AM signal has the carrier frequency in the center at frequency f_c, and for every frequency present in the message signal, f_m, there appears in the AM signal an upper sideband at $(f_c + f_m)$ and a lower sideband at $(f_c - f_m)$. You might note that the information in one sideband is redundant with that in the other since they are "mirror images" of each other about the carrier frequency.

The above concepts can be extended to include the case of a message signal which is nonperiodic. Suppose the message is given by $v_m(t)$; this nonperiodic signal has a Fourier transform which is bandlimited with a maximum frequency of $f_{m(\max)}$. This hypothetical function $v_m(t)$ is shown in Fig. 23.21(a) and its bandlimited Fourier transform in Fig. 23.21(b).

If this signal is used to AM modulate a carrier signal, the spectrum of the resulting AM signal appears as in Fig. 23.21(c). The upper and lower sidebands are equivalent to the Fourier transform of the original message, but shifted up in frequency and centered at the carrier frequency, f_c. Note that there is a delta or impulse function representing the carrier at $f = f_c$.

EXAMPLE 23.7

If the Fourier transform of a message signal has a maximum frequency, $f_{m(max)}$ = 30 kHz and it AM modulates a 5 MHz carrier signal, let us calculate the bandwidth, *BW*, of the AM signal.

Observe from Fig. 23.21(c) that the bandwidth of an AM signal is twice the highest frequency in the message signal. Therefore, BW = 60 kHz. ■

Amplitude modulation is used commercially today; the AM commercial radio band in the United States is authorized by the Federal Communications Commission (FCC) to operate from 525 kHz to 1.7 MHz. Each station is assigned a carrier frequency, and the regulations require that the carrier frequencies be separated by 10 kHz.

EXAMPLE 23.8

Given the regulations for commercial AM broadcast above, let us determine (a) how many "channels" or carrier frequencies can be placed in the assigned frequency range and (b) what is the maximum message frequency that can be transmitted without overlap.

(a) The frequency range allocated for AM broadcast is

$$f_{range} = 1700 \text{ kHz} - 525 \text{ kHz} = 1175 \text{ kHz}$$

This total bandwidth is partitioned so as to allow 10 kHz of separation between carrier frequencies; therefore, the total number of "channels," N is

$$N = \frac{1175 \text{ kHz}}{10 \text{ kHz}} \approx 118 \text{ channels}$$

(b) The carriers of two separate stations are separated by 10 kHz; let f_1 be one carrier and f_2 be the other ($f_1 < f_2$). We know that $f_2 = f_1 + 10$ kHz. If we let the maximum frequency of the message signal increase, the outer edges of both sidebands move away from the carrier frequency, increasing the total bandwidth of each AM signal. The maximum message frequency occurs when the top edge of the upper sideband of f_1 touches the lower edge of the lower sideband of f_2; this happens at the midpoint of the spacing between the carriers. Therefore, the maximum message frequency is half the frequency spacing between the carriers, that is

$$f_{m(max)} = \frac{10 \text{ kHz}}{2} = 5 \text{ kHz}$$

The FCC allows $f_{m(max)}$ to be slightly higher and further limits the number of stations in a listening area. ■

If you are familiar with audio systems, you know that an upper limit of 5 kHz on the audio frequency of transmission would noticeably degrade the quality of the signal, and is particularly noticeable in the quality of music reproduction. A high-quality audio signal should contain frequency components up to the highest range of human hearing, about

15 kHz. For this reason, the FM commercial broadcast band was established with a wider bandwidth per channel and the capability to transmit high-quality audio. FM will be discussed in the next section.

Modern communications systems sometimes use variations of AM to achieve a more efficient use of the transmitted energy and frequency spectrum.

Note that the carrier itself contains no information; all the information of the message signal is in the sidebands. Therefore, in some cases the carrier is eliminated from the transmitted signal, and only the sidebands are transmitted. This is called *double sideband* (DSB) transmission or *suppressed-carrier* transmission. The spectrum of a DSB signal appears exactly like that of a typical AM signal, such as that in Fig. 23.21(c), except the carrier is removed.

Also notice that the information content in the sidebands is redundant, since one sideband is the mirror image of the other. Therefore, a more efficient use of the spectrum can be accomplished by only transmitting one of the sidebands. *Single sideband* (SSB) transmission is commonly used by radio amateurs and other shortwave transmission applications. One can transmit either the *upper sideband* (USB) or the *lower sideband* (LSB).

The receiver used to detect these modified forms of AM must have special circuitry to recreate the missing carrier as a part of the demodulation process; the receivers for DSB or SSB signals are more complex than a conventional AM receiver.

There is one other form of modified AM in common usage, *vestigial sideband* transmission. This modulation scheme transmits one sideband completely and only a small portion (vestigial) of the other, thus reducing the total signal bandwidth. The video portion of television transmission uses this approach to reduce the frequency band required for each television channel, and thus allow more channels to be transmitted over the same frequency allocation.

DRILL EXERCISE

D23.6. If the commercial AM broadcast band were converted to SSB, with the same restrictions on message frequency, determine the number of separate channels that could be accommodated.

Ans: 236.

Frequency Modulation

Frequency modulation is easily visualized as a sinusoidal carrier signal which changes frequency in proportion to the amplitude of the message signal. This was introduced in Chapter 7, and illustrated in Figs. 7.8(a,b,c). A sinusoidal message signal, $v_m(t)$, is shown in Fig. 23.22(a); the frequency modulated carrier signal, $v_{FM}(t)$, is shown in Fig. 23.22(b).

The concept of FM was first proposed during the early 1920s; however, a mathematical analysis of FM done by John R. Carlson at Bell Telephone Laboratories indicated that FM modulation would produce an infinite number of sidebands; each FM signal would consume a very wide band of frequencies. This temporarily discouraged the development of FM; however, the subsequent successful demonstration of an FM system by E. H. Armstrong renewed interest. He showed that although the bandwidth required for each signal was wider than that for AM, there were advantages in terms of noise re-

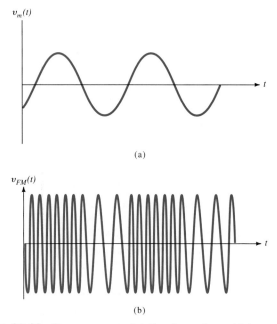

Figure 23.22   Frequency modulation by a sinusoidal waveform.

duction and immunity to interference. In effect there is a trade-off between bandwidth and improved signal quality (signal-to-noise ratio). The same message transmitted at the same power level using FM modulation will have greatly improved quality over AM systems, but at the expense of a much wider bandwidth.

In modern FM systems the transmitted signal is produced in a way to limit the bandwidth such that only a finite number of sidebands are transmitted. We will see there are two primary categories of FM: (1) *narrowband FM,* and (2) *wideband FM.*

A sinusoidal carrier signal is frequency modulated if its instantaneous angular frequency, $\omega_c(t)$, is linearly dependent upon the amplitude of the message signal, $v_m(t)$. Therefore

$$\omega_c(t) = \omega_o + k\,v_m(t) \tag{23.26}$$

where k is a constant in units of radians per second per volt, and ω_o is the frequency of the unmodulated carrier signal.

The time rate of change of phase angle is the angular frequency; or conversely, the instantaneous phase is the integral of instantaneous angular frequency. Thus, an expression for the FM modulated signal becomes

$$v_{FM}(t) = V_m\cos\left[\omega_o t + k\int v_m(t)\,dt\right] \tag{23.27}$$

An important consideration in FM systems is the maximum extent of frequency deviation from the nominal frequency, ω_o. If we define the maximum frequency deviation as $\Delta\omega$, then

$$\Delta\omega = k\,\big|\,v_m(t)\,\big|_{max} \tag{23.28}$$

In a narrowband FM signal the extent of frequency deviations is small. Normally the condition applied to define narrowband FM is that $\Delta\omega < 2\pi f_{m(max)}$; recall that $f_{m(max)}$ is the maximum frequency in the Fourier transform of the message signal. For a narrowband FM signal, it can be shown that the spectrum is approximately equivalent to AM. There is an upper and lower sideband which extends on either side of ω_o a frequency of $f_{m(max)}$. It is also true that the performance of narrowband FM systems is not that much improved over AM systems. The significant improvements in signal-to-noise ratio and rejection of interference is achieved with wideband FM.

A wideband FM signal can be shown theoretically to have an infinite number of sidebands; the analysis shows that the magnitude of the sidebands decreases at frequencies further from ω_o, and the amplitude of successive sidebands follows that of Bessel functions of the first kind. Fortunately the number of significant sidebands is limited, and a good approximation to the bandwidth covered by the significant sidebands is

$$BW = 2\,(\Delta f + f_{m(max)}) \qquad (23.29)$$

where $\Delta f = \Delta\omega/2\pi$

EXAMPLE 23.9

A wideband FM signal has a nominal carrier frequency of 0.1 MHz. If the message signal $v_m(t)$ is given by

$$v_m(t) = 10 \cos (2\pi 10^3\, t) \qquad (23.30)$$

and $k = 900$, let us determine the bandwidth of the modulated carrier and the band of frequencies occupied by this signal.

The values of $\Delta\omega$ and Δf are given by

$$\Delta\omega = k\,10 = 9000 \text{ rad/sec}$$

and

$$\Delta f = \frac{\Delta\omega}{2\pi} = 1432 \text{ Hz}$$

From the expression for the message signal

$$f_{m(max)} = 1000 \text{ Hz}$$

Therefore

$$BW = 2\,(\Delta f + f_{m(max)}) = 2\,(1432 + 1000) = 4864 \text{ Hz}$$

The band of frequencies occupied spans from

$$(100 \text{ kHz} - \frac{BW}{2}) \text{ to } (100 \text{ kHz} + \frac{BW}{2}) = 97.57 \text{ kHz to } 102.43 \text{ kHz} \qquad \blacksquare$$

In the United States the assigned band for FM commercial broadcast is from 88.0 MHz to 108.0 MHz; there are 100 possible assigned "channels" from 88.1 MHz to 107.9 MHz, each with an allowed bandwidth of 200 kHz. The wider bandwidth per channel than that allowed for AM offers higher quality audio since the allowed audio frequency range extends from 50 Hz to 15 kHz.

Most FM stations today transmit a stereo FM signal. *Stereo* involves the simultane-

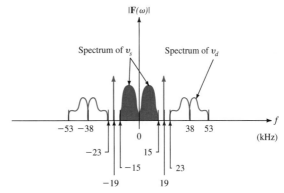

Figure 23.23   Spectrum of commercial stereo FM signal (centered on carrier frequency).

ous transmission of two signals, one designated as "left," $v_{mL}(t)$, and the other as "right," $v_{mR}(t)$. Typically these signals are derived from two microphones, one placed on the left side of the performer(s) and the other on the right side.

The commercial FM system was designed so that if you had a monoural FM receiver you would hear the sum of the left and the right signals. If you have a stereo receiver, you would be able to detect the two signals independently and direct them to two separate speakers in your room. To accomplish this, the two signals, $v_{mL}(t)$ and $v_{mR}(t)$ are added to create the sum signal, $v_s(t)$, that is

$$v_s(t) = v_{mL}(t) + v_{mR}(t) \tag{23.31}$$

There is also a difference signal created, which is

$$v_d(t) = v_{mL}(t) - v_{mR}(t) \tag{23.32}$$

From these two signals, a stereo receiver can add and subtract the two signals to recreate the original left and right signals.

The sum signal, $v_s(t)$, is transmitted with bandwidth of -15 kHz to $+15$ kHz about the nominal carrier frequency. The difference signal is transmitted using DSB modulation on a 38 kHz subcarrier. There is a pilot carrier at 19 kHz from center frequency which is used in the receiver's demodulation process. Therefore, the spectrum for a commercial FM broadcast station appears as in Fig. 23.23.

23.6

AM Demodulation: A Diode in a Radio Receiver

Recall that amplitude modulation (AM) is a process by which the amplitude of a sinusoidal wave, the carrier wave, is varied in proportion to the information to be transmitted. In Chapter 7, Fig. 7.8(b), an example of a ramp wave modulating a sinusoidal carrier wave was presented; other examples have been presented in this chapter. In this section a simple AM radio receiver is described.

A radio receiver tunes or selects the desired sinusoidal carrier wave by selecting its frequency with a resonant circuit of the type described in Chapter 6. It is generally an LC circuit tuned to the frequency of the carrier wave, f_c. This process of tuning one carrier frequency selects one station from all the many stations that are transmitting at various

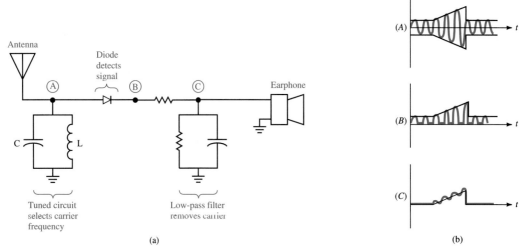

Figure 23.24(a)   Diagram of simple AM radio receiver.   **Figure 23.24(b)**   Waveforms in AM receiver circuit.

frequencies. The *selectivity* of a receiver is a measure of its ability to separate a signal at one frequency from a signal at an adjacent frequency.

This signal may then be connected to a diode which *rectifies* the signal, *detecting* or reconstructing the desired information. Finally the signal is filtered by an RC (low-pass) filter of the type described in Chapter 6, to remove the sinusoidal variation and leave only the desired information. The original information is thus extracted from the modulated signal. A diagram of such a radio receiver with these elements is shown in Fig. 23.24(a). The waveforms of the signal at various stages through the receiver appear as shown in Fig. 23.24(b).

A radio receiver such as that just described will actually operate and is called a "crystal set." The only problem is that the signal at the output is barely strong enough to drive an "earphone," a listening device you can put in your ear—like a hearing aid. It will not drive a speaker with enough power to be heard. What's needed is an amplifier to make the signal larger.

23.7
Time Division Multiplexing; Frequency Division Multiplexing

Multiplexing is used to transmit two or more separate messages via the same communication system. This is very common in most systems in order to fully utilize the capacity of the system. For example, a single communication medium between two major telephone central offices (such as a fiber optic link) may simultaneously carry thousands of separate phone calls. Signals can always be separated from each other if they are nonoverlapping in either frequency or time. Therefore, there are two fundamental methods of accomplishing multiplexing: (1) *frequency division multiplexing* (FDM), and (2) *time division multiplexing* (TDM).

Frequency division multiplexing was illustrated in the FM stereo system described in the previous section. The "sum" signal was transmitted in a 30 kHz bandwidth about the carrier frequency. However, a second signal, the "difference" signal, is transmitted by the same station with a spectrum centered at 38 kHz from the carrier; it does not overlap any of the frequency range utilized by the first signal. There are numerous other examples of frequency division multiplexing, one of the most familiar perhaps being the cable TV system. A large number (often 50 to 100) separate television channels are transmitted into your home over one cable; the signal for each channel is transmitted in a unique and nonoverlapping frequency band. The TV set then selects the range of frequencies corresponding to the channel you select, and the signal for the chosen channel is pulled from the clamor of all the other channels simultaneously present. Television will be addressed in more detail later, but as an example, television channel 2 is assigned 54–60 MHz, channel 3 is 60–66 MHz, etc. Television channels in the United States each occupy a 6 MHz bandwidth.

Time division multiplexing is a process by which more than one signal is simultaneously transmitted over the same channel by alternately switching from one signal to another. Time division multiplexing implies that each signal is sampled and the sample segments are transmitted serially. A simple illustration of a TDM system is shown in Fig. 23.25(a); switches S_1 and S_2 continuously cycle up and down together. In this system, two message signals $v_{m1}(t)$ and $v_{m2}(t)$ are alternately transmitted, and the transmitted message contains alternating segments of message 1 and 2. The key to reconstructing or separating both messages at the receiver is that the switch at the receiver operates in exact "lockstep" with that at the transmitter. In real TDM systems there are switching circuits to ensure this *synchronization.*

The received signals will be sampled data, and in order to reconstruct an approximation to the original signal, an averaging circuit such as an integrator with a fixed time constant is normally used. In general, we would like the switching or sampling rate to be much faster than the highest frequency components in the message signals. As we have shown earlier in this chapter, the minimum sampling or switching rate is twice the maximum frequency component of the message signal, according to the Nyquist criteria.

Time division multiplexing systems with more than two signals use the concept of a

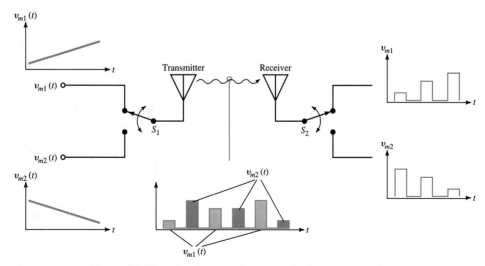

Figure 23.25   Simple time division multiplex (TDM) system.

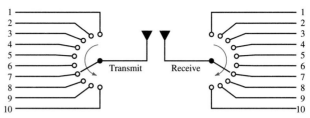

Figure 23.26   System for multiplexing 10 channels.

commutator or a rotating switch element at the transmitter and another similar unit in synchronization at the receiver. Figure 23.26 shows a system for the multiplexing of 10 channels.

EXAMPLE 23.10

Five audio channels are time division multiplexed on a single communications channel. If each of the audio signals has a frequency range of 50 Hz to 15 kHz, let us calculate (a) the minimum "cycle time" for the commutator, that is, time for the commutator to make a complete revolution covering all 5 signals, and (b) the pulse frequency on the communications channel.

(a) The system makes one sample of each signal every revolution, and according to the Nyquist criteria, 2×15 kHz $= 30$ kHz is the sampling rate, and therefore the commutator frequency required. The cycle time, T_c, is the inverse of cycle frequency, and hence

$$T_c = \frac{1}{f} = \frac{1}{30 \text{ kHz}} = 33.3 \ \mu s$$

(b) The commutator has a cycle frequency of 30 kHz and generates 5 pulses per revolution; therefore, the pulse frequency, f_p, is

$$f_p = 5 \ (30 \text{ kHz}) = 150 \text{ kHz} \qquad\blacksquare$$

23.8
Examples of Communications Systems

We have already described several common communications systems including AM and FM commercial radio, the telephone and the telegraph. In this section several additional important communication systems will be described.

Television

Television was introduced in the United States in the 1930s and found slow initial acceptance. The screens were small, 3″ to 4″ diagonally, and there was no regularly scheduled programming until the New York World's Fair in 1939. The rapid growth of TV occurred during the 1950s and by the 1960s virtually every home had a television set. The power of instantly transmitting visual images was recognized by the public. The standards for color television were developed and then adopted in 1954.

The television picture tube is much like the cathode ray tube in a laboratory oscilloscope. An electron beam is focused to a small diameter and directed toward the front center of the screen. The inside of the screen's glass is coated with a phosphor that glows

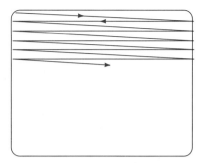

Figure 23.27   TV raster scan.

when hit by the electron beam making a spot of illumination. A color TV set actually has three electron beams, one for each of the three basic colors (red, green and blue) and they are used in combination to generate the other colors.

The image on the screen is generated by scanning the entire surface of the screen with the illuminated spot, and varying the color and intensity of the spot as it moves to make an image. It does this fast enough that the human eye sees only a completed image. The scanning process is called a *raster scan.*

The raster scan standard for television in the United States is a total of 525 lines/frame and 30 frames/second. Actually, only 495 lines are normally visible; the rest are used for special information like closed caption data. The raster scan starts at the top left of the screen and goes across, sloping down very slightly. When it hits the right edge, it quickly pops back to the left side and starts another line just beneath the first one. The process continues until the entire screen is covered as shown in Fig. 23.27. There is a small "trick" called *interlaced scanning* used to reduce the apparent flicker. The beam first traces out all the even numbered lines, top to bottom in 1/60 of a second; it then returns to the top and traces out the odd numbered lines in another 1/60 second. Each of the scans is called a *field,* and the two fields together create a *frame* in 1/30 second. There is a *horizontal sync pulse* transmitted at the end of every line which tells the TV receiver to begin a new scan line.

The amplitude of the video signal between horizontal sync pulses provides the information regarding intensity of the spot, from white to black. The intensity information is amplitude modulated on the video signal. A typical waveform segment for a video signal for one scan line is shown in Fig. 23.28. Note the horizontal sync pulses at each end of the line.

Also notice the burst of at least 8 cycles of a reference signal at 3.579545 MHz called the *color burst,* which appears at the end of the horizontal sync pulse. This signal provides a phase reference in the TV receiver for the color information which is a phase modulated subcarrier.

Each television channel in the United States is allocated a total bandwidth of 6 MHz as illustrated in Fig. 23.29. The sound signal is FM modulated and the FM sound carrier frequency is placed 4.5 MHz above the AM video carrier. If standard AM modulation were used, the video signal would exceed the 6 MHz allowed bandwidth; therefore, the video carrier is placed 1.25 MHz above the lower edge of the allocated channel. The upper video sideband extends above the carrier approximately 4.2 MHz and contains all the

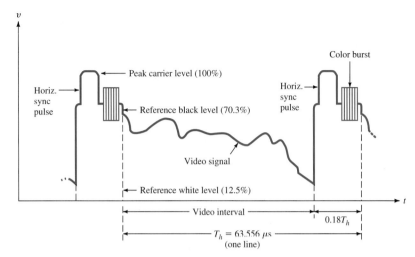

Figure 23.28   TV video signal—one scan line.

video information; the lower sideband is truncated so only part of it remains. This form of AM is called vestigial sideband AM which we discussed earlier. Also shown on the figure of the spectrum is the location of the phase modulated color subcarrier.

Figure 23.30 shows the frequency allocation for TV channels 2 through 69 in the United States. Channel numbers 13 and below are designated as VHF (*very high frequency*). Note the gap between channels 6 and 7; in this gap are the FM commercial radio, police communications, aircraft communications and electronic navigation signals, amateur radio, and a number of other radio services. There is also a gap between 13 and 14 which contains other activities including cellular telephone service. Channel numbers of 14 and above are designated as UHF (*ultra high frequency*).

EXAMPLE 23.11

Let us calculate the carrier frequency of the FM sound signal for television channel 6, and determine if this signal can be received by a regular FM radio.

From Fig. 23.30, channel 6 is assigned 82–88 MHz. From Fig. 23.29, the sound car-

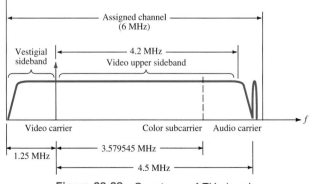

Figure 23.29   Spectrum of TV signal.

	Television channels in the United States			
Channel no.	Frequency band MHz		Channel no.	Frequency band MHz
2	54–60		36	602–608
3	60–66		37	608–614
4	66–72		38	614–620
5	76–82		39	620–626
6	82–88		40	626–632
7	174–180		41	632–638
8	180–186		42	638–644
9	186–192		43	644–650
10	192–198		44	650–656
11	198–204		45	656–662
12	204–210		46	662–668
13	210–216		47	668–674
14	470–476		48	674–680
15	476–482		49	680–686
16	482–488		50	686–692
17	488–494		51	692–698
18	494–500		52	698–704
19	500–506		53	704–710
20	506–512		54	710–716
21	512–518		55	716–722
22	518–524		56	722–728
23	524–530		57	728–734
24	530–536		58	734–740
25	536–542		59	740–746
26	542–548		60	746–752
27	548–554		61	752–758
28	554–560		62	758–764
29	560–566		63	764–770
30	566–572		64	770–776
31	572–578		65	776–782
32	578–584		66	782–788
33	584–590		67	788–794
34	590–596		68	794–800
35	596–602		69	800–806

VHF (channels up to 13), UHF (channels 14 and above)

Figure 23.30   Frequency assignments for TV channels in the United States.

rier frequency, f_{FM}, is

$$f_{FM} = 82 \text{ MHz} + 1.25 \text{ MHz} + 4.5 \text{ MHz} = 87.75 \text{ MHz}$$

The FM radio band begins at 88 MHz; because this is so close to the lower edge of the band, some receivers may be able to detect the sound from channel 6 at the lower end of their frequency dial. ■

DRILL EXERCISE

D23.7. A TV station operates on channel 27. Find the station's video carrier frequency.

Ans: 549.25 MHz.

Satellite Communications

The concept that a *communications satellite* could be placed in a particular earth orbit and that the satellite would appear stationary over a particular spot on the ground is widely attributed to the famous British science fiction writer, Arthur C. Clarke, the author of *2001: A Space Odyssey.* At a certain radius or altitude of orbit, the velocity of the space-craft is exactly the value required to remain over a spot on the rotating earth, as illustrated in Fig. 23.31. This so-called *geosynchronous* or *geostationary orbit* is over the equator, and is such that the time period for one orbit is exactly the time required for the earth to rotate once, or 24 hours.

In a normal mode of operation a signal is sent from a directional antenna on the earth, usually a familiar "dish" shaped antenna, up to the satellite; this is called the *uplink* and this transmission is at a particular range of carrier frequencies. The satellite receives the signal and retransmits the message back to earth, the *downlink,* on a different range of carrier frequencies. Typical uplink and downlink frequencies for INTELSAT VI are shown in Fig. 23.32 along with a typical system. The downlink signal can be received by any-one with a receiving antenna and equipment located in a cone of reception on the earth. This cone is optimized for the particular satellite application. In general, transmitted en-ergy is focused in areas of primary reception by directional antennae on the satellite.

DRILL EXERCISE

D23.8. The earth subtends what angle when viewed from geostationary orbit?

Ans: 17°.

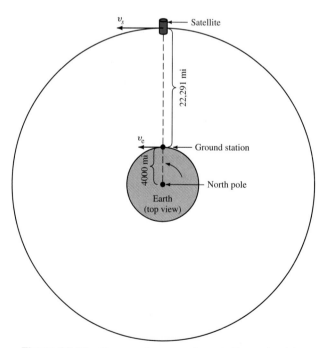

Figure 23.31   Geosynchronous (geostationary) orbit.

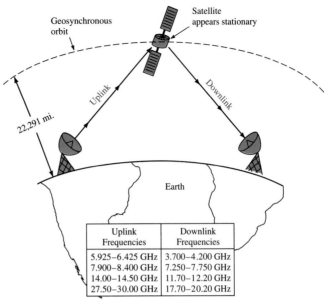

Uplink Frequencies	Downlink Frequencies
5.925–6.425 GHz	3.700–4.200 GHz
7.900–8.400 GHz	7.250–7.750 GHz
14.00–14.50 GHz	11.70–12.20 GHz
27.50–30.00 GHz	17.70–20.20 GHz

Figure 23.32   Uplink and downlink signals for INTELSAT VI.

EXAMPLE 23.12

If we assume the United States is 3,500 miles wide (west to east), let us calculate the angle of beam divergence (apex of the signal cone) required for a satellite downlink transmission to cover the entire country simultaneously from directly overhead.

   If we assume the geosynchronous orbit is at 22,291 miles, the situation is diagrammed in Fig. 23.33.

$$(0.5)\phi = \tan^{-1}\frac{\left[\dfrac{3500}{2}\right]}{22291} = 4.5°$$

Therefore

$$\phi = 9.0°$$ ∎

DRILL EXERCISE

D23.9. If we wish for a communications satellite in geosynchronous orbit to be separated by 4 degrees from its nearest neighbor when viewed from the earth's surface directly below the satellite, let us determine (a) the required separation between satellites, and (b) the maximum number of geosynchronous satellites possible worldwide.

   Ans: (a) 1555 mi. (b) 122.

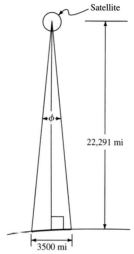

Figure 23.33   Diagram for Example 23.12.

EXAMPLE 23.13

Assuming there is no delay from the time a communications satellite receives a signal to the time it retransmits it, we wish to find the approximate time delay from earth transmission (uplink) to reception of the downlink signal.

Recall the speed of light (and radio) waves is 186,000 miles/second. The round-trip distance is

$$d = 2\,(22{,}291) \text{ mi} = 44{,}582 \text{ mi}$$

Therefore, the time required, t, is

$$t = \frac{d}{v} = \frac{44{,}582 \text{ mi}}{186{,}000 \text{ mi/s}} = 0.24 \text{ seconds} \qquad \blacksquare$$

There is a "built-in" delay of a minimum of approximately one-quarter of a second in any communications through a geosynchronous satellite; if the satellite is not directly overhead, the delay will be longer. You may have noticed some delay in speech reception of transcontinental phone conversations; these are often the result of the finite speed of light and the long distance signals must travel to and from geosynchronous orbit.

Satellites are equipped with multiple repeater (receive and retransmit) units called *transponders;* a satellite will generally have a number of transponders. For example, INTELSAT VI was designed with 45 transponders; newer satellites have more.

The bandwidth of each transponder depends on the particular satellite design, but generally they have ranged from bandwidths of 36 MHz to several hundred MHz. Therefore, a single transponder can simultaneously receive and retransmit multiple messages by frequency division multiplexing.

EXAMPLE 23.14

We wish to compute the number of television channels a satellite transponder with a 72 MHz bandwidth can carry simultaneously.

Recall that a television signal requires a 6 MHz bandwidth; therefore, the number of signals that can be accommodated in a 72 MHz bandwidth is

$$\frac{72 \text{ MHz}}{6 \text{ MHz}} = 12 \text{ channels} \qquad \blacksquare$$

Frequency Division Multiple Access (FDMA) is a variation of standard frequency division multiplexing in which portions of the bandwidth of a single transponder are assigned to different users. For example, a satellite which has a transponder with a 500 MHz bandwidth may assign a 250MHz segment of this transponder's bandwidth to Brazil for telephone transmissions and the other 250 MHz to Mexico for television use.

DRILL EXERCISE

D23.10. In the example of the previous paragraph, let us find (a) the number of telephone voice signals available to Brazil (assume a voice signal requires a bandwidth of 10 kHz), and (b) the number of TV channels available to Mexico.

Ans: (a) 25,000 (b) 41.

Time Division Multiple Access (TDMA) provides access to the transponder by multiple users using time division multiplexing. During a particular time segment, Δt_1, a ground-based station sending on the uplink may have the entire bandwidth of the transponder available. Then in the next time segment, Δt_2, user 1 must stop sending, and the use of the transponder is turned over to user 2. There may be many users time-sharing the transponder; after a rotation through all the users, user 1 transmits again and the cycle repeats. Using this system, the communications system transmits and receives information in bursts or frames when each user's time slot becomes available.

There are two main classes of communications satellites based on the way in which they achieve attitude control: (1) *spinners,* and (2) *three-axis stabilized.* A spinner is one which is typically cylindrical in shape, and the entire spacecraft body is designed to rotate in a range of 30 to 100 rpm. This provides an inherent gyroscopic action and maintains the satellite's orientation on its spin axis. The alternate approach is to provide three-axis stabilization of the entire spacecraft; such satellites are said to be "body stabilized." This is done with three small orthogonally mounted gyroscopes internal to the satellite. These gyroscopes establish a stable frame of reference, and then electronic controls periodically "fire" rocket or gas thrusters to maintain the proper satellite orientation. Examples of a spinner and three-axis stabilized satellite are shown in Fig. 23.34 (a) and (b).

The major subsystems of a communications satellite are as follows:

1. Attitude and Orbit Control System—This is the subsystem that maintains physical control and positioning of the satellite while it's in orbit. It consists of small rocket motors and thrusters that can provide bursts of rocket power to correct the satellite's drift from its proper attitude or orbit.
2. Telemetry, Tracking, and Command System—This subsystem electronically monitors the condition of the satellite, such as battery life, position, etc., and sends the signals for corrective action if needed.
3. Power System—This is a very important aspect of satellite design. Since it's vir-

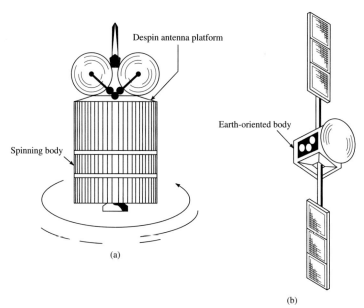

Figure 23.34   (a) Example of a spinner satellite; (b) Example of a three-axis stabilized satellite.

tually impossible to go replace the batteries if power fails, the power system must be designed for long life and high reliability. This system normally supplies electric power from a combination of batteries and solar cells. The uplink signals can be transmitted at high power, because energy is relatively inexpensive on earth; the downlink signal, transmitted from the satellite, must use as little power as possible and still provide adequate performance.

4. The Communications System—This is the heart of the satellite and consists of the group of transponders that actually provide the communications links and the antennae that receive and transmit the signals. This system is the satellite's reason for existence, but it is often a minor part of the weight and volume of a satellite.

Local Area Networks (LANs)

Local Area Networks, or LANs as they are called, are in essence a highway for data communications. They are used to interconnect computers and other information systems together within a small geographical area such as an office complex. They can be designed in various topologies, use different transmission schemes, employ a number of transmission media, and use different network access methods.

The network stations can be interconnected using any one of the topologies shown in Fig. 23.35. The *mesh topology,* shown in Fig. 23.35(a), is a point-to-point connection in which a direct, dedicated link is used to connect devices. This simple technique is viable only when the number of devices on the network is small.

The *star configuration,* shown in Fig. 23.35(b), is a pinwheel architecture employing a central switch. Time sharing is a good example of this type of network. However, if the central switch fails, total system failure occurs.

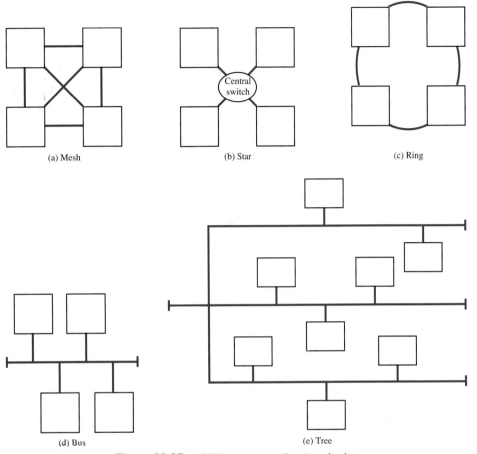

Figure 23.35   LAN interconnection topologies.

The *ring topology,* shown in Fig. 23.35(c), connects the network devices using point-to-point links in a closed loop. Traffic on the ring is unidirectional and repeaters are used to receive data from one link and pass it to the next. Reliability of the ring is a major issue in that an open circuit or repeater failure anywhere will disrupt traffic.

In the *bus connection,* shown in Fig. 23.35(d), every node is connected directly to the transmission medium. This topology is very flexible and additional stations can be easily added. Like the ring, this is a decentralized configuration and avoids the hazards of central switch failure.

The *tree configuration,* shown in Fig. 23.35(e), is simply a variation of the bus. However, problems are more easily isolated so that other portions of the network can function while repairs are being made.

LANs employ two types of *data transmission*: baseband and broadband. Both of these transmission schemes have been discussed earlier in this chapter.

The *transmission media* used in LANs is normally twisted pair, coaxial cable, or fiber optics. Twisted pair (two insulated conductors wrapped around one another) is generally the least expensive, coaxial cable is perhaps the most versatile, and fiber optic cable has the largest bandwidth and is completely immune to electrical interference.

The *access method,* which answers the question "who goes next?", is typically either polling, token passing, or contention. The polling method employs a master station which polls slave stations to determine access requirements. Token passing, normally used on a ring or bus, provides fast access by passing a bit position, called a token, along the highway giving each node permission to transmit at regular intervals. With the contention access method, each station decides when it will transmit, and obviously the general lack of decorum causes collisions. Problems with the earlier versions of this method led to the development of the carrier sense multiple access method with collision detection (CSMA/CD). This method vastly improved the success of data transmission.

In general, the three most popular LAN configurations are token rings, token buses, and CSMA/CD buses.

Cellular Telephones

This concept was developed at AT&T Bell Telephone Laboratories in 1947, and in 1962 the first tests were conducted to explore the commercial application of this technology. In 1970, the Federal Communications Commission (FCC) set aside radio frequencies for *cellular;* however, it was not until 1981 that the commission released its Final Report and Order in which it specified two competing licensees in every market, one of which would be the local telephone company. The markets are designed as the 306 largest metro markets and 428 rural service areas. Service was inaugurated in 1983. In 1985, there were approximately 100,000 users which skyrocketed by more than two orders of magnitude to approximately 14 million in 1993.

The FCC allocated a bandwidth of 50 MHz for cellular telephone service. The frequency bands are 824–849 and 869–894 MHz. Of the 999 channels in the band, 416 are reserved for each of the competing licensees. The dominant cellular technology is analog; however, a digital transmission system is emerging which improves sound quality and security and will launch the mobile computer and fax.

The basic cellular system is illustrated in Fig. 23.36. An area, for example, a city, is divided into cells which range from about one to 20 miles in diameter, depending upon terrain and required capacity. Each cell contains a low-power transmitter/receiver which

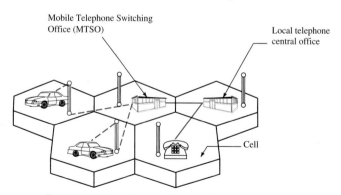

Figure 23.36   Basic cellular telephone system.

is connected to a mobile telephone switching office (MTSO), which in turn is linked to the telephone network serving homes and offices. By controlling transmitter power, the frequencies used within a cell can be confined to the cell. Therefore, and this is a key point, these same frequencies can also be used in another cell.

Suppose a call is made from a vehicle moving in a cellular area. When the call is initiated, it is received by the closest antenna in the cellular area, routed to a computerized switch which connects to a landline phone or another cellular phone. As the vehicle moves, the call's signal strength is constantly monitored. When the call's strength falls below some preset level, the cell controller signals the MTSO to "hand off" the call. The MTSO will transfer the call to the cell within the area that registers the highest signal strength. The MTSO also automatically switches the call's landline connection when the original cell acknowledges that the mobile unit has left its channel. All this happens so quickly that neither party to the call is aware that the switching is taking place.

With the introduction of digital technology, cellular systems will undoubtedly play an important role in the ever-expanding area of personal communication services.

23.9
Summary

- Electronic communication systems collect information, convert it into electric currents, fields, or light waves, and transmit these signals over electrical networks, optical fibers, or through space to another point at which they are reconverted into a form suitable for interpretation by the receiving station.
- The "message" signal is the original information which we wish to transmit. A carrier signal is a higher frequency signal to which we attach the message signal, a process called modulation. The message signal is termed the baseband signal.
- The reciprocal operation at a receiving station, that is, extracting the baseband signal from the modulated carrier signal, is called demodulation.
- Simple communication systems transmit the baseband signal directly, for example, the early telegraph or telephone system.
- Carrier transmission systems use RF (radio frequency) radiation or light waves to transmit information.
- RF radiation propagates through the atmosphere (or space) at the speed of light by several possible paths: (1) the direct wave (line-of-sight), (2) the sky wave, (3) the ground-reflected wave, and (4) the surface wave. The sky wave reflects off the ionosphere for a range of frequencies (approximately 3 to 30 MHz) making all long-distance short-wave communications possible.
- Coaxial cable is often used as the link for transmission of RF signals.
- Light waves have a high frequency and therefore have a very large bandwidth or capacity for information transmission. The optical fiber is the primary medium for optical transmissions.
- A signal can be represented as a function of time or as a function of frequency. There is a unique relationship between a signal expressed in the time domain and its expression in the frequency domain.
- For periodic signals the Fourier series may be used to convert from the time domain to the frequency domain.

- For nonperiodic signals the Fourier transform may be used to convert from the time domain to the frequency domain.
- The bandwidth of a signal is the range of frequencies which contain all its significant frequency components. For convenience, we often define the outer edges of the spectrum as the frequencies at which the magnitude of the signal's Fourier transform is reduced to $1/\sqrt{2}$ of its maximum value.
- The Nyquist criteria describes how often an analog signal must be sampled and stored in a sampled data system for accurate reconstruction of the original data. It states that if a signal has a maximum frequency component, f_m, the signal must be sampled at a rate of greater than twice f_m, that is, if the sampling frequency is f_s, $f_s \geq 2 f_m$.
- For amplitude modulation (AM), the amplitude of the carrier signal is varied in direct proportion to the amplitude of the message signal. The spectrum of AM signals is described.
- There are several modifications of AM that are used to reduce signal bandwidth or improve efficiency of transmission, for example, double sideband (DSB) or suppressed-carrier modulation, single sideband (SSB) or vestigial sideband modulation.
- For frequency modulation (FM), the frequency of the carrier signal is varied in direct proportion to the amplitude of the message signal. The spectrum of FM signals is described. There are two versions of FM, narrowband and wideband, depending on the extent of carrier frequency deviations allowed.
- A diode can be used for demodulation of an AM signal.
- Multiple signals can be transmitted over the same communication system using either time division multiplexing or frequency division multiplexing.
- Examples of communication systems are discussed including television, satellite systems, local area data networks, and cellular telephones.

PROBLEMS

23.1. (a) Define modulation. (b) Draw a sketch that illustrates the principle of AM modulation. (c) Repeat (b) for FM modulation.

23.2. Classify the first telegraph system and the first phone system as: (a) analog or digital, (b) baseband or carrier transmission systems.

23.3. Explain why transmission via the direct wave is not practical for distances more than a few hundred miles.

23.4. Repeat Example 23.1 if the receiving antenna was at ground level.

23.5. What factor(s) primarily affect the transmission of worldwide shortwave radio by the sky wave?

23.6. Sketch and label all parts in the cross section of (a) coaxial cable, (b) optical fiber.

23.7. What are the two main light sources for optical data transmission?

23.8. Plot the frequency spectrum of the signal, $v(t) = \cos(2\pi 10^4 t) + 0.8\cos(2\pi 10^3 t)$.

23.9. Compute the Fourier series for the signal shown in Fig. P23.9.

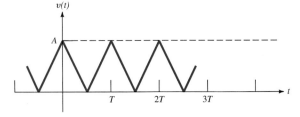

Figure P23.9

23.10. Compute the Fourier series for the signal shown in Fig. P23.10 (a) for the general case where τ is a variable, (b) for the case where $\tau = T/4$. (c) Plot the first four terms of the spectrum for (b).

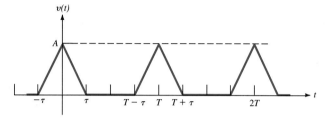

Figure P23.10

23.11. Compute the Fourier series for the signal shown in Fig. P23.11 (a) for the general case where τ is a variable, (b) for the case where $\tau = T/4$.

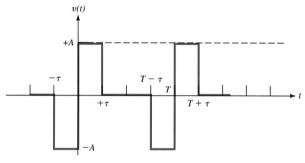

Figure P23.11

23.12. In problem 23.9 if $T = 1$ ms, plot the spectrum of this signal.

23.13. (a) Find the Fourier transform of the pulse shown in Fig. P23.13. (b) Sketch a plot of the Fourier transform as a function of frequency.

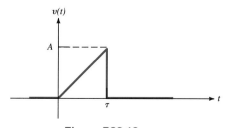

Figure P23.13

23.14. A single rectangular pulse of amplitude, A, has a width of 1 μs. Plot its Fourier transform as a function of frequency, f.

23.15. The signal described in problem 23.8 is to be sampled at a rate of f_s by an A/D converter. Determine the minimum value of f_s to ensure that information is not lost.

23.16. A square wave of frequency 1 kHz is filtered by a low-pass filter so that the only frequencies present are those up to and including the 5th harmonic. At what frequency must this signal be sampled by an A/D converter according to the Nyquist criteria?

23.17. (a) An AM transmitter has a carrier frequency of 1 MHz and is operating with a modulation index of $m = 1$. If the message transmitted is a 2 kHz tone, plot the frequency spectrum of this signal, and identify and label the carrier, the upper sideband and the lower sideband. Sketch the AM signal as a function of time. (b) Repeat (a) if $m = 0.5$.

23.18. Estimate the modulation index for the AM signal illustrated in Fig. 7.8(b).

23.19. If the analog signal described in problem 23.16 is used to AM modulate a carrier frequency, $f_c = 100$ kHz, sketch the spectrum of the AM signal.

23.20. A nonperiodic signal, $v(t)$, has a Fourier transform shown in Fig. P23.20; $v(t)$ amplitude modulates a carrier of frequency $f_c = 200$ kHz. (a) Sketch the Fourier transform of the AM signal. (b) What is the bandwidth of the AM signal?

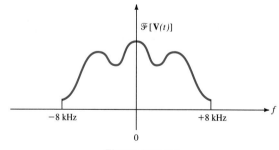

Figure P23.20

23.21. If the message signal with Fourier transform shown in Fig. P23.20 were transmitted by single sideband transmission, what would be the bandwidth of the signal?

23.22. The message signal with Fourier transform shown in Fig. P23.20 is transmitted by a wideband FM system. The total deviation of the carrier, $\Delta f = 12$ kHz. Determine the approximate bandwidth of this signal.

23.23. If narrowband FM were used to transmit the same signal as in problem 23.22, determine the approximate bandwidth of the FM signal.

23.24. Assuming a single television channel requires a bandwidth of 6 MHz, (a) how many channels may be simultaneously transmitted on a cable with a bandwidth of 100 MHz? (b) If you wish to provide a "guard band" or a frequency spacing of 1 MHz between each channel to minimize interference between channels, determine the new number of channels that can be transmitted. (c) This is an example of what type of multiplexing?

23.25. A time division multiplexing system is sequentially sampling 3 signals and each of these signals has a Fourier transform similar to that in Fig. P23.20. (a) At what frequency must each individual signal be sampled? (b) Determine the frequency of the pulses (samples) transmitted by this system.

23.26. Determine the center frequency for the audio transmission of a television signal on channel 18.

Complex Numbers

Complex numbers are typically represented in three forms: exponential, polar, or rectangular. In the exponential form a complex number $\mathbf{A}$ is written as

$$\mathbf{A} = Ae^{j\theta}$$

The real quantity A is known as the amplitude or magnitude, the real quantity θ is called the *angle*, and j is the imaginary operator $j = \sqrt{-1}$. θ is expressed in radians or degrees. The polar form of a complex number $\mathbf{A}$, which is symbolically equivalent to the exponential form, is written as

$$\mathbf{A} = A\,\underline{/\,\theta}$$

and the rectangular representation of a complex number is written as

$$\mathbf{A} = x + jy$$

where x is the real part of $\mathbf{A}$ and y is the imaginary part of $\mathbf{A}$.

The connection between the various representations of $\mathbf{A}$ can be seen via Euler's identity, which is

$$e^{j\theta} = \cos \theta + j \sin \theta$$

Using this identity, the complex number $\mathbf{A}$ can be written as

$$\mathbf{A} = Ae^{j\theta} = A \cos \theta + jA \sin \theta$$

which can be written as

$$\mathbf{A} = x + jy$$

Equating the real and imaginary parts of these two equations yields

$$x = A \cos \theta$$
$$y = A \sin \theta$$

From these equations we obtain

$$x^2 + y^2 = A^2 \cos^2\theta + A^2 \sin^2\theta = A^2$$

Therefore

$$A = \sqrt{x^2 + y^2}$$

Additionally

$$\frac{A \sin \theta}{A \cos \theta} = \tan \theta = \frac{y}{x}$$

and hence

$$\theta = \tan^{-1} \frac{y}{x}$$

The interrelationships among the three representations of a complex number are as follows.

Exponential	Polar	Rectangular
$Ae^{j\theta}$	$A \angle \theta$	$x + jy$
$\theta = \tan^{-1} y/x$	$\theta = \tan^{-1} y/x$	$x = A \cos\theta$
$A = \sqrt{x^2 + y^2}$	$A = \sqrt{x^2 + y^2}$	$y = A \sin\theta$

We will now show that the operations of addition, subtraction, multiplication, and division apply to complex numbers in the same manner that they apply to real numbers. The sum of two complex numbers $\mathbf{A} = x_1 + jy_1$ and $\mathbf{B} = x_2 + jy_2$ is

$$\mathbf{A} + \mathbf{B} = x_1 + jy_1 + x_2 + jy_2$$
$$= (x_1 + x_2) + j(y_1 + y_2)$$

that is, we simply add the individual real parts, and we add the individual imaginary parts to obtain the components of the resultant complex number.

Suppose we wish to calculate the sum $\mathbf{A} + \mathbf{B}$ if $\mathbf{A} = 5 \angle 36.9°$ and $\mathbf{B} = 5 \angle 53.1°$. We must first convert from polar to rectangular form.

$$\mathbf{A} = 5 \angle 36.9° = 4 + j3$$
$$\mathbf{B} = 5 \angle 53.1° = 3 + j4$$

Therefore

$$\mathbf{A} + \mathbf{B} = 4 + j3 + 3 + j4 = 7 + j7$$
$$= 9.9 \angle 45°$$

The difference of two complex numbers $\mathbf{A} = x_1 + jy_1$ and $\mathbf{B} = x_2 + jy_2$ is

$$\mathbf{A} - \mathbf{B} = (x_1 + jy_1) - (x_2 + jy_2)$$
$$= (x_1 - x_2) + j(y_1 - y_2)$$

that is, we simply subtract the individual real parts and we subtract the individual imaginary parts to obtain the components of the resultant complex number.

Let us calculate the difference $\mathbf{A} - \mathbf{B}$ if $\mathbf{A} = 5 \angle 36.9°$ and $\mathbf{B} = 5 \angle 53.1°$. Converting both numbers from polar to rectangular form

$$\mathbf{A} = 5 \angle 36.9° = 4 + j3$$
$$\mathbf{B} = 5 \angle 53.1° = 3 + j4$$

then

$$\mathbf{A} - \mathbf{B} = (4 + j3) - (3 + j4) = 1 - j1 = \sqrt{2} \angle -45°$$

The product of two complex numbers $\mathbf{A} = A_1 \angle \theta_1 = x_1 + y_1$, and $\mathbf{B} = B_2 \angle \theta_2 = x_2 + jy_2$ is

$$\mathbf{AB} = (A_1 e^{j\theta_1})(B_2 e^{j\theta_2}) = A_1 B_2 \angle \theta_1 + \theta_2$$

Given $\mathbf{A} = 5 \angle 36.9°$ and $\mathbf{B} = 5 \angle 53.1°$, we wish to calculate the product in both polar and rectangular forms.

$$\mathbf{AB} = (5 \angle 36.9°)(5 \angle 53.1°) = 25 \angle 90°$$
$$= (4 + j3)(3 + j4)$$
$$= (12 + j16 + j9 + j^2 12)$$
$$= 25j$$
$$= 25 \angle 90°$$

The quotient of two complex numbers $\mathbf{A} = A_1 \angle \theta_1 = x_1 + jy_1$ and $\mathbf{B} = B_2 \angle \theta_2 = x_2 + jy_2$ is

$$\frac{\mathbf{A}}{\mathbf{B}} = \frac{A_1 e^{j\theta_1}}{B_2 e^{j\theta_2}} = \frac{A_1}{B_2} e^{j(\theta_1 - \theta_2)} = \frac{A_1}{B_2} \angle \theta_1 - \theta_2$$

$\mathbf{A} = 10 \angle 30°$ and $\mathbf{B} = 5 \angle 53.1°$, we wish to determine the quotient $\mathbf{A/B}$ in both polar and rectangular forms.

$$\frac{\mathbf{A}}{\mathbf{B}} = \frac{10 \angle 30°}{5 \angle 53.1°}$$
$$= 2 \angle -23.1°$$
$$= 1.84 - j0.79$$

B

Computer Simulation Tools

There exists a wide variety of software tools that support both analysis and design in the field of electrical engineering. However, it is impossible in a comprehensive text of this type to cover these simulation tools in addition to the material that is contained in this book. Therefore, we will outline here some of the most popular software that is employed in electrical engineering and provide a source for obtaining it.

Circuits and Electronics

The most popular simulation tool for these areas is SPICE and its PC version PSPICE. These programs support a wide variety of analyses as well as the design of extremely large circuits. The recommended source for this software is

> Microsim Corporation
> 20 Fairbanks
> Irvine, CA 92718 USA
> 714-770-3022

Communications

A popular simulation program for this area is what is known as SIGNAL PROCESSING WORKSTATION (SPW). This program supports the simulation of both analog and digital communication systems as well as filter design. The recommended source for this software is

Alta Group
Cadence Design Systems Inc.
919 E Hillsdale Boulevard
Foster City, CA 94404
415-574-5800

Control

The two programs that appear to be the most popular simulation tools for the analysis and design of control systems are MATLAB and SIMULINK. These two software packages can be obtained from the following source.

The Math Works Inc.
24 Prime Park Way
Natick, MA 01760-1500
508-653-1415

Digital Systems

There are a wide variety of programs that support the simulation of digital systems. These tools include such things as schematic capture, IC layout and Very High Level Description Language (VHDL) for design and simulation. A recommended source for this software is

Mentor Graphics Corporation
27788 SW Parkway Avenue
Wilsonville, OR 97670-9215
503-685-7000

Similar packages are available from other sources such as Cadence, View Logic, and DAZIX.

Electric Machinery

The simulation tools employed in this area are typically SPICE, MATLAB and a software package konwn as the ELECTROMAGNETICS TRANSIENTS PROGRAM (EMPT). The best source for this latter simulation tool is

Electric Power Research Institute
P. O. Box 10412
Palo Alto, CA 94303
415-855-2411

APPENIDX

C

Fundamentals of Engineering (FE) Exam Review

Introduction

The engineering profession is regulated by boards in each of the 50 states and jurisdictions. Any individual who wishes to practice the profession of engineering by offering their professional services to the public must become registered as a professional engineer (PE).

The process of becoming a registered or licensed professional engineer is a multi-step process. First one must obtain the necessary education to be allowed to take the Fundamentals of Engineering (FE) examination; this was formerly called the Engineer-in-Training (EIT) exam. Engineering programs accredited by the Engineering Accreditation Commission (EAC) of the Accreditation Board for Engineering and Technology (ABET) are acceptable to all boards as qualifying education. Graduates of other programs should contact their appropriate registration board for the required education and experience prior to applying to take the FE examination. The FE exam is offered twice a year and is usually taken as a college senior or just following graduation. Those that successfully pass the FE exam are called an "Engineer-Intern" (EI), and are admitted to pre-professional status as a newly trained engineer.

After passing the FE exam the EI must obtain a minimum of 4 years acceptable experience before being qualified to take the Professional Engineering (PE) examination. This text serves as an excellent reference for both of these examinations; however, its use as a study guide for the FE exam will be emphasized here.

The FE examination consists of two parts. In the morning session there are 140 multiple choice problems (select answer A–E for each problem); approximately 14 of these will be in the area of electric circuits. The afternoon session consists of 70 problems; of these approximately 10 will be in the area of electrical circuits. The exam is prepared by the National Council of Examiners for Engineering and Surveying (NCEES).

In October 1993, the FE exam became a "supplied-reference" examination; candidates may not bring their own reference books to the examination. Instead candidates are given a booklet called the Reference Handbook containing relevant tables, formulae, or charts along with the exam. In this appendix we will outline the electrical engineering topics covered on the FE exam and identify the location of these topics in this text to facilitate study and review. In addition we will note important equations and concepts that are included in the FE Reference Handbook which you will be provided at the examination.

Examination Topics

Presented here is a combined topic list for the morning and afternoon sessions of the FE exam. Each of these areas will be addressed in the following sections.

1. DC Circuits
2. AC Circuits
3. Capacitance and Inductance
4. Transients
5. Diode Applications
6. Operational Amplifiers (Ideal)
7. Electric and Magnetic Fields

1. DC Circuits. Chapter 1 in this text provides an important introduction and definitions of quantities used in all subsequent chapters; quantities such as charge, voltage, current, etc., are defined in Section 1.2. Basic concepts, are illustrated with examples and drill exercises. Your time invested in thoroughly understanding these fundamental quantities will prove well worth the effort. This chapter uses numerous analogies to assist the newcomer in developing such understanding.

Equation (1.1) expresses the electrostatically induced force between two point charges; this equation is one of the first given in the Reference Handbook (the constant $k = 1/4\pi\epsilon$). Chapter 2 is titled "DC Circuits." This chapter develops Ohm's Law, explains Kirchhoff's Laws in Section 2.3, and illustrates their application with numerous examples to single loop circuits in Section 2.4.

Resistor combinations in series and parallel are expressed by Eqs. (2.17) and (2.18); these equations appear directly in the Reference Handbook. DC Nodal analysis presented in Section 2.7 and loop and mesh analysis presented in 2.8 provide a study of essential material for the systematic analysis of dc circuits. Thevenin's and Norton's theorems explained in Section 2.11 are provided in the Reference Handbook under the title *Source Equivalents.*

2. AC Circuits. Chapter 4 of this text provides a complete introduction to ac steady state analysis. Central to the analysis of ac circuits is the transformation of a sinusoidal time-varying quantity into the frequency domain as a phasor consisting of a magnitude and an angle. The phasor transformation defined by Eq. (4.17) shows that the amplitude of the time-varying function is equal to the magnitude of the phasor. One must use the same convention for the reverse transformation, i.e., the magnitude of the phasor transforms to the amplitude of the time-varying function.

The magnitude in the phasor transform can be defined alternatively as $1/\sqrt{2}$ times the amplitude of the time function, and thus the rms phasor can be expressed as

$$A \cos (\omega t \pm \theta) \leftrightarrow \frac{A}{\sqrt{2}} \angle \pm \theta \qquad (AC.1)$$

We can easily illustrate that consistent application of this alternative phasor definition results in exactly the same answers as when the phasor definition of Eq. (4.17) is utilized. For example let us rework the first part of Example 4.5 with the phasor definition of Eq. (AC.1).

The impedance $\mathbf{Z}$ is calculated exactly the same; therefore at f = 60 Hz, $\mathbf{Z} = 25 - j45.51\ \Omega$. We are given that $v(t) = 50 \cos (377t + 30°)$ V. Therefore using Eq. (AC.1), the rms phasor transform,

$$\mathbf{V} = \frac{50}{\sqrt{2}} \angle 30° = 35.4 \angle 30°$$

$$\mathbf{I} = \frac{\mathbf{V}}{\mathbf{Z}} = \frac{35.4 \angle 30°}{51.92 \angle -61.22°} = 0.679 \angle 91.22°\ A$$

Converting this back into the time domain again using Eq. (AC.1),

$$i(t) = \sqrt{2}(0.679) \cos(377t + 91.22°) = 0.96 \cos (377t + 91.22°)\ A$$

This is the same result as originally given in Example 4.5. The Reference Handbook provides the rms phasor definition given in Eq. (AC.1).

The important concepts of impedance and admittance are defined in Section 4.5 with several numerical examples. The important concept of effective or rms value is introduced in Section 5.4, the power factor in Section 5.5, and complex power in Section 5.6. The power factor (PF) is described in more detail in Section 5.7, entitled Power Factor Correction; this section describes an important practical example of ac circuits and a common type of exam question.

The algebraic manipulation of complex algebra which is necessary for calculations involving phasors, impedance, admittance, and complex power, can be assisted by the review of complex algebra; such a review is provided in Appendix A of this text. There is a brief review of the algebra of complex numbers in the Reference Handbook.

The formulae for resonance, including series resonance, quality factor, bandwidth and parallel resonance are described in Chapter 6 with a number of examples and drill exercises. Eqs. (6.21), (6.24), (6.26), and other relevant expressions are repeated in the Reference Handbook.

3. Capacitance and Inductance. The basic concepts of capacitance and inductance are first introduced in Chapter 1 and summarized in Figure 1.17. The beginning of Chapter 3 reviews these concepts in more detail and describes the series and parallel connection of capacitors in Eqs. (3.8) and (3.9), and the series and parallel combination of inductors in Eqs. (3.15) and (3.16). These equations are repeated in the Reference Handbook.

4. Transients. "Transient Analysis" is the title of Chapter 3. The transient analysis of RC and RL circuits is presented in Section 3.3 with four worked examples and three drill exercises. The Reference Handbook summarizes the response of a single loop series

RC circuit such as that illustrated in Fig. 3.7(a); equations similar to the following are presented:

$$v_C(t) = v_C(0)e^{-t/r} + V_s(1 - e^{-t/\tau})$$
$$i(t) = \{[V_s - v_C(0)]/R\}e^{-t/\tau}$$
$$v_R(t) = i(t)R = [V_s - v_C(0)]e^{-t/\tau}$$

where $v_C(0)$ is the initial voltage on the capacitor, and τ is the time constant, *RC*.

The set of analogous equations for a series *RL* circuit such as that presented in Figure 3.7(b) are also given; these include expressions for $i(t)$, $v_R(t)$ and $v_L(t)$ with the initial current expressed as $i(0)$ and the time constant τ as L/R.

5. Diode Applications. Chapter 8 provides a comprehensive review of the fundamentals of diode applications. The study of these applications begins in Section 8.6 with the diode circuit models; the ideal diode, the diode equation, and the piece-wise linear model. These basic diode circuit models are used in a wide variety of circuit applications in Sections 8.7 through 8.13. These applications include diode logic gates, power supply circuits including half-wave and full-wave rectification, filter circuits, clipper circuits, clamping circuits, zener diodes, photo diodes, LEDs, and the small-signal behavior of diodes.

6. Operational Amplifiers. Chapter 14 in this text reviews the fundamentals of operational amplifiers. The ideal op-amp is described in Section 14.2. In Section 14.3, and Sections 14.5 through 14.10 an array of ideal op-amp applications are discussed including the inverting amplifier, the non-inverting amplifier, the unity-gain buffer, the summing circuit, the integrator, the active filter, and the current-to-voltage converter. This chapter is independent and may be studied at any point after dc circuits.

7. Electric and Magnetic Fields. The definition of a uniform electric field between parallel plates is presented in Eq. 8.2. Chapter 18 is titled "Magnetically Coupled Circuits and Transformers" and introduces the fundamental definitions required for analysis of magnetic fields, magnetic circuits, and transformers. Faraday's Law is illustrated by the expression (18.11) and other related equations in Chapter 18. Chapters 19 and 20 discuss the application of these principles to dc and ac machines (motors and generators).

8. Additional Topics. The Reference Handbook presents an expression for the resistance of a bar of cross-sectional area, A; this equation is given in the text as Eq. (13.1). The handbook also describes the change in resistance resulting from temperature changes, which is described by a temperature coefficient, alpha; these concepts are illustrated by problems 7.7 and 7.8 in the text.

For additional or updated information regarding the FE examination contact directly the NCEES, PO Box 1686, Clemson, SC, 29633-1686.

Index

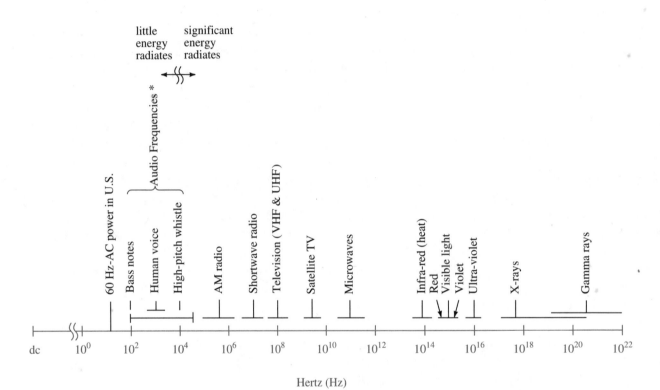

little significant
energy energy
radiates radiates

60 Hz-AC power in U.S.

Bass notes

Audio Frequencies *

Human voice

High-pitch whistle

AM radio

Shortwave radio

Television (VHF & UHF)

Satellite TV

Microwaves

Infra-red (heat)

Red

Visible light

Violet

Ultra-violet

X-rays

Gamma rays

dc 10^0 10^2 10^4 10^6 10^8 10^{10} 10^{12} 10^{14} 10^{16} 10^{18} 10^{20} 10^{22}

Hertz (Hz)

Frequency Spectrum